DEVOLUTION, PORT GOVERNANCE AND PORT PERFORMANCE

RESEARCH IN TRANSPORTATION ECONOMICS

Series Editor: Martin Dresner

RESEARCH IN TRANSPORTATION ECONOMICS VOLUME 17

DEVOLUTION, PORT GOVERNANCE AND PORT PERFORMANCE

EDITED BY

MARY R. BROOKS

Dalhousie University, Nova Scotia, Canada

KEVIN CULLINANE

University of Newcastle, Newcastle upon Tyne, UK

ELSEVIER
JAI

Amsterdam – Boston – Heidelberg – London – New York – Oxford
Paris – San Diego – San Francisco – Singapore – Sydney – Tokyo
JAI Press is an imprint of Elsevier

JAI Press is an imprint of Elsevier
The Boulevard, Langford Lane, Kidlington, Oxford OX5 1GB, UK
Radarweg 29, PO Box 211, 1000 AE Amsterdam, The Netherlands
525 B Street, Suite 1900, San Diego, CA 92101-4495, USA

First edition 2007

British Library Cataloguing in Publication Data
A catalogue record for this book is available from the British Library

ISBN-13:   978-0-7623-1197-2
ISBN-10:   0-7623-1197-5
ISSN: 0739-8859 (Series)

For information on all JAI Press publications
visit our website at books.elsevier.com

Printed and bound in The Netherlands

07 08 09 10 11 10 9 8 7 6 5 4 3 2 1

CONTENTS

v

LIST OF CONTRIBUTORS

Alfred J. Baird	Maritime Research Group, Transport Research Institute, Napier University, Edinburgh, Scotland, UK
Ramon Baltazar	Dalhousie University, Halifax, Canada
Khalid Bichou	Centre for Transport Studies, Imperial College, London, UK
Mary R. Brooks	School of Business Administration, Dalhousie University, Halifax, Canada
A. Guldem Cerit	School of Maritime Business and Management, Dokuz Eylül University, Izmir, Turkey
Kevin Cullinane	School of Marine Science & Technology, University of Newcastle upon Tyne, Newcastle upon Tyne, UK
Peter W. de Langen	Department of Port, Transport and Regional Economics, Erasmus University, Rotterdam, The Netherlands
Soner Esmer	School of Maritime Business and Management, Dokuz Eylül University, Izmir, Turkey
Sophia Everett	The Australian Centre for Integrated Freight Systems Management, The University of Melbourne, Melbourne, Australia
James A. Fawcett	University of Southern California, Los Angeles, USA

ix

Hakki Kisi   School of Maritime Business and
                           Management, Dokuz Eylül University,
                           Izmir, Turkey

Jasmine S. L. Lam   School of Civil and Environmental
                           Engineering, Nanyang Technological
                           University, Singapore

Maria Lamonarca   Department for the Study of
                           Mediterranean Societies, University of
                           Bari, Bari, Italy

Sung-Woo Lee   Korea Maritime Institute, Seoul, Korea

Hilde Meersman   Department of Transport and Regional
                           Economics, University of Antwerp,
                           Antwerp, Belgium

Theo Notteboom   ITMMA – University of Antwerp,
                           Antwerp, Belgium

Ersel Zafer Oral   School of Maritime Business and
                           Management, Dokuz Eylül University,
                           Izmir, Turkey

Athanasios A. Pallis   Department of Shipping, Trade and
                           Transport, University of the Aegean,
                           Chios, Greece

Paola Papa   Department of Economics, University of
                           Bari, Bari, Italy

Ross Robinson   The Australian Centre for Integrated
                           Freight Systems Management, The
                           University of Melbourne, Melbourne,
                           Australia

Ricardo J. Sánchez   Austral University, Buenos Aires,
                           Argentina,

                           ECLAC/UN, Santiago, Chile

Dong-Woo Song   Centre of Urban Planning and
                           Environmental Management, The
                           University of Hong Kong, Hong Kong

Wayne K. Talley	Department of Economics, Old Dominion University, Norfolk, USA
Okan Tuna	School of Maritime Business and Management, Dokuz Eylül University, Izmir, Turkey
Vincent F. Valentine	Economist
Marisa A. Valleri	Department of Economics, University of Bari, Bari, Italy
Eddy Van de Voorde	Department of Transport and Regional Economics, University of Antwerp, Antwerp, Belgium
Larissa M. van der Lugt	Department of Port, Transport and Regional Economics, Erasmus University, Rotterdam, The Netherlands
Thierry Vanelslander	Department of Transport and Regional Economics, University of Antwerp, Antwerp, Belgium
Teng-Fei Wang	Plymouth Business School, University of Plymouth, Plymouth, UK
Gordon Wilmsmeier	Osnabrück University, Bielefeld, Germany
Wei Yim Yap	Maritime and Port Authority of Singapore, Singapore

ACKNOWLEDGMENTS

This book is a product of five years of effort and a labour of love on the part of many. In the summer of 2001, a group of researchers interested in port governance and devolution, and their impact on port performance, met in Hong Kong. The research content of the book evolved over subsequent meetings in Panama in November 2002, Belgium in July 2003, Izmir in July 2004 and Cyprus in July 2005. The meetings explored the complexity of the issues and planned a research agenda that was exceptionally ambitious. The content of these chapters reflects the work to date and should only be construed as a beginning.

We are happy to acknowledge on the next page those reviewers who assisted with shaping the contributions of network members. The reviewers read, and in some cases, reread the work of the authors presented here, providing advice and guidance along the way. This volume would not have been realized without the invaluable support of Margaret Sweet and Janet Lord, both at Dalhousie University, and without the funding assistance of the Centre for International Business Studies at Dalhousie University and the School of Marine Science and Technology at the University of Newcastle.

I would like to dedicate this work to Richard O. Goss, for his keen interest in port policy, and in particular Canadian port policy, and for his challenging insights into all things maritime. He has always been supportive of my research efforts, and a good colleague and friend.

Mary R. Brooks

I would like to dedicate this work to Prof. Phil Goodwin in recognition of his boundless enthusiasm for transport issues that matter, the inspiration he instills in those who have worked with him and the continued support he has provided to me over many years.

Kevin Cullinane

REVIEWERS

Alf Baird
Ramon Baltazar
Khalid Bichou
Kenneth J. Button
Young-Tae Chang
Claude Comtois
Peter de Langen
John Dinwoodie
Michael Dooms
Sophia Everett
David Gillen
Devinder Grewal
Elvira Haezendonck
Trevor Heaver
Marc Hershman
Jan Hoffmann
Richard Horn
John Mangan

Bob McCalla
David Menachof
Steve Meyrick
Peter Nijkamp
Theo Notteboom
Thanos Pallis
Photis Panayides
Stephen Pettit
Ross Robinson
John Rowcroft
Dong-Keun Ryoo
Tomás Serebrisky
Brian Slack
Dong-Wook Song
Jose Tongzon
Vince Valentine
Larissa van der Lugt
Gordon Wilmsmeier

PART I:
BACKGROUND

CHAPTER 1

INTRODUCTION

Mary R. Brooks and Kevin Cullinane

1. INTRODUCTION

The relationship between ports and governments has changed profoundly over the past quarter of a century. Governments have moved to extract themselves from the business of port operations and, for the most part, have focused their efforts on the monitoring and oversight responsibilities associated with the concept that the role of government is to provide a safe and secure environment for citizens, and a level playing field for commercial activities. This trend for ports has not been universal and it has not occurred in isolation. It has been driven by two larger global trends affecting many industries – the globalization of production and distribution of manufactured goods and the movement towards new public management in government.

There can be no doubt that globalization has changed the world economy. The last 25 years have seen a dramatic restructuring of the production and consumption locations of goods and services and of the distribution network that delivers them to consumers. Economies of scale in manufacturing and the resultant specialization of national economies have been most obvious. Not as obvious outside the transport community has been the development of transportation networks to deliver parts for assembly and re-export in order to service a burgeoning demand for goods from an emergent and growing middle class in many developing countries, as well as the traditional and mature consumer demand that exists in the developed

Devolution, Port Governance and Port Performance
Research in Transportation Economics, Volume 17, 3–28
Copyright © 2007 by Elsevier Ltd.
ISSN: 0739-8859/doi:10.1016/S0739-8859(06)17001-8

world. Coupled with the rise of Southeast Asia, China and India on the back of their low labour costs, this highly efficient transport logistics network has fundamentally led to pressure on the shipping industry for greater economies of scale in transport activities and on the port industry to effectively manage operations at a lower cost.[1]

The trend towards new public management has also been ubiquitous. In the mid-1980s, thinking about government reform as a means to improve efficiency and responsiveness began to take hold in a number of developed countries, particularly those with strong central governments like the United Kingdom, Canada, Australia and New Zealand (Manning, 2000). Experiments in creating semi-autonomous agencies to deliver services previously provided by government began to emerge, marking the beginning of an earnest effort on the part of governments to improve the delivery of services, including port services, as an important element in enhancing national competitiveness. The new public management movement gained significant traction following the publication of Osborne and Gaebler's (1992) tome *Reinventing Government*. The opportunity to refocus government attention on its core business (i.e., from operation and legislation to policy-making centred on delivering the public interest) held the additional appeal of providing an opportunity for incorporating performance incentives into contracts for services (Manning, 2000). In particular, following the fall of the Berlin Wall, the appeal of efficient service delivery spread to developing countries. As interest in the new public management philosophy spread, it was only natural that large infrastructure-based service enterprises began to be examined for ways to make them more efficient. At a time when ports (and governments) were being pressured to make investments to accommodate the larger containerships that were servicing the new global trade patterns, the spotlight of many governments came to rest on port activities. The hype of new public management made governments more aware of alternative approaches to achieving new port investment at a time when new ways of thinking about investment were needed. The terms concessions, franchising and public–private partnership became commonly used, but remained poorly defined. The same can also be said of the words privatization, commercialization and corporatization.

This book examines the changed port management environment, focusing particularly on government policies such as devolution, regulatory reform and newly imposed governance models, all of which have exerted a significant influence over the nature of that changed environment. This change is illustrated in Fig. 1.1. The impact that these policies have had on port strategies and port performance is also analysed. Before outlining the

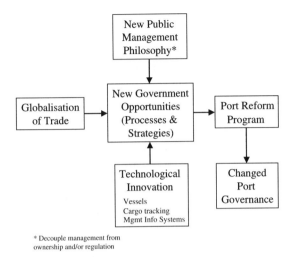

Fig. 1.1. Drivers of Port Reform.

structure of this book, however, it is critically important to provide some greater clarification as to what these terms – devolution, governance and performance – actually mean within the context of this book.

2. WHAT IS DEVOLUTION?

Rodal and Mulder (1993, p. 28) define devolution as "the transfer of functions or responsibility for the delivery of programs and services from the federal government to another entity," which may be "another order of government or a non-governmental organization, community group, client association, business or industry." While this might appear to be rather an all-encompassing definition, we actually perceive devolution as broader than Rodal and Mulder's control-consultation-partnership continuum with its decreasing financial and administrative involvement from government, and simultaneous increase in the other party's commitment and responsibility. As well, this definition clearly implies that the *decentralization* of government responsibility and accountability (from central, national control to local responsiveness) is one form of devolution. Rodal and Mulder stop short of explicitly recognizing the transfer of ownership to the private sector as the most extreme form of devolution, perhaps because their focus is exclusively on public sector responsibilities (and thus once these responsibilities are removed from the public sector, they no longer exist in the public sector management framework). For the purposes of this book, however, we agree with Brooks, Prentice, and Flood (2000), as illustrated by Brooks (2004), that privatization is indeed considered to be a form of devolution, as the transfer of ownership and/or management activities must form part of the same continuum of allocation of responsibility. The distinction between the two approaches is illustrated in Fig. 1.2.

The term privatization has been popularized; it is often freely used but seldom in a specific sense or underpinned by a specific definition. Generally, it is taken to mean the direct transfer of the ownership of assets from the public to the private sector. Broader usage of the term often includes what are actually deregulatory measures that are intended to open up former public sector monopolies to private sector competition – a process that is not actually privatization, but rather *deregulation*. Some confusion in the use of terminology has arisen because policies of deregulation are sometimes implemented concomitantly with a transfer in the ownership of assets from the public to the private sector (i.e., privatization) and necessitate what was previously the public sector entity adopting a new set of objectives and approaches to management. The latter is an inevitable corollary not only of its mere participation, but also of its ambitions to survive and prosper within a newly emergent competitive environment.

Equally, there are *franchising* arrangements that involve private companies tendering for transport services that were formerly directly provided by local or central government. Button and Rietveld (1993) have categorized this franchising of services to the private sector as falling within the remit of privatization. Although this does indeed constitute a mechanism for stimulating greater private sector participation in an entity or activity, we

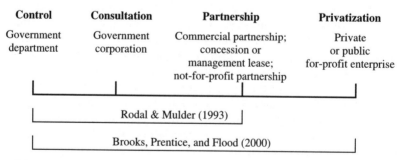

Fig. 1.2. Devolution Continuum. *Note:* Typology of Devolution Implies Governance Choice by Regulator. *Source:* Brooks (2004).

consider this to be simply a specific form of "contracting out," as within the literature on business and economics *franchising* is more properly regarded as a specific type of outlet replication method. For the purposes of the analyses contained in this book, therefore, it is treated as being clearly distinct from privatization.

Any new "contracting out" arrangements can extend beyond simply providing operating services. They can include, for example, the construction of major pieces of infrastructure. In concession[2] agreements, for example, publicly determined infrastructure is built and operated by a private company. At the end of the agreement, the assets will revert to public ownership under pre-agreed terms, even though the investments made by these private sector companies are often sunk and some agreements may contain provision for the recovery of the investment from sources other than revenue streams. Although management concessions are often regarded as examples of privatizations, from the perspective of the terminological definitions used within this book, they are actually considered as examples of commercialization, or partnerships as defined by Rodal and Mulder (1993).

The nature and division of public and private sector participation is central to the debate about what constitutes privatization. True privatization is the full transfer of ownership, even where this applies only to a certain proportion of an entity's assets or shareholding as is the case with a *partial privatization*. Privatization is not just a transfer of temporal rights, however long-lived. It is certainly the case that concession agreements are considered to be privatizations by some. Within the context of this work, however, we do not regard concessions as examples of privatization since the duration of the contracts confers merely the rhetoric of ownership, although not the reality of it. For the purposes of this analysis, therefore, concession agreements are treated as a specific form of commercialization, rather than as a form of privatization.

Commercialization entails the state withdrawing from the operation of transportation infrastructure while retaining ownership of it. The commercialized entity is managed and operated by a separate, non-recourse entity,[3] which makes lease payments to the landlord (the state) for the use of the infrastructure. Commercialization may be accompanied by the decentralization of oversight responsibility from the national to the local level, increasing the flexibility of the entity to respond to local needs and situations. It is very clear that commercialization is built on the fundamental tenet that ownership remains with the government and that land (or water) assets in particular are not transferred to the operator, and cannot therefore be sold. Operating assets like cranes (superstructure) may be transferred, depending on the negotiated arrangement. The ownership of new capital assets is determined by the nature of the terms of the agreement between the legal entity responsible for their management and the government transferring responsibility for the performance of the port.

Corporatization is a specific form of commercialization that involves the creation of a separate legal entity, a corporate form, which takes on legal responsibility to provide the functions or activities mandated in its charter or by-laws. In many countries, corporatization is used synonymously with privatization when the government creates a new legal entity to take on the management of the port assets and provides it with a commercial lease. In many instances, the difference between straightforward commercialization and corporatization specifically is vague; we see the primary distinguishing feature of corporatization to be whether there is the creation of a legal entity[4] with share capital and a profit-making mandate.

To explain the differences between the terms, Table 1.1 provides a synopsis of the terms and their key characteristics. Critical questions that differentiate the concepts revolve around whether there has been the creation of a new legal entity for the management of the operation, and whether there has been a transfer of both rights (obligations) and risks to an existing or new legal entity. Each of these approaches results in a different governance model.

In sum, *decentralization* has a differing set of objectives from other forms of devolution. It is intended to secure local responsiveness on the part of government in its provision of services. However, privatization, corporatization and commercialization are all intended to bring the *operations and management* of assets onto a more commercial footing based on either (a) the desire to extract greater profits from the assets under management, or (b) to limit the impact of loss-making assets on government finances.[5] The distribution of the benefits from these processes is a separate issue from the extraction of the benefits. As defined by Rodal and Mulder (1993), devolution included transfers to other levels of government (decentralization) and, therefore, throughout this book the word devolution is used as an all-encompassing term to imply the full range of activities government might undertake in reforming its governance of port activities.

3. WHAT IS GOVERNANCE?

Since the 1980s, many governments have devolved responsibility for transport support industries from the public to the private sector. For example,

Table 1.1. Characteristics of Types of Devolution.

Characteristics	Decentralization	Commercialization	Corporatization	Privatization
Changed government objectives	From executing national policy to local responsiveness	To improve efficiency and responsiveness	To improve efficiency and responsiveness	Private sector-led efficiency; market competition
Changed organizational structure (new lines of accountability)	Yes	Yes	Yes	Yes
Establishment of a legal entity	No	Yes; there is no share capital	Yes; share capital may be owned in part or in full by government	Yes (if not sold to an existing private entity)
Control of operations and management	Local government	Transferred from government	Transferred from government	Transferred from government
Ownership of existing capital assets[a]	May be transferred to another level of government	Not transferred	Usually not transferred	Transferred
Ownership of new capital assets[a]	Dependent on negotiated arrangement	Dependent on negotiated arrangement	Resides with new entity	Privately owned by new entity
Responsibility for risk[a]	Remains with public sector	Transferred to new entity; there may still be recourse to government, depending on contract terms	Transferred to private sector	Transferred to private sector
Right to borrow money[a]	May require national approval	Yes, there may be caps imposed	Yes	Yes
Ability to sell the assets[a]	Usually not without national government approval	Usually limited, depending on contract terms	Usually limited, depending on contract terms	Yes

[a]Critical control elements.

governments have privatized ports in the United Kingdom (Thomas, 1994; Baird, 1995), put concessions out to tender for the management of facilities (Kent & Hochstein, 1998) or transferred the management of ports or terminals to other government departments. In some cases, transfer arrangements have been unique. In Canada, for example, a three-pronged approach was implemented that saw ownership of the largest ports retained by government within non-recourse agencies but their management commercialized. At the same time, if not transferred to a regional or municipal government, ports with a regional or local focus were privatized and remote ports were retained so that their ownership and operation remain in the public interest. Meanwhile, in the US there has not been any obvious transfer of ownership, management or control. There is substantial variety in the institutional arrangements for the governance of ports in the US. It is, therefore, important that we understand what is meant by the term governance.

While governance is generally not defined in many economics dictionaries or texts, free-market economists use the term to encapsulate the adoption and enforcement of rules governing conduct and property rights. Governance is often confused with government. It may be imposed by governments or adopted voluntarily by groups or associations. As such, governance is a notion that can be applied to more than just corporations. While governance principles are applicable to all relationships between businesses and their shareholders, they can also be suitably applied to relationships between governments and their voters and taxpayers, between public/private agencies and their stakeholders, between organizations and those who establish them to undertake activities on their behalf. This broad applicability is reflected in the World Bank Institute's (2000) definition of governance as

> The traditions and institutions by which authority in a country is exercised for the common good. This includes (i) the process by which those in authority are selected, monitored and replaced, (ii) the capacity of the government to effectively manage its resources and implement sound policies, and (iii) the respect of citizens and the state for the institutions that govern economic and social interactions among them.

In essence, therefore, the systems, structures and processes that organize groups of individuals to a common purpose can be perceived as constituting the governance structure of the group, society or voluntary organization. At the other end of the continuum, the legislation and regulations that the government imposes on a business or not-for-profit entity also shape the governance structure within which such organizations operate. The structures and processes put in place by national laws, such as the requirements for open procurement processes, do form part of the governance of government. Most frequently, however, the term *governance* is used to refer more specifically to *corporate governance*, an area that is widely researched. A wealth of information on how it is implemented in various countries and under various business models is contained in Keasey, Thompson, and Wright (1999).

The OECD (1999) defines *corporate governance* as

> ... the system by which business corporations are directed and controlled. The corporate governance structure specifies the distribution of rights and responsibilities among the different participants in the corporation, such as the board, managers, shareholders and stakeholders, and spells out the rules and procedures for making decisions on corporate affairs. By doing this, it also provides the structure through which corporate objectives are set, and the means of obtaining those objectives and monitoring performance.

Therefore, corporations are not expected to deliver the social welfare or public policy objectives of government. Sternberg (1998, p. 20) provides a much shorter definition, whereby corporate governance constitutes "ways of ensuring that corporate actions, assets and agents are directed at achieving corporate objectives established by the corporation's shareholders." She argues that it is a mistake to criticize corporations for a failure to achieve public policy objectives or to accord greater importance to their stakeholders, on the assumption that stakeholder participation will provide for better governance.[6] When an organization is devolved, stakeholders raise concerns that their interests will no longer be considered, and that society's interests will be ignored in the pursuit of greater efficiency and profits. In consequence, it is often argued that stakeholders enjoy greater protection when an organization is managed by government.

Therefore, in the private ownership model, corporate governance is the structure, roles and responsibilities that provide the means by which the organization is managed as an economic entity, based upon the objectives of the corporation. The corporation is in business to meet the objectives set by the Board for the benefit of owners, not specifically to meet the objectives of government, regulators or other stakeholders. In the fully public model, the government is accountable to its citizens, voters and taxpayers. In mixed models, the governance issues become muddied; it is not always clear where responsibilities and accountability lie.

There are two aspects of governance that exert an influence over transportation. First, transport companies must operate within the restrictions that their Boards of Directors impose on their activities and processes; these boards may represent family owners (in family-owned companies), shareholders (in publicly traded companies) or owners in private firms. In the private sector, corporate governance guidelines are relatively straightforward; within each corporation, the management is free to deliver the strategic vision of the entity in a manner deemed best to deliver the strategic intent of the organization, within the governance guidelines established by applicable laws.

Second, from the broader perspective, governments impose governance systems on public and private corporations, associations and societies. Whether the transport entity is owned by the private sector, by government or by associations, governance rules and processes exist to ensure that the Board meets its fiduciary responsibilities, and is both responsible and accountable to the appropriate party. Under the rules imposed by government, the Board is also required to warrant that the corporation is in compliance with the laws applicable to the corporation.

To fully understand the importance of the particular governance model as it applies to any transport operation, it is not sufficient to be aware of the applicable regulations; it is important to know who is responsible for the activities of the enterprise, who will be held accountable for the outcomes of those activities (and by whom), what constitutes a conflict of interest, and what processes and safeguards exist for the protection of the public interest.

Throughout the book, the phrase "port reform" is intended to mean the restructuring of the governance of ports as part of a government's devolution program.

4. WHAT IS PERFORMANCE?

Those in favour of the privatization of government services argue that free-market competition improves services, lowers costs and generates revenue for the government from real estate sales and corporate income taxes (Goldsmith, 1997). Its detractors argue that benefits accrue as a result of wage cuts and greater use of part-time labour, with the private sector catering only to those already having a better standard of living. Starkie and Thompson (1985) have expressed their concern that privatization may result in cross-subsidization and the inefficient allocation of resources. In other words, the implementation of devolution raises questions about the performance of the devolved entity and who benefits from the reallocation of obligations, responsibilities, risks and accountability. Devolution implies a change in governance model, which itself raises a question about consequent performance. To assess performance outcomes, be they the performance of new governance models or the performance of devolution programs, we first need to define what is meant by performance.

Performance is generally defined as an accomplishment, and so context becomes critical to the view held of what constitutes "good," "satisfactory" or "acceptable" performance. The modifiers associated with performance are as important as the context because they define the intent or objectives that are sought. In the context of port performance, the focus of academic research to date has largely been on efficiency. However, not all ports have set economic performance objectives; the assessment of performance as an outcome, therefore, is measured against the objectives of the entity providing port services and/or of the government that has instituted a program of port reform. This leads therefore to the overall purpose of this collection of papers: to examine the paths of devolution taken in the port industry, to explore the resulting governance models and to begin the examination of outcomes. As will be seen from the final chapter of the book, there is more work to do in the area of performance evaluation.

5. STRUCTURE OF THE BOOK

To recap, the past two decades have witnessed a period of port reform that has been driven by governments preoccupied with the prevailing economic thinking, in particular the theories of new public management. This port reform has been implemented primarily through a process of devolution. Part I of this book contains this chapter and one by Notteboom (Chapter 2). This chapter provides a framework for defining the basic concepts involved in devolution, while that of Notteboom paints a picture of the current port environment, its likely future evolution and the expected impact this will have on the functioning of ports, and specifically container ports, as logistics nodes in global networks. Key economic developments that will significantly affect the sector overall, as well as the individual competitiveness of ports, are identified as being: the globalization of production and consumption; the changing structure of supply chains and the emergence of supply networks; the growth in the outsourcing of logistical decisions (including port choice) to specialist third-party logistics providers; and the increased level of concentration within the international logistics industry that has come about as the result of a move towards the greater integration of logistics systems. Notteboom concludes that the global port marketplace is characterized by the presence of powerful and relatively footloose customers and providers, extensive business networks and complex logistics systems that all combine to create a high degree of uncertainty in the industry. Since major port customers no longer concentrate their service packages solely on the sea-to-land interface, but on the quality and reliability of the entire logistics chain, ports can no longer expect to attract cargo simply because they are natural gateways to rich hinterlands. The appropriate response to this changing competitive environment, the author asserts, is for port managers to reassess existing organizational and management structures, as well as their port strategy. Although the competitive position of a port does not depend solely upon its administrative structure, Notteboom does point to the need for the establishment of an appropriate legislative and administrative framework that facilitates the achievement of the efficiency objectives that support a port's competitiveness. These two chapters serve to provide an introduction and background to the chapters that follow.

The remaining three parts of the book are intended to correspond to the three intersecting themes contained within it. Part II examines the port industry in 14 countries or administrations, and presents the thinking behind any devolution programs that have been implemented. Each country (or administration) has followed its own unique path, with some building on experiences seen elsewhere, and others initiating their own novel experiments in new public management approaches. The content focuses on the theme of devolution, but not all devolution has occurred in similar ways. The most extreme example of devolution is the United Kingdom, and so it is discussed first.

Unlike devolution programs elsewhere, the Thatcher experiment often involved full-scale privatization and the policy decision to allow the industry, for the most part, to self-regulate its activities. This experiment was therefore bolder than that seen in other countries, where the decision went a shorter distance along the devolution continuum presented in Fig. 1.2. As a consequence, politicians incorporated greater oversight provisions into the governance model. Baird and Valentine (Chapter 3) present a number of approaches that were taken in the UK and summarize the methods by which the major sales were completed. While Goss (1999) argued that the role of government is to prevent a single entity from intercepting the economic rents at the expense of society as a whole, Baird and Valentine conclude that the UK approach to privatization has actually encouraged this to happen. In their view, seaports should be public assets managed in the public interest. Perhaps the reason why the UK model of port governance attracts so much attention is that there is neither a port policy statement by government nor a continuing oversight role built into the privatization legislation or subsequent regulation. The government is dependent on irregular reviews arising from stakeholder discontent to act as a catalyst for re-examination of its decisions (e.g., Gilman, 2003).

In contrast to the bold experiment of port devolution in the United Kingdom, Chapters 4–8 present the approaches to port devolution adopted by other European countries.

In Chapter 4, Meersman, Van de Voorde and Vanelslander describe Belgium's unique geographical and political context, within which the port governance issue should be considered. An approach to port governance has emerged that encompasses a diverse variety of governance structures. While some municipal authority ports have been corporatized, others have been privatized, either fully or partially. The authors address where the different responsibilities for investment and the setting of port policy lie, but point to the potential complications that exist as the result of the influence exerted at regional level by both the Benelux nations and the European Union. The chapter concludes by highlighting the dramatic level of competition that exists between ports in the Hamburg–Le Havre range and the relationship that exists between port governance and the outcomes of that competition. In so doing, the critical importance of ensuring the optimum design of port governance structures is alluded to, especially in ensuring a level playing field between competing ports. Given the comparative success achieved by its ports, particularly with respect to container handling at Antwerp, it would seem that thus far Belgium's approach to governance has proven reasonably successful. There are clear concerns, however, that this might be undermined by conflicting or contradictory policies set at supranational levels: the EU and/or Benelux.

In Chapter 5, the governance of the three major ports in the Netherlands (Rotterdam, Amsterdam and Zeeland) is analysed by de Langen. Changes in port governance in the Netherlands have been prompted by considerable changes to the environment in which ports operate. The traditional Dutch model of port governance has been that ownership and control are both vested at the local level of a municipality or region, with national government responsible solely for major investments in port infrastructure. Consequently, while the three major Dutch ports remain owned by local governments, they operate relatively autonomously. Decisions on private sector participation in ports, therefore, have been at the discretion of the regional/municipal port authorities alone and, as a result, different models and approaches have emerged. Rotterdam has been corporatized, although subsequent discoveries of shortcomings in its governance have prompted the central government to go so far as to take a minority shareholding; prompted not only as a means of ensuring future financing requirements for expansion, but also by the desire to secure the national interest. On the other hand, Amsterdam and Zeeland have refrained from any form of commercialization, with the former engaging in periodic debates on the issue and the latter unequivocally ruling out this option. The chapter concludes that the prevailing governance model influences a port's strategy and vice versa. The local ownership of ports explains a strategic focus on creating prosperity for the "local" port cluster, while reducing (local) negative effects. The strategy required to perform in the North European port environment also influences the governance model. Thus, all three ports engage in cooperation with other regional port authorities in their vicinity.

The Italian case is quite interesting because, unlike the previous countries, it is one where the central government has a strong regulatory control over the transport system but has attempted to implement new public management principles without changing the existing bureaucratic structure. Valleri, Lamonarca, and Papa (Chapter 6) examine the restrictions such a bureaucratic structure imposes on the ability of the port authorities to respond appropriately to a changing commercial environment. While recognizing that authorities need to move towards more innovative approaches, they conclude that rigid regulation, financial restrictions and bureaucratic structures hamper the development of an efficient port system. This is particularly disconcerting given the geographic advantage that has accrued to southern Italian ports in the recent global restructuring of trading activities.

In Chapter 7, Pallis examines the corporatization and privatization of Greek ports as they seek to overcome past inefficiencies. Greek port reform is a relatively recent effort and is still in progress. While the purpose of Greek port reform was to improve port efficiency, modernize services and instill a commercial ethos, Pallis concludes that the corporatization of 10 ports and the public listing of two trans-European ports did not alter the situation as expected. Government maintained a majority of the shares in the two publicly listed companies and continued to interfere in the decisions of the ports. He notes, in particular, the problems experienced in the summer of 2005 and the extent of bureaucratic interference that prevents the reform process from achieving the benefits that are sought. He concludes that, if future benefits from reform are to be realized, there is a need to take the process further and for government to reconsider the political practices at play.

A similar phase of port reform is underway in Turkey. In Chapter 8, Oral, Kissi, Cerit, Tuna, and Esmer describe the current complex and bureaucratic governance structure, which is the object of the port reform process. They point to a number of inadequacies with an industry where the vast majority of the country's cargo throughput falls under the purview of the public sector, highlighting specifically the onerous procedures for the approval of investments to increase port capacity. Although describing the current port reform process as "privatization," the authors make explicit the fact that there will be no transfer of ownership from the public to the private sector. As such, the process is actually one that, for the purposes of a wider analysis, we have defined as commercialization and which involves solely the transfer of operating rights under a lease or other methods. The chapter concludes that, irrespective of the current program of port reform, in order to achieve a truly substantive improvement in port performance, an overarching organization should be established in Turkey to formulate a national port policy and to coordinate and oversee port investments, development and competition. The need to match appropriate institutions to the purpose of reform is clearly underlined.

From Turkey, this section moves to the Americas, beginning with Chapter 9 by Sánchez and Wilmsmeier on the impact of the reform program on two ports in the Plate River basin – Buenos Aires and Montevideo. Although located in different countries (Argentina and Uruguay, respectively), these ports do compete in the same market. What makes this chapter particularly interesting is that both countries have faced similar catastrophic economic crises, and have responded with devolution programs (across a range of different industries, including ports) that are intended to ease the financial crisis facing the two national governments and to make the two ports more competitive in servicing the region. Despite similarities in what Baltazar and Brooks (Chapter 17) define as the "remote" environment, the two ports had dramatically different outcomes to their devolution programs. Sanchez and Wilmsmeier trace the root causes of this performance gap to the governance models imposed by the two governments in their devolution programs and to the general codes of governance and the business climate that exist within each country. The chapter draws forward some interesting concepts from later chapters in this book and makes a strong case for the dynamic nature of the three themes of this book.

From the tip of South America, the next chapter discusses the port industry in perhaps one of the most complex governance environments, that of the United States of America. In Chapter 10 Fawcett provides a detailed examination of the ports and port policy found in the US. Owing to its historical development as a federation of autonomous states, the US is characterized by port governance that varies by state. Over its history, the country has moved from a purely private model to one where the largest seaports are commonly owned by public agencies, governed by public boards, and follow a landlord model, one where operations are undertaken by the tenants who lease the facilities. In contrast, smaller seaports have often remained in private hands. It is perhaps best summarized as a situation where the absence of a national port policy, coupled with a philosophy of local responsiveness, has meant that infrastructure is financed by state or municipal governments in a climate where the "squeakiest wheel gets the grease," and channel maintenance is supplied by the US Army Corps of Engineers. Resisting the pressure exerted by the new public management ethos, the US has not engaged in programs of port reform that encourage privatization, instead preferring to continue the public control that has enabled regional or local governments to make investments that are deemed to be in the public interest.

North of the United States is a culturally similar situation. However, Canada's port reform process could not be further from that found across its southern border. Brooks (Chapter 11) notes that the devolution program in Canada was implemented in waves beginning in the 1980s (although the paper provides some earlier history) and was founded to some extent on the principles espoused by Goss (1983) in work conducted for the Canadian government. The second wave was driven by new public management principles and a desire to remove loss-making transport activities from being a drain on the fiscal purse. While the UK government would have recommended a course of privatization as the solution to the Canadian government's financial dilemma, its former colony embarked on a unique experiment in not-for-profit commercialization for the largest and most strategic ports. The thinking was that different types of ports required different governance models and, in consequence, three types of governance models resulted. This wave of commercialization became embodied in the *Canada Marine Act 1998*, a piece of legislation that, as concluded by both Brooks (in Chapter 11) and Baltazar and Brooks (in Chapter 17), has not yet satisfactorily resolved outstanding port governance issues in Canada.

While both Canada and Australia are former colonies of the United Kingdom, port reform processes unfolded quite differently. In Chapter 12, Everett and Robinson present a summary of port devolution in Australia, a process that took place under state, rather than national, jurisdiction. Although some ports were privatized, the dominant model was that of corporatization, with each state government (the owners of the ports) enacting legislation aimed at implementing its own corporatization model. The authors point out that these policies of port reform were instituted in the hope of increasing port efficiency, generating profitability and reducing the burden on public debt that accrued from the public sector ownership and operation of ports. The authors also suggest that another important motivation was the prevailing economic/political/ideological environment at the time the policy was implemented. Everett and Robinson point to the inherent difficulties in attempting to evaluate the impact of such policies pre- and post-implementation, but assert that Australia has generally failed to transform its ports into commercial businesses along the lines of private sector operations; in effect, the port reform process has failed to fulfill its objective. The authors lay the blame for this on the inadequacy and inappropriateness of the nature and form of the legislation that was generally enacted to facilitate corporatization. They point to the comparative success achieved in Tasmania, where the legislation underpinning port corporatization differed from that implemented in most of the rest of Australia. They also allude to the possibility that port reform in Australia did not go far enough and that privatization may have been a better option for achieving the objective of engendering a truly commercially oriented port sector in Australia.

As the chapters have now reached the Pacific Rim, Chapters 13–16 focus on port devolution and reform in three Asian countries – Singapore, Hong Kong (treated separately), China and Korea.

The corporatization of the former Port of Singapore Authority (PSA) in 1997 transformed the port from its previous status as a government body to one where, although it remains entirely government-owned, it is independent of government. Cullinane, Yap and Lam (Chapter 13) point out that this corporatization was motivated by a desire to enhance the commercial flexibility of the national port operator, not only with respect to responding to the demands of the port's bigger customers within what is a fiercely competitive regional environment that revolves around a somewhat transient transshipment trade, but also in facilitating PSA's strategy of global diversification. The authors suggest that the most significant impact of the changes in governance has been to allow PSA to focus on port-related business (particularly in an international context), while the Maritime and Port Authority pursues the policy of developing Singapore as an International Maritime Centre, with the intention of attracting activities with higher added value than those associated with the port. The authors conclude that the strategic importance of the port to the Singaporean economy and the inherent danger of the government losing control over its ability to meet long-term objectives for the economy as a whole will ultimately preclude any further significant devolution of decision-making authority.

In Chapter 14, Song and Cullinane detail the governance structure for the port of Hong Kong. They point out that the main port facilities are privately owned and operated and have been for a very long time, with the government's role limited to undertaking long-term strategic planning for port facilities and providing the land and necessary supporting infrastructure. The port of Hong Kong has not really engaged in any radical port reform in recent years. Instead, the focus has been on implementing changes to the governance of the port at the macro level, with the main impact being that there is now much greater involvement of Hong Kong's shipping industry in providing government with advice on port-related matters. The authors assert that while both the port and its shipping industry are pivotal to the success of Hong Kong's wider logistics policies, recent changes in the governance structure of the port have actually divorced port and shipping industry representation from the wider logistics debate. It is concluded that this may prove to be a strategic disadvantage within the context of the intense cross-border competition that Hong Kong faces from ports in the south of mainland China.

In Chapter 15, Wang and Cullinane point out that, concomitant with wider national economic development and reform, the governance structure for China's ports has also changed dramatically over the last few decades. The main avowed objectives for China's port reform are to free the port from the bureaucracy of government, and to increase the competitiveness and efficiency of the port industry. As such, the port governance structure has evolved from one characterized by a high degree of centralization towards a form where authority and management decision making is much more decentralized, and from a wholly planned economic system to one that is much more market oriented. The authors point out that one of the most important linchpins of port reform in China has been the privatization policy that has been pursued, despite the fact that the political ideology underpinning government and its institutions regards this term as anathema – so much so that it is frequently avoided in official documents in favour of "corporatization" or "marketization." New legislation in 2004 explicitly states that China's central government will no longer be involved in the ownership of ports, with its main responsibility now resting solely with the strategic planning of the port network for the whole country. The same legislation overcomes the serious drawback that the port authority was simultaneously the regulator and a market player, by replacing the original port authority with two different entities to reflect these different activities. Finally, in compliance with China's obligations on accession to the World Trade Organization (Li, Cullinane, & Cheng, 2003), this legislation abolishes the previous ceiling on stakes in ports that may be held by foreign investors. While port reform in China may be described as rather a passive and slow-moving process, Wang and Cullinane conclude that the underlying objective and progress towards that objective are quite clear. They point out, however, that what remains unclear is to what extent the emergent governance structure will influence the future development of the port industry in China.

Song and Lee (Chapter 16) describe the historical and ongoing process of decentralization that has occurred in the governance of Korean ports over recent years, with particular emphasis placed on the nation's critically important container port sector. As a consequence of this process, Korea has moved from a position where, until 1996, central government exerted exclusive control over ports, having direct responsibility for everything from planning and development down through to operating ports. The authors suggest that this centralized governance model was necessary because of Korea's government-focused economic development policy and trade-oriented economy. However, it did lead to problems with respect to inefficient management and operations, longer response time to a rapidly changing business environment and bureaucratic administration. Ownership, authority and responsibility for port matters are now increasingly vested at the level of the region or city in which a port is located. Although many of the smaller ports remain under central control, the decentralization process for larger ports is just about complete. Changes to the port governance structure were instigated through a mix of approaches. For example, national and then regional authorities and agencies responsible for port matters have been corporatized in order to introduce a "private spirit" into port management and development. To create greater competition in the market, private sector participation in container terminal operations was facilitated by the granting of concessions to both domestic and overseas investors. The latter needs, of course, the reassurance that comes from the more transparent governance model that the authors suggest is now in place. Song and Lee conclude that the policy changes that have been implemented in recent years have created a complex set of governance structures within which Korea's ports operate and that, although central government is still to some extent involved in even operational matters, that interference is becoming less as the overall governance structure continues to evolve.

These 14 chapters provide a rich insight into the thinking of government, the resulting devolution programs and the outcome or impact the programs had. These chapters also set the scene and provide content for the cross-cutting chapters on governance and performance that follow.

The next section of the book focuses on port governance and devolution generally. It begins with a presentation of the Matching Framework proposed by Baltazar and Brooks (Chapter 17). Using configuration theory, they conclude that there is a conceptual framework for looking at a firm's strategic responses to changes in the environment and that the governance structures that are chosen to address these changes will have an influence on a port's performance. This conceptual approach links performance to the strategic intent of the port organization. They note that not all ports have profit-motivated objectives, as found in a survey of ports by the Port Performance Research Network,[7] and conclude that the Matching Framework provides a tool for practitioners to assess strategic fit, in order to optimize performance outcomes by minimizing the potential for misalignment between the strategy, structure and environment in which the port finds itself. They also conclude that the framework is useful for academic researchers interested in case study approaches, most suitable for studies that compare and contrast the performance and configuration characteristics of different organizations over a common time frame, or in studies that examine the configuration and performance of single organizations over different time frames. This work provides a framework useful for the longer-term study of port performance.

The second paper in this section (Chapter 18) focuses on the governance models in place in ports surveyed by the Port Performance Research Network. The 42 ports responding to the survey have a variety of strategic objectives (as noted by Baltazar and Brooks in Chapter 17). Brooks and Cullinane examine the existing typologies of governance models for ports and hypothesize that the range of activities in which ports engage is so varied that existing models cannot hope to capture the complexity of the situation. The results of an analysis of the survey returns from 42 ports worldwide support this hypothesis, thereby suggesting that existing port governance models constitute drastic oversimplifications of reality. The port governance situation is one of such enormous complexity that all existing governance models are rendered invalid. The nature of this complexity is graphically illustrated by the number of activity groupings that emerge from an analysis of the correlation between the governance approaches of the different ports that make up the sample. Although the findings from the analysis of survey returns would seem to suggest the existence of a plethora of potential port governance models, in practice the range of possibilities is much more limited.

Notteboom (Chapter 19) examines the use of concession agreements as a specific tool of governance. Concession agreements are widely used in the port industry and Notteboom defines the types of agreements and the processes by which the successful bidder is chosen. He notes the wide variation in agreements in Europe alone and discusses the effectiveness of specific approaches. He identifies that the process and the awarding of port concession agreements offer port authorities the opportunity to structure the relationship between the authority and the operator, and hence the prospect of improving the performance of the port to meet the intent of the authority. He concludes that the area of concessions and their effectiveness is ripe for further economic research, given the wide adoption of the approach and the diversity of its implementation.

While governance models may be imposed on a particular port authority by a government, the existence of stakeholders (with their own agendas) is clearly a driver of government policy. Port clusters raise new issues in governance and de Langen's Chapter 20 on governance in port clusters discusses how conflicts of interest may be addressed in port communities. The chapter develops a framework for analysing conflicts of interest, a problem of particular importance in the port industry as global transport players are not "embedded" in one port but serve globally diverse constituencies. De Langen argues that cluster governance is a potential source of competitive advantage for a port, a conclusion that many ports currently developing stakeholder management processes will find of interest. He concludes with a research agenda that will be of interest to academics and practitioners in ports with large maritime clusters and a predisposition to stakeholder management.

The section on governance and devolution closes with the examination of a relatively unusual situation, the one of supranational governance as found in Europe (Chapter 21). The European Commission (EC) has raised port policy from the level of national debate to a trans-European discussion of what ports may or may not do about supporting particular port activities and port investments. Pallis provides a comprehensive overview of the state of Europe-wide port policies and notes that the spirit of the policies is in line with the EC's focus on integration but that this imposition of supranational governance will require port authorities to adopt new roles and redefine their relationships with their port users and service providers.

These discussions of governance provide considerable food for thought. What they have not done, however, is tie devolution and governance to performance. The final section of the book discusses the current state of performance evaluation in ports and looks at what needs to happen in a future research agenda to link performance and governance.

The section on performance begins with a paper by Talley (Chapter 22), which addresses the thorny issue of how a port's performance should properly be evaluated. The author asserts that in a competitive port environment, time-related costs are important determinants in port selection for both shippers and carriers. It follows, therefore, that the evaluation of port performance should be based on a comparison of actual with optimum throughput, contingent on achieving an economic objective (e.g., profit maximization). This underpinning assumption facilitates the analysis of port performance indicators which, adopting an economics perspective, constitute choice variables (i.e., variables whose values are under the control of port management, wherever that responsibility may lie) for optimizing the port's economic objective. Such an analysis will ultimately yield standards (or benchmarks) against which actual performance may be compared. Talley presents a basic economic model of a port that provides the theoretical foundations for understanding both the single-port approach and the multi-port approach to port performance evaluation. While the former is based on a comparison of optimum and actual performance indicators over time, the latter revolves around a cross-sectional or panel data analysis of a sample of ports to determine best practice optima. Talley concludes that this conceptual model provides a holistic understanding that allows the economic performance of a port to be evaluated from the standpoints of technical efficiency, cost efficiency and effectiveness.

In Chapter 23, Cullinane and Wang examine the most widely used method for evaluating port performance in a multi-port context – data envelopment analysis (DEA). The chapter begins by justifying the importance of measuring efficiency and the inadequacy of the more elementary engineering methods based on partial productivity measures. By reviewing the many applications of DEA to the port sector, the versatility of the method is revealed by the great diversity of analytical objectives that have been achieved through its application. The ease of use of DEA is then portrayed through an illustrative case study.

Ports are but a link in the global supply chain. Bichou (Chapter 24) notes that, despite the variety of performance models and measurement systems reported today, an integrative benchmarking approach is seldom adopted and performance measurements are often fragmented and biased towards sea access. He believes that ports need to examine their performance in the broader context of the total supply chain, and that by doing so, they will be better able to develop more appropriate measurement tools. He concludes that more research is needed with respect to the quantification of the performance model and its associated concepts.

These chapters all raise questions about performance and how it should be measured. Based on the adage that you cannot manage what you have not measured, Brooks (Chapter 25) looks at performance from a strategic management perspective and discusses port performance from both a firm level (for its own internal management) and a government program perspective. The chapter also provides some food for thought on the potential to learn from the airport industry, examining its benchmarking activities and looking for transferable lessons. In particular, airport performance measurement has moved into the domain of third-party customer satisfaction measurement, and the chapter calls for a similar approach for ports. It draws on performance measurement as it currently occurs in the 42 ports examined by the Port Performance Research Network to draw conclusions about which internal measures ports have found useful, which measures they report externally, and how governments may choose to measure the success of devolution programs and hence monitor the progress of port reform (devolution).

Finally, Brooks and Cullinane (Chapter 26) draw conclusions on the content of these chapters and set out a future research agenda for those interested in the issues of devolution, governance and port performance. Having derived reasonably conclusive evidence as to the inadequacy of existing models of port governance, a more appropriately complex model is proposed, based on the analysis reported in Chapter 18. This model is characterized not only by greater detail in the definition of groups of activities upon which ports must take governance decisions, but also by a much more refined definition of the range of governance typologies that may be selected by decision makers. In terms of defining the future research agenda in this field, the key objective must lie with investigating, and possibly identifying, the nature of the relationship between this newly expanded range of port governance configurations and port performance. Since this must be achieved for each of a variety of contexts that impact not only port performance, but also the definition of performance per se, it will not be a simple task. With the achievement of this objective, the opportunity exists for the port governance model developed and presented in Chapter 18 to be transformed from a mere portrayal of available decision alternatives (in the sense of positive economics) to a much more normative portrayal where specific governance models (or sets of them) are directly and explicitly linked to performance outcomes that are flexibly defined.

The road map for the book is presented in Fig. 1.3. It is intended to provide a quick reference to the three themes and where they are discussed.

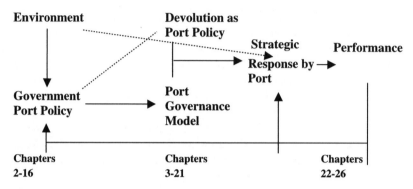

Fig. 1.3. A Road Map for the Book.

6. CONCLUDING REMARKS

There is no shortage of research on port performance and port efficiency. This book, however, was motivated by the absence of an effort to integrate the various pieces into a whole, one that would provide guidance to ports and governments. Given the sheer multiplicity of governance models implemented in the global rush towards new public management, it remains unclear if governments have accomplished that which they desired. There has been significant unhappiness with port reform activities on the part of taxpayers and citizens, and yet there is evidence that today's port network is more commercially responsive and efficient than was the case two decades ago. This book lays out the facts of what happened in a number of countries, presents assessments of the current state of play from both economics and strategic management perspectives, and from the viewpoints of ports, governments, stakeholders and supply chain partners. Taken all together, there is a rich tapestry of detail for those who would look back and a research agenda for those who would move forward without reinventing that which has already been done.

NOTES

1. Juhel (2001), for example, discussed the changing global environment and its impact on ports, institutions and regulations, while Peters (2001) documented the changing global shipping environment.

2. Button and Rietveld (1993) noted that the term *concession* is a derivation of the French word *concessionaire*.

3. Whether the commercialized body has recourse to the government is dependent on the nature of the commercialization. Commercialization involves the establishment of a new legal entity, owned by some level of government and reporting to a Finance or Transport Ministry. Perhaps a key element of the commercialization process, for those seeking the objective of an improved fiscal situation for the government, would be to remove ports from the purview of the Minister of Transport and transfer them to the Minister of Finance. This is a double-edged sword; Finance Ministers may see ports as cash cows, leaving them starved for investment when it is time to grow the business. On the other hand, Transport Ministers may have the tendency to interfere if public policy objectives are not being met. What a government decides to retain as responsibilities and what it decides to relinquish are the central governance issues in port reform.

4. In the parlance of UK company law, this amounts to the legal entity possessing a "corporate personality."

5. For Canada, Brooks (2004) noted that commercialization was primarily driven by the fiscal imperative of an unmanageable deficit. Everett (2003) indicated that Australia, for the most part, chose corporatization for two reasons: to improve port efficiency and to separate operations from government.

6. Stakeholders are groups affected by the decisions of the corporation, e.g., employees, customers, the greater community and advocates for the environment or for product safety.

7. The Port Performance Research Network was established in 2001 as a special interest group within the International Association of Maritime Economists. Its members are interested in the issues discussed in this book, and have contributed data to a number of its chapters. The network is chaired by Mary R. Brooks.

REFERENCES

Baird, A. J. (1995). Privatisation of trust ports in the United Kingdom: Review and analysis of the first sales. *Transport Policy, 2*(2), 135–143.

Brooks, M. R. (2004). The governance structure of ports. *Review of Network Economics: Special Issue on the Industrial Organization of Shipping and Ports, 3*(2), 168–183. http://www.rnejournal.com

Brooks, M. R., Prentice, B., & Flood, T. (2000). Governance and commercialization: Delivering the vision. *Proceedings, Canadian Transportation Research Forum, 1*, 129–143.

Button, K. J., & Rietveld, P. (1993). Financing urban transport projects in Europe. *Transport, 20*, 251–265.

Everett, S. (2003). Corporatization: A legislative framework for port inefficiencies. *Maritime Policy and Management, 30*(3), 211–219.

Gilman, S. (2003). Sustainability and national policy in UK port development. *Maritime Policy and Management, 30*(4), 275–291.

Goldsmith, S. (1997). Can business really do business with government? *Harvard Business Review, 75*(3), 110–122.

Goss, R. (1983). *Policies for Canadian seaports.* Ottawa: Canadian Transport Commission.

Goss, R. O. (1999). On the distribution of economic rent in seaports. *International Journal of Maritime Economics*, *1*(1), 1–9.

Juhel, M. H. (2001). Globalisation, privatisation and restructuring of ports. *International Journal of Maritime Economics*, *3*(2), 139–174.

Keasey, K., Thompson, S., & Wright, M. (1999). *Corporate governance*. Cheltenham, UK: Edward Elgar.

Kent, P. E., & Hochstein, A. (1998). Port reform and privatisation in conditions of limited competition: The experience in Colombia, Costa Rica and Nicaragua. *Maritime Policy and Management*, *25*(4), 313–333.

Li, K. X., Cullinane, K. P. B., & Cheng, J. (2003). The application of WTO rules in China and the implications for foreign direct investment. *Journal of World Investment*, *4*(2), 343–361.

Manning, N. (2000). *The new public management and its legacy*. Washington: World Bank, http://www1.worldbank.org/publicsector/civilservice/debate1.htm

OECD. (1999). OECD Principles of Corporate Governance (SG/CG(99)5). Paris: Organisation of Economic Co-Operation and Development, April.

Osborne, D., & Gaebler, T. (1992). Introduction: An American perestroika. *Reinventing government: How the entrepreneurial spirit is transforming the public Sector*. Reading, MA: Addison-Wesley.

Peters, H. J. F. (2001). Developments in global seatrade and container shipping markets: Their effects on the port industry and private sector involvement. *International Journal of Maritime Economics*, *3*, 3–26.

Rodal, A., & Mulder, N. (1993). Partnerships, devolution and power-sharing: Issues and implications for management. *Optimum, The Journal of Public Sector Management*, *24*(3), 27–48.

Starkie, D. N. M., & Thompson, D. J. (1985). The airports policy white paper: Privatisation and regulation. *Fiscal Studies*, *6*(4), 30–41.

Sternberg, E. (1998). *Corporate governance: Accountability in the marketplace*. London: Institute of Economic Affairs.

Thomas, B. J. (1994). The privatization of United Kingdom seaports. *Maritime Policy and Management*, *21*(2), 135–148.

World Bank Institute. (2000). http://www.worldbank.org/wbi/wbigf/governance.html

CHAPTER 2

STRATEGIC CHALLENGES TO CONTAINER PORTS IN A CHANGING MARKET ENVIRONMENT *

Theo Notteboom

ABSTRACT

The global market place, with powerful and relatively footloose players, extensive business networks and complex logistics systems creates a high degree of uncertainty in the port industry and leaves port managers with the question of how to respond effectively to market dynamics. The focus of port competition is gradually changing and so are the roles of the various stakeholders involved. This chapter provides a bird's eye view on the economic and logistics market developments affecting (container) ports. It identifies key market developments in trade and logistics and analyses how the economic and logistics trends affect seaport authorities.

---

*This chapter is partly based on the study 'Factual report – Work Package 1: Overall market dynamics and their influence on the port sector' commissioned by the European Sea Ports Organization (Notteboom & Winkelmans, 2004).

Devolution, Port Governance and Port Performance
Research in Transportation Economics, Volume 17, 29–52
Copyright © 2007 by Elsevier Ltd.
ISSN: 0739-8859/doi:10.1016/S0739-8859(06)17002-X

1. INTRODUCTION

The market environment in which seaports operate is changing. Ports are confronted with changing economic and logistics systems. The global market place, with powerful and relatively footloose players, extensive business networks and complex logistics systems have a dramatic impact on the raison d'être of seaports. The logistics environment creates a high degree of uncertainty and leaves port managers with the question of how to respond effectively to market dynamics. Port authorities and port management teams, whose objectives are significantly economic, are forced to re-assess their role and related governance structures.

This chapter focuses on the port market environment and its impact on the functioning of ports as logistics nodes in global networks. It gives a bird's eye view of the economic and logistics market developments affecting seaports. The paper mainly deals with containerised cargo logistics. In the first part key market developments in trade and logistics are identified. The second part analyses how the economic and logistics trends described in the first part affect seaport authorities. The chapter is targeted at institutions that govern and manage seaport operations and whose objectives are economic.

2. TRADE ROUTES

International trade represents a growing share of global output, and growth in trade is expected to outstrip overall growth in output for the foreseeable future. On the basis of current trends, international trade may grow to the equivalent of 30 percent of world output by 2010 (from its current level of around 15 percent). The rising significance of trade is a consequence of the increasing integration of the global economy. Legal and cultural obstacles to trade are diminishing at the same time as the motivation to trade is increasing. Integration is occurring both at the regional level, through initiatives such as NAFTA and the European Union (EU) Single Market, and at the global level, supported by the continuing evolution of WTO.

The last three decades have seen important modifications in international trading flows. The bulk of international trade occurs within economic blocks, especially the EU and NAFTA. Other significant flows are between Asia-Pacific and North America (especially the United States), between Europe and North America and between Europe and Asia-Pacific.

The unprecedented economic growth of East Asia transformed the patterns of world trade. The Asian financial crisis in the summer of 1997 meant a temporary setback for a number of countries. China, in particular, now attracts a lot of international attention. As one of the world's most rapidly growing economies, China has achieved an average GDP growth of 9 percent, which it has been able to maintain since 1979 (UNCTAD, 2002). The Chinese economic boom is reflected in the liner service schedules of major shipping lines. The liner trade speaks of the China effect. Shipping lines are dedicating higher capacities and deploying larger vessels to cope with increasing Chinese containerised imports and exports (Yap, Lam, & Notteboom, 2003). The China effect has also resulted in changes to the ranking of the world's largest container ports (Table 2.1). Average annual growth at container ports in mainland China (excluding Hong Kong) is elevated (Table 2.2). The rising throughput figures appear to justify investment efforts to keep capacity in line with growing demand, with most major ports either earmarked for expansion, or already undergoing massive construction, e.g. the enormous terminal development at Yangshan near Shanghai.

Structural changes are also taking place in other regions of the world. For instance, Western European markets are becoming mature. The total

Table 2.1.　World Container Port Ranking in 2004 (in Million TEU).

Rank	Port	Country	mTEU
1	Hong Kong SAR	**China**	21.93
2	Singapore	Singapore	21.33
3	Shanghai	**China**	14.55
4	Shenzhen	**China**	13.66
5	Busan	South Korea	11.43
6	Kaohsiung	Taiwan	9.71
7	Rotterdam	The Netherlands	8.22
8	Los Angeles	United States	7.32
9	Hamburg	Germany	7.00
10	Dubai	United Arab Emirates	6.42
11	Antwerp	Belgium	6.06
12	Long Beach	United States	5.78
13	Port Kelang	Malaysia	5.24
14	Qingdao	**China**	5.14
15	New York/New Jersey	United States	4.47
16	Tanjung Pelepas	Malaysia	4.02
17	Ningbo	**China**	4.01
18	Tianjin	**China**	3.81
19	Laem Chabang	Thailand	3.62
20	Tokyo	Japan	3.58

Source: Porter (2005).

Table 2.2. Container Throughput in Chinese Ports.

Million TEU	1990	1998	2002	2004	Average Growth per Year	
					1998–2002 (%)	2002–2004 (%)
Shanghai	0.46	3.07	8.61	14.55	45	34
Shenzhen	0.03	2.06	7.61	13.66	67	40
Qingdao	0.14	1.21	3.41	5.14	45	25
Tianjin	0.29	1.02	2.41	3.81	34	29
Guangzhou	0.08	0.85	2.17	n.a.	39	–
Ningbo	0.00	0.35	1.86	4.01	107	58
Xiamen	0.03	0.65	1.75	n.a.	42	–
Dalian	0.13	0.53	1.35	n.a.	39	–
Jingmen	0.00	0.00	0.49	n.a.	–	–
Fuzhou	0.00	0.06	0.48	n.a.	176	–

Source: Porter (2005).

market volume in Europe's most important countries and in traditional market sectors, such as consumer goods or automotive, are showing moderate growth rates, which contrast with the boom in these markets in the 1970s and 1980s. Economic development in Central and Eastern Europe is expected to grow significantly in the future, with forecast annual GDP growth of between 4 and 4.8 percent until 2009 (European Commission, 2001). Northern ports, in particular Hamburg, are likely to benefit the most from the recent EU enlargement, whereas new development opportunities might arise for secondary port systems in the Adriatic and the Baltic Sea. Both in the Baltic and the Mediterranean, extensive hub-feeder container systems and short sea shipping networks came into existence in the last decade to cope with the increasing volumes and to connect to other European port ranges (the Hamburg–Le Havre range in particular).

As the Russian market has become a close neighbour of the new EU due to the latter's enlargement, Russia might develop a transit function for goods flow between Asia-Pacific and the EU. Plans even exist to connect the Eurasian land-bridge system to the North-American land bridges by creating a fixed link over the Baring Straits. The feasibility of such a link remains uncertain due to a high investment cost, harsh climatic conditions and the low economic potential along the new link. Another project having a high potential impact on the routing of trade flows is the expansion of lock capacity in the Panama Canal. The design of the new locks would allow Aframax and Suezmax vessels to pass through the Canal. It would also allow post-Panamax container ships to make the passage from the Pacific to the Atlantic, thereby opening opportunities for container lines to introduce new round-the-world service concepts.

3. TRENDS IN LOGISTICS

3.1. Globalisation

One of the main driving forces for change in the port industry emerges from the globalisation of production. Multinational enterprises (MNE) are the key drivers of global production networks and associated distribution networks. A shift has taken place from capital-intensive activities – such as ownership and management of a large number of manufacturing sites, distribution centres and sales outlets – towards another type of activity, which is far less capital intensive and focuses more on developing a strong brand. Branding forms a key concept in the new business model of MNEs. This involves a strong focus on customers and product innovation, whereas production is outsourced to a network of suppliers (Christopher, 1992). Many of the world's largest MNEs manage extensive networks of globally dispersed inputs. Global sourcing as such is a major driver of world trade and has deeply affected transportation and distribution patterns and related cost structures. Advances in IT have made it possible for MNEs to promote their brands and to exert control over the quality and reliability within supply chains, without having ownership of supply factors.

3.2. Shifts in Supply Chains

Logistics models evolve continuously as a result of influences and factors such as globalisation and expansion into new markets, mass customisation in response to product and market segmentation, lean manufacturing practices and associated shifts in costs. Service expectations of customers are moving towards a push for higher flexibility, reliability and precision. In many industries, product innovation has become a large competitive factor. As a result, average product life cycles and supply chain cycles have decreased. The number of products to be shipped and the shipment frequency has increased, whereas batch sizes are becoming smaller. There is a growing demand from the customer for 'make-to-order' or 'customised' products, delivered at maximum speed, with supreme delivery reliability, at the lowest possible cost. The focus is on supply chain excellence, with superior customer service and lowest cost to serve.

As a result, international supply chains have become complex and the pressure on the logistics industry is increasing. In 2003, logistics costs amounted to 9.6 percent of sales. The worldwide logistics industry is expected to grow by 4–5 percent every year (IBM, 2003). The supply chains need to be supported by a wide range of advanced communication tools and new powerful, reliable and cost-effective transportation networks need to be set up and operated by IT-supported logistics service providers.

3.3. Outsourcing: Third Party Logistics and Fourth Party Logistics on the Rise

Leading-edge companies are taking a broader view of the parts of their business they seek to control and manage. The re-engineering of supply chain processes (including customer order management, procurement, production planning, distribution, etc.) to enhance performance typically results in collaborative networks with logistics partners. Many companies have acknowledged that warehousing and transportation are not part of their core business and as a result these operations are outsourced to logistics service providers. A study by CGE&Y (2002) indicates that in Europe 94 percent of companies have already outsourced part of their warehousing and transportation operations to logistics service providers. IBM (2003) reported that the share of outsourced logistics would grow worldwide by 15–20 percent, and in Europe by 9 percent every year.

Increasing customer demands drive the Third Party Logistics (3PL) service industry forward. Customers' need for a wider array of global services and for truly integrated services and capabilities (design, build and operate) triggered a shift from transportation-based 3PLs to warehousing and distribution providers and at the same time opened the market to innovative forms of non-asset-based logistics service provision, i.e. Fourth Party Logistics (4PL). Whereas a 3PL service provider typically invests in warehouses and transport material, a 4PL service provider restricts its scope to IT-based supply chain design. Consultants and IT shops help 3PLs and 4PLs expand into new markets and become full-service logistics providers. The competence of 4PLs lies in the selection, linking and bundling of service providers as well as the alignment of interests of all concerned in the supply chain.

Notwithstanding the emergence of non-asset-based 4PLs in parts of Europe, Asia and North America, they will not take over the role of key player in the worldwide logistics market from the asset-based 3PLs. Hence, asset-based full service providers such as the express integrators DHL and FedEx increasingly install the IT control systems themselves. Moreover, many logistics users prefer to keep control of the design of the supply chain in-house instead of being totally dependent on 4PLs. The distinction between 3PL and 4PL is not always straightforward. An existing 3PL chain player may act also as a 4PL in the same chain.

3.4. Logistics Integration and Consolidation in the Logistics Service Provider Industry

Globalisation and outsourcing open new windows of opportunities for shipping lines, forwarders, terminal operators and other transport operators. Manufacturers are looking for global logistics packages rather than just straight shipping or forwarding. Most actors in the transport chain have responded by providing new value-added services in an integrated package, through a vertical integration along the supply chain.

The vertical integration in the logistics industry has enabled many freight forwarders to take control of larger segments of the supply chain. For instance, European companies like Danzas (since 1999 part of Deutsche Post), Schenker/BTL (the merger between Schenker Logistics and Scansped), Frans Maas and Kuehne + Nagel evolved from basic forwarders or road haulage companies to full logistics service providers (Fig. 2.1). With an increasing level of functional integration, many intermediate steps in the transport chain have been removed. Mergers and acquisitions have permitted the emergence of large logistics operators that control many segments of the supply chain: the megacarrier. The megacarriers meet the requirements of many shippers to have a single contact point on a regional or even global level (the 'one-stop shop'). Technology has also played a particular role in this process, namely in terms of IT (control of the process) and intermodal integration (control of the flows).

Mergers and acquisitions shape the contemporary business environment. Mergers and acquisitions are not only driven by companies searching for takeover candidates, but also by companies which have decided to divest aspects of their businesses and are consequently looking for buyers of these businesses. For instance, in the last decade a large industry consolidation has started in the European logistics service provider industry, triggered by companies such as ABX Logistics and Deutsche Post.

Consolidation and vertical integration strategies have created a logistics market consisting of a wide variety of service providers ranging from megacarriers to local niche operators. Not only does the geographic coverage of the players differ (from global to local), but also major differences can be observed in the focus (generalist versus specialist), in the service offering

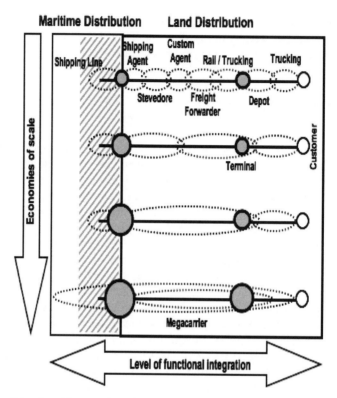

Fig. 2.1. Functional Integration of Supply Chains. *Source:* Notteboom & Rodrigue
(2005), adapted from Robinson (2002).

(from single service to one-stop shop) and in asset-orientation (asset-based
versus non-asset-based).

3.5. New Approaches Towards Logistics Networks

Logistics networks are being reconfigured. Two main developments concern
the decoupling of order and delivery and the transition from chains to
networks (see Fig. 2.2). The decoupling of order and delivery coincides with
a stronger role for IT and web-based platforms. The transition from chains
to networks implies that the chain manager now has a number of alternatives available to route cargo through the logistics chain, ranging from delivery via a main distribution centre (MDC) or via a regional distribution
centre (RDC) to various solutions based on the inclusion of cross-dock

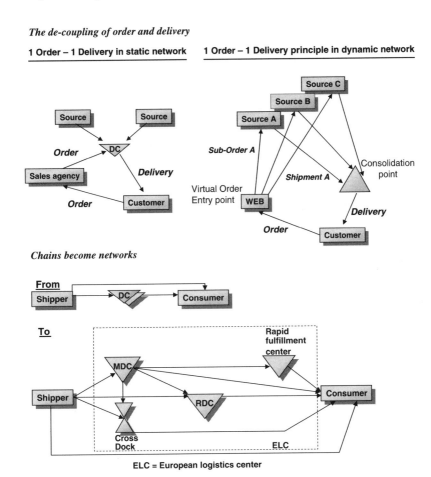

Fig. 2.2. The Reconfiguration of Logistics Networks. *Note:* DC = distribution centre, MDC = Main DC, RDC = Regional DC. The principle of cross docking means that the products are almost immediately transferred from the discharge area to the load area (no temporary storage). *Source:* Buck Consultants International (1997).

facilities or a rapid fulfilment centre. Distributions systems have to adapt to the new requirements.

Supply chains are being redesigned to respond to varying customer and product service-level requirements. The variables that affect site selection are numerous and quite diverse and can be of a quantitative or qualitative nature (cf. centrality, accessibility, size of the market, track record regarding reputation/experience, land and its attributes, labour (costs, quality, productivity), capital (investment climate, bank environment), government policy and planning (subsidies, taxes) and personal factors and amenities). Many companies fall back on intuition and rules of thumb in selecting an appropriate site. Logistics service providers have developed powerful tools to assist shippers in selecting an appropriate network configuration and in site selection.

Many products need to be made country- or customer-specific (labelling, kitting, adding manuals in local languages, etc.) before they can be delivered to the customer. Historically these country- or customer-specific activities were mostly done in the factory, and this led to high inventory levels. Owing to the increasing variety of products and shorter product life cycles, many companies have chosen to move their country- and customer-specific kitting or assembly operations as close to the customer as possible. This implies that the traditional storage and distribution functions of many distribution centres are supplemented by semi-industrial activities such as the 'customising' and 'localising' of products, adding components or manuals, product testing, quality control or even final assembly. These activities are referred to as value-added logistics services (VAL). While setting up their logistics platforms, logistics service providers favour locations that combine a central location (i.e. proximity to the consumer market) with an intermodal gateway function. Seaports and sites along hinterland corridors typically meet these requirements.

4. TRENDS IN LINER SHIPPING AND THE CONTAINER TERMINAL INDUSTRY

4.1. Scale Increases in Vessel Size

Lines operate regular, reliable and frequent services and incur high fixed costs. Once the large and expensive networks are set up, the pressure is on to fill them with freight. Throughout the 1990s a great deal of attention was devoted to larger, more fuel-economic vessels and this indeed produced substantial reductions in cost per twenty-foot equivalent unit (TEU) of capacity provided (Table 2.3). Larger ships typically have a lower cost per TEU-mile than smaller units with the same load factor (see, e.g. Cullinane & Khanna, 1999; Drewry Shipping Consultants, 2001; Gilman, 1999). Given that there seem to be no technical reasons preventing containerships from getting larger, it will be economic and operational considerations that will act as the ultimate limit on post-Panamax vessel sizes and designs of the future.

Table 2.3. Scale Increases in Vessel Size: Evolution of the World Cellular Fleet 1991–2006.

	January 1991	Shares (%)	January 1996	Shares (%)	January 2001	Shares (%)	January 2006	Shares (%)
> 5000 TEU	0	0.0	30,648	1.0	621,855	12.7	2,355,033	30.0
4000/4999 TEU	140,032	7.5	428,429	14.4	766,048	15.6	1,339,978	17.1
3000/3999 TEU	325,906	17.6	612,377	20.6	814,713	16.6	892,463	11.4
2000/2999 TEU	538,766	29.0	673,074	22.6	1,006,006	20.5	1,391,216	17.7
1500/1999 TEU	238,495	12.8	367,853	12.3	604,713	12.3	719,631	9.2
1000/1499 TEU	329,578	17.7	480,270	16.1	567,952	11.6	596,047	7.6
500/999 TEU	191,733	10.3	269,339	9.0	393,744	8.0	438,249	5.6
100/499 TEU	92,417	5.0	117,187	3.9	132,472	2.7	114,976	1.5
Total	1,856,927	100.0	2,979,177	100.0	4,907,503	100.0	7,847,593	100.0

Note: Projection in January 2006 as compiled with existing fleet and order book as on 15 June 2003.
Source: BRS Alphaliner (2003).

4.2. Cooperation, Mergers and Acquisitions in Liner Shipping

Horizontal integration in liner shipping comes in three forms: trade agreements such as liner conferences, operating agreements (i.e. vessel sharing agreements, slot chartering agreements, consortia and strategic alliances) and mergers and acquisitions (M&A).

The merger of P&OCL and Nedlloyd in 1997 to become P&O Nedlloyd and the takeover of SeaLand by Maersk in 1999 are two of the well-known M&A in liner shipping. In 2005, a new level was reached by the takeover of P&O Nedlloyd, the third largest container shipping line, by Maersk, the world's number one, and the takeover of CP Ships by the mother company of Hapag-Lloyd. The economic rationale for M&A is growth to achieve economies of scale, market share and market power. Other motives for M&A in liner shipping relate to gaining instant access to markets and distribution networks, obtaining access to new technologies or diversifying.

The top 20 carriers controlled 26 percent of the world slot capacity in 1980, 41.6 percent in 1992 and about 58 percent in 2005 (see also Table 2.4). More important than which carriers are in the top 20 is the fact that only a few container carriers outside the top 20 operate post-Panamax vessels and most of the top 20 carriers are involved in multi-trade strategic alliances. The most important alliances are the Grand Alliance (P&O Nedlloyd, OOCL, Hapag-Lloyd, NYK and MISC), the Cosco/K-Line/Yangming Alliance, the United Alliance (Hanjin and Senator) and the New World Alliance (APL, Hyundai and Mitsui OSK Lines).

Strategic alliances provide their members with easy access to more loops or services with relative low cost implications and allow them to share terminals and cooperate in many areas at sea and ashore, thereby achieving cost savings in the end. Mergers and acquisitions have led to the reshuffling of partners across alliances. Midoro and Pitto (2000) and Graham (1998) argue that despite these advantages of alliance formation, strategic alliances have not become a stabilising factor in liner shipping due to the organisational complexity of the alliance and perceived intra-alliance competition that undermines trust between the carriers involved.

4.3. The Emergence of Global Terminal Operators

In a response to the concentration trend that is unfolding in container shipping, a number of terminal operators have opted for scale increases. P&O Ports is set to join Hutchison, PSA and APM Terminals at the head of the global port operator league table. These companies have established a truly

Table 2.4. Slot Capacity Operated by the Top 20 Shipping Lines.

	January 1980 Carrier	Slot Capacity	September 1995 Carrier	Slot Capacity	January 2000 Carrier	Slot Capacity	May 2005 Carrier	Slot Capacity
1	SeaLand	70,000	SeaLand	196,708	AP Moller-Maersk	620,324	AP Moller-Maersk	1,051,350
2	Hapag-Lloyd	41,000	Maersk	186,040	Evergreen	317,292	MSC	687,621
3	OCL	31,400	Evergreen	181,982	P&O Nedlloyd	280,794	P&O Nedlloyd	464,769
4	Maersk	25,600	COSCO	169,795	Hanjin/DSR senator	244,636	Evergreen Group	442,962
5	NYK line	24,000	NYK line	137,018	MSC	224,620	CMA/CGM Group	411,723
6	Evergreen	23,800	Nedlloyd	119,599	NOL/APL	207,992	NOL/APL	316,389
7	OOCL	22,800	Mitsui OSK Lines	118,208	COSCO	198,841	COSCO	299,452
8	Zim	21,100	P&OCL	98,893	NYK line	166,206	China shipping	298,589
9	US line	20,900	Hanjin shipping	92,332	CP Ships/Americana	141,419	Hanjin/Senator	294,532
10	APL	20,000	MSC	88,955	Zim	136,075	NYK line	287,250
11	Mitsui OSK lines	19,800	APL	81,547	Mitsui OSK lines	132,618	OOCL	239,274
12	Farrell lines	16,400	Zim	79,738	CMA/CGM	122,848	CSAV Group	215,336
13	NOL	14,800	K-Line	75,528	K-line	112,884	K line	213,266
14	Trans freight line	13,900	DSR-Senator	75,497	Hapag-Lloyd	102,769	Hapag-Lloyd	212,890
15	CGM	12,700	Hapag-Lloyd	71,688	Hyundai	102,314	Zim	211,107
16	Yang Ming	12,700	NOL	63,469	OOCL	101,044	Mitsui OSK lines	205,724
17	Nedlloyd	11,700	Yang Ming	60,034	Yang Ming	93,348	CP Ships Group	195,177
18	Columbus Line	11,200	Hyundai	59,195	China shipping	86,335	Yang Ming	185,758
19	Safmarine	11,100	OOCL	55,811	UASC	74,989	Hamburg-Süd	162,158
20	Ben Line	10,300	CMA	46,026	Wan Hai	70,755	Hyundai	148,681
Slop capacity of top 20		435,000		2,058,063		3,538,103		6,544,008
C4-index (%)		38.6		35.7		41.4		40.4
Share top 5 in top 20 (%)		44.1		42.3		47.7		46.7
Share top 10 in top 20 (%)		69.1		67.5		71.7		69.6

Source: Compiled from various issues of **BRS** Alphaliner and Containerisation International (1996, 2001 and 2006).

global presence, collectively operating in over 90 ports in 37 countries. In developing a global expansion strategy, HPH, PSA Corp, APM Terminals and P&O Ports try to keep a competitive edge by building barriers to prevent competitors entering their domains or succeeding if they do. For example, PSA Corporation first built a stronghold at its home base of Singapore before taking the step towards global scale and coverage. Once the company established itself as an international benchmark, its ambitions went global through a mixed strategy of organic growth (new terminals) and acquisitions (e.g. HesseNoordNatie in 2002) backed up by a sound financial status.

Smaller terminal operators have not been successful in neutralising the power of these giants. Many of them avoid direct competition by concentrating on market niches, for example in the short sea market. By 2008, the top four operators will control over one-third of total world container port capacity (Drewry Shipping Consultants, 2003). For instance, in Europe the top six leading operators handled nearly 70 percent of the total European container throughput in 2002 (Table 2.5), compared to 53 percent in 1998, illustrating the mature and consolidated nature of this market. These figures are expected to rise as consolidation continues and the big players plan new terminals.

4.4. An Increased Focus on Landside Logistics

From the viewpoint of a shipping line, inland logistics is one of the most vital areas left in which to cut costs. In a typical intermodal transport operation, inland transport now accounts for a much larger component of the cost than running the vessel. Inland costs constitute 40–80 percent of the total costs of container shipping.

Carriers that traditionally have been concerned only with the transportation of goods from one point to another are now seeking logistics businesses in the area of just-in-time inventory practices, supply chain integration and logistics information system management. With only a few exceptions, however, the management of pure logistics services is done by subsidiaries that share the same mother company as the shipping line but operate independently of liner shipping operations, and as such, also ship cargo on competitor lines (Heaver, 2002).

Some shipping lines such as Maersk Sealand have gone rather far in door-to-door services and integrated logistic packages (i.e. Maersk Logistics), managing the container terminal operation (i.e. APM Terminals with a network of dedicated terminals that has been opened to third-party users as well) and inland transport (e.g. European Rail Shuttle in joint venture with

Table 2.5. Global Terminal Operators' Presence in Europe.

	Worldwide Throughput 2002	European Throughput 2002	European Throughput 1998	Annual Growth Europe 1998–2002 (%)
Hutchison Port Holding (HPH) – China Felixstowe (UK), Thamesport (UK), Harwich (UK), ECT-Rotterdam (The Netherlands)	36.70	6.90	7.75	-2.7
PSA Corp – Singapore Voltri-Genoa (Italy), Sines (Portugal), VECON-Venice (Italy), HesseNoordNatie-Antwerp/Zeebrugge (Belgium)	26.20	5.44	0.60	201.7
APM Terminals – Denmark Bremerhaven (Germany), Rotterdam (The Netherlands), Algeciras (Spain), Gioia Tauro (Italy, 10% stake)	17.20	3.24	1.00	56.0
P&O Ports – UK Antwerp (Belgium), Marseille/Le Havre (France, joint-venture CMA-CGM), Southampton (UK), Tilbury (UK)	12.80	2.76	1.25	30.2
Eurogate – Germany Eurokai-Hamburg (Germany), BLG-Bremen (Germany), La Spezia (Italy), CICT-Cagliari (Italy) Medcenter-Gioia Tauro (Italy), Liscont-Lisbon (Portugal), Livorno (Italy), Salerno (Italy)	9.59	9.59	5.73	16.8
HHLA – Germany Hamburg (Germany)	4.00	4.00	2.35	17.6
Total of six major European container terminals operating companies	106.49	31.93	18.68	17.7
Grand total	275.00	46.50	35.06	8.2
Share six operators in grand total (%)	38.7	68.7	53.3	

Note: Figures include all terminals in which non-minority shareholdings were held.
Source: Based on terminal operator data and Drewry Shipping Consultants (2003).

P&O Nedlloyd) and bypassing the freight forwarder by developing direct relationships with the shipper. Other shipping lines stick to the shipping business and try to enhance network integration through structural or ad hoc coordination with independent inland transport operators and logistics service providers. A last group of shipping lines combines a strategy of selective investments in key supporting activities (e.g. agency services or distribution centres) with subcontracting of less critical services. Shipping lines generally do not own inland transport equipment. Instead they attempt to use trustworthy independent inland operators' services on a (long-term) contractual basis.

The formation of global alliances has taken intercarrier cooperation to new heights, with members sharing inland logistics information, techniques and resources as well as negotiating collectively with suppliers (terminals, rail operators, feeders, barge operators, etc.). By extending to the landside, alliances clearly differ from older forms of operating agreements.

Inland and container logistics thus constitute an important field of action for shipping lines. Lines that are successful in achieving cost gains from smarter management of inland and container logistics can secure an important cost savings advantage.

Also, the leading terminal operating companies have developed diverging strategies towards the control of larger parts of the supply chain. The door-to-door philosophy has transformed a number of terminal operators into logistics organisations. The services offered include warehousing, distribution and low-end, value-added logistical services (e.g. customising products for the local markets). The recent focus of Hutchison on inland logistics in China is an example.

Not every terminal operator is integrating by acquiring or setting up separate companies or business units. In many cases, effective network integration is realised through better coordination with third-party transport operators or logistics service providers.

4.5. Changes in Liner Service Network Design

Liner service network design has tended to move from a pure cost-driven exercise to a more customer-oriented differentiation exercise, as the optimal network design is not only a function of carrier-specific operational factors, but more and more of shippers' needs (for transit time and other service elements) and of shippers' willingness to pay for a better service. The reality of deep-sea operations is that even the largest ships operate on multi-port itineraries. Alliances and consolidation have created multi-string networks on the major trade routes and both shippers and liners are used to it. As liner service network design has become a more customer-oriented differentiation exercise, this could very well introduce a tendency towards less transhipment and more direct ports of call (even for the bigger vessels).

Asian carriers such as APL, Hanjin, NYK, China shipping and HMM typically focus on intra-Asian trade, trans-Pacific trade and the Europe–Far East route, partly because of their huge dependence on export flows generated by their Asian home bases (Frémont & Soppé, 2003). MOL and Evergreen are among the few exceptions frequenting secondary routes such as Africa and South America. Many of these carriers have allocated 70–80 percent of their slot capacity to a strategic alliance. Maersk Sealand, MSC, CMA-CGM and P&O Nedlloyd are among the truly global liner operators, with a strong presence also in secondary routes.

4.6. Concluding Remarks

The essence of shipping lines' existence is gradually shifting from pure shipping operations to integrated logistics solutions. Through various forms of integration along the supply chain, shipping lines are trying to generate revenue, streamline sea, port and land operations and create customer value. For the time being, container terminal operators are mainly focused on increasing the scale of operations. Global terminal operators clearly have shifted their mindset from a local port level to a port network level, albeit that the terminal network effects still have to be exploited to the full.

5. THE IMPACT ON SEAPORTS AND SEAPORT AUTHORITIES

The above sections underline that ports are confronted with everchanging economic and logistics systems. The global market place, with powerful and relatively footloose players, extensive business networks and complex logistics systems, have a dramatic impact on the raison d'être of seaports. Port authorities and port management teams whose objectives are significantly economic, are forced to re-assess their role and specify competencies that should lead to competitive advantage and should position the port for growth. This part attempts to provide a bird's eye view of the way the economic and logistics trends described in earlier sections affect ports and how seaport authorities can deal with related challenges.

5.1. Port Competition and Logistics Chains

The integration strategies of the market players have blurred the traditional division of tasks in the logistics chain and, as such, have created an environment in which ports are increasingly competing, not as individual places that handle ships but within transport chains or supply chains. The logistics chain has become more than ever the relevant scope for analysing port competitiveness. This also implies that a port's competitiveness has become increasingly dependent on external coordination and control. Port choice has become more a function of network costs and port selection criteria are related to the entire network, in which the port is just one node. The ports being chosen are those that will help to minimise the sum of sea, port and inland costs.

Port authorities are forced to approach port development and management issues in the broader framework of logistics chains. This observation demands closer coordination with logistics actors outside the port perimeter and a more integrated approach to port infrastructure planning and concession policy. For instance, terminals are not an end in themselves. Efficient cargo handling facilities contribute to the industrial and logistics development in the port area and the hinterland.

5.2. Dealing with Powerful Port Users

Logistics integration in the transport industry results in a concentration of power at the port demand side. Seaports increasingly have to deal with large port clients who possess strong bargaining power vis-à-vis terminal operations and inland transport operations.

Ports and port authorities essentially have to deliver value to each and every customer (shippers or their representatives, shipping lines and other stakeholders) and capture value. Given the market environment, value creation is not an easy task. It demands the identification of what the various types of customers really want and how port managers can play a role in the value creation process.

In the contemporary logistic-restructured port environment it has become more difficult to identify the port customers who really exert power in the logistic chain or who are driving port selection. In some cases, the chain manager is situated at the end of the chain. For instance, supermarket chains (Wal-Mart, Carrefour) exert power over the supply lines of food products. In these high-volume logistics chains, a seaport is seen as a bundling point, a buffer within the scope of inventory management and/or a fast transit point. In other logistics chains, commodity traders have a major impact on the routing of cargo. Large forwarding agencies negotiate rates with shipping lines and route the cargo they manage according to a combination of determinants such as price, transit time and reliability. Large shippers often have direct contracts with one or more shipping lines for their worldwide shipments. The applicable terms of sale and terms of trade play a key role in the controlling power in the logistics chain (buyer side or seller side). As such, the question of who really decides which port to choose depends on the type of cargo involved, the cargo generating power of the shipper, the characteristics related to specific trade routes and the terms of trade and terms of sale. Identifying the real chain manager/decision maker is a key challenge to port managers. In some cases, market players are both port user and port service supplier at the same time (e.g. a shipping line operating a dedicated terminal).

While cooperation at the operational level between the actors in the supply chain may have increased, this has not necessarily resulted in increased commitment to a long-term future relationship with the port. The purchasing power of the large intermodal carriers, reinforced by strategic alliances between them, is used to play off one port or group of ports against another. The loyalty of a port client cannot be taken for granted. Ports face the constant risk of losing important clients, not because of deficiencies in port infrastructure or terminal operations, but because the client has rearranged its service networks or has engaged in new partnerships with other carriers. Because of the sheer size of the port users, the loss or the acquisition of a customer may in some cases imply losing or acquiring 10 to even 20 percent of the port's container traffic.

Port authorities should facilitate private port companies to 'capture' some shippers and carriers that control huge cargo flows and which are in a good position to generate value-added for the port region. If a seaport wants to attract or retain some of the megacarriers, it has to position itself as an efficient intermodal hub and distribution service centre acting within extensive transport and communications networks. The success of a port (authority) will depend on the ability to integrate the port effectively into the networks of business relationships that shape supply chains. In other words, the success of a seaport no longer exclusively depends on its internal weaknesses and strengths. It is being more and more determined by the ability of the port community to fully exploit synergies with other transport nodes and other players in the logistics networks of which they are part. To be successful, port authorities have to think along with the customer, trying to figure out what his needs are, not only in the port but also throughout the logistics chains and networks. This demands the creation of a platform in which various stakeholders (carriers, shippers, transport operators, labour and government bodies) are working together to identify and address issues affecting logistics performance. Port authorities can be a catalyst in this process, even though their direct impact on cargo flows is limited.

5.3. Seaports as Habitats for Logistics Services

Seaports are key constituents of many supply chains and their pre-eminent role in international distribution is unlikely to be challenged in the fore-seeable future. The gateway position of major seaports offers opportunities for the development of value-added logistics (VAL). A seaport can evolve from a pure transhipment centre to a complex of key functions within a logistics system. In cooperation with other parties involved, port authorities can actively stimulate the logistics development of port areas through the enhancement of flexible labour conditions, smooth customs formalities (in combination with freeport status) and powerful information systems.

Warehouses are traditionally located just behind the terminals. At present, logistics activities can take place on the terminal itself, in a logistics park where several logistics activities are concentrated or, in the case of industrial sub-contracting, on the site of an industrial company. While there is a clear tendency in the container sector to move away from the terminal, in other cargo categories an expansion of logistics on the terminal itself can be witnessed. As such, a mix of pure stevedoring activities and logistics activities occurs.

In the new logistics market environment, the following logistics activities typically find a good habitat in seaports:

- Logistics activities resulting in a considerable reduction in the transported volume.
- Logistics activities involving large volumes of bulk cargoes suitable for inland navigation and rail.
- Logistics activities directly related to companies that have a site in the port area.
- Logistics activities related to cargo that needs flexible storage to create a buffer (products subject to seasonal fluctuations or irregular supply).
- Logistics activities with a high dependency on short sea shipping.

Moreover, port areas typically possess a strong competitiveness for distribution centres in a multiple import structure and as a consolidation centre for export cargo.

The changing logistics environment poses new challenges in the relations between seaports and inland ports. A large number of port authorities promote an efficient intermodal system in order to secure cargo under conditions of high competition. This includes, for example, involvement in the introduction of new shuttle train services to the hinterland, together with the respective national railway companies, rail operators, terminal operators, shipping companies and/or large shippers.

5.4. New Port Dynamics

In the old port model, neighbouring ports of comparable size competed for cargo to and from a shared hinterland. However, changes in liner service networks and larger ships have forced previously non-competing ports into head-to-head competition. This has led to changes in the port hierarchy.

First, seaports located far from each other are now to some extent competing. Second, the new requirements related to deep-sea services do not necessarily make the existing large container ports the best locations for setting up hub operations. That is why the position of the large load centres is to some extent threatened by medium-sized ports and new hub terminals. New terminal facilities might give shipping lines and alliances more opportunities to use their bargaining power to play off one port against another.

The consolidation process in the container handling industry also has a large impact on individual ports. First, the large terminal operators are becoming more footloose as the network approach loosens their former strong ties with one particular seaport. Second, competition is shifting from port authorities to private terminal operators who are trying to establish terminal networks. Third, the influx of overseas capital in seaports, together with the consolidation in the cargo handling business, has created circumstances in which some stevedoring companies have acquired a very strategic position in a port's future. The key position of such terminal operators inevitably attracts a lot of attention from the local port community because it wants to ensure that the economic rents of these terminal operations stay local.

5.5. Increased Focus on Containers in Port Investments

Containerisation has become a must for ports, as the provision of container facilities is considered to be one of the prerequisites for success in the newly logistics-restructured environment. In many cases the capacity of a port authority to respond to the challenges is determined by the extent to which it can secure financial support from its local community or national government. In the past, many governments around the world have funded the majority of large infrastructure works in ports. The gradual withdrawal of governments from the financing of container terminal infrastructure might confront even the largest and most prosperous port authorities with severe financial pressures in attempting to keep their competitive edge in an industry too often guided by the belief that the best workable strategy to defeat competitors is building new highly efficient terminals.

5.6. Seaports and the Community

A large part of the community takes the port and maritime industry for granted and is ignorant of how the industry is organised and operated and to what extent it contributes to global trade and local economies. More attention is given to the fact that these industries generate negative effects, such as road congestion in and around ports, the use of scarce land, pollution (oil spills) and a lack of safety. Environmental and safety considerations are very prominent in community groups' strategy. The economic value of a port development project now tends to be taken as given, so the argument concentrates on the environmental criteria (e.g. dredging and dredge disposal, loss of wetlands, emissions into the air, water pollution, congestion, loss of open space, light and noise externalities, potential conflicts with commercial fishing and recreational uses of water areas). Port authorities and port companies must demonstrate a high level of environmental performance in order to ensure community support. However, environmental aspects also play an increasing role in attracting trading partners and potential investors. A port with a strong environmental record and a high level of community support is likely to be favoured. In this respect, port authorities have a leading and catalytic role to play in the 'greening' of port operations and management.

6. CONCLUSIONS

The current issues faced by port managers are multiple and complex. The global market place, with powerful and relatively footloose players, extensive business networks and complex logistics systems, creates a high degree of uncertainty in the port industry and leaves port managers with the question of how to respond effectively to market dynamics. The focus of port competition is gradually changing, as are the roles of the various stakeholders involved.

The market trends described in this chapter are a clear invitation to port managers to reassess existing organisational and management structures in seaports, as well as port strategy. The traditional concept of a seaport being a landlord or a total organisation with single and/or multiple facilities in a single location is no longer straightforward. The modern port concept leads to a comprehensive organisation taking care of multiple services across multiple locations. Containerisation and intermodality revolutionised modern shipping as well as hinterland transportation so deeply that there is definitely a need to re-assess the role and functions of (container) ports.

The competitive position of a seaport does not depend solely on its administrative structure; it is more a matter of commercial attitude and mentality. Port economics has indeed become more a matter of management style. Port management objectives nowadays are much more directed towards efficiency than to distributional equity. Commercialised and/or corporatised ports might find difficulties in avoiding politicisation of the so-called technocratic port organisations, as they often rely on external political decisions, especially in the case of government-funded port investments. The establishment of an appropriate legislative framework that guarantees an efficiency-oriented approach is one of the main challenges to port policy makers.

Modern ports must be capable of accommodating larger port clients, who possess strong bargaining power vis-à-vis terminal and inland transport operators. As such, port authorities must not expect to attract cargo simply because they are natural gateways to rich hinterlands. Major port clients concentrate their service packages not on the ports' sea-to-land interface, but on the quality and reliability of the entire logistics chain. Capturing and keeping important footloose clients on a sustainable basis requires integrated services characterised by a high level of reliability and flexibility, short time-to-market, as well as non-market conditions such as transparency within efficient governance structures.

ACKNOWLEDGMENT

The author would like to thank ESPO for the permission granted for making use of the report in preparing this chapter.

REFERENCES

BRS Alphaliner. (2003). Fleet report.
Buck Consultants International. (1997). *Europese distributie en waardetoevoeging door buitenlands bedrijven (European distribution and value added logistics by foreign companies)*. Nijmegen: Buck Consultants International.

CGE&Y. (2002). *EU enlargement: European distribution centres on the move?* Available as a PDF from http://www.nfia.com/logistics/pdfs/EUreport.pdf

Christopher, M. (1992). *Logistics and supply chain management: Strategies for reducing costs and improving services.* London: Pitman Publishing.

Containerisation International. (1996, 2001 and 2006).

Cullinane, K. P. B., & Khanna, M. (1999). Economies of scale in large container ships. *Journal of Transport Economics and Policy, 33*(2), 185–208.

Drewry Shipping Consultants. (2001). *Post-Panamax containerships – the next generation.* London: Drewry.

Drewry Shipping Consultants. (2003). *Annual review of global container terminal operators.* London: Drewry.

European Commission. (2001). *The economic impact of enlargement.* EC, DG EcFin, Office for Official Publications of the European Communities, Brussels.

Frémont, A., & Soppé, M. (2003). *The service strategies of liner shipping companies.* Research seminar: Maritime transport, globalisation, regional integration and territorial development, Le Havre.

Gilman, S. (1999). The size economies and network efficiency of large containerships. *International Journal of Maritime Economics, 1,* 5–18.

Graham, M. G. (1998). Stability and competition in intermodal container shipping: Finding a balance. *Maritime Policy and Management, 25,* 129–147.

Heaver, T. (2002). The evolving roles of shipping lines in international logistics. *International Journal of Maritime Economics, 4,* 210–230.

IBM Business Consulting Services. (2003). *Opportunities and challenges for logistics service providers in Europe.* Brussels: IBM.

Midoro, R., & Pitto, A. (2000). A critical evaluation of strategic alliances in liner shipping. *Maritime Policy and Management, 27,* 31–40.

Notteboom, T., & Rodrigue, J.-P. (2005). Port regionalization: Towards a new phase in port development. *Maritime Policy and Management, 32,* 297–313.

Notteboom, T., & Winkelmans, W. (2004). *Factual report – Work Package 1: Overall market dynamics and their influence on the port sector.* Study commissioned by the European Sea Ports Organisation (ESPO).

Porter, J. (2005). China extends top 20 presence. *Lloyd's List*, March 16.

Robinson, R. (2002). Ports as elements in value-driven chain systems: The new paradigm. *Maritime Policy and Management, 29,* 241–255.

UNCTAD. (2002). *Review of maritime transport.* Geneva: UNCTAD.

Yap, W. Y, Lam, J. S. L., & Notteboom, T. (2003). Developments in container port competition in East Asia. *Proceedings of IAME 2003 conference, International Association of Maritime Economists*, Busan (South Korea) (pp. 715–735).

PART II:
PORTS AND PORT POLICY

CHAPTER 3

PORT PRIVATISATION IN THE UNITED KINGDOM

Alfred J. Baird and Vincent F. Valentine

ABSTRACT

This chapter considers developments and changes in port policy and regulation occurring in the United Kingdom over the last few decades. This provides relevant contextual background relating to the issues and arguments surrounding the sale of former state-owned ports in the UK, the different methods of disposal employed, the associated liberalisation of dockworking and the legislative framework. It is argued that there were two distinct phases of port privatisation in the UK, with phase I involving the sale of state-owned ports and railway ports in the early 1980s, and phase II the disposal of major trust ports. However, it remains that the method and approach used to privatise ports in the UK differs markedly from the privatisation process for ports in most other countries, very few actually selling off port land. Port privatisation in the UK was never about creating new and improved port infrastructure and facilities to benefit the economy, which was the aim in other countries; it was simply a mechanism used to remove port assets from public ownership. This raises issues related to the aims and objectives of seaports and the role of government in this regard. The wider purpose of seaports in facilitating trade and generating economic and social

Devolution, Port Governance and Port Performance
Research in Transportation Economics, Volume 17, 55–84
Copyright © 2007 by Elsevier Ltd.
ISSN: 0739-8859/doi:10.1016/S0739-8859(06)17003-1

benefits still tends to be stressed by public-owned (but private operated) ports in most other countries, whereas the narrower profit-making goal of private enterprise is paramount in the UK.

1. INTRODUCTION

Privatisation can encompass many meanings. For example, privatisation may involve the sale of existing state-owned assets, the use of private financing and/or management or the contracting out of certain tasks to the private sector (Gomez-Ibanez & Meyer, 1993). This implies that there may be, in fact, wide variations in the actual extent and processes of privatisation adopted in any given situation.

The modern word 'privatisation' came into being during the late 1960s and was later attributed to the UK government's reforms on ownership and operation of numerous companies managed by the state. Chapman (1990) has accredited Drucker (1969) as the author of the word 'privatization,' in its American spelling. The actual process of implementing privatisation is not however a 20th Century concept, nor can it be said to have originated in the UK. Historians have shown that private port concessions have apparently been around since Roman times (Grosdidier de Matons, 1997).

It was this 20th Century christening of an established process, a renaissance of an earlier idea on company ownership that brought the word 'privatisation' into everyday vocabulary. Although the UK privatised much of its industry in the late 20th Century it does not mean that all industry capable of being privatised has been. Yet in order to understand privatisation more fully it is necessary to examine how the concept has evolved during recent times. This chapter considers in some detail the privatisation experience in the UK during the last two decades, and ideas and concepts surrounding it, before focusing more specifically upon privatisation of UK seaports.

Since 2000, the UK government has raised the equivalent of US$7.2 billion from privatising airports, motorways and other former state-owned assets. The total revenue produced from all UK privatisations between 1980 and 2005 stands at over US$121 billion from 186 separate issues.[1] In Europe, the cumulative value for all privatisations to 2005 stands at over US$641 billion. Gibbon (1998, 2000) states that all revenue raised through government privatisations worldwide exceeded US$1 trillion sometime during the second half of 1999, at which time Europe accounted for approximately half. Gibbon (2000) also states that the amount raised (from privatisation) was approximately US$140 billion worldwide.[2]

2. PRIVATISATION

2.1. From Nationalisation to Privatisation

Before considering the privatisation process it is necessary to look briefly at how major industries first came into the hands of governments. In the UK, most of the nationalised industries were created during the post Second World War, 1945–1951, Attlee government (Parker, 1997). In essence, the reasons for nationalisation were twofold. First, to take into the hands of government, and thereby the people, those industries and utilities that were viewed as essential services (i.e. water, gas, electricity and telecommunications). Chick (1987) stated that one of the main aims of nationalisation was to secure 'improved coordination' in sectors such as fuel, power, transport and communications. Second, during the 1950s the Labour Party's aim was to take over the 'commanding heights of the economy' so that private entrepreneurs would not be able to make exorbitant profits from the people (Chapman, 1990).

Nationalisation principally occurred in two stages. The first occurred in the late 1940s under the post-war Labour government and consisted of those industries that were of a public utility nature such as coal, electricity, gas and public transport. The notion prevalent at this time was that these industries had suffered from a lack of investment during the period between the two world wars, 1918–1939, and that it would be best if they were brought under the wing of a centrally planned government. The second stage again came under a Labour government in the 1970s, this time mainly consisting of transport-related industries such as British Aerospace, British Shipbuilders, British Leyland and International Computer Limited. The exception to these two phases is British Steel, whose ownership constantly changed from being a nationalised industry during the 1940s, to a privatised one in the 1950s, to become renationalised in the 1960s and subsequently reprivatised in the 1980s (Hare & Simpson, 1996). In addition, British Airways, together with other airlines, were nationalised in 1939 to form the British Overseas Airways Corporation, but in 1974 reverted back to its former name, and then in 1987 was privatised.

One of the key arguments for nationalisation was a belief that if nationalised industries had the competition element removed from their business, this would enable them to concentrate on their main purpose, which was to provide a service to the consumer. The historian Keegan (1999) gave four reasons for the Labour government's decision to nationalise. First, the extinction of internal credit to finance production combined with the collapse of exports forced industry to seek government help. Nationalising these industries therefore seemed an obvious way of controlling how this assistance was spent. Second, the Labour Party was committed by Clause 4 of its constitution to ownership of the means of production. Although the policy was not implemented to the extent expected by its full Marxist origins, the Party felt that some degree of control was necessary in order to support the voters who placed it in power because of its policies. Third, during the years following the First World War, the owners of many industries sought assistance from the government because of what was seen as their own incompetence and inefficiency in handling their businesses. These owners of the means of production were the descendants of the original Victorian entrepreneurs who were now viewed by the middle classes as no more capable, if not worse, at running their enterprises than government. Fourth, the decline of the British Empire brought those that had managed the means of production within industry in places like India the opportunity to apply their skills in the homeland's ready-made institutions, e.g. the Milk Marketing Board, the Transport Commission and the Gas Council.

The post-war years saw a marked increase in the size of the public sector, from 25 percent of UK GDP in 1946 to 52 percent in 1970 (Jackson & Price, 1994). The turmoil and suffering experienced during the Second World War were to be put aside for a new epoch in which the state controlled much of industry, and where people were listened to through the voice of unions. By 1979, 13 million of a total workforce of 20 million belonged to unions. Salaries rose by 15, 19 and 23 percent in each year between 1973 and 1975 (Keegan, 1999). Despite this, workers benefiting from wage rises were also becoming disenchanted with union disruption. By the end of 1978, there were 48 public corporations, including nationalised industries as well as partially owned enterprises such as British Petroleum and bodies like the National Enterprise Board, the latter set up to provide government backing for private companies (Semple, 1979). According to Gough (1979), all the advanced capitalist countries exhibited the same trends in the late post-war period, namely rising public expenditure as a share of GDP, and social expenditure rising as a share of public expenditure.

2.2. Towards a New Era in Industry Ownership

In 1974, Margaret Thatcher was appointed leader of the Conservative Party, and in 1979 under her leadership the party won the general election. From an historical point of view, this marked a turning point in the debate on nationalised industries, for the process that followed heralded a reversal of the political objectives of much of the previous three and a half decades. The process the Conservatives decided to adopt was one of privatisation. This was not a new idea and had been included in early 1970s' political manifestos (see Howell, 1970) but not, notably, in the 1979 manifesto. In fact, the previous Labour government had privatised 32 percent of British Petroleum in 1977 but what the new 1979 Conservative government did was to accelerate the process. Thatcher, in describing privatisation, called it one of the "central means of reversing the corrosive and corrupting effects of socialism" (Thatcher, 1993, p. 676); regulation that in the public sector had been "covert" had through privatisation become "overt and specific," thus providing a clearer and better discipline (p. 677).

The economic cost in terms of subsidies was growing and the profits, if any, of these companies were decreasing. The government's social obligations were expanding and required growth in taxes to fund them. British Airways, for example, suffered a deficit of £544 million in 1981–1982. This deficit had to be funded by the government through revenue collected in taxes. During the 1970s, the financial and industrial troubles of British Leyland (automobiles), British Steel and British Coal dominated the Cabinet's agenda. Their borrowing and subsidy requirements were a constant drain on public finances. In 1979, losses from nationalised industries were costing each taxpayer £300 a year at 1995 prices (Anonymous, 1995). Punnett (1994) stated that by the 1970s the public sector employed over a quarter of the country's workforce yet accounted for only 10 percent of GDP. Therefore, it became imperative for the government to reduce this burden by selling industries, and to instead receive income from corporation tax on future profits. This combined increase in tax and income from industry sales in exchange for an asset that was costing money seemed like an attractive solution for the government. However, in order to privatise a company it first had to be profitable, otherwise nobody would want to buy its shares. It was also sometimes necessary to implement an element of deregulation within industries in order that the market could be liberalised and competition introduced.

Privatisation in Britain can thus be divided into two phases. The first phase, 1979–1983, involved the sale of companies with no real characteristics that would justify their continued retention within the public sector (e.g. Amersham International, Britoil, British Rail Hotels, International Computer Limited, Ferranti Cable & Wireless, British Aerospace, British Freight and British Transport Docks Board). The British Transport Docks Board (BTDB), which was subsequently renamed Associated British Ports (ABP), thus came at the end of this first phase of privatisations with a 51.8 percent listing on the stock exchange. During the second phase of privatisations, the government sold the remaining 48.2 percent of ABP. The second phase began with the sale of British

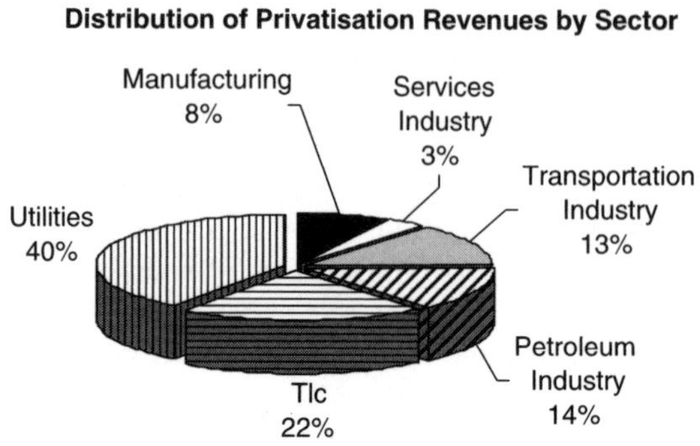

Distribution of Privatisation Revenues by Sector

Fig. 3.1. UK Privatisation Revenues by Sector. *Source:* www.privatizationbarom-
eter.net (2004).

Petroleum in 1984 and continues to the present day, even though Labour
ousted the Conservative government in May 1997. Previously, the Labour
Party had stated that it would renationalise all of the industries privatised by
the Conservatives, thus 1984 was the beginning of a new era that saw the
flotation of some of the largest corporations in the world, the first time this had
ever been attempted anywhere (i.e. national 'public utility industries' such as
telephone, gas, water and electricity), and in which total output for the entire
economy is produced by one organisation (see Fig. 3.1 and Table 3.1).

3. UK PORT POLICY AND REGULATION

3.1. Operational Framework

According to Goss (1997), there has been no single uniform port policy in
the UK since 1945, but instead a movement from central planning to privatisation. Although a National Ports Council (NPC) was established in
1965, its role was not accepted by government or by the ports themselves.
The Department of Transport even increased its own staffing dramatically
during the years of the NPC in order to second-guess the NPC's actions
(Wilson, 1983). By the time the NPC was abolished in 1981, no study had
been published on the efficiency of UK ports. This was strange given that
one of the roles of the NPC was to administer a national ports policy that
surely by its very nature would require an investigative study to establish
how the ports industry as a whole was performing.

Table 3.1. UK Privatisation Revenues.

Year	Value of Transaction	Public Offers (US$million)	Private Sales (US$million)	Public Offers	Private Sales	Total Transactions
1977	972	972	—	1	0	1
1978	—	—	—	0	0	0
1979	694.6	694.6	—	1	0	1
1980	—	—	—	0	0	0
1981	667.6	667.6	—	2	0	2
1982	1,021.90	1,021.90	—	2	0	2
1983	1,281.60	1,281.60	—	3	0	3
1984	5,508.90	5,508.90	—	4	0	4
1985	2,072.80	2,072.80	—	3	0	3
1986	7,943.94	7,610.40	333.54	1	4	5
1987	17,751.82	17,715.30	36.52	4	4	8
1988	5,120.31	4,673.40	446.91	1	7	8
1989	6,947.83	6,909.50	38.33	9	3	12
1990	9,137.51	8,558.50	579.01	10	1	11
1991	19,644.92	18,761.10	883.82	3	5	8
1992	643.81	51.3	592.51	1	10	11
1993	8,263.19	8,075.70	187.49	2	10	12
1994	787.93	—	787.93	0	16	16
1995	9,455.35	7,573.70	1,881.65	3	15	18
1996	7,863.66	5,357.50	2,506.16	4	21	25
1997	862.31	—	862.31	0	15	15
1998	6,661.22	117.8	6,543.42	1	6	7
1999	917.48	761.2	156.28	2	2	4
2000	166.31	154.5	11.81	1	2	3
2001	1,438.48	—	1,438.48	0	3	3
2002	10.93	—	10.93	0	1	1
2003	5,778.70	—	5,778.70	0	3	3
Total	121,615.10	98,539.30	23,075.80	58	128	186

Source: www.privatizationbarometer.net, 2004.

It was suggested that individual ports became so opposed to the NPC that they maintained that each port was so different from all the others that no comparable efficiency indicators could possibly be devised (Goss, 1997). This view, although disputed by Livesey (1997), no doubt contributed to the government's desire for change and what the Conservative government often referred to as the idea of 'rolling back the frontiers of the state.' Wilson (1983) maintained that efficiency figures tended to be deliberately withheld by ports because of their poor efficiency rating when compared to continental ports, thus preventing the NPC from making adequate plans.

The deregulation policy pursued by the Conservative government was designed to remove obstacles hindering the operation of market forces, and to set a level playing field for trading activities. Similar deregulation policies in Italy, Spain and Portugal had prompted other European countries such as Greece to consider its options in the context of the European Union's policy on economic and regulatory harmonisation (Goulielmos, 1999).

The UK port privatisation process began with the closure of the NPC in 1981 followed by the privatisation of the newly formed ABP in 1983 and 1984. In 1989, the government then repealed the Dockworkers Act of 1946 (Regulation of Employment) and abolished the National Dock Labour Scheme (NDLS). The NDLS applied to many of the UK ports, with the notable exception of Felixstowe which, due to its good employee relations and because it was considered not to be a significant port (in the early 1960s), was never included in the scheme (Baird, 1998). Prior to containerisation, Felixstowe simply consisted of a relatively small tidal dock. It was only in the early 1970s, after the US carrier Sea-Land Service Inc. had helped to develop a dedicated container terminal at the Suffolk port, that Felixstowe began to expand. By the late 1970s Felixstowe was starting to attract some of the new Asian container lines, most of which had no tradition of port calls in the UK and simply selected the most cost-effective port alternative available at the time.

The Scheme, which operated in some 60 ports, legally prevented anyone other than a registered dock worker from performing dock work, and was said to have resulted in highly excessive manning levels throughput the ports industry (Ross, 1995). The ambiguity of the term 'dock work' kept lawyers in business for some 40 years, according to Finney (1990). Juhel and Pollock (1999), quoting from an unnamed study, stated that 79 percent of former registered dockworkers became redundant after the scheme was abolished, of which 19 percent wished to remain active but could not find work. Some 55 percent found employment elsewhere while 25 percent re-entered the port industry.

The abolition of the NDLS removed restrictive and archaic employment regulations and helped to create an environment for the introduction of a range of new and flexible employment practices, which was urgently required due to fundamental technological changes affecting the industry. Turnbull and Weston (1993) argued that abolition of the NDLS resulted in excess capacity through over-investment in the UK ports industry, citing the Isle of Grain (Thamesport Container Terminal) as an example of excess container facilities in the South East of England that merely attracted customers from nearby Felixstowe, rather than creating 'new business'. However, such a claim appears far less relevant today given the serious lack of deep-water container port capacity in the UK.

Thompson (1987), in a review of UK transport privatisation, concluded that there was more to the process than just removing restrictive legislation. He argued for a greater division of the newly privatised industries so as to create more companies and thereby promote efficiency through increased competition. Another challenge concerned the valuation of state assets; generally speaking, the government managed to dispose of ports at what subsequently proved to be very cheap, heavily discounted prices, representing a substantial loss of value to the state (Baird, 1995; Goss, 1999; Saundry & Turnbull, 1997; Sidery, 1994).

3.2. Legislative Framework

After the UK Government abolished the NPC, whose main function was to develop a national plan for ports, the emphasis from this period onwards was on privatisation and increased competition to ensure greater efficiency, with the private sector promoting its own port expansion schemes in line with demand. The focus of Government policy was on the need to regulate the sector, though today for the most part competent port authorities are themselves regulators within port areas, and many (competent port authorities) are now privately owned. At the time of privatisation, the relevant port authority was dissolved in most cases, with the private successor company thereafter assuming all statutory duties and responsibilities formerly held by the authority. Curiously, and unlike all other privatised utilities such as energy, water, railways, aviation and buses, the UK ports industry has never been given its own specific regulator (e.g. Of-Port, or Office of the Port Regulator). In addition, government policy remains separated from commercial and managerial decision-making, allowing the private sector to determine its own port investment priorities.

The UK Government's non-interventionist policy approach was confirmed in its 2000 ports policy paper *Modern Ports* (DETR, 2000), the first

UK ports policy paper in 30 years. While a number of initiatives arose from this work, such as a project appraisal framework for ports, environmental best practice and improved statistics, there was to be no Government funding for ports (unlike virtually all other transport modes).

The Government's policy for ports is mainly focusing on guidance as opposed to direct intervention (DETR, 2000). Government will

> support sustainable port projects for which there is a clear need, with each looked at in detail on its merits; however, this support relates only to the planning process and not to any financial support. (DETR, 2000, p. 5)

The Government takes full account of the need for good access to ports in developing policies and programmes for the various forms of transport, and encourages the use of ports by coastal and short sea shipping services. The absence of public sector investment in ports nevertheless seems discriminatory in the sense that infrastructure for other major transport modes, and road and rail in particular, remains heavily subsidised. Whereas government expects the market to deliver sustainable sea transport and coastal/short sea shipping solutions, at the same time the state itself invests heavily in competing land transport modes, resulting in a distorted market.

The main legislative Acts applying to UK ports today include the *Harbours Act 1964*, the *Pilotage Act 1987* and the *Ports Act 1991* (ESPO, 2004). The increasing environmental responsibilities of harbour authorities and their role in coastal management extend the range of legislation affecting statutory harbour authorities. Harbours are for the most part founded under individual Acts of Parliament, which often embodied the provisions of the 1847 *Harbours, Docks and Piers Clauses Act*. Under these Acts, ports developed sets of bye-laws to provide their own legislative framework. Some ports also have General Powers of Direction to regulate shipping movements and provide safe navigation within designated harbour waters. In some cases, these are used in place of bye-laws and are generally viewed as a more flexible alternative. Laws relating to a particular harbour authority are therefore contained in their local Acts and orders, as well as in general law.

In the case of privatised and privately owned ports, this does of course mean that private actors have now become regulators within their own defined areas of jurisdiction, which in turn raises questions to do with free and fair competition, and the potential for monopolies arising, and so on. Although it is the case that the UK has gone further than any other country in transferring port regulatory powers to private port successor companies, it remains that there is no specific state regulatory body for ports. The UK government maintains that port users can object to matters like port charges by making representations to the Minister concerned. That this option is seldom used implies port users are either not always aware that it exists, or simply do not have the time to follow it up. There appears to be a need for research into this issue in order to ascertain if there are abuses of market power and privately held regulatory powers. The alternative is simply to consider, as the government appears to do, that because there are few complaints, the system must therefore be assumed to function well.

3.3. Responsibilities and Decision-Making

As harbour authorities are formed under specific Acts of Parliament, this can by implication impose a duty on the entity concerned to establish and maintain the service in question. The main responsibilities and functions of harbour authorities can be described as follows:

- To provide and maintain harbour facilities
- To ensure safe navigation within harbour waters by providing lighting and buoys, removing wrecks and maintaining approach channels of sufficient depth through dredging
- To regulate vessel movements and berthing in the harbour
- To license construction works within the harbour area
- To provide pilotage service and other harbour operations such as cargo handling

Private ports are responsible to shareholders, trust ports to a range of stakeholders and municipal ports to local communities via their local council. One of the 'Modern Ports' initiatives was to review the structure, governance and accountability of the trust port sector and this work is now complete (DETR, 2000). A similar review of municipal ports started in early 2004 and should be completed in 2005. However, there has been no review of the private ports sector, and there is unlikely to be one, as the Government assumes the private sector already acts efficiently and in the interests of port users, based on the notion that port users have a choice of port supplier. (This is not so evident in the context of major port activities such as deep-water container terminals, RoRo ferry terminals, and port-connected oil refineries.) Given that the private sector now accounts for the vast majority of the UK ports industry, one might expect government from time to time to take a more informed look at its activities and workings. For the last 15–20 years, most of the UK's major ports have been operated as private companies, yet there has never been any government-sponsored research to find out precisely how effective these ports are, in a local, regional, national as well as an international context.

The UK Government therefore takes a very hands-off approach to port management and investment decisions. Decisions on investment are made by individual port authorities/port owners and are approved by their board of directors/trustees based on the commercial viability of the proposal. Companies are required to raise capital for port investments, even for major infrastructure such as dredged channels, navigation aids, harbour protection and increasingly local road and rail access improvements.

New port developments normally require a Harbour Revision Order (HRO). The harbour authority applies to the Secretary of State for the Order. Objections are then invited, and if they are not withdrawn a Public Inquiry is held. The Inspector who conducts the Inquiry then submits a report to the Secretary of State in the light of which the Secretary of State decides whether or not to make the Order. Applications for a HRO have to be supported by an environmental impact assessment and a special assessment if a Natura 2000 site is involved. Environmental issues usually feature prominently at a Public Inquiry.

Over the last few years there have been several Public Inquiries held in the UK to consider major port development schemes. In 2003–2004, a Public Inquiry was held into the proposed development of a major new container terminal at the ABP-owned Port of Southampton (Dibden Bay). The Inspector and subsequently the Government decided that this scheme should not proceed on environmental grounds, and permission was therefore refused (ABP, 2005). Similar Public Inquiries were held during 2004 into proposed new container terminals at London Gateway (Thames), Harwich (Bathside Bay) and Felixstowe (Felixstowe South). The London Gateway project has been given approval by the Secretary of State, subject to agreement on necessary landside investments, and a decision on the other two schemes is expected before the end of 2005.

Five further major deep-water container port schemes are being proposed in the UK. Three of these are in England – Teesport, Liverpool and Bristol – and two in Scotland – Hunterston on the Firth of Clyde and Scapa Flow in Orkney (Lewis, 2005). Scottish Ministers in Edinburgh are expected to play a role in regard to the latter two proposals given that the Scottish Executive also has responsibility for ports.

3.4. Cargo Handling

Cargo-handling services may be carried out by the port entity itself, by port subsidiary companies or by companies specialising in its provision. Regardless of who carries out cargo handling, such activities receive no government or other subsidy. In practice, ports often undertake some cargo handling themselves and in many cases all or nearly all of it. UK ports now see their role not just as modal interchange points but as part of the integrated supply chain. For certain types of traffic, ports are increasingly making arrangements with cargo owners for a range of logistics services to be offered within port areas, and which may include warehousing, packaging and landward distribution as well as cargo handling. Some ports have also integrated forward to the extent that they now operate short sea shipping services (e.g. Clydeport, MDHC).

Since the UK Government abolished the NDLS, normal employment legislation has applied to employer and employees, who are free to establish their own contractual employment relationship. These agreements can be negotiated at individual, port or company level, whereas previously under the Scheme employment negotiations were undertaken at national level. Today, earnings of port employees are subject to local negotiation. Training is provided by port employers at their own expense. Port users have argued that the most important change in UK ports in recent years was actually abolition of the NDLC (John, 1995; Ross, 1995) rather than privatisation per se; in this sense, who owns the docks is apparently of rather less relevance to a shipowner as long as cargo handling is efficient, reliable and affordable.

Self-handling (i.e. ship's crew handling cargo) is largely confined to the RoRo sector, but it can also include the lashing of vehicles, although the practice varies from port to port. Some ports do not permit self-handling of cargo, on the grounds that this would amount to cherry-picking and would not be in the overall interests of the port. The need for expensive equipment and high investment generally militates against self-handling of cargo, especially for lift-on lift-off containers, though this is rather less the case for RoRo traffic, and especially driver-accompanied vehicles. In a number of countries, truck drivers are permitted to load and unload RoRo ferries (e.g. Turkey and Japan), and this appears to be an increasing trend, as reflected in ongoing attempts by the European Commission to bring forward its Port Services Directive, which is intended to further liberalise the sector (*Containerisation International*, 2005).

In the case of cargo handling, the most obvious approach for a would-be stevedore is to purchase or take over a lease on a terminal. In common-user terminals, the service provider would need to approach the owner of the terminal (who may or may not also be the port authority, and which may or may not be privately owned). Some ports operate a licensing system for cargo handling, but most do not and there is no legal requirement to do so. For towage, a special authorisation is needed for safety reasons, albeit there is no standard regime for operator selection or appeal procedures.

With regard to contracts, there is no standard duration, but terminal concessions of 15–25 years are not unknown (Institute of Shipping & Logistics, 2005). In individual cases, the length of contract will tend to reflect the scale of investment involved. The largest investments can result in concessions of more than 35 years. Many terminals are either owned outright or operated on long leases; this includes, for example, oil terminals and many of the numerous wharves on the Thames and Humber. Extensions are frequently negotiated, often in association with a proposal for new investment. Early terminations usually arise only when the operator runs into financial difficulties, in which case the port owner sometimes takes over the operation itself. There is no specific legal framework governing these arrangements. In certain ports and terminals, the number of cargo handlers is restricted by the port authority/owner up to the maximum number that the traffic volume is thought able to sustain. In other cases the port, as a matter of policy, undertakes all handling of certain types of cargo. In several UK ports that are owned and administered by the one port company, and within which there may be a number of different port and terminal facilities, there is just the one cargo-handling supplier, and usually this is under the control of the port-owning company itself (e.g. Forth, Clyde).

3.5. Pilotage

The 1987 Pilotage Act placed responsibility for marine pilotage on Competent Harbour Authorities (CHAs) who are usually the port authorities (irrespective whether public or private). Pilots may be either employees of the CHA or self-employed, but the CHA sets the rules for pilotage and the charges. It also issues Pilotage Exemption Certificates (PECs) to suitably qualified regular users of a port (e.g. ferries) and UK legislation provides that regular users of a port are entitled to apply for a PEC and these are widely issued. Prior to 1987, pilotage was administered by independent pilotage authorities; the 1987 changes were a simplification that the government considers has generally helped to reduce the costs of pilotage without jeopardising safety. The 1987 Act provides that competent harbour authorities are responsible for provision of pilotage services: they authorise the pilots, determine the rules and set the charges. Pilots may be employees of the port, self-employed or employed by a company that contracts for the provision of the service.

Responsibility for towage within defined port areas is fulfilled in most ports by private towage undertakings, and in some instances by the port itself. Towage services are, in most cases, provided by companies specialising in their provision. However, in some cases towage companies may be owned or partly owned by port authorities. There are no restrictions on the number of providers, but in practice in many ports there is only enough towage business to support one operator.

4. UK PORT SECTOR TODAY

4.1. Major UK Ports

There are approximately 1,000 ports and terminal/wharf facilities in the UK. Of these, over 650 have statutory powers, of which about 120 are commercially active. The industry is today far more consolidated and much of the trade is concentrated in the largest ports; in 2003, the top 20 ports handled 85 percent of all UK port traffic (Table 3.2). Out of a total 555.7 million tonnes, the top 20 ports handled 470.7 million tonnes. Fifteen of the top 20 ports are privately owned, while three are trust ports and two are municipal ports.

London, Southampton and Liverpool are large multi-purpose ports, each handling a very wide range of traffic types. Felixstowe's primary focus is on containers, although the port also handles hundreds of thousands of RoRo trailers. The main focus of five of the top 20 ports is on a single traffic-type, either dry or wet bulk. Overall growth in dry/wet bulk traffic is only around one percent per annum, according to Department for Transport statistics, whereas growth in the unitised freight sector is higher, at between five and seven percent per annum.

Fig. 3.2 shows that 66 percent of UK port volumes are controlled by private ports among the top 20, 12 percent by trusts and 7 percent by municipal ports; collectively 85 percent of UK port volumes are handled by these top 20 ports The remaining 15 percent of tonnage is handled by a myriad of smaller ports outside the top 20, of which some are also privately owned. This implies that approximately 70 percent of the UK ports industry by tonnage is today handled by privately owned ports (see Fig. 3.3 for the location of the major UK ports).

Grimsby & Immingham was the UK's leading port in 2003, handling 55.9 million tonnes, followed by Teesport with 53.8 million tonnes and London with 51.0 million tonnes. Fourth and fifth largest were Forth with 38.8 million tonnes and Southampton with 35.8 million tonnes. These five ports accounted for 42 percent of all UK port traffic. Eleven other ports each handled in excess of 10 million tonnes in 2003, and collectively these ports accounted for a further 37 percent of all UK traffic.

Table 3.2. Major Ports of United Kingdom, 2003 (in Million Tonnes).

Rank	Port	Million Tonnes	Ownership	Main Focus
1	Grimsby & Immingham	55.9	Private	Dry/wet bulk; RoRo; vehicles
2	Tees port	53.8	Private	Dry/wet bulk; RoRo; vehicles
3	London	51.0	Private (1)	Dry/wet bulk; containers; RoRo
4	Forth	38.8	Private	Wet bulk; containers
5	Southampton	35.8	Private	Wet bulk; containers; vehicles; cruise
6	Milford Haven	32.7	Trust	Wet bulk
7	Liverpool	31.7	Private	Containers; RoRo; wet/dry bulk; passengers
8	Sullom Voe (Shetland)	26.4	Municipal	Wet bulk
9	Felixstowe	22.3	Private (1)	Containers; RoRo
10	Dover	18.8	Trust	RoRo; passengers
11	Medway (Sheerness)	15.6	Private	Vehicles; dry bulk
12	Orkney	14.4	Municipal	Wet bulk
13	Belfast	13.2	Trust	Containers; RoRo; passengers
14	Bristol	11.4	Private (2)	Vehicles; dry bulk
15	Hull	10.5	Private	RoRo; passengers; containers
16	Rivers Hull & Humber	10.0	Private	Wet/dry bulk
17	Clyde	9.2	Private	Wet/dry bulk; containers
18	Port Talbot	7.8	Private	Dry bulk
19	Manchester	6.1	Private	Dry/wet bulk
20	Glensanda	5.3	Private	Dry bulk
	Total top 20	470.7	85%	
	Other UK ports	85.0	15%	
	Total	555.7	100%	

Notes:
(1) Public trust is responsible for navigation/pilotage only.
(2) Port is leased to private operator by the municipality.
Source: 2003 Maritime Statistics, UK Department for Transport

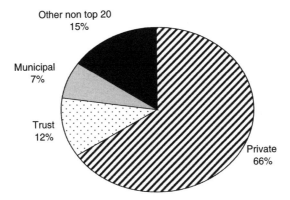

Fig. 3.2. Top 20 UK Ports (Plus 'Other' Ports, by Share of Tonnage/Ownership Type) (2003).

Looking at the historic trends, total tonnage handled by UK ports has risen by three quarters since 1965. However, the number of freight units (containers and trailers) has increased more than fivefold since 1970, with the number of road trailers rising from 0.4 to 6.3 million in 2003, and the number of containers rising from 1.5 to 4.5 million units. About 70 percent of UK ports' volumes are concerned with bulk and general cargoes such as liquid bulk, grain, steel, cars and agricultural produce. Since the 1990s, volumes in these cargoes have declined in favour of lighter and higher value container and RoRo traffic. Liquid bulk is considered to have the lowest value-added potential as it is generally pumped through pipelines, whereas potential for port value-added in the unitised sector tends to be much greater.

The industry therefore mostly consists of private ports, in addition to trust and municipal ports, which compete with each other and operate as stand-alone, self-financing commercial enterprises.

4.2. Company and Privatised Ports

The former British Transport Docks Board (BTDB) was a publicly owned nationalised industry (like British Railways) and subject to normal constraints on investment and borrowing applying to all publicly owned industries. The BTDB was privatised via a public flotation on the stock market in 1983 and is now known as Associated British Ports plc. Currently,

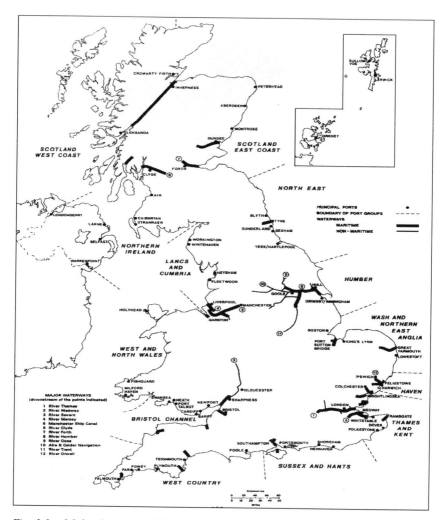

Fig. 3.3. Major Ports and Inland Waterways in the UK. *Source:* Department for
Transport.

ABP owns and operates 21 ports in the UK, including the major ports of
Hull, Immingham and Southampton. ABP also has a number of overseas
port investments, including subsidiary terminals and stevedore companies in
the USA. (In 2004, ABP's proposal for a new deep-water container terminal

at Dibden Bay (Southampton) was rejected by government over concerns about its effect upon the environment.)

The privatisation of ABP in 1983 was followed with the privatisation, via negotiated sale, of several British Rail Sealink ports and associated ferry assets to James Sherwood's Sea Containers Group (Thomas, 1994). Sea Containers subsequently sold most of the former Sealink ferry and port assets, the ferry division went to Sweden's Stena Line together with some ports, including Holyhead and Stranraer; subsequently Stena Line sold the port of Harwich to Hutchison Ports in 1998, and Sea Containers sold Heysham to Mersey Docks & Harbour Company.

Several UK ports have always been in private ownership, are generally known as company ports and were created by their own Act of Parliament. Company ports include the Manchester Ship Canal Company (acquired by Peel Holdings in 1991); Felixstowe Dock & Railway Company Ltd. (FDRC), now owned by Hong Kong's Hutchison Ports Holdings; and Mersey Docks & Harbour Company (MDHC), acquired by Peel Holdings in 2005. MDHC (owner of the Port of Liverpool) has historically had a significant Government shareholding at different times in its existence. In 1971, MDHC was given financial assistance by the government due to its near bankrupt state and, in return, the government took a 21 percent share of the company, which later was diluted to 13.9 percent upon acquisition of the Port of Medway (Sheerness). In 1982, MDHC was close to declaring bankruptcy when, for the second time, it depleted its financial reserves. However, in the past decade cargo throughput has tripled, turnover has more than doubled and profits soared more than 15 times. This increase in profitability enabled MDHC to acquire the Port of Medway (Sheerness) in October 1993 for £104 million, followed recently by purchase of the port of Heysham. In a little published move in March 1998, the government sold its stake in MDHC on the open market. MDHC was acquired in 2005 by Peel Holdings, which also owns the ports of Manchester and Clydeport.

The *1991 Ports Act* paved the way for further privatisation and seven former trust ports transferred to the private sector between 1992 and 1998. This included Clydeport, Dundee, Forth, Ipswich, Sheerness, Teesport and Tilbury. A number of these ports have subsequently changed hands since they were privatised, giving further evidence of a rather fluid market for ports as traded businesses in the UK. Teesport was instrumental in bringing about the introduction of the *1991 Ports Act* through lobbying parliament, but was unfortunately unsuccessful in its own management buy-out bid. Instead, the government accepted a bid of £180 million from Powell Duffryn PLC in 1992 (Anonymous, 1992). However, Powell Duffryn itself was purchased by

Prestige Acquisitions, a buyout vehicle backed by Japan's Nikko Securities for £570 million towards the end of 2000 (Anonymous, 2000). In 2004, the new company owner, PD Ports, was listed on the Alternative Investment Market (stock market) with Nikko Principal Investment Ltd retaining 29.9 percent.

Other acquisitions included Forth Ports' purchase of both Tilbury and Dundee, MDHC's acquisition the Port of Medway, ABP's purchase of Ipswich and Peel Holdings' acquisition of Clydeport and more recently MDHC.

The port of Bristol, which is owned by the local authority, is leased on a long-term basis to the privately owned Bristol Port Company. The latter approach is a rather more international or 'Hanseatic' model of privatisation as opposed to what has become known as the more typical 'Anglo-Saxon' (i.e. UK) approach whereby port land is simply sold, rather than leased by the public sector to a private port operating company via a concession agreement (Suykens, 1996).

An important reason given by government for promoting the privatisation process was the need to make it easier for ports to attract loan capital for financing port investment from private funds on a commercial basis. Government maintains that privatised companies have invested substantially, although at many ports such investment has very often taken the form of diversification into other fields such as property development (e.g. the Edinburgh/Leith, Glasgow, Cardiff waterfronts). However, in more recent years UK privatised ports have undertaken what might be regarded as significant investments in new port capacity as their existing facilities operate close to or beyond designed limits due to trade growth (e.g. Southampton, Liverpool, Tilbury, Hull, Immingham, Forth and Teesport). Expansion has also continued at Felixstowe and Thamesport container terminals, both owned by Hutchison Ports, though these two facilities have always been privately owned and were, therefore, not subject to the privatisation process.

The small Port of Boston in Lincolnshire was privatised prior to the *1991 Ports Act*, and this port has had a number of ownership changes since. Located on England's east coast, the port handles mainly agricultural products, paper and steel. It was owned by the local borough council until 1989 when a new initiative by the council resulted in privatisation of the port. Sold in 1990, John Sutcliffe & Son took a 25 percent stake and Budge Construction Group the remaining 75 percent. Budge Construction took complete control in 1992 with the backing of 3i and Charterhouse Development Capital (Smith, 1995). In 1996, Cleveland Trust acquired a 75 percent stake in the Port of Boston and listed it on the stock exchange.

Later, Cleveland Trust took full ownership before itself being taken over by Ashtenne Holdings PLC in May 1999. In 2004, Ashtenne Holdings sold the port to Victoria Wharf Ltd. for £5 million.

Table 3.3 provides a summary of the major port privatisations in the UK and from this it can be seen how various port companies have changed hands.

Has UK port privatisation been successful? From the first port privatisation in 1983 (ABP) to the last (MDHC), port sales yielded some US$824 million in revenue for the state. While this may seem positive, and it ignores the increase in revenue gained from taxation on port profits, it needs to be looked at in the context of what it actually costs to create a port in the first place. For example, the new container terminal proposed for London (e.g. P & O Ports London Gateway Terminal) is alone expected to cost approximately US$1.1 billion (£650 million). In this context, it does appear that a great many of the ports sold in the UK have not really raised that much in the way of revenue.

Indeed, there are a number of examples that illustrate the fact that port sales were heavily discounted. Medway Ports, for instance, was privatised in March 1992 to a management buy-out team for £13 million. Just 18 months later, the buy-out team sold the port to another recently privatised port, MDHC, for £104 million. The chief executive of Medway alone made a personal profit of £12 million in cash and shares, with other senior managers also becoming millionaires overnight (Anonymous, 1993). The ABP share sell-off was 34 times oversubscribed, so great was the demand. By 1995, shares in ABP had increased 20-fold in value between 1983 and 1997 (Anonymous, 1997a, 1997b). Other gains, also highly disproportionate to the initial investment, were made by shareholders in Forth Ports, Tilbury and Clydeport.

UK privatised ports, like most other privatised utilities, tended to be highly profitable businesses; indeed, many of the trust ports sold were profitable before privatisation (Baird, 1995). Subsequent high profits were mainly due to the disposal of valuable assets at what were undoubtedly heavily discounted sale prices, successor companies starting life with artificially low debt levels. In addition, a number of major privatised ports were given virtual monopolies over large port areas, in which there were significant captive port users (e.g. oil refineries served by pipeline/tanker jetties and so on). Private companies were also allowed to assume former port authority regulatory powers (including powers that may be used to prevent competition arising in a given port's area of jurisdiction), further strengthening their competitive positions.

Table 3.3. Summary of Major UK Port Privatisations (Excludes Sealink British Rail Ports).

Date Privatised	Company Name	% Sold	Value of Transaction in US$ Million	Method of Sale	Subsequent Sales
1 February 1983	Associated British Ports Holdings	51.5	33.40	Public offering	No change
1 April 1984	Associated British Ports Holdings	48.5	74.80	Public offering	No change
31 January 1992	Teesport	100	328.23	Private sale to Powell Duffryn	Bought by Nikko Securities in 2000, renamed PD Ports and listed on the stock exchange in 2004
11 March 1992	Port of Tilbury	100	55.17	MBO followed by private sale to Forth Ports	No change
18 March 1992	Forth Ports Plc	90	51.30	Public offering	No change
24 March 1992	Clyde Port Authority	100	44.82	Private sale (management buy-out, MBO)	Bought by Peel Holdings in 2003
30 June 1992	Medway Ports Authority	100	51.20	Private sale (management buy-out, MBO)	Bought by MDHC in 2003
17 November 1995	Dundee Port Authority	100	28.31	Private sale to Forth Ports	No change
20 March 1997	Ipswich Ports Authority	100	39.08	Private sale to ABP	No change
2 March 1998	Mersey Docks & Harbour Co	13.9	117.80	Public offering	Bought by Peel Holdings in 2005
Total			US$824.11		

Source: Baird and Valentine (2005).

4.3. Trust Ports

Historically, trust ports are largely an English concept that goes back in the main to the early part of the Industrial Revolution, and then was applied to the other parts of the UK (i.e. Scotland) as well as to former British colonies (such as India and Pakistan). Trust ports are independent statutory bodies (each established by its own Act of Parliament), governed by a board of Trustees and charged with promoting the well being of the port to meet the needs of users and stakeholders. Any trading surplus is ploughed back into improving port facilities. When introducing the *1991 Ports Act*, the Government acknowledged that trust ports were neither public nor private, being in a kind of limbo, making this form of port ownership quite unusual (Baird, 1995).

Prior to port privatisation, trust ports were a key category of port administration/ownership in the UK. In 1989, 26 (37 percent) of the top 70 ports were trust ports. The top 70 ports accounted for 468 million tonnes of cargo throughput, some 97 percent of the total UK traffic, and the trust ports' share of this was 196 million tonnes or 42 percent (Wild, Fells, & Dearing, 1995). Following the privatisation of Teesport, Medway, Tilbury, Forth and Clyde in the early 1990s, and the sales of Ipswich and Dundee shortly thereafter, the remaining trust ports today account for less than one quarter of the UK ports industry by tonnage.

One of the remaining trusts, the Port of Tyne, is covered by its own Act of Parliament, which was passed in 1989 giving it greater powers to sell land and enter into joint ventures. It has already sold 'vast tracts' of former port land to Tyne and Wear Development Corporation (Guest, 1995). In 1996, there was a House of Commons majority of 12 on the issue, with 234 members voting for privatisation of the port, and 222 against (Anonymous, 1996). The Port of Tyne, which was to be privatised by the former Conservative government, saw these plans dropped following the Labour Party's victory in the 1997 General Election, and it remains a trust port today. This was not a reversal of previous government policy by a new government; the Labour government that gained power from May 1997 continued the previous Conservative government's privatisation programme.

Another trust, the Port of Dover, is the UK's largest ferry port, handling 57 percent of the UK's international seaborne passenger traffic (14.6 million passengers) in 2002 and 41 percent of international road goods vehicles (1.8 million trailers). The Port of Dover has been notably anti-privatisation, arguing that it sees no benefit from a change in ownership for what is regarded as an already efficient operation, and with the ferry terminals already privately operated. In a report on the Port of Dover by management consultants Deloitte & Touche (referred to by Anonymous, 1997b), it was concluded that privatisation would not be in the long-term interests of the port or its users. Dover is one of the most successful of the trust ports in that it has a long trading relationship with its mainland European counterparts and this has enabled it to operate profitably. However, its main trading partner, the Port of Calais, currently suffers from congestion and this subsequently has a knock-on effect at Dover.

Of the remaining UK trust ports, 20 have an annual turnover above £1.0 million, and a further eight record annual turnover of between £500,000 and £1.0 million. Several port trusts now register negligible income, derived in some instances from activities such as tourism and car parks. Nevertheless, a number of trust ports are highly important in specific markets. Aberdeen, for instance, is the UK's main port serving the North Sea oil sector as well as supporting diverse general cargo and island ferry markets.

Something of an oddity remains in respect of trust ports at Harwich and London (Thames), both of which are non-port owning/operating and mainly exist to provide conservancy (e.g. dredged navigation channels and aids to navigation) and pilotage services to users of privately owned ports and wharves in their respective areas. Other trust ports in the UK include Bridlington, Brightlingsea, Dumfries, Falmouth, Gloucester, Great Yarmouth, King's Lynn, Lerwick, Milford Haven, Newport (Gwent), Newlyn, Poole, Shoreham, Teignmouth and Ullapool.

4.4. Municipal Ports

A few commercially significant ports are municipally owned. Sullom Voe in Shetland and Scapa Flow in Orkney have privately owned oil terminal installations with public municipal harbour departments managing port conservancy, VTS and pilotage for these terminals. Both are highly significant oil ports by volume, benefiting from close proximity to North Sea and Atlantic oil fields. Other municipal ports of varying sizes include Aberystwyth, Bristol, Colchester, Exeter, Penzance, Perth, Preston, Ramsgate, Southend, Sunderland, Tenby, Weymouth and Workington. There are also many smaller municipally owned harbours and piers situated on the numerous isles of Scotland.

The municipal ports sector is currently undergoing a review, partly to examine financing arrangements but also to assess whether municipal structures are compatible with efficient ports capable of competing with trust and privatised ports. At the beginning of 2000, the government, concerned about the running of the nation's municipal ports, issued a White Paper recommending that all the UK's ports, with the exception of Northern Ireland, review their activities with a view to making changes (DETR, 2000). The recommendations arose because of concerns that some municipal ports were being run by inexperienced locals as opposed to industrialists or those with knowledge of the business requirements. The White Paper listed the existence of some 88 municipal ports and harbour installations in the UK. This is also against the background of changes to financing whereby Supplementary Credit Approvals (SCAs), which require ports to make applications for funding for emergency repair work, have been disbanded and now municipal ports may borrow on the open market but within limits set by local authorities. A further complicating factor is that municipal ports, as a part of government, have to undergo reviews of their effectiveness, the results of which are often misleading as the reviews are aimed at assessing local services rather than commercial operations. Potentially, municipal ports could convert to trust or privatised status although there are no current plans to do so in advance of the results of the review. The Port of Bristol is an example (unusual in the UK) of a local authority contracting out port operations on a long lease to a private consortium. This is along the lines of the corporatisation of local authority airports and is another possibility for future operations at some ports.

Under the new finance system, local authorities have more responsibility for taking on new borrowing and meeting additional financing costs. Locally set financial indicators will be an important part of self-regulation, building on the principles of best value and strategic asset management planning. The aim is to enable authorities to have a proper debate about the potential consequences of their proposed investment policies, and transparency and accountability will hopefully be enhanced as a result.

In the Highlands and Islands of Scotland, local authorities manage over 200 other mostly minor harbour and pier facilities including many ferry terminals. Many of these ports are regarded as essential public transport facilities serving remote local/island communities, often supporting subsidised ferry services (provided via public service obligations). Some examples included the state-owned Caledonian MacBrayne on Clyde and Western Isles services, the partially state-owned Northlink for Northern Isles services, plus various inter-island ferry services in Orkney, Shetland and the Western Isles. Most of these facilities tend to be publicly funded by the Scottish Executive, supported in some cases by the local authorities and EU grants from the European Regional Development Fund.

In addition to the above categories, the British Waterways Board (BWB), the only remaining nationalised sector of the ports industry, owns four relatively minor ports and operates two of them. BWB is a UK-wide, public corporation 'sponsored' by the UK Department for the Environment, Food and Rural Affairs in England and Wales as well as by the Scottish Executive.

5. CONCLUSIONS

The port privatisation process in the UK can be divided into two eras: the first occurring from the early 1980s with the formation of ABP and the negotiated sale of Sealink railway ports, and the second occurring after the *1991 Ports Act*. Deregulation of the labour market, which followed several years after the first privatisations in 1989 with the abolition of the NDLS, was a critical element in allowing ports to become more competitive. Indeed, port users maintain that abolition of the Scheme was far more important than who actually owns the port.

The method of port privatisation adopted in the UK remains highly unusual, as it involved the wholesale transfer of all three core functions of the port – landowner, utility (i.e. port operations) and regulator (Baird, 1997). It is the transfer of this last function, namely the port authority as regulator within designated port areas, which has attracted particular attention of other countries studying port privatisation. Elsewhere, the public port authority is viewed as a fixed entity that always remains in situ, and virtually all other countries that have privatised their ports (i.e. terminals, cargo-handling operations etc.) have retained a public port authority which continues to function as the regulatory body for the port as a whole (also acting as landlord in many instances).

Internationally, the far more universal preference for privatisation solutions encompasses: (a) a set time period for terminal operating leases/concessions and/or BOT schemes (with port land reverting to the state at the end of the concession) and (b) retention of some form of public port authority. International evidence confirms that the state does not need to sell off port land in order to generate private sector investment in its seaports. In other words, selling off port land (and dissolving the public port authority!) was not, and is not, a necessary precondition for securing private sector investment in (UK) ports.

A great many countries have embraced port privatisation, but they have invariably tended to walk away from the outright sale of port land, or what Suykens (1996) termed the 'Anglo Saxon' model of port privatisation as practised in the UK. In the main, most countries have preferred instead to operate via terminal lease/concession schemes, where the private sector is brought in to build or modernise ports as required by the economy concerned, and to manage them; conversely, the UK method of privatisation was more concerned with the transfer of ownership of what were mature port estates and related assets (often sold initially to former port officials), most holding vast urban property development potential.

UK port privatisation was not concerned with the modernisation and enhancement of seaport capacity that would aid the flow of trade and benefit the economy, which tends to be a key objective of port privatisation internationally (Baird, 2000). Nowhere in the privatisation of UK ports was there a specific requirement made, as a condition of sale, for specific investments to be made (by the new private owners) in new or enhanced infrastructure that could aid the economy; private port successor companies were thus free to do as they wished with port land, subject to the normal planning process.

In the UK, the state has therefore more or less withdrawn from its ports industry and indeed today the state does not even have a specific port regulatory body. Internationally, this is a highly unusual stance to take with regard to something normally considered to be as strategically important as a nation's seaports, through which virtually all trade flows. Whether via a port authority, marine department or some other body, in the vast majority of countries the public sector today still retains a central role with respect to seaport planning, regulation, development and investment (Baird, 2002).

Goss (1999) suggested that the new private owners of UK ports were simply intercepting a disproportionate share of the economic rents generated by their strategic position in the market, and of course this must inevitably be at the expense of other stakeholder groups (e.g. port users, the wider economy, society as a whole). He argued that the role of government is to prevent the interception of economic rents by any single stakeholder, yet in the UK the form of privatisation adopted has actually promoted such interception.

The UK government has undertaken reviews of trust and municipal ports; it would be well advised to also undertake a review of private ports, the latter now accounting for approximately 70 percent of the UK ports industry. There is no evidence as to the real effectiveness of privatised ports in the UK, other than the fact they make a great deal of money and help move rising levels of trade (the latter achieved with increasing difficulty). Such a review might ideally focus on the views and needs of port users, taking into account forecast trade patterns, technological developments and demand for new port capacity.

Perhaps the process and extent of privatisation decided upon comes down to what is considered to be the ultimate objective of a seaport. Ask any private port owner what the key objective of his port is and the response will inevitably be 'to make money' (the authors have tried this and that was the answer!). Ask a public port (i.e. publicly owned/regulated, but privately operated) the same question and the response will more likely be along the lines of 'to help facilitate trade and generate economic benefits for the region or country concerned' (see Cass, 1996). While the pure profit motive assumes that a port is just like any other business in the economy, trade development and related economic (and environmental) benefits, in our view, must ultimately be the raison d'être of any seaport. Responsibilities of this nature logically fall into the realm of the public sector.

NOTES

1. Derived by the authors from data collated at www.privatizationbarometer.net (official provider of privatization data to OECD).
2. The *caveat* when reading global figures relates to fluctuations in currencies, inflation and possible local variances that may exist.

REFERENCES

ABP. (2005). *ABP Interim Report 2005*. London: Associated British Ports.

Anonymous. (1992). Outside bidders pull hard in tug-of-war. *The Guardian*, 4 January, p. 13.

Anonymous. (1993). How the boat came in at private dock. *The Times*, 23 September, p. 25.

Anonymous. (1995). Disgusted. *The Economist*, 11 March, p. 11.

Anonymous. (1996). Port of Tyne privatisation gets go-ahead. *Lloyd's List*, 7 December, p. 12.

Anonymous. (1997a). Success of privatisations setting pace for world-wide trend. *Lloyd's List*, 10 April, p. 6.

Anonymous. (1997b). *Fairplay*, 2 October.

Anonymous. (2000). Nikko buys Powell Duffryn for £507m. *Lloyd's List*, 6 November, p. 1.

Baird, A. J. (1995). Privatisation of trust ports in the United Kingdom: Review and analysis of the first sales. *Journal of Transport Policy*, 2(2), 135–143.

Baird, A. J. (1997). Port privatisation: An analytical framework. *Proceedings of the International Association of Maritime Economists Conference (IAME)*, London.

Baird, A. J. (1998). The Port of Felixstowe. In: T. Kreukels & E. Wever (Eds), *North Sea ports in transition: Changing tides* (pp. 99–132). Assen, The Netherlands: Van Gorcum.

Baird, A. J. (2000). Privatisation and deregulation in seaports. In: B. Bradshaw & H. Lawton-Smith (Eds), *Privatization and deregulation of transport* (pp. 397–412). London: Macmillan Press Ltd.

Baird, A. J. (2002). Privatisation trends at the world's top-100 container ports. *Maritime Policy & Management*, 29(3), 271–284.

Cass, S. (Ed.) (1996). *Port privatization – process, players and progress.* London: IIR Publications/ Cargo Systems.

Chapman, C. (1990). *Selling the family silver.* London: Hutchinson Business Books Limited.

Chick, M. (1987). Privatisation – the triumph of past practice over current requirements. *Business History, 29*(4), 104–116.

Containerisation International. (2005). EU ports' package still left in doubt. *Containerisation International,* October, p. 41.

DETR. (2000). *Modern ports: A UK policy.* London: HMSO Department of the Environment, Transport & Regions.

Drucker, P. (1969). *The age of discontinuity: Guidelines to our changing society.* London: Heinemann.

ESPO. (2004). *ESPO factual report on the port sector.* Brussels: ESPO (European Sea Ports Organisation).

Finney, N. (1990). *Thatcherite plan to destroy National Dock Labour Scheme.* A speech by the former director of the National Association of Port Employers, Australia. Retrieved 23 April 2000 from www.labournet.org.ul/ndls.htm

Gibbon, H. (1998). Worldwide economic orthodoxy. *Privatisation International, 123,* 4–5.

Gibbon, H. (2000). Editor's letter. *Privatisation yearbook.* London, *Thomson Financial,* p. 1.

Gomez-Ibanez, J., & Meyer, J. R. (1993). *Going private: The international experience with transport privatisation.* Washington, DC: The Brookings Institution.

Goss, R. O. (1997). British ports policies since 1945. *Journal of Transport Economics and Policy, 32*(1), 51–57.

Goss, R. O. (1999). On the distribution of economic rent in seaports. *International Journal of Maritime Economics, 1*(1), 1–9.

Gough, I. (1979). *The political economy of the welfare state.* London: Macmillan.

Goulielmos, A. M. (1999). Deregulation in major Greek ports: The way it has to be done. *International Journal of Transport Economics, 26*(1), 121–148.

Grosdidier de Matons, J. (1997). Is a public authority still necessary following privatisation? *Proceedings of the Cargo Systems Port Financing Conference,* London, 26–27 June.

Guest, A. (1995). Sell-off call cost Tyne port major investment. *Lloyd's List,* 12 September, p. 1.

Hare, P., & Simpson, L. (1996). *UK economy: Performance and policy.* Hemel Hempstead, UK: Prentice-Hall/Harvester Wheatsheaf.

Howell, D. (1970). *A new style of government.* London: Conservative Central Office pamphlet.

Institute of Shipping & Logistics. (2005). *Public financing and charging practices of seaports in the EU.* Interim report for the European Commission DG TREN. Institute of Shipping & Logistics, Bremen.

Jackson, P. M., & Price, C. M. (1994). *Privatisation and regulation – a review of the issues.* Harlow, UK: Longman.

John, M. (1995). Port productivity: A user's view. *Proceedings of the UK Port Privatisation Conference,* Edinburgh, 21 September.

Juhel, M. & Pollock E. (1999). Whatever happened to ports? *Urban Age, 6* (4). www.wds.worldbank.org

Keegan, J. (1999). The British century. *The Daily Telegraph* (weekend supplement), 13/14 November.

Lewis, C. (2005). Regions eye deepsea box trade. *Cargo Systems,* July/August, pp. 30–37.

Livesey, R. C. (1997). British ports policies since 1945 – a comment. *Journal of Transport Economics and Policy, 32*(3), 405–406.

Parker, D. (1997). *Caveat emptor at privatisation? Reflections on the regulatory contract in the UK.* Working Paper 9734. Aston Business School, Birmingham, UK.

Punnett, R. M. (1994). *British government & politics* (6th ed.). Aldershot, UK: Dartmouth Publishing.

Ross, H. (1995). Ports and innovative shipowners: Encouraging enterprise. *Proceedings of the UK Port Privatisation Conference*, Edinburgh, 21 September.

Saundry, R., & Turnbull, P. (1997). Private profit, public loss: The financial and economic performance of UK ports. *Maritime Policy & Management, 24*(4), 319–334.

Semple, M. (1979). Employment in the public and private sectors 1961–78. *Economic Trends, 313*, 90–108.

Sidery, R. (1994). Understanding the importance of the valuation process. *Proceedings of the World Port Privatisation Conference*, London, 27–28 September.

Smith, L. (1995). Boston port owner sees acquisition as first step in UK east coast expansion. *Lloyd's List*, 27 June, p. 1.

Suykens, F. (1996). The future of European Ports. In: L. Bekemans & S. Beckwith (Eds), *Ports for Europe*. Brussels: European Interuniversity Press.

Thatcher, M. (1993). *The Downing Street years*. London: Harper Collins.

Thomas, B. J. (1994). Privatisation of UK seaports. *Maritime Policy & Management, 21*(2), 135–148.

Thompson, D. J. (1987). Privatisation in the UK. *European Economic Review, 31*, 368–374.

Turnbull, P., & Weston, S. (1993). The British port transport industry: Part one, Operational structure, investment and competition. *Maritime Policy and Management, 20*(2), 109–120.

Wild, P., Fells, H., & Dearing, F. (1995). *International ports*. London: Lloyd's of London Press Ltd.

Wilson, G. K. (1983). Planning: Lessons from the ports. *Public Administration, 61*, 265–281.

CHAPTER 4

FIGHTING FOR MONEY, INVESTMENTS AND CAPACITY: PORT GOVERNANCE AND DEVOLUTION IN BELGIUM

Hilde Meersman, Eddy Van de Voorde and Thierry Vanelslander

ABSTRACT

Any economic analysis of seaport activity requires a clear definition of the concept of seaport and an understanding of the institutional context. Policymaking in ports, including Flanders, is largely controlled by an administrative authority. The reason for public authorities' great interest is that ports are of enormous economic significance to the area governed. The degree of involvement and the relevant government level differs among Flemish ports. Port investment projects are financed according to predefined rules, but for specific projects, exceptions are observed. European port policy has a clear impact, although port competition is still prone to competitive distortion.

Devolution, Port Governance and Port Performance
Research in Transportation Economics, Volume 17, 85–107
Copyright © 2007 by Elsevier Ltd.
All rights of reproduction in any form reserved
ISSN: 0739-8859/doi:10.1016/S0739-8859(06)17004-3

1. INTRODUCTION

The starting point for any economic analysis of seaport activity is a clear definition of the concept of a seaport and an understanding of the institutional context. As to the definition of a seaport, reference is made to the nature of the vessels entering the port: if sea vessels can reach the port, it is considered to be a seaport; if only inland vessels can reach the port, it is considered to be an inland port (Blauwens, De Baere, & Van de Voorde, 2002). Alternative but less suitable approaches would be the locational approach to defining a seaport (Henk, 2003) or the hinterland-size approach (Branch, 1986).

Broadly speaking, issues in port policymaking are most often related to the necessity of creating additional docks and/or terminals, and the appropriate timing and location for such expansion operations. In other words, port policymaking is concerned with the manner in which the scarce production factors of 'time' and 'space' can, or shall be, assigned to port activities.

Irrespective of whether a port is in private or public hands, policymaking is largely controlled by an administrative authority. The reason why public authorities in particular tend to show great interest in seaports is that ports are of enormous economic significance to the area governed: depending on the administrative level, they generate significant benefits in the shape of direct and indirect employment, added value and international trade (Meersman, Van de Voorde, & Vanelslander, 2003). On the other hand, there are costs to consider, including public investment in maritime access, port infrastructure, hinterland connections, as well as negative externalities such as congestion and air pollution.

This form of public administrative control due to the national economic importance of ports also exists in Belgium. In terms of surface area and population, Belgium is one of the smaller member states of the European Union. However, because of Belgium's central geographical location in the vicinity of important production and market centres, and because of its dense population, it has four seaports; all are situated in Flanders and together are the motor of the country's economy.

In 2003, the direct added value created by the four seaports rose by 3.6 percent to EUR 11.5 billion. The indirect added value amounted to EUR 10.5 billion (National Bank of Belgium, 2005).[1] Also in 2003, the port of Antwerp accounted for 5.3 percent of the added value generated by the Belgian economy and 9.3 percent of that of the Flemish economy. As to employment, 105,419 jobs are created directly and 133,457 jobs indirectly at

Flemish ports. About 3.6 percent of Belgian and 6.2 percent of Flemish employment is generated directly or indirectly by the port of Antwerp. There are considerable differences between the four ports, though Antwerp accounted for 64.4 percent of the added value, compared to 25.7 percent in the case of Ghent and 6.1 percent for Zeebruges (National Bank of Belgium, 2005).

It should be noted that the level of port industrialisation is an important determinant of a port's contribution to the national economy. Although the port of Ghent, for instance, handles less traffic than the port of Zeebruges (23.5 versus 30.6 million tonnes), the former's added value is higher. The type of activities inside the port perimeter is therefore of crucial importance.

Institutionally speaking, Belgium is a federal state, consisting of three Regions (Flanders, Wallonia and the Brussels Capital Region) and three Communities (the Flemish, the French- and the German-speaking communities). It is the Regions that have full decision-making power with regard to seaport policy. The Flemish Region sets the rules with respect to the ports of Antwerp, Zeebruges, Ghent and Ostend, the Brussels Region does so for its seaport, and the Walloon Region is in charge of seaport policy at the port of Liège. Distinguishing among the regions is important, as policies diverge among them and as the European Community in its documents and data treats ports in the different Regions separately. In addition, it is the Regions themselves that are in charge of contacts with the European Commission, for instance with respect to state aid to ports.

This paper deals in greater depth with port governance in Flanders. It considers consecutively: port management, i.e. the manner in which daily operations are run; seaport policy, i.e. the role that the higher authorities play; and provides an assessment of the port governance issue. The great economic significance of seaports and the resulting toughness of port competition have, all too often, caused port policy and the role of government to be preoccupied with a struggle for money, investments and capacity, as we already indicate in the title of this chapter.

2. FLANDERS' PORTS IN A NUTSHELL

The principle of the derived nature of transport demand also holds in the port sector. A period of strong economic growth, further internationalisation and globalisation, and ever-growing trade flows automatically implies increased throughput in the large port ranges. However, whatever holds at aggregate level may not ring true for each port separately. Within the Hamburg–Le Havre range, port competition is stiff. This means that international and national influences can affect the competitive position of individual ports. Port management and policy are therefore crucially important factors to a port's comparative success or failure. Figs. 4.1–4.5 provide an overview of the relative size of Flanders' ports in 2004.

Over the decades, one has become accustomed to Flemish seaports reporting good growth figures year after year. In 2004, Antwerp once again

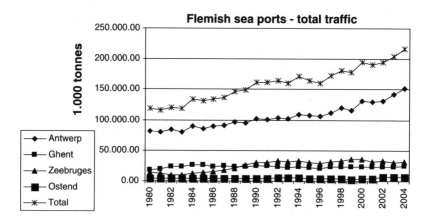

Fig. 4.1. Flemish Seaports: Total Traffic. *Source:* Flemish Port Commission (2005).

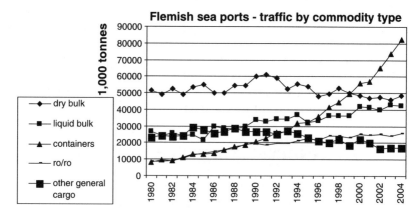

Fig. 4.2. Flemish Seaports: Traffic by Commodity Type. *Source:* Flemish Port Commission (2005).

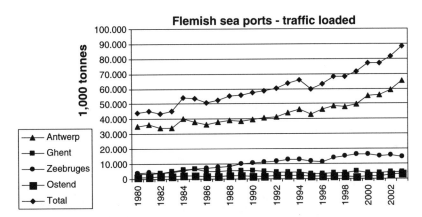

Fig. 4.3. Flemish Seaports: Traffic Loaded. *Source:* Flemish Port Commission (2005).

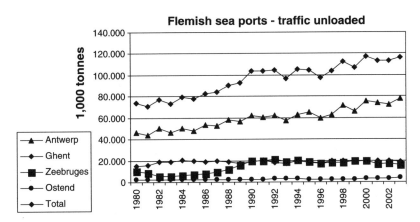

Fig. 4.4. Flemish Seaports: Traffic Unloaded. *Source:* Flemish Port Commission (2005).

broke its own traffic record. Total throughput rose above 152 million tonnes, beating the previous record year, 2003, by 6.6 percent. Between 2000 and 2004, traffic in Antwerp grew by approximately 30 percent.

Aggregate figures can, however, hide divergent trends. This is also the case for the port of Antwerp. Handling of liquid bulk remained stable in 2004, but traffic in other goods categories rose by several percent. The big

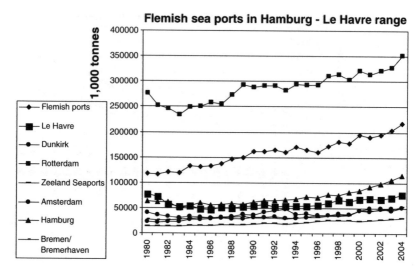

Fig. 4.5. Belgian Seaports in Hamburg–Le Havre Range. *Source:* Flemish Port
Commission (2005).

success story, though, is undoubtedly container throughput: in 2004, a total
of 6,063,744 TEU was achieved, representing 11.4 percent growth over the
previous year. This corresponds to a tonnage figure of over 68 million.
Antwerp's container capacity has, for that matter, been increased by a further 6.4 million TEU with the opening in 2005 of a brand new tidal dock
called Deurganckdok.

Equally interesting is the distinction between loaded and unloaded tonnage. More goods are unloaded than loaded at Flemish ports (116,200,000
versus 88,001,000 tonnes in 2003). Taking into account flows that almost
never generate backhaul cargo, shipping companies do on average have a
reasonable chance of attracting cargo for the return trip. This enhances a
port's competitiveness.

Another important aspect is the respective market shares. In 2004, the
port of Antwerp achieved a market share of 16.4 percent of total freight
throughput in the Hamburg–Le Havre range. This means the port ranks
second, behind Rotterdam's market share of 37.9 percent. Hamburg, by
comparison, achieved a 12.3 percent share. As far as non-containerised
general cargo is concerned, Antwerp is the largest port within the range,
while in terms of container throughput it ranks third behind Rotterdam and
Hamburg.

3. PORT GOVERNANCE IN FLANDERS

Belgium's seaports are not all governed in the same way. Moreover, there is no uniformity within Europe in terms of the financial and organisational structure of ports. This is not without consequence for their competitive position, as becomes apparent if one considers the various managerial structures in greater detail. This is especially the case with respect to how financing specifications and the degree of commercial freedom is determined.

3.1. The Broad European Framework

Within Europe, we can distinguish between various types of port governance. In Fig. 4.6, the emphasis is on the distinction between pure private ownership on the one hand and mixed private–public management on the other. In Fig. 4.7, on the other hand, the focus is on the distinction between central and decentralised management.

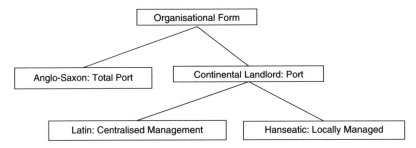

Fig. 4.6. Types of Port Management by Distinction Fully Private/Mixed Structure. *Source:* Suykens (1995).

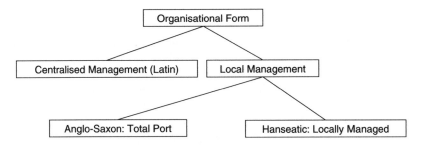

Fig. 4.7. Types of Port Management by Distinction Centralised or Local. *Source:* Jansson and Shneerson (1982, p. 4).

In practice, we can thus distinguish between state-run ports, municipal ports and privately owned ports (Grammenos, 2002). The first type has a centralised management, while the other two are locally run. The main objective of centralised governance is to arrive at an equitable distribution of means and to enhance the implementation of national port policy, while a local management can respond to specific needs of the port more swiftly. Furthermore, the first two types generally have a mixed (private–public) structure, while ports of the third port type are typically fully privatised. The current situation in Europe is summarised in Table 4.1.

Table 4.1 takes no account of ports belonging to a hybrid class, with a mixed state ownership[2] but private-style daily operations organised in an autonomous body, as is the case in certain French ports (the so-called *Ports Autonomes*). Another form of French port organisation is locally oriented and combines the benefits of a mixed management (no dominance of private interests) and private ownership (no public or political interference). The latter type is either run by the local 'Chambre de Commerce et d'Industrie' or by the regional authority.[3]

Particularly in the Hanseatic port type, but also in the Latin and the Anglo-Saxon systems, capital goods (i.e. port infrastructure) are usually in public hands, while labour (i.e. goods-handling) and specific superstructure are usually privately owned. In other sectors of the economy, capital and labour are often in the same hands.

3.2. Evolution of Port Governance Types in Flanders

In Flanders, port governance is typically decentralised. Ports are either run at the municipal level (Antwerp, Ghent and Ostend) or by an autonomous body (Zeebruges). There are no fully privatised seaports in Belgium.

3.2.1. Ghent and Antwerp
The ports of Ghent and Antwerp were transformed into municipal companies on 1 January 1979 and 1 January 1988, respectively. The ports are managed outside the general municipal services, yet on the basis of a regulation that is drawn up by the municipal council.

Previously, the situation had been rather ambiguous. A municipal port authority is, after all, not an autonomous body. Management and supervision were in the hands of the alderman for the port or another member on the local council. The port thus remained subject to decisions taken by the municipal council and to managerial guardianship. A separate budget and business account used to be drawn up, and expenses and revenue fell

Table 4.1. Organisational Structure of European Ports.

	Public		Private	
	Type I	Type II	Type III	Type IV
Ownership	Public	Public	Mixed	Private
Autonomy of Port Management	Very restricted	Limited	High	Complete
Responsibility of Port Management	State-operated/'Tool port'[b]/'Landlord port'[c]	'Landlord port'[c] (predominant) 'Tool port'[b]	'Full Service port'[d]	'Full Service port'[d]
External public funding	Extensive	Important	Very limited	No public aid
Cost-recovery practices	Not principal objective	Partial recovery predominant	Full services, some infrastructure investments	Full cost recovery
Access to provision of services	Open tender/direct agreement	Direct agreement predominant	Direct agreement	Normally closed
Relative importance in traffic terms[a]	Limited 8%	Very important 75%	Limited 7%	Limited 10%
EU states employing organisation types I–IV	Dk, Gr, F, P, D, I	B, Dk, Fin, F, D, Gr, NL, P, E, S, I	Dk, Ir, S, UK	Mostly UK, but also in other member states

Source: Commission of the European Communities (2001).

[a]Traffic estimates based on EU member states replies and best evidence available.

[b]A port where the public authority is providing not only basic infrastructure but also (some) facilities to port operators.

[c]A port where the public authority is coordinating port development and manages only basic infrastructure.

[d]A port operating company runs the port entirely. This company is very often established in a mixed holding between public and private operators.

under the competence of a company treasurer rather than the municipal treasurer.

The port of Antwerp was transformed into an autonomous municipal port company. Ghent held on to its specific form of mixed management, but in consequence of the new port decree, it was eventually transformed into a municipal autonomous port company in late September 1999. The port decree, which prescribes uniform rules for the management of Flemish seaports, contains a number of stipulations to which port companies must conform, one of which is that they must possess legal personality. The port of Ghent's mixed public–private form of governance did not possess legal personality, so that a reform imposed itself. The ports of Antwerp, Ostend and Zeebruges, by contrast, already fulfilled this condition.

In the case of Antwerp, one must take account of the fact that the port area on the left bank of the Scheldt (LBS) lies in East Flanders territory, while the area on the right bank lies in Antwerp. A 1978 act, amended in 1987, prescribes that land and industrial management of LBS should be in the hands of a so-called intercommunal company,[4] which leases the land under a leasehold system, while the management of docks and channels is in the hands of the Port of Antwerp, which gives them in concession. The act is designed to strike a compromise between unity of port governance between the two 'riverbanks' and the sovereignty of the local authorities to which the area on the left bank belongs. At the same time, however, it is stipulated that the port dues and other port-related revenues on LBS should have the same structure and be of the same order as those on the right bank.

3.2.2. Zeebruges

The port is managed by *Maatschappij der Brugse Zeevaartinrichtingen* (M.B.Z.), which by law has been a public utility company since 1954. The shares of the company were initially divided between the Flemish Region (65.4 percent), the city of Bruges (31.0 percent) and private shareholders (3.6 percent). In 2001, the Flemish government transferred its shares to the city of Bruges.

The port is managed on the basis of the statutes of the company. The highest body is the meeting of shareholders. The company is managed by the Board of Directors, which is made up of 15 members, including the Regional Port Commissioner. The accounts are supervised by a commissioner appointed by the Board of Directors.

The M.B.Z. held an exploitation concession until 1997, which has since been extended. Should M.B.Z. ever be disbanded, the port and its equipment shall be returned in part to the Flemish Region and in part to the city of Bruges.

3.2.3. *Ostend*

The port of Ostend used to be fully integrated into the municipal council, as was the case in Ghent and Antwerp.[5] This meant that port services were subject to the same municipal regulation as any other municipal service. This structure brought with it a number of specific problems that stood in the way of the port's future development: the fragmentation of port management, the absence of a port authority and the further absence of formal deliberation and cooperation structures between port managers and port users.

A solution needed to be found in the short run. The available potential of the port of Ostend required a coordinated and dynamic commercial approach, so that its possibilities could be exploited fully and more efficiently. One could not continue either with negotiations over nor the approval of a renovation plan, if a structure was lacking that assured the drawing up of an adequate corporate plan, an integrated port management and, last but not least, an appropriate commercial approach.

The general starting points for the new organisational structure were a unified port management for the entire port area, coupled with managerial autonomy, and supported by a port authority. Eventually, the port of Ostend was transformed into an autonomous port company on 1 January 1998.

4. FLEMISH SEAPORT POLICY

Regional government still plays a significant role in Flemish seaports. First and foremost, it remains the primary financier of port infrastructure. Government is also closely involved in outlining seaport policy, which is moreover embedded in the policy of the Benelux Economic Union as well as the European Union.

4.1. *Investments in Port Infrastructure*

Impulses for investments in Flemish port infrastructure and in new maritime access routes come from two sources: initiatives on the part of the regional authorities and initiatives taken by the port authority.

In the first case, the regional authority may finance the infrastructure works to be executed and have them carried out under its own direct management. After completion, ownership or exploitation of the port infrastructure can be transferred to the port authority. Examples of this procedure are plenty: the expansion of the port of Zeebruges, the further development of the port of Antwerp on the left bank of the Scheldt, including the recent completion of Deurganck container dock, the construction of the quay walls along the Ghent to Terneuzen canal, and the Kluizen docks in Ghent.

The port authority can also take the initiative to have infrastructure work carried out under its own management. To this end, it can apply for subsidies from the Flemish Region.

From 1988, under the so-called St Anna Plan, a subsidy scheme was adopted, even for investments made by the authorities themselves:

- the basic infrastructure, such as maritime locks, harbour dams, dredging works, road networks and the like, is 100 percent financed by the Flemish Region;
- additional infrastructure, such as quay walls, is 60 percent financed by the Flemish Region and 40 percent by the port management, irrespective of whether the Region has taken the initiative or not. In the case of renovation works and smaller seaports, the proportion is 80 percent/20 percent; and
- superstructure[6] is not subsidised, with the exception of superstructure that the port authority requires to fulfil its public duty.

As regards the maritime access routes, a distinction needs to be made between access to the ports of Ostend and Zeebruges on the one hand and to those of Ghent and Antwerp on the other. Ostend and Zeebruges are located on the coast. The necessary dredging of the access channels is carried out at the expense of the Flemish Region. The ports of Ghent and Antwerp, on the other hand, are accessible to sea-going vessels via the Scheldt, a tidal river. Significantly, a large part of the river (known as the Western Scheldt) lies in Dutch territory. The status of the Scheldt was fixed under the Belgian–Dutch Separation Treaties of 1839–1843. Table 4.2 provides an overview of who takes which decisions and who bears the costs.

Pilotage, concreting and maintenance of the fairway on the Western Scheldt are supervised by a council of permanent commissioners made up of two Dutch and two Belgian members. A radar system was constructed along the Scheldt and its maritime access channels, together with an information processing system. The Flemish Region is responsible for 90 percent of the construction and operational costs on Dutch territory and the full cost on Flemish territory.

The port of Ghent is connected with the Western Scheldt via the Ghent to Terneuzen canal. The Belgian section of the maritime canal to Ghent is regional property, so that it is maintained and improved at the expense of

Table 4.2. Decision-Making Regarding the Scheldt.

Item	Who Does What?	Costs Borne by ...
Dutch section (Western Scheldt)		
Concreting and beaconing	The Netherlands	Mostly by the Flemish Region
Dredging Western Scheldt (keeping depth and maintenance)	Flemish Region (Dutch permission required)	Flemish Region
Hydraulic engineering on the Western Scheldt	Deliberation within Technical Scheldt Commission	
Belgian section (Maritime Scheldt)		
Maintenance and improvement of the river (regional patrimony)	Flemish Region	Flemish Region
Concreting and beaconing	Flemish Region	Flemish Region
Radio Telephony	Flemish Region	Flemish Region
Maritime access channels		
Wielingen, Scheur	Flemish Region	Flemish Region

the Flemish Region. The Dutch section is owned by the Dutch state. Improvement works, including the construction of locks at Terneuzen, are subject to negotiation between Belgium and the Netherlands. On the basis of the treaty of 20 June 1960, Belgium contributes 80 percent of the costs of works in the Netherlands, while Belgium contributes an annual sum to maintenance.

Fig. 4.8 gives an overview of the amounts of capital the Flemish Region has invested in Flemish ports over the period 1989–2004.

4.2. How has Flemish Seaport Policy Taken Shape?

Belgian, and subsequently Flemish, seaport policy has always been a rather ambiguous affair. On the one hand, it used to be the case that port management was decentralised, while on the other government made a significant financial contribution to the maritime access route and subsidises port infrastructure. Moreover, actual transport policy, including port policy, fell within the area of competence of the Minister of Traffic and Transport, while the development of transport and port infrastructure was the responsibility of the Minister of Public Works.

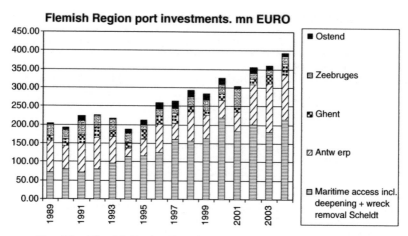

Fig. 4.8. Flemish Region Port Investments. *Source:* SERV (2005).

This situation necessitated a level of coordination and, in a later phase, organised deliberation on seaport-related policy. In this context, two important dates should be mentioned:

- 2 April 1963: the establishment of the National Commission for the Enhancement of Port Interests.
- 3 January 1978: reform of the above commission into the National Commission for Port Policy.

The purpose was to create a consultative body for representatives of the ports and the most important ministries involved. Port undertakings were, however, clearly underrepresented.

Since 1988, Belgian seaport policy has been completely devolved to the Flemish Region, so that now the regional authorities have decision-making power in many seaport matters. After devolution of seaport policy, the Flemish Executive decided on 15 November 1989 to set up a Flemish Port Commission to prepare port policy. The text defines the core purpose of the commission as follows: "... in order to advise the Flemish Executive in outlining Flanders' port policy, taking into account a number of pressing issues in the field of investments as well as the competitive position vis-à-vis foreign ports, and the operational and managerial structure."

The Flemish Port Commission's assignment is to make a general contribution to the preparation of port-related policy. This entails, among other things:

- working out general policy objectives for the infrastructure and exploitation of the ports;
- the formulation of proposals regarding the managerial and operational structures of ports, as well as the conditions under which competition takes place between ports (funding, subsidising, cooperation agreements, port dues, ...);
- the drafting of proposals to enhance the complementary nature of the seaports through deliberation, greater specialisation in particular kinds of traffic and closer cooperation between ports;
- the formulation of proposals regarding all hinterland connections to the ports and their role in export policy;
- the preparation of interregional and international consultation with regard to seaport policy;
- developing these general policy objectives into concrete infrastructure plans and projects for ports as a whole;
- stimulating the creation and organisation of scientific policymaking tools for supporting the economic aspects of seaport policy;[7] and
- organising consultation between all parties directly involved in all bottlenecks in the implementation of port-related policy with a view to enhancing inter-port cooperation.

Moreover, the Commission is supposed to formulate recommendations for socioeconomic studies of newly planned projects that are to be proposed in the following budget year and whose cost over several budget years exceeds EUR 10 million.

The Port Commission is made up of a chairman and 30 members, who are appointed for a four-year term. Mandates are renewable. Table 4.3 provides an overview of the composition of the Commission.

The advice and recommendations of the Commission are issued for the benefit of the Flemish Minister of Public Works. Approval is by simple majority of members present. The minister can only depart from a unanimous recommendation if such a step is motivated by self-interest.

The Port Commission can call on the assistance of external experts who do not have seats on the commission. Working parties may be established to carry out preparatory research into specific aspects.

Since its creation in 1989, the Flemish Port Commission has given advice on various matters of major regional and economic importance. In 1991–1992, the Flemish Port Commission advised on the first regional port decree,

Table 4.3. Composition of the Flemish Port Commission.

Representatives	Appointed by	Number
Employees	The Flemish Government, at the suggestion of their representative body in the SERV[a]	8
Employers	Idem	8
Port authorities	The Flemish Government, at the suggestion of the Flemish Minister of Public Works (five members for Antwerp, two members for the other ports)	11
Rail, road, inland navigation	Flemish Government, at the suggestion of the Flemish Minister of Public Works	3
Chairperson	Idem (not entitled to vote)	1

Source: SERV, 2005.
[a]SERV is the Socio-Economic Council of Flanders.

which in the end was rejected. In 1999, a new port decree was accepted and approved by the Flemish government. Six principles dominate this decree (SERV, 2005):

• More autonomy of local port authorities in management and operations.
• Uniform operating conditions for all seaports.
• Greater flexibility for port authorities in labour matters.
• Mandatory legal status for all port authorities.
• A clear and transparent relationship between port authorities and the Flemish Region.
• A more objective Flemish financing policy.

Various executive statements have been issued by the Flemish Port Commission since 1999, dealing with topics outlined in Table 4.4.

In general, the Port Commission deals with following topics (SERV, 2005):

• Financing
• Maritime and hinterland access
• Environment
• European policy.

Table 4.4. Flemish Port Commission Executive Statements.

Date	Topic
2/3/1999	Lock maintenance financing
12/1/2001	Organisation and operation of sub-regional concertation forums
27/4/2001	Sale rights of port authorities
13/7/2001	Provisionary delineation of port areas
13/7/2001	Subsidies to ports for port captainery servicing traffic, safety and the environment
13/7/2001	Delineation of maritime access routes and port infrastructure
13/7/2001	Conditions and procedures for project subsidies
13/7/2001	Co-financing of maritime access mooring locations
30/10/2002	Set-up of sub-regional concertation forum for the Port of Ghent
21/5/2004	Revision of port delineation
31/12/2004	Environmental effect reporting
31/12/2004	Port security
31/12/2004	Services on the internal market
31/12/2004	Access to the market for port services

Source: SERV (2005).

4.3. The Level of the Benelux Economic Union

Within the Benelux Economic Union, there are a number of ports attracting globally significant traffic volumes. Historically speaking, the divergent development of the Belgian and the Dutch states has resulted in different port structures in the two countries, as well as different forms of port exploitation and management.

The economic significance of the Benelux seaports is reflected in a number of treaties. First and foremost, there is the 'Treaty Establishing the Benelux Economic Union' (3 February 1958), article 69 of which stipulates that:

The above signatory parties to the treaty commit themselves to directing their common policy towards enhancing a harmonious development of and an active cooperation between their seaports.

Furthermore, there is the Belgian–Dutch Treaty concerning the new Scheldt–Rhine connection (13 May 1963), which explicitly mentions seaports.

In 1971, the Dutch Minister of Transport and Waterways took the initiative for the so-called "Benelux seaport deliberation", with a view to optimising the utilisation of the potential that seaport development represented to the Benelux. The three countries should aim their common policy at enhancing a harmonious development of, and active cooperation between, their seaports.

In 1991, within the framework of the Benelux seaport deliberation, a meeting was held at the level of the executive committee and the coordination committee. The executive committee of the Benelux Seaport Deliberation is made up of the chairperson and the secretary of the Flemish Port Commission (*Vlaamse Havencommissie*) and the Dutch Port Council (*Havenraad*) on the one hand and civil servants on the other. For Belgium, these are representatives of the national and the Flemish authorities, as well as one representative each from the regional governments of Wallonia and Brussels Capital.

The activities of the executive committee are prepared at the level of a coordination committee, which is made up of the secretaries of the port commissions and two Dutch and two Flemish civil servants.

Important topics to have been discussed at this level include European mobility policy, European transport policy and the Benelux seaports, and pilot services and pilotage. The Benelux also lent support to the Rhine–Scheldt–Delta Cooperation (RSD). The ports of the Rhine–Scheldt delta have united in this organisation with a view to harmonising the management and policies of the ports concerned and in order to protect their mutual interests.

4.4. The Impact of European Port Policy

The significance of seaports to the European Union is clear to see. It is therefore all the more surprising that the Treaty of Rome Establishing the European Economic Community makes no mention of seaports. However, there is a subsequent judgment by the European Court of Justice (4 April 1974), in a dispute between the European Commission and the French government, according to which the general stipulations of the Treaty are applicable to maritime transport. Consequently, many port-related issues (e.g. rules of competition, subsidising) may be approached from the perspective of these general stipulations. With the 1992 reform of the Treaty with a view to the creation of the European single market, it was stipulated that maritime transport was subject to the terms of the Treaty.

In addition, seaport policy is also a function of industrial policy. Whatever the European Commission decides in that field has direct consequences for port policy (e.g. energy policy, agricultural policy, social policy, taxation, transport policy, maritime policy).

In recent time, the European Commission has devoted much greater attention to transport in general and seaports in particular. On 10 December 1997, the European Commission published a 'Green Paper on Seaports and

Maritime Infrastructure'. The purpose was to launch a debate on seaports and their efficiency, their integration into multimodal networks and the rules of competition that should apply. The Green Paper had been prompted by the following observations:

- competition between seaports is becoming increasingly fierce as a result of a liberalisation of the global economy, technological progress and the development of Trans-European Networks (TENs);
- maritime transport has not succeeded in reducing the modal share of road transport within Europe; and
- safety of shipping needs to be guaranteed; to this end, port services such as pilotage, towage and mooring must be of the highest possible quality.

In early 2001, the European Commission issued a draft guideline concerning access to the market of port services. The purpose was to ensure the right to free entrepreneurship in the port services sector, in accordance with the basic treaties of the European Union. However, in November 2003, the European Parliament rejected the proposed compromise. In 2004, an amended guideline was put forward that strove to regulate goods-handling, towage, pilotage and mooring and unmooring.

5. PORT GOVERNANCE ISSUES: AN ASSESSMENT

The stiff competition that exists between the different ports of the Hamburg–Le Havre range, within and beyond the national boundaries of the various EU member states, implies that good port governance has become crucially important to a port's chances of success. The overall package of goods to be traded within the range is known, and price elasticity across the range is low, at least for dry and liquid bulk. Among others, this is illustrated by Van de Voorde (2005). This elasticity does, however, become significant when it concerns the choice for a specific port. Port authorities, terminal operators and other players who are active within the port perimeter are all too aware of this, which explains why scientific research into port competition has received growing attention over the past decade or so (see among others Heaver, Meersman, & Van de Voorde, 2001; Huybrechts et al., 2002; Leggate, McConville, & Morvillo, 2005; Notteboom, 2002). At the same time, efforts are made at the Flemish, as well as the European, policymaking level to create a level-playing field in response to mutual accusations of distortion of port competition.

At the present moment, there are a number of important focal points that require urgent solutions and/or decisions. Invariably, these aspects relate to port management, port policy and, most certainly, also port governance.

A first important issue concerns the most appropriate definition of a seaport. Two rather opposed views are held. The first takes as broad a view as possible of seaports, both functionally and geographically, so that the notion of 'seaport' also encompasses port industrialisation. This view is generally held in Flanders and the Netherlands, for example. The second definition is much narrower and describes seaports primarily as locations for loading and unloading of sea-going vessels. This perspective dominates in Germany, among other places. Obviously, the definition applied will determine what falls within the port perimeter and will therefore also affect port management and policymaking.

Moreover, there is great diversity within Europe in terms of port governance. Ports are seen partly as a public service (cf. the infrastructure) and partly as a commercial undertaking (cf. goods-handling). As a result of the public service aspect, government influence remains considerable in most European ports.

We also observe great differences between EU member states in terms of port operations. This become apparent when one considers the competence of the port authorities in relation to the various aspects of a port: maritime access routes, hinterland connections, port infrastructure and superstructure and goods-handling in the port itself. Suykens (1995, p. 19) distinguishes two main trends in this respect. In the case of the Continental trend (including Belgium), ports are managed and operated by a port authority to a limited extent. The maritime access routes and hinterland connections are the responsibility of the central authorities and goods-handling is in the hands of the private sector. On the other hand, there is a trend towards managing ports as a totally integrated organisation (e.g. in the UK). Maritime access routes, the port and goods-handling are then all managed by a single body responsible for all port functions. Under this model, one often sees cross-subsidising.

Possible distortion of port competition is, to an extent, a consequence of these different approaches to port governance, so that charges and forms of management can diverge. See, for instance, Trujillo and Nombela (2001) and Van de Voorde and Winkelmans (2002). This gives rise to the issue of financial intervention on the part of government and, in consequences, the issue of possible traffic diversion. Only if the latter occurs can one speak of distortion of competition. The differences in port dues and charges are sometimes greater between ports within the same country than between ports in different countries.

On the basis of the foregoing, a number of critical aspects relating to port governance can be derived for the Belgian and the European port sector:

- taxation and subsidisation policies
- the land and quay concession policies of port authorities
- joint venture activities of port authorities (e.g. in private companies) and potential conflict of interest
- state support for maritime access and effect on port dues
- operational deficits
- privileged loans
- privileged hinterland transport rates.

Each of these issues requires further theoretical and empirical research. This research should provide the scientific underpinning for a comparison between port governance rules followed by Belgian and European ports. More important still: to what extent do differences in port governance rules affect the competitive position of ports? And what are the economic consequences? But we should not restrict ourselves to researching port governance: research should also be conducive to an evolution towards *good* port governance.

6. CONCLUSIONS

In most countries, growing attention is paid to the issue of port governance. This is also the case in Flanders, a region that possesses a number of important seaports and has a long tradition of port activity. It is also a country where internal, as well as external, port competition is fierce, and where, previously, reference has been made to possible distortion of competition.

Strikingly, in Flanders, port management is decentralised. Ports are run at municipal level or by an autonomous body. Fully privatised ports do not exist. If one considers the role of the higher authorities in port policymaking, then one notices that this role is concerned mostly with investments in port infrastructure and the manner in which seaport policy is shaped. More recently, one notices the growing influence of the European level.

In Flanders, scientific research has, in recent times, focused mainly on port competition and the manner in which port competition is influenced. As far as the latter is concerned, the authorities clearly play an important role, besides a battery of other important variables, like hinterland connections, labour productivity, etc. However, as far as pure 'port governance' is concerned, that same research has only reached the stage of describing the state of affairs. Further study should make available an analytical framework for comparative empirical research. In this manner, one can prevent unnatural influences on port competition, and thus also prevent mutual accusations of dishonest practices and even the distortion of competition.

NOTES

1. The National Bank's calculations of these indicators are based on the balance sheet of the various undertakings in the four ports.
2. The Director and the Chief Accountant ('Agent Comptable') of a Port Autonome are appointed by the State.
3. A recent decree states that all Ports of National Interest, which were State-owned (e.g. Calais, Nice) but mainly run by the local Chamber of Commerce by means of long-term concessions, will be handed over to the Regions, which will lead a bidding process by (consortia of) local authorities.
4. Participants in this intercommunal company called IMALSO are the Flemish Region (10 percent), the municipality of Beveren (14.6 percent), the Intercommunal Company of the Country of Waas (48.7 percent), the city of Antwerp (25 percent) and the municipality of Zwijndrecht (1.7 percent) (Nieuwsbank, 1999).
5. Until 1998, a distinction used to be made in Ostend between the commercial port and the quays of the outport.
6. Superstructure includes buildings as well as all movable equipment.
7. Specifically, this involves the making of medium- and long-term traffic forecasts, research into the competitive position of the Flemish ports, studies into capacity and capacity utilisation, sectoral analyses of specific goods categories, research into the most appropriate assessment tool for port projects, studies into the social and economic contribution that the ports make.

REFERENCES

Blauwens, G., De Baere, P., & Van de Voorde, E. (2002). *Transport economics.* Antwerp: De Boeck Ltd.
Branch, A. E. (1986). *Elements of port operation and management I.* London: Chapman & Hall.
Commission of the European Communities. (2001). *Public financing and charging practices in the community seaport sector.* Working Document 14/02/2001. Brussels: Commission Staff.
Flemish Port Commission. (2005). *Jaaroverzicht Vlaamse Havens 2004.* Brussels: SERV.
Grammenos, C. Th. (Ed.). (2002). *The handbook of maritime economics and business.* London: LLP.
Heaver, T., Meersman, H., & Van de Voorde, E. (2001). Co-operation and competition in international container transport: Strategies for ports. *Maritime Policy & Management, 28*(3), 293–305.

Henk, R. (2003). Safe and efficient shipping – joint research creates a guidebook for inland port developers. *Texas Transportation Researcher, 39*(1), 13.

Huybrechts, M., Meersman, H., Van de Voorde, E., Van Hooydonk, E., Verbeke, A., & Winkelmans, W. (Eds). (2002). *Port competitiveness – an economic and legal analysis of the factors determining the competitiveness of seaports.* Antwerp: De Boeck Ltd.

Jansson, J. O., & Shneerson, D. (1982). *Port economics.* Cambridge, MA: MIT Press.

Leggate, H., McConville, J., & Morvillo, A. (Eds). (2005). *International maritime transport: Perspectives.* London: Routledge.

Meersman, H., Van de Voorde, E., & Vanelslander, T. (2003). The industrial economic structure of the port and maritime sector: An attempt at quantification. In: Cetena/Atena (Eds), *NAV2003: International conference on ship and shipping research, conference proceedings* (Vol. III, pp. 2.3.1–2.3.19), Palermo.

National Bank of Belgium. (2005). *Economisch belang van de Vlaamse zeehavens.* Working Paper no. 69. NBB, Brussels.

Nieuwsbank. (1999). *Persmededeling van de Vlaamse Regering.* Retrieved from www.nieuwsbank.nl/inp/1999/04/0428F060.HTM

Notteboom, T. E. (2002). Consolidation and contestability in the European container handling industry. *Maritime Policy & Management, 29*(3), 257–269.

SERV. (2005). *Annual Report 2004.* Retrieved from www.serv.be/vhc/

Suykens, F. (1995). *Controlling your port's future: The role of the private sector, government and the port authority.* American Association of Port Authorities, Annual Convention, 18 October, New Orleans, LA.

Trujillo, L., & Nombela, G. (2001). Multiservice infrastructure. Privatizing port services. *Public policy for the private sector.* Note 222. World Bank Group Private Sector and Infrastructure Network, Washington, DC.

Van de Voorde, E., & Winkelmans, W. (2002). A general introduction to port competition and management. In: M. Huybrechts, H. Meersman, E. Van de Voorde, E. Van Hooydonk, A. Verbeke & W. Winkelmans (Eds), *Port competitiveness – an economic and legal analysis of the factors determining the competitiveness of seaports* (pp. 1–16). Antwerp: De Boeck Ltd.

Van de Voorde, E. E. M. (2005). What future the maritime sector? Some considerations on globalisation, co-operation and market power. In: A. Kanafani & K. Kuroda (Eds), *Global competition in transportation markets: Analysis and policy making, research in transportation economics, 13* (pp. 253–277). Amsterdam: Elsevier JAI.

CHAPTER 5

GOVERNANCE STRUCTURES OF PORT AUTHORITIES IN THE NETHERLANDS

Peter W. de Langen and Larissa M. van der Lugt

ABSTRACT

This chapter describes the governance structures of the port authorities of the major Dutch seaports. This description is based on a framework to assess the fit between the environment, governance model, strategy and capabilities of the port authority. Changes in the port environment have triggered changes in the governance model and strategy of seaports. For the three major Dutch seaports, regional cooperation with ports in the vicinity and efforts to professionalize the port authority are important issues. Furthermore, all three ports develop new activities to play a more active role in enhancing the competitiveness of the port.

1. INTRODUCTION

This paper describes recent changes and plans for future changes in the governance structures of the port authorities (PAs) of the three major Dutch ports. In the next section, a framework to analyse the strategic fit between environment, strategy, governance and capabilities of PAs is presented and discussed. This framework is used for the description of (changes in) the

Devolution, Port Governance and Port Performance
Research in Transportation Economics, Volume 17, 109–137
Copyright © 2007 by Elsevier Ltd.
ISSN: 0739-8859/doi:10.1016/S0739-8859(06)17005-5

governance structure of PAs in the Netherlands. Section 3 briefly summarizes changes in the international and national environment in which the Dutch seaports operate. Then, national port policies are discussed followed by descriptions of the governance structures of the PAs of Rotterdam, Amsterdam and Zeeland Seaports. The description of these three PAs deals with three issues: institutional position and structure, responsibilities and activities as developed by the PAs, and the jurisdiction of the PA, including regional cooperation. Of all these aspects, the most relevant changes over the last decade and possible foreseen changes are described. A concluding section finalizes the paper.

2. A FIT BETWEEN ENVIRONMENT, GOVERNANCE MODEL, STRATEGY AND CAPABILITIES

In this section, a framework to analyse how the strategies, governance model and capabilities of PAs change over time is presented. The framework is based on the following principles/assumptions:

- Organizations operate effectively and efficiently when there is a *fit* between the environment of the organization, the governance model of the organization, the strategy of the organization and the resources and capabilities of this organization (see Mintzberg, 1990; Venkatraman, 1989, for an overview of studies using the construct of fit[1] and Baltazar and Brooks (Chapter 17 of this book) for an application on ports).
- Fit can be defined as "the degree to which characteristics of one component are consistent with the characteristics of another component" (see Nadler & Tushman, 1980, p. 40).
- Inconsistencies, e.g. a huge public involvement in an international environment increasingly dominated by private *global players*, lead to inefficiency and/or ineffectiveness and consequently pressure for the realignment of the four components in the model.
- The more inconsistencies, the larger the pressure to adapt. However, since most PAs are public and there is no market for the ownership of PAs, there is no process of natural selection that leads to the disappearance of ineffective PAs.[2] If realignment does take place, a considerable time lag is frequently involved.
- The global environment can hardly be influenced by PAs. Thus, changes in the environment are a major trigger for changes of governance models, strategies and the capabilities of PAs.

- The local port environment influences the strategy, governance structure and capabilities of the PA. The PA also influences the port environment, e.g. through port planning.
- The governance structure, strategy and capabilities of the PA all influence each other. There is no dominant line of influence: e.g. existing capabilities in some cases determine the strategy of the PA, while in other cases PAs deliberately develop new capabilities.
- The local port environment, the governance model, the strategy and capabilities of the PA, together with the global environment, determine the performance of the PA. This performance can be assessed at the port level, for instance with performance indicators such as throughput volumes and value-added, and for the PA, for instance with indicators such as return on investments. Performance at both levels is interrelated. Fig. 5.1 shows the analytical framework.

Fig. 5.1 shows the embeddedness of port governance with the environment, strategy and the resources/capabilities. A two-level approach, distinguishing between port cluster level and PA level, is needed for the analysis.

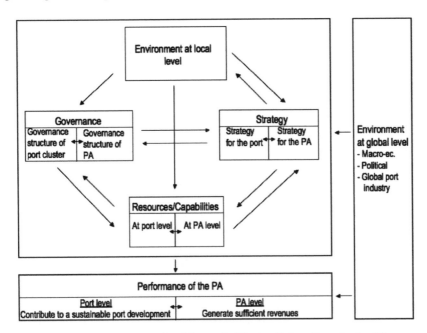

Fig. 5.1.　The Framework of Strategic Fit Applied to Port Authorities.

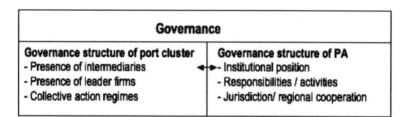

Fig. 5.2.   Specific Aspects of Port Governance at Two Levels.

Central in this paper is the governance structure of the PAs in the main Dutch ports. The specific aspects of governance both at port cluster and PA level are outlined in Fig. 5.2.

In the analysis of the three selected main Dutch ports we first focus on the specific relevant aspects of the governance structure of PAs, followed by an analysis of the influences of environment, strategy and capabilities on the governance as indicated by the thick arrows in Fig. 5.1. Examples from the three ports are used to clarify the framework and demonstrate its value added for analysing governance in seaports.

3. THE PORT INDUSTRY IN THE NETHERLANDS

The largest three ports in the Netherlands are Rotterdam, Amsterdam and Zeeland Seaports. Apart from these three ports, the Netherlands has about 10 other ports. The throughput figures of all Dutch ports are given in Table 5.1. Fig. 5.3 shows the location of the Dutch ports.

Table 5.1 shows that the three major Dutch ports described in this paper together account for over 90 percent of the total port throughput. The other relatively large port (Velsen/Ymuiden) is a privately owned single-user port that handles iron ore for Corus,[3] a steel producer. The other smaller ports serve specific local hinterlands, such as Delftzijl/Eemshaven, Harlingen and Dordrecht or, in the case of Scheveningen, handle only RoRo traffic.

Rotterdam is by far the largest Dutch (and European) port. Amsterdam and Zeeland Seaports are substantially smaller, but are still among the 10 largest ports of the Hamburg–Le Havre Range, Europe's most important port range in throughput volume. Ports in this range have a significant contestable hinterland and compete for cargo to destinations such as South Germany and Middle and Eastern Europe. The interport competition is relatively strong in this port range. The throughput figures and market

Table 5.1.   Throughput Volumes of All Dutch Ports.

Port	Throughput (× 1000 ton)	Market Share (%)
Rotterdam	327,531	75.0
Amsterdam	44,518	10.2
Velsen/Ymuiden	20,163	4.6
Vlissingen (part Zeeland Seaports)	15,149	3.5
Terneuzen (part Zeeland Seaports)	12,862	2.9
Scheveningen	4,686	1.1
Moerdijk	4,244	1.0
Delftzijl/Eemshaven	3,080	0.7
Dordrecht	2,430	0.6
Harlingen	1,129	0.3
Zaanstad	530	0.1
Beverwijk	253	0.1
Den Helder	146	0.0

Source: Nationale Havenraad (2004).

shares of the largest ports in this range, for the most important commodities are given in Table 5.2.

Table 5.2 shows that the three large Dutch ports have – at least for some commodities – substantial market shares in the relevant port range. All three Dutch ports have a role in international trade.

These throughput figures are one indicator of the size and importance of a port. An additional indicator that reflects the fact that ports are also concentrations of all kinds of port-related activities (de Langen, 2004; Dutch Ministry of Transport, 2004) is the *value-added* generated in the port areas. This is probably a more important indicator of the importance of a port for the national and local economy (de Langen, 2004).

Table 5.3 shows the value-added generated in all Dutch seaports, the relative importance of the Dutch ports in terms of value-added and the average value-added created per ton throughput. These figures are calculated based on a methodology developed by the Dutch Port Council. Included is all value-added generated by port-related firms that are located in the port area. These firms include cargo handling, but also transport, logistics and manufacturing activities. The latter are very important in terms of value-added (roughly 60 percent of all value-added generated in the Dutch seaports). The value-added is calculated based on reliable corporate tax data that is also used in the 'national accounts'. Indirect effects, either inside or outside the port area, are not included in the calculations.

Fig. 5.3.   Location of Dutch Ports.

Table 5.3 also shows that the three major Dutch seaports account for the majority of all port-related value-added in the Netherlands. However, the smaller Dutch ports have a relatively large value-added per ton; the value-added general in Zaanstad and Den Helder is particularly striking. The explanation for Zaanstad is the presence of high-value processing activities for cocoa, while in Den Helder, the economic activities of the Navy are included.[4]

The relatively low value-added per ton in Rotterdam is also striking. This is explained by two factors. First, Rotterdam handles a relatively large proportion of commodities that generate a relatively low value-added, such as coal and iron ore (cf. the weighting rules as discussed by Haezendonck, 2001). Second, Rotterdam is a *transit port*, and these transit volumes generate a relatively low value-added.

Table 5.2. Market Shares of Dutch and Other Large Ports in the Hamburg–Le Havre Range.

	Dry Bulk	Subtotal Liquid Bulk	Containers	RoRo	Other General Cargo	Total Cargo
Total throughput (2003)						
Rotterdam	85,944	152,509	70,606	10,489	8,251	327,799
Antwerp	25,912	35,127	61,350	6,046	14,440	142,875
Hamburg	27,841	11,576	64,279	0	2,587	106,283
Le Havre	4,855	44,618	19,040	2,792	80	71,385
Amsterdam[a]	44,641	13,625	663	719	5,816	65,464
Dunkirk	25,786	13,225	1,537	7,944	1,596	50,088
Bremen	8,698	1,932	31,802	0	6,541	48,973
Zeebrugge	1,661	4,869	12,271	11,107	662	30,570
Zeeland Seaports	10,787	9,032	352	2,277	3,791	28,001
Ghent	16,871	3,081	243	1,425	1,918	23,538
Total	252,996	289,594	262,143	42,799	45,682	894,976
Market shares (2003) (%)						
Rotterdam	34.0	52.7	26.9	24.5	18.1	36.6
Hamburg	11.0	4.0	24.5	0.0	5.7	11.9
Antwerp	10.2	12.1	23.4	14.1	31.6	16.0
Le Havre	1.9	15.4	7.3	6.5	0.2	8.0
Amsterdam	17.6	4.7	0.3	1.7	12.7	7.3
Dunkirk	10.2	4.6	0.6	18.6	3.5	5.6
Bremen	3.4	0.7	12.1	0.0	14.3	5.5
Zeebrugge	0.7	1.7	4.7	26.0	1.4	3.4
Zeeland Seaports	4.3	3.1	0.1	5.3	8.3	3.1
Ghent	6.7	1.1	0.1	3.3	4.2	2.6

Source: Port of Rotterdam (2004) and Nationale Havenraad (2004).
[a]These data are for the whole North Sea Canal Area.

Table 5.3. Value-Added and Value-Added Per Ton Created in the Dutch Seaports.

Port	Throughput (× 1000 ton)	Value-Added (Million Euro)	Share Value-Added (%)	Value Added Per Ton (Euro)
Rotterdam	327,531	6,346	50.8	19
Velsen/Ymuiden	20,163	1,198	9.6	59
Terneuzen (part Zeeland Seaports)	12,862	1,161	9.3	90
Amsterdam	44,518	1,083	8.7	24
Vlissingen (part Zeeland Seaports)	15,149	635	5.1	42
Moerdijk	4,244	527	4.2	124
Zaanstad	530	430	3.4	811
Delftzijl/Eemshaven	3,080	417	3.3	135
Dordrecht	2,430	276	2.2	114
Scheveningen	4,686	125	1.0	27
Den Helder	146	107	0.9	733
Beverwijk	253	116	0.9	458
Harlingen	1,129	81	0.6	72

Source: Nationale Havenraad (2004).

Following this general overview of the Dutch ports, the port environment for the Dutch ports is briefly discussed. This environment influences the governance of the PAs (see Fig. 5.1).

4. A CHANGING PORT ENVIRONMENT

Ports – like all organizations – operate in changing environments. These changes pose challenges to PAs, and trigger new initiatives such as port expansion, deepening of the port, privatization or changes in the governance model of the PA. Table 5.4 provides a brief overview of these changes, based on a review of the relevant literature.

In Table 5.4, a distinction is made between three relevant networks, each with an impact on ports. First, the port is part of a *transport network*. Second, the port is part of a *logistics network*. Ports need to fit perfectly in a logistics network to be competitive (Paixão & Marlow, 2003). Third, ports are part of an international *manufacturing and trade network*. In general, the dominant influence is 'outside in'; developments in trade and manufacturing influence the requirements for logistics systems and consequently influence the demand for transport and finally port services.[5] The role of ports in

Table 5.4. Relevant Changes in the Port Environment.

Network	Trends
Trade and manufacturing	Trade liberalization, leading to an increased international division of labour
	International standardization
	Global sourcing of MNCs and global supply chain management
	Internationalization of firms, loose ties between regional institutions and MNCs
Logistics	Outsourcing of activities, increased importance of logistics networks
	Advanced ICT systems, enabling collaborative planning
	Standardization and unitization of logistics flows and procedures
	Shorter product life cycles and more attention for flexibility of logistics systems
	Various firms aiming to provide global logistics services
	Increased pressure to reduce time-to-markets and stock levels
Transport	Increased safety/security levels
	Declining costs of international transport
	Liberalization of transport markets
	National and international policies to promote sustainable transport
	Evolving structure of liner shipping networks, increasingly complex network structures
	Yield maximization of slots and containers, better (computer-aided) planning
	Consolidation in the transport industry

Source: de Langen and Chouly (2004), Drewry Shipping Consultants (2002), Van Klink and de Langen (1999), Notteboom (2004) and de Souza, Beresford, and Petit (2003).

these three networks requires ports to be responsive or agile (Paixão & Marlow, 2003). The roles of different port users, such as shippers, shipping lines, forwarders, logistics service providers and manufacturing firms and their port choice criteria change over time.

These changes in the port environment have led to considerable changes in the governance models of PAs. Some important consequences of the changes in the environment of PAs for the issue of port governance are (see Brooks, 2004):

1. Ports face increasingly powerful clients (shipping lines, shippers and forwarders) with a decreasing interest in one particular port, because they operate on a global level (Notteboom, 2004).

2. Local stakeholders such as residents and environmental groups are well organized and put pressure on ports to improve safety and sustainability (Gilman, 2003).
3. The port community becomes more international, forcing a change in port management styles, towards more openness and transparency (Slack & Fremont, 2005).
4. Ports are increasingly regarded as regular services instead of as public goods (Baird, 2004). Thus, the principle of cost recovery is increasingly applied to ports. Furthermore, market access is an increasingly important policy objective, especially in the European Union.
5. The general trend in the political environment towards privatization and liberalization has also had its influence on the port industry.

The impact of these changes on the governance models of Dutch PAs is described in the following sections.

5. THE EFFECT OF DUTCH PORT POLICIES ON PORT GOVERNANCE

The Dutch seaports have historically been governed by regional or municipal public organizations (Kreukels & Wever, 1998). National involvement is relatively limited, although large-scale infrastructure investments and port expansion projects are funded by the national government. The latest port policy document provides clear guidelines on investment policies of national government: investments need to have sufficient benefits to justify investment costs. A social economic impact study with forecasts accepted by national government is thus mandatory. In such an analysis, advantages for Dutch consumers and for importing and exporting firms are centre stage. Employment effects are hardly taken into account because it is argued that new jobs stemming from investments in ports do not lead to lower unemployment but mainly *replace* other jobs. This argument is especially valid in periods when the level of structural unemployment is relatively small. National government also favours co-funding by regional authorities and private organizations.

When various projects competing for public funding have the same cost/benefit ratio, preference will be given to investments in 'mainport Rotterdam' (Dutch Ministry of Transport, 2004, p. 54), because of the economic and social benefits of concentrating port activities in one location. The importance of this policy guideline should not be overstated though; the cost–benefit ratio is the main determinant.

National government does not have a clear and strict policy with regard to governance models of Dutch seaports: "It is up to the regional and local governments to decide whether or not to commercialize port authorities. When PAs are commercialized, public interests need to be secured" (Dutch Ministry of Transport, 2004, p. 44). Thus, the commercialization of Rotterdam's PA (Port of Rotterdam, PoR) was not enforced by national government. Other Dutch ports operate as municipal or regional ports; this variety of port model is not regarded as ineffective. National government has decided to participate in PoR, to secure effective spending of public resources. This participation is *not* aimed at securing public interests: "National government regards participation not as an effective means to secure public interests. Participations are solely meant to secure effective management of (partially) public investments" (Dutch Ministry of Transport, 2004, p. 45).

National government supports cooperation between PAs to improve the quality and cost-effectiveness of public responsibilities, such as safety. National government also encourages commercial cooperation between Dutch PAs, but regards this as a decision of the PAs. A similar approach is taken on cross-border cooperation, for instance between Zeeland Seaports and adjacent Belgian ports (Gent, Antwerpen, Zeebrugge) or between Groningen Seaports and adjacent German seaports (Dutch Ministry of Transport, 2004).

6. ROTTERDAM

6.1. Institutional Position and Structure

The port of Rotterdam is managed by one PA, PoR. In 2004, the institutional position of PoR changed from a municipal department to a public corporation. The municipality of Rotterdam is still the only shareholder of PoR (summer 2005), but it has been agreed that the national government will purchase a minority share of PoR, as a means of providing funds for the port expansion project Second Maasvlakte. Owing to the corporatization, the management (Executive Board) of PoR is no longer controlled directly by the municipality, but by a Supervisory Board.

The profile of the supervisory board members is such that the board "must be capable of working in the hybrid setting of a private company with public accountability" (Port of Rotterdam, 2005). Thus, supervisory board members have experience in managing public corporations, the transport industry and stakeholder management. The shareholders (the municipality of Rotterdam and, from 2006 onwards, national government) have a formal influence at the annual shareholders' meeting. In this meeting, they can influence strategies and investments and appoint new members of the supervisory board. Other than that, the supervisory board is the sole body involved in monitoring the executive board.

Prior to the corporatization, the management of PoR had successfully claimed more autonomy, arguing that the PA is fundamentally different from other municipal departments. Reasons for this increased autonomy were the need to sign confidential agreements (that cannot be publicly discussed in the city council), and the need to act quickly and settle deals in negotiations. Given these special characteristics, the municipality agreed to give the executive directors autonomy to sign deals of up to 15 million euros. Furthermore, prior to the corporatization, the municipality had agreed to establish a subsidiary of PoR, Mainport Holding Rotterdam (MHR), to administer all the participations of PoR. This gave PoR substantial freedom for participating in new ventures. At the end of 2003, just prior to the corporatization, participations of MHR included: a participation in ECT, the largest container terminal operator in Rotterdam, in which Hutchinson Port Holdings, a multinational terminal operator from Hong Kong, had acquired a majority share; KSD Dirkzwager, a private firm based in Rotterdam that provides all kinds of port-related information services; European Inland Terminals, an organization owning inland terminals and over 10 other participations. The municipality was hardly involved in setting the strategy of MHR.

Thus, in practice, PoR was run as an autonomous corporation prior to the formal corporatization. The monitoring and control of PoR was limited, partly because of the granted autonomy and partly because both the aldermen and city council did not possess the right incentives and capabilities (because these capabilities are not part of the selection criteria for these functions) to monitor such a complex organization as PoR.

In the autumn of 2004, after the corporatization, it had become clear that this lack of monitoring had led to increasingly risky participations of PoR, which were all decided upon by the CEO alone. One of these decisions, to give international banks guarantees for huge loans to a private company, turned into a disaster when this company went bankrupt. It turned out that the CEO had arranged these deals without informing other directors or the supervisory board or city aldermen. This led to the resignation of the CEO. Afterwards it became clear that the lack of monitoring had led to more risky decisions, some without the expected benefits.

The overall conclusion from this case is that tight and capable monitoring of PoR is required to prevent excessive risk-taking or other decisions not in the interest of the principal, while safeguarding the flexibility needed to act in an international and changing environment. Monitoring of PoR by a city council is probably not sufficiently effective in the increasingly complex port environment. In any case, most major stakeholders (Chamber of Commerce, port business association) supported the commercialization, and it is widely argued that the new governance structure will prevent excessive risk-taking in the future.

6.2. Responsibilities and Activities

PoR positions itself as a port manager with an active involvement in the port. The PA invests substantially in activities besides the ones required for the safe and efficient management of vessel and cargo flows, and which have benefits for the port community (see de Langen, 2004). Examples of such investments include:

- *ICT infrastructure.* Through the establishment of PortInfolink a subsidiary of PoR (see www.portinfolink.nl).
- *Port marketing and promotion.* PoR invests in 'Rotterdam Representatives' in important markets, such as China, Japan and the USA. Furthermore, PoR contributes financially to Rotterdam Port Promotion Council (RPPC, see www.rppc.nl) an organization aiming to attract cargo to Rotterdam.
- *Port-related training and education.* PoR invests in, among others, Rotterdam Transport Schools a network of the largest providers of port-related education (see www.rdamtransportschools.nl) and port-related research (see www.actransport.nl).
- *Hinterland access.* Through participation in hinterland terminals.
- In the past the PoR also had its own consultancy department with the aim to commercialize its knowledge of port development and port management that it had built up within the organization. A few years ago this department was withdrawn; the main reason given was that this activity did not match the corporate view on the scope of activities of the PoR and, besides, it yielded a poor rate of return.

This positioning with active involvement in the port can also be found in the mission of the PoR. The mission of PoR is "to strengthen the position of the Rotterdam port and industrial complex at the European level, now and in the long term" (Port of Rotterdam, 2004). In order to achieve this mission "PoR positions itself as a port authority and an international service provider that:

- stimulates and facilitates economic activities in the port and industrial complex;
- cooperates with relevant parties in an open and transparent manner;
- makes clear choices that support the favourable development of the port and industrial complex with a entrepreneurial attitude."

A consequence of the increased involvement in the port is that the scope of activities of the PoR and the division of responsibilities between PoR and the port cluster have become less clear. Discussions arise often, both within and outside the PA organization, about the logic of certain activities and involvements and the potential negative external effects that these involvements might cause.

The internal organization of the PoR reflects the core activity of the organization: to develop, maintain and support a safe and sustainable port. The organization scheme is outlined in the appendix.

6.3. Jurisdiction/Regional Cooperation

Even though PoR is (still) 100 percent owned by the municipality of Rotterdam, PoR manages the entire port complex, including facilities in the municipalities of Maassluis and Vlaardingen. Furthermore, PoR has a partnership with the nearby port of Dordrecht. Finally, PoR has a 50 percent share in Exploitatiemaatschappij Schelde Maas (ESM), an organization in charge of port development in Zeeland Seaports (see later). ESM is a joint venture of PoR and Zeeland Seaports. Both have a 50 percent share and aim to act jointly in major port development projects, including the Westerschelde Container Terminal, a newly planned container terminal with a projected throughput of about 2 million TEU, and the Axelse Vlakte, a port-related industrial site in Terneuzen. The advantages of this partnership for PoR are some control over port tariffs in Zeeland, to prevent destructive price competition between both ports for the same customer. Furthermore, PoR broadens its supply of port sites. However, the cooperation has not been without problems: Zeeland Seaports regards the fact that Rotterdam knows all details of offers to potential customers – that may also be attractive for Rotterdam – as a step too far. The future of the cooperation is thus somewhat uncertain.

6.4. Foreseen Changes of the Governance Model

Two further changes of the governance model are expected or considered. First, it has been agreed that in January 2006 the national government will become a minority shareholder. This transaction is directly related to the port expansion project Second Maasvlakte,[6] but also justified by the national interest issues associated with the port of Rotterdam. It is still unclear whether national government also demands the right to nominate members to the supervisory board, especially given the general policy of national government to refrain from active involvement in minority shareholdings.

Second, the further integration of PAs in the south-west of the Netherlands is under debate. Rotterdam has stated its ambition to manage all ports in this region (Rotterdam, Dordrecht – an inland shallow draft port, Moerdijk – a medium-sized industrial port with a relatively shallow draft and Zeeland Seaports – a medium-sized European port). Rotterdam has developed partnerships with Zeeland Seaports and Dordrecht (see above). Further integration seems unlikely in the short run, but may be viable in the long run. As a matter of fact, even the issue of cooperation with Antwerp, a major and nearby competitor, arises from time to time, prompted especially by chemical industry users of both ports who regard such cooperation as the way ahead.

7. AMSTERDAM

7.1. Institutional Position and Structure

The port of Amsterdam is managed by Amsterdam Port Authority (APA), a department of the municipality of Amsterdam. APA became more independent from the municipality in 1998:

> We will therefore [...] increasingly position ourselves side by side with the companies to develop new, innovative projects. This requires a new approach from the Port Authority [...]. It was with its decision in 1998 to realize our internal independence that the city council gave us the opportunity to take on this role. (Amsterdam Port Authority, 1999, p. 4)

This "internal independence" means APA can more easily engage in contractual relations with third parties and operates relatively autonomously. However, all annual accounts and major investments have to be approved by the city council. Thus, the city council in Amsterdam has more tools to monitor APA than the city council in Rotterdam prior to its commercialization.

7.2. Responsibilities and Activities

The mission of APA also reflects its ambition to be engaged in activities beyond its pure landlord function: "Together with other port managers and companies, the APA manages the industrial and logistics complex along the North Sea Canal. Together they are responsible for an innovative and diversified port region" (Amsterdam Port Authority, 2004).

The core of the mission is quite similar to Rotterdam's mission: both stress the importance of the whole economic complex. Amsterdam also emphasizes the ambition to be an innovative and diverse port. This ambition is reflected in the targeted economic activities. Traditionally, PAs do not invest in real estate or provide consultancy services. For a number of years Amsterdam has gone beyond this traditional landlord role of PAs, for instance by investing in warehousing and all-weather terminal facilities. These investments are a commercial success. The activities are reflected in the organization schedule of the APA, which is outlined in the appendix. This organization structure is similar to that of PoR. Perhaps the most interesting feature of the organization is the presence of a real estate business unit and a commercial consultancy department.

7.3. Jurisdiction/Regional Cooperation

The port of Amsterdam is the largest port in the 'Noordzeekanaalgebied', the area from the North Sea to Amsterdam (see Fig. 5.4).

The various PAs in this area started cooperating in 1998, by signing a 'North Sea Canal Area Partnership protocol'. This partnership is institutionalized through the 'governing platform North Sea Canal Area', consisting of the following organizations: the municipalities of Amsterdam, Beverwijk, Velsen, Zaanstad, Haarlemmerliede, the province of Noord-Holland, the public infrastructure department of Noord-Holland, Zeehaven IJmuiden NV (a private port facility, mainly handling fish), the Amsterdam Chamber of Commerce, APA, Amsterdam city district 'Noord' and Corus (a steel firm with a private port in Velsen).[7]

These partners have set strategic guidelines for long-term development and have developed a masterplan for the whole area. On top of that, in November 2004 a spatial economic plan for the area up to 2020 was

Fig. 5.4. The North Sea Canal Area.

presented. This plan describes rather detailed goals for the various parts of the area. In the vision for 2020, several potential growth clusters are identified and the growth clusters of "containers and value added logistics, automotive activities, energy and recycling, and short sea activities" are identified as among the most attractive opportunities for further development of the area.

Even though the above-mentioned partnership may provide a sound basis for effective joint long-term planning, operational cooperation between all partners is also required to ensure the long-term plan is also jointly implemented. For that reason, the 'Regionale Ontwikkelingsmaatschappij Noordzeekanaalgebied' (Regional Development Agency North Sea Canal Area, RON, see www.nzkg.nl) was established in 1999, to carry out all commercial activities resulting from the masterplan. These activities include: restructuring outdated port areas, developing new port areas and managing joint marketing and acquisition efforts to attract new investments to the area. RON should also operate as a 'one-stop shop' for firms considering investing in the area. It has funds for investments and should be self-sustaining. The shareholders of RON are the municipalities of Amsterdam, Beverwijk, Velsen and Zaanstad, the province of Noord-Holland, ABN AMRO, Corus, Zeehaven IJmuiden NV and ORAM (the association of firms in Amsterdam). RON carries out various restructuring and development projects in the area, in some cases through participation, in other cases autonomously.

The governing platform and RON enable joint governance of the whole port area. Nevertheless, such integrated governance still depends on the commitment of the municipal organizations involved. The good election results of political parties with 'localist' agendas have been a drawback for regional cooperation, but, overall, the various stakeholders are satisfied with the results achieved so far.

7.4. Foreseen Changes of the Governance Model

In the late 1990s, it was decided to grant APA more internal independence, but not to take the step to commercialization. The current governance structure of APA is not widely regarded as the best model, thus discussions on the most appropriate model arise from time to time. Commercialization, as in Rotterdam, is debated at times and is regarded as a valid option. Further integration of the port management activities in the North Sea Canal Area is also anticipated, even though no major steps can be expected in the short run.

8. ZEELAND SEAPORTS

8.1. Institutional Position and Structure

Zeeland Seaports is the product of various institutional changes. The PAs of Terneuzen and Vlissingen were legally established in 1970 and 1971, respectively, both with the aim of unifying the governance of port facilities, which was previously dispersed over various municipal and regional organizations. National, provincial and municipal authorities participated in both PAs. This decision provided a basis for a relatively fast growth of both ports.[8]

In 1994, the Dutch government decided to discontinue participation in the two PAs and provided a financial incentive for the creation of one PA for both ports. Since 1 January 1998 the ports of Vlissingen and Terneuzen have operated under one PA: Zeeland Seaports.

Zeeland Seaports still has separate financial accounts for both ports, but they are expected to be integrated in the medium run. All office activities of Zeeland Seaports were integrated into one office, in Terneuzen. Zeeland seaports has a Managing Director who reports to the board of governors, consisting of one representative each of the municipalities of Vlissingen, Terneuzen and Borsele, and one representative from the province of Zeeland[9] (Zeeland Seaports, 2004).

8.2. Responsibilities and Activities

The mission of Zeeland Seaports reflects its responsibilities and is "to generate prosperity in the region by developing and managing the ports of Terneuzen and Vlissingen" (Zeeland Seaports, 2004). This mission clearly emphasizes the role of the port in regional economic development, as was the case in Rotterdam and Amsterdam. This is at least partially explained by the 'owners' of Zeeland Seaports; three municipalities and the province. The organization chart of Zeeland Seaports is given in the appendix.

While Zeeland Seaports has divisions that operate in Vlissingen and Terneuzen, these ports are still separate business units with separate financial accounts.

8.3. Jurisdiction / Regional Cooperation

Zeeland Seaports is the PA for all port facilities in Vlissingen and Terneuzen, the two principal ports in the province of Zeeland (other ports are predominantly fishing ports).

As discussed in the section on Rotterdam, Zeeland Seaports and PoR have cooperated since 1995, through the joint venture ESM. In 2003, cooperation was intensified and major new projects, especially the Axelse Vlakte (Terneuzen) and the Westerschelde Container Terminal (WCT, in Vlissingen), are now administered by ESM. The advantage of this partnership for Zeeland Seaports is that it benefits from the vast international acquisition network of Rotterdam. However, the involvement and commitment of PoR in the projects of ESM is (still) limited. For instance, PoR was hardly involved in the plans for the WCT and the political discussion on the benefits and effects of this new container terminal. Zeeland Seaports is not fully satisfied with the cooperation, and the future of the cooperation is uncertain.

Zeeland Seaports signed a cooperation agreement with the PA of Ghent in 2002. This agreement was rather far-reaching, including issues such as joint marketing and promotion, encouraging contacts between companies and the promotion of interests regarding infrastructure and port projects (Zeeland Seaports, 2004). So far, however, this partnership has not led to strong cooperation. More analysis of advantages and modes of cooperation is required. Such a study started in 2005 and is expected to provide the basis for a decision whether or not it is worthwhile to cooperate more strongly, including commercial cooperation.

Finally, Zeeland Seaports signed a 'Memorandum Scheldehavens' with the ports of Antwerp, Zeebrugge and Ghent in 2003. All parties agreed to cooperate in areas such as security, maritime training and labour market initiatives. This is a rather loose form of cooperation, leaving all commercial issues in the hands of the individual ports.

8.4. Further Changes of the Governance Model

Zeeland Seaports has been through a major change in governance structure, especially the integration of two separate PAs and the establishment of ESM. Unlike in Amsterdam, where the commercialization of the PA is regarded as a possible option for the future, Zeeland Seaports has explicitly stated, in its vision for 2005–2015 (Zeeland Seaports, 2004) that it will not move towards more commercialization. An important argument in this respect is the fact that the already established ESM can act commercially. Zeeland Seaports has decided to allocate various commercial activities to ESM and thus does not need to be commercialized itself. Furthermore, the current governance model of Zeeland Seaports is effective in promoting coordination between all relevant public organizations (three municipalities and the province). This coordination is important because, without support from these organizations, port planning and development is impossible. Thus, the current model is regarded as suited to the role of Zeeland Seaports.

Integration of the two separate business units for the distinctive port areas is expected. This separation is explained by the strong track record of Terneuzen (consistent profits) and weak track record of Vlissingen (mostly losses) prior to the merger. Almost 10 years after the first steps towards cooperation, this imbalance has changed and, more importantly, separate business units are not cost-effective from an organizational point of view and do not promote effective decision-making. Thus, full integration is expected (Zeeland Seaports, 2004).

9. CONCLUSIONS

In this section, the changes in the governance model of the PAs of the three major Dutch ports are summarized. Even though there are differences between the three ports discussed, similarities prevail. Tables 5.5–5.7 summarize the relevant developments in PA governance in the Netherlands.

Earlier in this paper the interrelationship between port governance and the factors of environment, strategy and resources/capabilities were explained and illustrated by the framework in Fig. 5.1. We stated that when

Table 5.5. Summary of Relevant Changes in Governance Structure of PoR.

Port of Rotterdam	Relevant Changes	Reasons	Results/Consequences
Institutional position and structure of PA	1995–2004: obtaining more autonomy from municipality 2004: corporatization	Increase of flexibility and responsiveness More autonomous operation	Increase of risky undertakings Not clear yet, to some extent 'back to core business' mentality
Responsibilities and activities	Investments beyond landlord function ICT infra Port marketing Training and education Hinterland transport facilities	Better monitoring by capable Board Easier engagement in contractual relationships To solve collective action problem To defend the regional interest	Diffusion of division of responsibilities between private actors and PoR Discussions about scope of activities of PoR
Jurisdiction/regional cooperation	1995: joint venture with Zeeland Seaports in EMS, organization in charge of port developments in Zeeland	To avoid destructive price competition To broaden the service area	Limited possibilities because of reserved attitude of Zeeland Seaports

Table 5.6. Summary of Relevant Changes in Governance Structure of APA.

Amsterdam Port Authority	Relevant Changes	Reasons	Results/Consequences
Institutional position and structure of PA	1998: more independence from municipality	More autonomous operation; Easier engagement in contractual relationships	Monitoring capabilities required by municipality
Responsibilities and activities	Ambition to be engaged in activities beyond landlord function; Innovation; Diversification of the port	Importance of the whole economic complex; Need for support of collective action	Diffusion of division of responsibilities between private actors and APA; Discussions about scope of activities of APA
Jurisdiction/regional cooperation	1998: North Sea Canal Area Partnership Protocol; 1999: establishment of Regional Development Agency North Sea Canal	Improving long-term development of area; Joint governance of the whole area; To ensure joint implementation through operational cooperation; Joint governance of the whole area	Satisfaction of stakeholders, but, required commitment of municipalities involved may slow down decision processes

Table 5.7. Summary of Relevant Changes in Governance Structure of Zeeland Seaports.

Zeeland Seaports	Relevant Changes	Reasons	Results/Consequences
Institutional position and structure of PA	1970/1971: establishment of two port authorities with involvement of national, regional and municipal government	To unify the governance of various port facilities to improve performance	Fast growth of both ports
	1998: integration of two port authorities into one	Economic integration of both ports	Synergies in port management
		Policy of devolution from national to regional governments	Increased 'reputation'
Responsibilities and activities	Activities mainly aimed at regional economic development	Large port users do not 'demand' involvement of PA through additional investments	Limited staff of port authority
	Focus on core activities: safety and stable environment for firms		
Jurisdiction/regional cooperation	1995: signature of cooperation agreement with PoR 2003: intensification of cooperation	Broadening of international acquisition network	Limited satisfaction of Zeeland Seaports
	2003: cooperative agreement with port of Ghent	Integration of port activities Ghent and Terneuzen	No results yet

the different components are coherent (strategic fit), good performance is expected. In this paper, we focused on the governance structure of PAs in the Netherlands. The interrelationships between governance and the factors of environment, strategy and resources/capabilities is explained below.

9.1. Governance and Environment

The three Dutch ports operate in more or less the same environment: they compete with each other and other ports in the Hamburg–Le Havre Range for cargo and economic activities. Some port activities are relatively captive, but, overall, competition is fierce. The regulatory environment is also very similar: the three Dutch ports face the same European and Dutch regulation. Important elements of this regulation include the principles of cost recovery, promotion of sustainable transport and market access for port services.

The regulatory environment mandates transparent financial accounting and autonomous port management. Consequently, the three Dutch ports operate relatively autonomously. The institutional environment also explains the ownership of the PAs by local governments: traditionally, municipalities and provinces are involved in managing transport infrastructure (roads, waterways, etc.) and national government is only involved with infrastructure of national importance.

All three ports engage in cooperation with other PAs in their vicinity. This is mainly explained by the fact that the port complex surpasses the regional jurisdiction of one individual PA. Thus, integrated port planning is only viable when PAs cooperate.

The environment, especially European port policies, also forces ports to enhance market access to seaports. Consequently, the PAs have to reconsider their tender and land allocation procedures. Strategies of strong cooperation with local port service providers are increasingly under pressure and PAs have started to promote intra-port competition.

The environment (regulatory and market) affects the responsibilities and activities of the three PAs. Competitive factors such as port facilities, sufficient draft and available land are not sufficient to satisfy port users. Additional competitive factors, such as a high-quality labour market, hinterland access (see de Langen & Chouly, 2004) and ICT infrastructure, are in the collective interest of all firms in the port cluster, but are generally not provided without an active role of the PA (see de Langen, 2004). Thus, PAs are increasingly under pressure to invest in such competitive factors. In all three cases, the PAs now pay a lot of attention to such investments, as demonstrated by their annual reports (Amsterdam Port Authority, 2004; Port of Rotterdam, 2004; Zeeland Seaports, 2004).

9.2. Governance and Strategy

The governance model of PAs influences the strategy and vice versa. The fact that ports are locally owned explains the strategic focus on creating prosperity for the 'local' port cluster, while reducing (local) negative effects. The reduction of local negative externalities is an important objective in all three Dutch ports.

The strategy of the three ports also influences the governance model: PAs should be able to create partnerships beyond the municipal level. Thus, all three PAs have claimed more autonomy and have developed partnerships with other regional PAs and other relevant organizations, both public and private. The pressure resulting from European regulation and a more demanding and less embedded port community has led to a change in the traditional governance models. Following these changes, the strategic fit between environment and governance model/strategies has improved.

9.3. Governance and Capabilities

The capabilities have to fit with the strategy and with the selected governance model. *Networking capabilities* and *investment selection* are especially important for the Dutch PAs. The specific function of a PA in supporting a port complex that consists of a multitude of companies and operates in an international port network requires the development of networking capabilities. Interaction and cooperation with the ports' logistics and industrial companies, local and national governments and port users is crucial.

Dutch ports also need to develop capabilities for effective stakeholder management. Given the governance structure and regulatory environment, PAs need to invest in stakeholder relations.

All three ports also seek to exploit their port management capabilities overseas. PoR does so by managing a port complex (Oman), while the other two provide consultancy and training services to ports.

One important difference between the three PAs is that, because of the size of the port complex, the tough stakeholder environment and the pressing challenge of effectively managing the economic and spatial transformation of the port area, Rotterdam's PA is frequently a front-runner. Consequently, PoR has a substantial staff engaged in new activities and engages in experiments more often than the other two PAs. For this reason,

PoR, in particular, needs to develop into a learning organization. This is expected to become an increasingly important capability for Rotterdam, as well as other PAs.

NOTES

1. In fact, these authors are quite critical of the value of the 'fit' model for strategic management research, for different reasons. It is an effective tool for a description and analysis of organizations.

2. The increasing globalization of the port environment does not necessarily lead to complete convergence of PAs (Hall, 2003) but different PAs respond in similar ways to the changing environment. Baird (2002), for instance, shows that most of the top 100 container ports operate as landlords, while many others restructure towards the landlord model.

3. This explains why Velsen/Ymuiden is not included in this study even though it is a substantial port: it is a user-owned facility where 'port governance' is hardly relevant. The terminal does not have separate accounts or a separate judicial entity but is fully integrated in the steel company.

4. These activities hardly generate throughput volumes. Consequently, value-added per ton throughput is high.

5. 'Inside-out' influences can be observed as well. The most important example is the introduction of the container in the transport system, which has triggered a boost in international trade.

6. Such a contribution can be justified from two perspectives: first, the economic benefits of the expansion project are larger than the costs (Central Planning Bureau, 2004). Second, government funding is necessary for the breakwater, an investment that can be regarded as a public good, since it provides protection against flooding of the coastal zone.

7. The Ministries of Economic Affairs and Spatial Planning are advisory members of the platform.

8. The fast growth of Terneuzen in this period is to a large extent explained by the fact that the municipality was designated a 'growth municipality', a status with substantial advantages (subsidies) for investors. Companies like Ovet and Air Products arrived in this period.

9. This board of governors is selected from a 'general board of governors' with *two* representatives from the three municipalities and the province. This general board of governors meets a couple of times a year and is formally the highest governing body.

REFERENCES

Amsterdam Port Authority. (1999). *Annual report*. Amsterdam: APA.
Amsterdam Port Authority. (2004). *Annual report*. Available at www.portofamsterdam.com

Baird, A. J. (2002). Privatization trends at the world's top-100 container ports. *Maritime Policy and Management, 29*(3), 271–284.

Baird, A. J. (2004). Public goods and the public financing of major European seaports. *Maritime Policy and Management, 31*(4), 1–17.

Brooks, M. R. (2004). The governance structure of ports. *Review of Network Economics, 3*(2), 68–183.

Central Planning Bureau. (2004). *Welvaartseffecten van Maasvlakte 2*. Kosten-batenanalyse van uitbreiding van de Rotterdamse haven door landaanwinning. The Hague: CPB.

de Langen, P. W. (2004). *The performance of seaport clusters; a framework to analyze cluster performance and an application to the seaport clusters in Durban, Rotterdam and the lower Mississippi*. Ph.D. thesis, ERIM, Rotterdam.

de Langen, P. W., & Chouly, A. (2004). Hinterland access regimes in seaports. *European Journal of Transport and Infrastructure Research, 4*(4), 361–380.

de Souza, G. A., Beresford, A. K. C., & Petit, S. J. (2003). Liner shipping companies and terminal operators: Internationalization or globalization? *Maritime Economics and Logistics, 5*(4), 393–412.

Drewry Shipping Consultants. (2002). *Global container terminals: Profit, performance and prospects*. London: Drewry.

Dutch Ministry of Transport. (2004). *Zeehavens: Ankers van de economie*. The Hague: SDU.

Gilman, S. (2003). Sustainability and national policy in UK port development. *Maritime Policy and Management, 30*(4), 275–291.

Haezendonck, E. (2001). *Essays on strategy analysis for seaports*. Leuven: Garant.

Hall, P. V. (2003). Regional institutional transformation: Reflections from the Baltimore waterfront. *Economic Geography, 79*(4), 347–363.

Kreukels, T., & Wever, E. (Eds) (1998). *North Sea ports in transition: Changing tide*. Assen: Van Gorcum.

Mintzberg, H. (1990). The design school: Reconsidering the basic premises of strategic management. *Strategic Management Journal, 11*(3), 171–195.

Nadler, D., & Tushman, M. (1980). A diagnostic model for organizational behavior. In: J. R. Hackman, E. E. Lawler & L. W. Porter (Eds), *Perspectives on behavior in organizations* (pp. 83–100). New York: McGraw-Hill.

Nationale Havenraad. (2004). *Port statistics*. Available at www.havenraad.nl

Notteboom, Th. E. (2004). Container shipping and ports: An overview. *Review of Network Economics, 3*(2), 86–106.

Paixão, A. C., & Marlow, P. B. (2003). Fourth generation ports – a question of agility? *International Journal of Physical Distribution & Logistics Management, 33*(4), 355–376.

Port of Rotterdam. (2004). *Port statistics, mission statement and company goals*. Available at www.portofrotterdam.com

Port of Rotterdam. (2005). *Governance structure*. Available at www.portofrotterdam.com

Slack, B., & Fremont, A. (2005). Transformation of port terminal operations: From the local to the global. *Transport Reviews, 25*(1), 117–130.

Van Klink H. A., & de Langen, P. W. (1999). *Scale and scope in mainport Rotterdam*. Final report. Rotterdam: ETECA.

Venkatraman, N. (1989). The concept of fit in strategy research: Toward verbal and statistical correspondence. *Academy of Management Review, 14*(3), 423–444.

Zeeland Seaports. (2004). *Annual report*. Available at www.zeelandseaports.com

APPENDIX ORGANIZATION CHARTS OF THE THREE PORT AUTHORITIES

The internal organization of PoR has been changed in 2004 and is given in Fig. 5.5. Figures 5.6 and 5.7 show the organization charts of Amsterdam Port Authority and Zeeland Seaports.

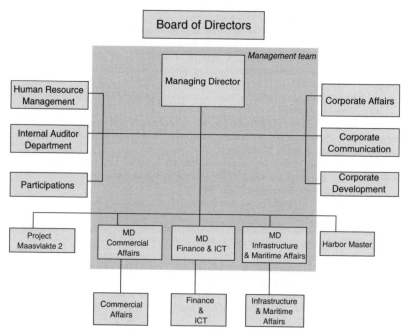

Fig. 5.5.   Organization Chart of PoR. *Source:* Port of Rotterdam (2004).

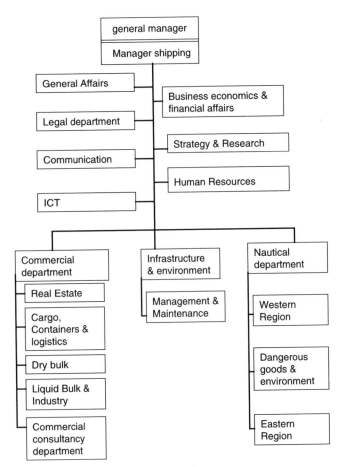

Fig. 5.6. Organization Chart of Amsterdam Port Authority. *Source:* Amsterdam Port Authority (2004).

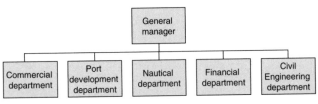

Fig. 5.7. Organization Chart of Zeeland Seaports. *Source:* Zeeland Seaports (2004).

CHAPTER 6

PORT GOVERNANCE IN ITALY

Marisa A. Valleri, Maria Lamonarca and Paola Papa

ABSTRACT

*After an overview of the administrative evolution of Italian port govern-
ance, this chapter examines the main three variables – environment,
strategy and structure – and their relationships, according to the Matching Framework. The main outcomes show that the recent privatisation
process is still far from an optimisation of port management, since some
structural and bureaucratic constraints prevent Italian ports from benefiting from the positive environmental dynamics. Some Italian ports,
indeed, are taking advantage of the recovered centrality of the Mediterranean area in the maritime flows and from strategic foreign investments.*

1. INTRODUCTION

The recent changes in maritime economics and, in particular, in port economics, modified the role of ports which are now considered fundamental
links in the global logistic chain, able to produce innovative transport services in a more or less competitive way. These changes also focused attention
on the institutional and managerial assets, with the aim of detecting governance structures that enable ports to be more competitive and to acquire
the right amount of flexibility to adapt to market changes, in order to seize
the arising opportunities (Brooks, Button, & Nijkamp, 2002).

Devolution, Port Governance and Port Performance
Research in Transportation Economics, Volume 17, 139–153
Copyright © 2007 by Elsevier Ltd.
ISSN: 0739-8859/doi:10.1016/S0739-8859(06)17006-7

In particular, this chapter aims at analysing the evolution of laws and of the management of assets at Italian ports. Such analysis will focus on the recent legislative and managerial developments in the Italian port policy, using the Matching Framework (Baltazar & Brooks, 2001) as a reference scheme. This permits us to relate the three fundamental variables (environment, strategy and structure) of an organisation as a port, with the aim of drawing conclusions about the current performance of the devolution efforts in the Italian port sector.

2. THE ITALIAN PORT SYSTEM

Italy's irregular 8,000 km coastline favoured the development of a number of ports. Nonetheless, the prevalence of mountains and hills longitudinally crossing the peninsula has always constituted an obstacle to the implementation of easy connections between ports and their hinterland.

There are over 300 ports in Italy, considering also exclusively tourist and fishery ports. Maritime ports, which originate an activity of goods and/or passenger traffic, no matter how little, and are therefore functional to the national transport system, number about 100 (see Fig. 6.1).

Italian Law no. 84/94 (Italian port reform law no. 84 of 28 January, 1994) divided ports into categories and classes. Category I includes ports for the military defence and security of the Nation; category II includes commercial ports, which, in turn, are divided into three classes: class I (ports of international economic relevance), class II (ports of national economic relevance) and class III (ports of regional and interregional economic relevance).

The same law determined the prominence of some ports belonging to one of the first two classes of category II, there constituting port authorities. Initially, there were 18 ports (Ancona, Bari, Brindisi, Cagliari, Catania, Civitavecchia, Genoa, La Spezia, Leghorn, Marina di Carrara, Messina, Naples, Palermo, Ravenna, Savona, Taranto, Trieste and Venice) but the number increased to 24 (including also Piombino, Gioia Tauro, Salerno, Olbia – Golfo Aranci, Augusta and Trapani), following further Decrees of the President of the Republic, as paragraph 8 of art. 6 of the above-mentioned law determines that other port authorities could be constituted if an increase in the volume of the traffic of goods not lower than 3 million tons per year, or 200,000 TEUs, was recorded over three years.

Table 6.1 shows the recent evolution of goods and container transport in the main Italian ports. In particular, in the four-year period under consideration, a total growth in the sector of goods was recorded, amounting to

Fig. 6.1. The Italian Port Authorities.

slightly less than 6 percent, due to the fact that most ports under evaluation showed a positive trend.

It is the container sector, however, that shows the most significant growth (amounting to 26 percent between 2000 and 2003), with the dominant position of the terminals in Gioia Tauro, which in 2003 moved about 3.1 million TEUs, Genoa 1.6 million TEUs, and La Spezia 1.0 million TEUs, followed by Taranto 0.7 million TEUs. It is evident that the positive trend is mainly due to the development of predominantly transshipment ports (Gioia Tauro, Taranto and Cagliari).

Table 6.1.   Goods and Container Transport in the Major Italian Ports (2000–2003).

Ports	Goods (million tons)				Container ('000 TEUs)			
	2000	2001	2002	2003	2000	2001	2002	2003
Ancona	11.15	13.72	12.51	9.57	84	90	94	76
Augusta	31.30	31.31	31.59	29.28	–	–	–	–
Bari	3.45	3.50	3.61	3.93	1	2	12	24
Brindisi	8.05	8.92	8.74	10.17	7	6	1	2
Cagliari–Sarroch	30.32	29.04	31.60	34.07	22	35	74	314
Catania	2.64	2.65	2.74	4.13	13	11	13	14
Civitavecchia	9.85	8.84	9.35	8.43	13	16	21	25
Genoa	50.80	50.18	51.75	53.71	1,501	1,527	1,531	1,606
Gioia Tauro	30.82	29.61	25.59	25.45	2,653	2,488	3,009	3,149
La Spezia	16.52	15.85	18.20	19.79	910	975	975	1,007
Leghorn	24.58	24.66	25.33	25.73	501	502	520	541
Marina di Carrara	3.38	3.15	3.27	3.06	11	9	10	9
Messina-Milazzo	10.62	13.95	15.73	17.56	–	–	–	–
Naples	14.78	16.72	18.63	19.41	397	431	444	433
Olbia–G.Aranci	5.76	5.20	5.70	5.92	–	–	–	–
Palermo	5.22	5.27	4.91	5.41	17	15	11	15
Piombino	10.38	9.00	8.171	8.67	–	–	–	–
Ravenna	22.68	23.81	23.93	24.91	181	158	161	160
Salerno	3.83	4.45	4.97	7.08	276	321	375	417
Savona-Vado	13.20	13.27	13.15	13.41	37	51	55	54
Taranto	33.88	34.53	34.67	37.51	3	198	472	658
Trieste	47.61	49.14	47.17	46.00	17	19	17	13
Venice	28.18	28.81	29.55	30.13	206	201	185	120
Total	419.00	425.58	430.861	443.33	7,051	7,280	8,226	8,908

Source: Federtrasporto. (2004). *Statistiche – Trasporto marittimo.*

All in all, the national network of goods flow indicates three large geographic areas having different functions and vocations: the Northeast one (Northern Adriatic), the Northwest (Northern Thyrrenian) and South.

Northeastern ports are distant from the main ship routes between the Suez Canal and the Straits of Gibraltar. The prevailing traffic in these ports, therefore, is trans-Adriatic and intra-Mediterranean. On the other hand, Northwestern ports, although also distant from the main routes of crossing of the Mediterranean basin enjoy the attraction effect of the tight port network formed by Marseille, Barcelona and Valencia. The weak point here seems to be the link with other transport modes, complementary to the maritime mode. In fact, road mobility is hampered by the location of the port inside towns or in their vicinity, whereas railway mobility is limited by the nearness of the Alps or the Appennines. Southern ports, however, have an optimal location compared with the Gibraltar–Suez direction. Nonetheless, they are heavily affected by the lack of internal links with the national and European infrastructure network.

To increase its share in the European port industry, the Italian port system therefore needs to overcome limitations concerning ports such as infrastructure (in relation to the services offered, the equipped areas, etc.) and links with the hinterland.

3. THE EVOLUTION OF THE ITALIAN PORT LAWS

Italian laws have historically considered port services and activities as public goods. The 1942 code of navigation achieved the maximum functionality towards public interest in the activities in this sector and confirmed a high level of intrusion of administrative power on those interested in maritime navigation and related activities.

On the one hand, it emphasised ports as belonging to the maritime state and, on the other, the availability of port services to the public. The character of state property is related to both the function of port areas and to the ownership of the port by the Public Administration, with the consequent imposition of bonds and organisational models also on the private actors operating in the port.

In time, the overlapping of competences of the various local and central administrations interested in port activities caused government to look for autonomous forms of management that allowed for a better exploitation and development of the port. With this aim, it was necessary to speed up the adoption of provisions, and this only seemed possible through the centralisation of many functions and activities into a single, local and autonomous body – the State. The need for such a new organisation was first recognised in 1903 with the creation of the Consortium of the Port of Genoa (Law 12 February 1903, no. 50). It was an autonomous consortium whose aim was, through the special funds it had been endowed with, to see to the implementation of public works, management and the coordination of port services.

The constitution of the Consortium of Genoa, just like the successive constitutions of other port organisations in some of the major ports, met the need to transfer management tasks, implementation and maintenance of port works, rather than the real will to decentralise functions and powers.

When the Fascists were in power, port organisations were dissolved, in line with the centralising principle, typical of the regime. Only after World War I were these organisations resumed, having wider functions ranging from administration (planning of port works, concession of maritime state property, discipline of port works, authority over tariffs) to entrepreneurship, aimed at managing port operations and service supply. Where such organisations were not present, the Minister of Merchant Marine entrusted various actors with the management of ports, such as volunteer consortia among local entities, special branches of the Chambers of Commerce, businesses and mixed consortia, and private businesses. The remaining ports were directly administered by the state through local maritime authorities.

The level of autonomy was different in different ports. If the ports administered by maritime authorities were directly controlled by the state, those managed by port consortia could enjoy a certain level of financial autonomy.

Important support for the process of liberalisation of the port sector (Cullinane & Song, 2002) came from the substantial changes that occurred in the international maritime economy, in particular, the development of container traffic and the growth of intermodal transport. Compared with such revolutions, the organisational assets of Italian ports appeared more and more inadequate. The demand for port services was changing, requiring better and better performance. The need to overcome the prevailing institutional model's difficulty in adapting to new demands brought about an intense and prolonged debate about the possible lines of reform of port regulation. In particular, during the 1980s, many reorganisation proposals for the port sector were formulated. Among these, a document by Confindustria dated 1984 must be underlined; it featured the separation of the activity of control carried out by the port organisation from the organisation of port services, which had to be entrusted to business-like managers. The document also suggested creating public–private mixed societies for the activities of planning, execution and management of port areas in order to promote an increase in competitiveness for Mediterranean ports (Valleri & Van de Voorde, 1992).

Summarising, Italian ports before 1994 were characterised by

- The bureaucratic presence of the Public Administration;
- A strong connection between administrative and entrepreneurial activities by port organisations;
- A regulation of port activities imposing bonds and organisational models also to private operators, in virtue of concessions and/or authorisations;

- The copresence of different models of management in different ports;
- The monopoly on the supply of manpower in port operations in favour of the workers of port companies (art. 110, last paragraph of the code on navigation).

4. THE PORT REFORM

The basic principle of the Framework Law no. 84/1994 on the "Reorganisation of legislation on ports" is to separate the development of port operations and the task of controlling and directing port activities. Actually, the former are carried out by authorised private port enterprises, whereas the latter are provided by port authorities, as constituted in the main Italian ports, which discipline and monitor the carrying out of port operations.

Briefly, *port authorities* replace port organisations, whereas *maritime authorities*, which are present in every port in accordance with the Italian Navigation Code and other special laws, carry out a policy and safety role as well as those administrative duties that do not come under the port authorities' competence. Maritime authorities have specifically the following functions:

- coast guard, search and sea rescue;
- safety of navigation;
- maritime police (the regulating of traffic in and out of the port's waters, berthing and maritime traffic; organisation of ancillary services and relevant tariffs; inquiries and investigations on navigation accidents);
- naval property and administrative regulation of vessels; and
- marine environment safety and protection.

Where port authorities were not constituted, only local maritime authorities discipline and survey on the carrying out of port functions. Finally, a small number of ports are managed by *special port enterprises*, which are peripheral structures of the Chambers of Commerce, as expressly indicated by law 84 (Table 6.2). There are four Special Enterprises of the Chambers, two of which (Porto Torres and Agrigento) are exclusively dedicated to activities of promotion, whereas the other two (Monfalcone and Chioggia) carry out production services, drawing income from managing stores and mechanical structures under convention, as well as from the rents of port buildings and services.

The Port Authority is a public body endowed with legal status, with administrative autonomy that is subjected to restrictions imposed by the

Table 6.2.   Port Structure in Italy.

Type	Criteria
Port authorities	24 major ports, a number likely to increase if the parameters of traffic set by the law are met (for three years, traffic volume not lower than 3 million tons per year or 200,000 TEUs); these are non-economic public bodies endowed with juridical personality and administrative, budget and financial autonomy within the limits set by the law
Maritime authorities	Lesser ports, which do not meet the parameters necessary for the constitution of a Port Authority; these are decentralized structures of the central government without decision-making autonomy
Special port enterprises	These are present in the ports of Chioggia, Monfalcone, Porto Torres and Agrigento; such facilities are constituted and financed by the Chambers of Commerce

Ministry of Transport and with budgetary and financial autonomy subjected to restrictions imposed by the law. It is composed of the Chairman (appointed by the Minister of Transport in concert with the Regional Governor), the Port Committee (composed of representatives of the local authorities), the General Secretariat and the Auditor's Committee.

Each Port Authority has the following functions:

- policy and planning function (through strategic management and land-planning); and also coordination, marketing and control over port operations and activities performed within the port area;
- maintenance; and
- landlord (owning the port land and granting terminals through concessions to private operators) with the possibility of directly providing utilities within the port.

Therefore, port authorities play a major role in port territory planning; such a role was confirmed by the law decree dated 31 March 1998, no. 112 on the attribution of administrative functions and tasks to the regions and to local bodies. Port planning documents are adopted by the Port Committee, following an agreement with the interested municipalities. The activities of port promotion and development are up to the port authorities. The Port Authority President and the "Comitato Portuale" (Port Committee) are responsible for finding out and programming the infrastructure and superstructure works to improve the port efficiency. In order to do this, they prepare and approve a "Piano Triennale" (three-year plan) of port development. The building of infrastructure and the purchasing of superstructure depend on private funds, different financial sources (Interregg Programme, European Programme, Regional Programme, etc.) and Ministerial transfer funds. Capital and routine maintenance are up to both port authorities and maritime authorities. The latter are harbour offices of the Minister.

As for the issuing of concessions, a concession may last 30 years at the most, with the exception of large investments in infrastructure charged to the concessionary, for which, with prior authorisation of the appropriate Ministry, port authorities may grant longer concessions on the grounds that it is consistent with the port's development strategy.

As already indicated, according to the principle of separation of tasks, the law does not allow port authorities to directly perform port operations and activities; they can only provide ancillary activities, based for example on supporting intermodal and logistic development. Entrepreneurial activity within the port area cannot be performed by the port authority, only by competitive private enterprises.

5. ENVIRONMENT, STRUCTURE AND STRATEGY OF ITALIAN PORTS

The utilisation of the Matching Framework proposed by Baltazar and Brooks (2001) permits the analysis of the performance of a port organisation on the basis of three variables: environment, strategies and structure. Such a scheme is based on the classic paradigm of business economy, which considers the environment, strategies and structure as critical variables for the success of an enterprise, by adopting it to port systems.

With reference to the Italian situation (Table 6.3), all the external sectors that, either directly or indirectly, affect the port are considered as environment (industrial sector, economic situation, social and cultural environment, human resources, financial resources, political context, etc.). The variable "structure," on the other hand, represents the model utilised to implement strategy, which, in turn, consists of policies aimed at facilitating the integrated and coordinated development of Italian ports.

The starting point to analyse the variable "environment" is the consideration of all the factors affecting the organisation's ability to achieve its own objectives, starting from the idea that the environment is characterised by a given degree of uncertainty deriving from the lack of perfect

Table 6.3.   The Matching Framework Applied to Italian Ports.

	Configuration 1	Configuration 2
Environment		High uncertainty
		High complexity and dynamism
Strategy	Efficiency-oriented	
	Focus on delivery of the basic product and service	
Structure	Mechanistic	
	Centralised decision-making characterised by procedural standardisation	

Source: Adapted from Baltazar & Brooks (2001).

information which, in turn, brings about high levels of complexity and dynamism. In terms of a good performance of the organisation, therefore, it is necessary for the structure to be adequate to the level of complexity and dynamism of the environment. In other words, a highly dynamic environment requires a decentralised structure characterised by higher flexibility and lighter bureaucratic procedures. Moreover, vis-à-vis possible strategies, an organisation may decide to focus its attention either on the production of basic products and services or on innovation and differentiation policies. More generally, we could say that the higher the "degree of fitness" among the variables taken into account, the better the result expected in terms of the economic performance of the port organisation.

With regard to the variable environment, and taking into account the evolution of Italian ports in time, it appears that the public authority has historically governed the port sector through a very centralised port policy. The state, under a system of monopoly, controlled the port land, infrastructure, equipment and also the financing of new investments without stimulating any private entrepreneurial development. Only after the 1994 reform, with the accompanying decentralisation of powers and the introduction of port authorities in the main Italian ports, did the important process of privatisation of port services get started.

The reform process coincided with the modification of international maritime networks, which brought the Mediterranean area within main international traffic flows linking Asia, Europe and North America (Fig. 6.2), as a result of the growth in the importance of Asia in global trade.

The Mediterranean basin is becoming one of the most dynamic European areas, especially for container transshipment, and the Italian ports can

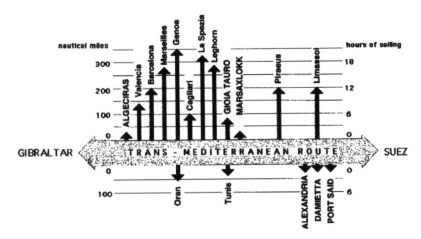

Fig. 6.2.   Deviation of Mediterranean Ports from the Main Gibraltar–Suez Shipping Lane. *Source:* Ridolfi (1999).

acquire importance compared with the past (Valleri & Venezia, 1997). Expanding transshipment ports, such as Gioia Tauro and Taranto, have therefore gained a foothold in the market. Moreover, during the last few years, some Italian ports, taking advantage of the port reform and the privatisation process still in progress, are increasing their volume of traffic, especially due to growing investments by the big international shipping companies (e.g., in Gioia Tauro, Taranto and Cagliari) (Table 6.4).

Singapore's PSA Corp., the world's second largest terminal operator, gained a strong foothold in the European port sector by agreeing to take over a 60 percent shareholding in Sinport, the FIAT logistic subsidiary, which holds majority stakes in Genoa Voltri Terminal Europa (VTE), Venice and Civitavecchia. Indeed, in the port of Taranto, a company belonging to Evergreen Marine Corporation of Taipei (Taiwan) today manages a terminal container, with a 60-year concession agreement with the Taranto Port Authority. Moreover, the lead Mediterranean transshipment port, Gioia Tauro, is experiencing positive growth due to both its strategic position within the main trans-Mediterranean routes and the investments of Contship and other companies belonging to the German Euro Eurokai and Eurogate group.

On the basis of the developments just described, it appears that the external environment is becoming more complex and dynamic and consequently quite uncertain. The changed environmental scenario stimulated the

Table 6.4. Terminal Operators' Investments in the Main Italian Ports
(2003).

Port	Terminal	Operator
Gioia Tauro	Medcenter Container Terminal (MCT)	Contship Eurogate (1996–) Maersk (10%)
La Spezia	La Spezia Container Terminal (LSCT)	Contship Eurogate (1996–)
Cagliari	Mediterranean International Transshipment Hub (MITH)	P&O (2001–)
Taranto	Taranto Container Terminal (TCT)	Evergreen (2001–)
Trieste	Molo VII	ECT (1997–2001)Koper (2001–)
Genoa	Voltri Terminal Europa (VTE)	PSA (1997–)
Venice	Venice Container Terminal (Vecon)	PSA (1998–)
Leghorn	Terminal Darsena Toscana	Eurogate (2001–)
Ravenna	Terminal Container Ravenna (TCR)	Eurogate (2001–)
Civitavecchia	Roma Terminal Container (RTC)	PSA/Evergreen (1998–)

revision of port regulation that, as stated above, brought about the 1994 reform (Musso, 2001). Although there was a decentralisation of power in major ports, with a consequent higher management autonomy compared with the past, the reform still left some structural bonds unchanged. In fact, Italian ports are still penalised by a very bureaucratic administrative structure, provided with little operational flexibility vis-à-vis planning procedures and implementation modes of port infrastructures and political interferences, particularly in the process of defining the bodies of port authorities and composition of port committees. This restricted managerial autonomy often limits the port strategic development, leading ports to focus on the delivery of basic products and services and on efficiency-oriented strategies, rather than letting them commit to supplying added-value services, such as logistics activities and IT supports.

At the same time, financial restrictions still remain. The port authority's financial statement, completely separate from that of the municipality and prepared in accordance with civil accounting principles, shows revenues deriving from port dues, concessions of port areas and income from the provision of the same port services. However, trends in traffic have an almost imperceptible effect on the financial statement of port authorities. This is because a large part of harbour dues charged by the Italian port authorities to port users is directly channeled into state coffers, while port authorities retain only around 15 percent. This percentage is quite low, particularly when compared with the percentage of other ports abroad

(Moglia & Sanguineri, 2000). Therefore, the idea to devolve a great portion of port revenue to the port governing body could fill one of the principal limits of the reform law and offer more competitiveness to Italian ports (Sciutto, 2002).

All in all, the need arises to restructure the port legal setting, to revitalise strategy and port structure, as well as the role and functions of port authorities (De Monie, 2004), with an institutional reform that can allow authorities to increase their range of action and limit the intervention of the central government and of the political interference.

Using the Matching Framework, the Italian port system is still far from a complete alignment among the organisational external environment, strategy and structure. Even if the environment is becoming extremely complex and dynamic for Italian and other Mediterranean ports due to their strategic position in new international trade networks, there is a strong need to restructure the port legal setting and revitalise port strategy and structure. If port authorities intend to move towards innovation-oriented strategies, based on the development of adequate industrial nodes, hinterland connections, logistic activities and modern technologies, more organic structures and decentralised decision-making processes must be urgently implemented.

5. CONCLUSIONS

Up to the beginning of the 1990s, the Italian port sector was characterised by centralised policy and organisation of the port maritime activities. As a consequence, the entrepreneurial activities carried out in ports did not feel the need for productivity improvements and a focus on competitiveness. Following a revolution in technology and the opening of markets, and in order to increase port productivity and competitiveness, it is now paramount to focus attention on the models of governance that may influence the port's life cycle, keeping in the market those ports that appear as more "desirable" in terms of time and quality of services. In such a general view, the Italian ports were subjected to rigid regulation that hampered the adaptation of the organisation to the new, pressing needs of the world market, resulting in a management that did not help Italian port development. This contributed to the beginning of a radical process of reform in the port sector, which involved the organisation of major ports, with the introduction of some autonomous port authorities, which now support the already existing maritime authorities.

Nonetheless, the existing administrative and managerial assets do not seem to be completely adequate to sustain a competitive port development; as a matter of fact, the application of the Matching Framework detected some discrepancies among the variables considered.

The present port environment is more dynamic, due to the new centrality of the Mediterranean area with respect to the international maritime routes, thanks to new foreign investments. However, the Italian ports are still anchored to bureaucratic structures and slow decision-making procedures. Even if, on the one hand, the port reform law promoted terminal privatisation, on the other, it failed the objective of giving Italian ports full decision-making and financial autonomy that is a fundamental element to bring ports towards competitive development.

Such inadequacy also resides in the fact that there is still no Italian "port range": as a consequence, the port environment suffers from the competition among small ports, which wage war among themselves rather than specialising. A greater attention to the processes of port governance is, therefore, necessary just as it is necessary for port authorities to implement infrastructure or service investments, according to the strategic aims that each port deems more important, pending a general reorganisation of both local development and port systems.

REFERENCES

Baltazar, R., & Brooks, M. R. (2001). *The governance of port devolution: A tale of two countries.* World Conference on Transport Research, Seoul, Korea, July.

Brooks, M. R., Button, K., & Nijkamp, P. (Eds) (2002). *Classics in transport analysis: Maritime transport.* Cheltenham, UK: Edward Elgar.

Cullinane, K. P. B., & Song, D.-W. (2002). Port privatization: Principles and practice. *Transport Reviews, 22*(1), 55–75.

De Monie, G. (2004). *Mission and role of port authorities after 'privatisation'.* ITTMA-PPPP Seminar, Antwerp, November.

Federtrasporto. (2004). Statistiche – trasporto marittimo. http://www.federtrasporto.it

Italian port reform law no. 84 of 28 January (1994). *Riordino della legislazione in materia portuale.*

Moglia, F., & Sanguineri, M. (2000). The new models of port governance. *Proceedings of the International Association of Maritime Economists.* Conference, Naples.

Musso, E. (2001). *Regulation and deregulation nei porti: che cosa, come e quando regolamentare. Giornate di studi superiori sulla organizzazione dei trasporti nell' integrazione economica europea.* Trieste: ISTIEE.

Ridolfi, G. (1999). Containerisation in the Mediterranean: Between global ocean routeways and feeder services. *GeoJournal, 48*, 29–34.

Sciutto, G. (Ed.) (2002). *I porti italiani e la sfida di mercati.* Genoa: Sciro Edizioni.

Valleri, M. A., & Van de Voorde, E. (1992). Do Mediterranean ports have a future? *Journal of Regional Policy, 2,* 411–430.

Valleri, M., & Venezia, E. (1997). Regional policy and maritime transport: The Italian case. *Proceedings of the 8th Congress of International Maritime Association of the Mediterranean* (IMAM), Istanbul, Turkey.

CHAPTER 7

PORT GOVERNANCE IN GREECE

Athanasios A. Pallis

ABSTRACT

This chapter examines the development and current forms of port governance in Greece. The national port policy reforms that have been underway since the late 1990s have resulted in the conversion of 12 ports of national interest to government-owned port corporations, thus directing the analysis into a devolution context. The emphasis is on the configuration of three key variables of port governance, namely environment, strategy, and structure. The chapter concludes that in the early post-reform years the absence of a matching configuration of these interrelated variables contributes to the apparent difficulties that Greek ports continue to experience in enhancing competitiveness.

1. INTRODUCTION

Since the late 1990s, the port sector in Greece has changed its management model through the conversion of 12 ports of national interest from 'public law undertakings' to government-owned port corporations. These reforms attempt to facilitate the adjustment of the sector to contemporary trends and to overcome the deficiencies of the pre-existing port structures, by transferring the management and the responsibility for port services provision to port-level entities. To address the resulting port-performance potentials, this

Devolution, Port Governance and Port Performance
Research in Transportation Economics, Volume 17, 155–169
ISSN: 0739-8859/doi:10.1016/S0739-8859(06)17007-9

chapter follows the path of port governance (Baltazar & Brooks, 2001; Notteboom & Winkelmans, 2001, 2002; Wang, Koi-Yu Ng & Olivier, 2004).

The emphasis is on the principles and the institutions by which authority over ports in Greece is exercised, and on the process by which ports select, monitor and replace strategies and the capacities of those in authority to manage port resources, implement policies and govern economic interactions.

2. BACKGROUND TO PORT REFORM

Although Greece is a traditional shipping nation, national administrations have placed little importance on infrastructure for the development of maritime transport. This is due to the small position that the country holds in the global marketplace. The principal state agent (Ministry of Mercantile Marine – MMM) focuses on 'flag-state' policies rather than on policies for providing port services ('port-state' policies). The potential of Greek ports as important nodes connecting the Far East with Europe has been neglected.

Greek ports, like other Mediterranean ports, were organised as comprehensive state-controlled port organisations, with the state acting both as regulator and port service provider. Ports were 'public law undertakings' ruled according to the general regulatory regime of public entities. In this governance model, a state-controlled port authority owns and maintains the infrastructure and superstructure and provides all port services. The exclusive provider of each service is either the port authority's personnel, or the local federations of port-workers. The private sector is involved in the provision of these services solely when port authorities lack the capacity or the equipment (i.e. handling cranes and towing) to provide them, while some services (e.g. pilotage) are provided by the MMM.

The first reform discussions took place in the early 1990s. Trade expansion and restructuring had exposed port deficiencies. International ports failed to improve their image and the users' criticism was intense. Common complaints were the absence of long-term vision, the insensitivity towards users' demands, absence of port facilities and inland connections and lack of investment. The criticism for unproductive decisions and the absence of innovative ideas by managers who lacked sector experience but whose appointment facilitated clientele political practices were also explicit, as were the conflicts between vested interests and port administrations (INU, 1995).

An inappropriate legislative regime had produced port structures that resulted in controlling boards with little business competence. Politicians used ports to solve employment problems, even exercise social policies. In

1997, the Port of Piraeus was obliged (under a 1961 Law) to spend 20 percent of its annual income to provide pensions to 1,800 retired port workers, delaying projects required to meet market developments. World-wide trends transformed ports to a capital-intensive sector but in Greece the outcome was low productivity levels, generally poor operating conditions and infrastructure inefficiencies (Chlomoudis & Pallis, 1997). Ineffective pricing mechanisms (Psaraftis, 2005), sub-optimal performance and lack of long-term strategies prevented ports from fulfilling stakeholder expectations and resulted in frequent loss of business. When changes in transportation demanded specialisation and post-Fordist methods of operation (Chlomoudis, Karalis, & Pallis, 2003), Greek ports experienced 'political management' structures that impeded progress in responding to structural changes, as was happening elsewhere in Europe (Notteboom & Winkelmans, 2001).

The country's commitment to be part of the European Monetary Union resulted in economic policies to reduce public deficit and inflation, and constrained investments in 'public law undertakings', furthering limitations. Despite the EU initiatives to improve port infrastructure via public/private partnerships (Pallis, 1997), there was no significant private capital mobilisation and the situation deteriorated. For some this was a policy of 'privatisation through bankruptcy' (Goulielmos, 1999).

A new government in 1993 sought ways to transfer the management of certain ports to local authorities. This ideologically driven move was initiated for the purpose of decentralisation, but had nothing to do with relaxing state ownership, control or management (Goulielmos, 1999). The deregulation of utilities became part of the agenda, in view of the Single European Market. Still, there were no immediate waves of change. The resistance on the domestic front was apparent and structural reforms in the Greek public sector were limited. The publication of an EU Green Paper stating the intention to create a level playing field between and within EU international ports (CEU, 1997) contributed to a wait and see policy that minimised political costs. Nonetheless, the pace towards reforms accelerated as the formation of EU policies proved to be remarkably slow (Chlomoudis & Pallis, 2002) and the dynamics of the economy-posed adjustment pressures.

3. PORT REFORM

Port policy restructuring has been underway in Greece since 1999. The objectives are to increase the participation of Greek ports in global maritime transport, enhance the EU emphasis on the role of its members as port-states (Pallis, 2002) and promote "the greatest possible participation in the provision of port services" (MMM, 2002, p. 7). The national administration acknowledged that this policy should be based on the circumstances of Greece, the new volumes of international trade, new technologies and organisational structures that attract high-yield investments.

This strategy has five objectives:

- bilateral maritime relations with countries exporting significant cargo volumes;
- port competitiveness in light of the international economic environment;
- sustainable and integrated port development in order to meet social and environmental needs;
- social cohesion of the island area and populations; and
- safeguarding cargoes transported through Greek ports.

3.1. The New Statutory Formation

The first port-reform initiatives (1999) considered the two trans-European port organisations of the country. The ports of Piraeus and Thessaloniki became corporations, at that time wholly owned by the Greek State (Law 2688/1999). Following a later decision, these ports were listed on the Athens Stock Exchange (Thessaloniki since 2001 and Piraeus since 2003), with the state retaining 75% of their shares. Two years later (Law 2932/2001) 10 other ports of national interest (Table 7.1) were converted from Port Funds to limited companies. Each company has one share owned by the state. These ports are supposed to operate as private-sector businesses with the objective of developing infrastructure and offering quality and competitive services. Along with Piraeus and Thessaloniki, they participated in the self-governed Hellenic Ports Association (HEPA) that was established in 2003 in order to ensure collaboration between ports.

As for the remaining ports, local and municipal authorities have undertaken the management of the remaining Greek ports, replacing managing boards directly appointed by the central administration, with the objective being better utilisation of public funds and other resources, and better service for residents and the local tourist industry. Part of the reform was also the establishment of a Port Planning and Development Committee whose members are representatives of eight ministries. This Committee is responsible for the general planning, monitoring and implementation of port programmes, and the allocation of financing for the adaptation of

Table 7.1. The Greek Port System.

	Port	Goods Throughput (in '000 Tonnes)	Passenger Traffic (in '000 Persons)	Sea-going Vessels (in Number)
Major trans-European ports	Piraeus	21,425	12,537	26,333
	Thessaloniki	14,899	201	2.855
	Patra	3,399	1,263	81,581
	Heraklion	3,350	1,652	1,879
	Elefsina	3,250	800	5,003
	Kavala	1,633	1,454	7,330
	Volos	1,241	392	3,494
National ports	Lavrio	825	213	1,815
	Corfu[a]	642	2,146	9,899
	Alexandroupoli	639	169	1,757
	Igoumenitsa	495	1,193	12,227
	Rafina[a]	97	1,761	4.305
	Total of the 12 ports	51,894	23,782	158,478
Peripheral ports	1,250 Peripheral ports, marinas, fishing harbours subject to the jurisdiction of 188 port authorities			
Municipal ports	53			

Source: Hellenic Ports Association (HEPA) (2003); all other data are for 2003 and provided by port authorities.
[a]2002 Data.

infrastructure. In total, the gross weight of goods handled in all Greek major and small ports is 111.1 million tonnes, including almost 2 million TEUs container cargo (ESPO, 2005) (Fig. 7.1).

4. EFFICIENCY-ORIENTED POST-REFORM STRATEGIES

Public control remains and all ports are under the supervision of the MMM. The ownership of the ports' assets has passed (from port funds) to the State, which signed 40-year concession agreements with the respective port authorities for the exclusive right of use and exploitation of land buildings and infrastructure within port zones, in order to provide all users with port services. The concession fee is 2% of the operating income. The scope is to generate a degree of autonomy regarding port management and operation, while a special governmental secretariat administers the national port system. This secretariat is expected to operate as an *external state*, which

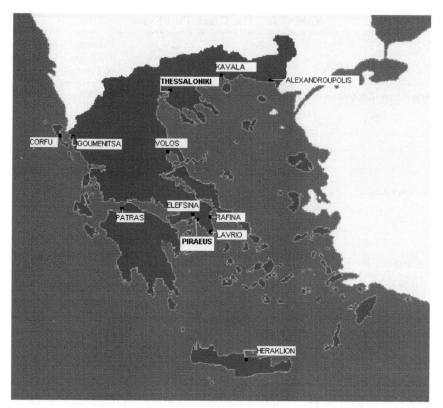

Fig. 7.1. The Geography of the Greek Port System: Ports of National Interest.

intervenes from a position outside and above port operations and with a view to maximising the common good, that is, developing strategic objectives (i.e. sustainability). This strategy matches considerably the changes that had taken place in other Mediterranean EU countries (Italy, France and Portugal). The tradition that led to a distinct Mediterranean port region within the EU (CEU, 1997) is alive, albeit its features are structurally different from those of the past.

Using the World Bank (2000, module 3, p. 46) taxonomy, in the case of ports of national interest the reform follows a corporatisation process

> in which a public sector undertaking ... is transformed into a company under private corporate law that conducts the port's business and holds its assets, although the shares are issued and ... owned entirely by the government .... The main objective is to decrease direct government control over the company and to make it more responsive to market forces.

As regards Piraeus and Thessaloniki, public listing was a move to secure private funds and limit the fiscal burden of port modernisation. For those advocating the privatisation of ports, this remains a first step towards this direction.

The 2004 change of the political party in power was accompanied by inertia. This had commenced in the pre-election period and ended a year later with the appointment of entirely new Boards of Directors and CEOs in all ports of national interest. Before that, a 'first generation' of autonomous port managers had been appointed via a supposedly transparent and independent process. This approach resulted in further delays, as the 'second generation' decided to follow a public sector tradition and review the existing business plans. Three months later, the inexperienced (as regards ports) CEO of the Port of Piraeus resigned. Within the last six years, the ports of Piraeus and Thessaloniki have been managed by five and three different CEOs, respectively. In mid-2006, the ministry dissolved the self-governed Hellenic Ports Association, only to establish a National Centre of Port Development that is governed by the MMM itself.

The ultimate structures of port devolution might take several forms (Brooks, 2004; Cullinane & Song, 2002) with the balance between public and private sectors standing as the critical issue (Juhel, 2001). In Greece, neither the re-allocation of duties and responsibilities between public and private actors, nor the re-distribution of risks and rewards has developed yet. Ports have not implemented an organisational restructuring, although in the aftermath of the reforms most ports defined strategic and business plans.

This is partly attributed to the lack of labour reform. There was neither a new labour statute, nor any personnel retraining to increase commercial orientation and improve managerial procedures. Many of the traditional regulations (operational practices and dockers' payment scheme) are still the same, while redundancies are not allowed. While modern container terminals employ approximately 10 people per crane, Greek ports employ almost 20. The attitude in pushing reforms was to let the real reforms be made by the ports themselves, but the central government continues to control several aspects (i.e. the process of hiring employees), which brings into question how substantive the reforms have been in practice.

Besides, the regulatory framework limits concessions and intra-port competition. The absence of such competition, along with the lack of business culture and the inexperience of the newly formed port entities in long-term planning, contribute to the existing lack of efficiency (Barros & Athanassiou, 2004). Port authorities remain the sole providers of core services (e.g. handling) while the provision of some nautical services by private firms (e.g. towage) is heavily regulated. The interest from shipping and other private companies for investing in both cargo and passenger terminals and the provision of port services is explicit (i.e. Cosco, China Shipping) national legislation does not allow for free market entry.

Given that most Greek ports have not yet become competitive third-generation ports, port authorities direct their efforts towards efficiency-oriented strategies. Strengthening investments and enhancing the uninterrupted handling of all kinds of cargoes are major targets. Owing to the lack of intermodal connections, delivery of core port services is a key theme. However, national ports have not secured funding for implementing recently defined business plans. In 2005, the government expressed its intention to tender part of the port of Piraeus. This movement has the potential to become the biggest substantive reform of all, yet it is still unclear how this will become reality. As in other cases, neither public nor private funds have been secured.

The greater flexibility in the making of decisions induces additional themes in port strategies. Port authorities attempt to decide whether to focus on transhipment or origin-destination traffic. They are also in the process of choosing between searching for the benefits of competition (Elefsina vs. Piraeus) and endorsing cooperation and port-networking strategies, either between them (Piraeus and Heraklion) or with other ports in the region (Cyprus and Syria). At port level, this implies planning rationalisation to reach optimal utilisation levels. Comparing actual TEU throughput with throughput capacity, Piraeus and Thessaloniki have a utilisation rate (78.9%) that is not lower than the ratios observed in other European ports. Thus, the emphasis is on operational efficiency (berthing systems and new technologies) and port tariff restructuring, so as to better attract cargoes and achieve simplification in service provision.

Greek ports are also interested in developing user-oriented adaptability. This is based on the assumption that the biggest ports could be assisted by synergies with shipping and cargo-handling companies, in order to satisfy user demand. Signing contracts with container carriers (such as MSC, Norasia and China SCL) were among the priorities following the transformation of Piraeus into a corporation. The potential of storage facilities and equipment concessions to port users are strongly promoted in Thessaloniki. Smaller ports pursue a niche market strategy, as they lack both the experience and the capability to develop multi-services provision strategies.

In this vein, CEOs of Greek ports seek developments that could benefit them by enhancing their autonomy, or by allowing more flexible strategies, i.e. an EU-imposed port services directive that would open access to the port services markets (Pallis & Vaggelas, 2005). With geographical powers erod-ing, the vital issue is the presence of commercial ethos rather than the nature of competition. In conditions of limited private sector involvement, they seek the disappearance of public monopolies and/or the generation of con-ditions for labour reform. Notably, given the pro-privatisation intentions of the current government, the strategic planning that is followed anticipates the highest support for 'the imminent privatisation. According to the plan-ning of the MMM, this will be completed in two phases, with 51 percent being owned by the public... (and) will take place with the consent of the Port's personnel' (ThPA, 2005, p. 1).

5. ECONOMIC CONTEXT: UNCERTAINTY, COMPLEXITY, DYNAMISM

In the 1990s container handling in Greek ports increased from 480,000 to 1.3 million TEUs. In the period 2000–2004, this volume continued to rise, mainly due to the incoming cargoes from the Far East and China. Similar to other Mediterranean ports, Greek ports handle mostly trans-oceanic traffic and represent the least integrated part of the industry in intra-EU traffic. Piraeus attempts to act as the major gateway in the Southeast Mediterra-nean region and serves transhipment trade, a strategy that has been quoted as successful. Transhipment increased by 20% annually and in 2004 rep-resents 63% of the containers handled.

The situation in the Mediterranean is not static, particularly regarding containers. The successful evolution of the Medcenter in Gioia Tauro, which started operations in 1995 and turned into a major container terminal, pro-vides a major example (Table 7.2). The mid-1990s decision of two lines (Evergreen and Lloyd Triestino) to move their transhipment hub from Piraeus to Gioia Tauro produced a temporary but significant reduction of container traffic through Piraeus, although traffic of imported and exported containers continued to grow. With shipping lines developing hub and spoke systems, which ports establish themselves as hubs in the Mediterranean and thus ac-commodate the traffic on the route connecting the Far East with Europe and which establish themselves as markets in the Black Sea is up for grabs.

Then, port users press for service modernisation. The users of Piraeus represent a case of pressure for value-added services and labour adaptations. As operational deficiencies continue, some of them are considering redirec-tion of cargoes to nearby ports. The main user of the Piraeus container terminal, MSC (700,000 TEUs per annum), is considering using Limassol to

Table 7.2.   Ranking of the Major Mediterranean Ports: 1990–2002 ('000 TEUs Handled).

Port	Country	2002	Rank	2000	Rank	1990	Rank	Percentage of Increase 1990–2000
Gioia Tauro	Italy	2,883	1	2,653	1	N/A	–	–
Valencia	Spain	1,826	2	1,308	5	387	7	237
Algeciras	Spain	1,732	3	2,009	2	553	1	263
Genoa	Italy	1,531	4	1,501	3	310	8	383
Piraeus	Greece	1,405	5	1,161	6	426	5	172
Barcelona	Spain	1,122	6	1,388	4	448	4	209
Marseilles	France	811	7	726	8	482	2	51
La Spezia	Italy	780	8	910	7	450	3	102
Leghorn	Italy	547	9	501	9	416	6	20
Naples	Italy	444	10	397	10	133	9	198
Thessaloniki	Greece	240	11	230	12	54	11	326

Note: The first year of port operation was 1995.
Sources: Containerisation International Yearbook; various port websites.

accommodate traffic expansion and transhipment, unless operating conditions improve. The users of Thessaloniki port have stated their dissatisfaction with the inactivity of the port authority regarding the upgrade of services. Some companies have recently decided to shift to ports in neighbouring countries, but circumstances are no better there.

The EU port policy developments and the strategies of the competitors add to the complexity. The EU decision (2001) to include ports in the Trans-European Transport Networks (TEN-T) plans permits the financing of ports fulfilling one of the following criteria:

• International ports, whose annual traffic exceeds 1.5 million tonnes or 200,000 passengers, which have established intermodal links with the TEN-T.
• Community ports, whose annual traffic exceeds 500,000 tonnes or varies between 10,000 and 199,000 passengers, which have established intermodal links with the TEN-T.
• Local ports that do not fulfil criteria A or B but are located on islands or remote inland areas and are considered necessary for the provision of steady connections with specific areas.

The concept of high-speed maritime corridors (motorways of the sea), the Marco Polo programme supporting multimodal transport, the EU

Table 7.3. TEN-T Category A (International Ports) Greek Ports.

Ports of National Interest		Other Greek Ports	
Piraeus	Lavrio	Mytilene	Kyllini
Thessaloniki	Kavala	Halkida	Tinos
Patra	Igoumenitsa	Paros	Mykonos
Elefsina	Rafina	Skiathos	Rhodes
Heraklion	Corfu	Naxos	Chios
Volos		Souda	

contribution in infrastructure modernisation, R&D frameworks and the promotion of short-sea shipping provide further challenges and opportunities. The status quo might change as ports can effectively exercise their rights and enhance their performance (e.g. Gioia Tauro benefited noticeably by EU funds). The Greek ports classified in Category A are numerous (Table 7.3), so the impact of TEN-T plans might be substantial.

If adopted, the EU port services directive (CEU, 2004) would establish intra-port competition in the provision of services in eight ports of national interest (Piraeus, Thessaloniki, Elefsina, Patras, Igoumenitsa, Iraklion, Rafina and Corfu). At the moment, the direct private sector involvement in port services provision is not allowed. In September 2003, attempts by a private company to offer port services (operate a car terminal in Elefsina) were blocked as illegal. One company (Astakos SA) has gained the right to serve cargoes in western Greece, since there is no alternative provider; it still has to limit its activities to the development of an industrial and commercial free zone that provides logistics solutions. Given the consolidation in the European container handling market (ESPO, 2005), both national and international firms and consortia are expected to express their interest in providing port services.

Regional competitors have introduced reforms that allow freer access to port services provision. Spain has allowed concessions, while international groups currently control large container terminals in Italy. Cyprus is developing a concession programme while Marsaxlokk's success was followed by a 30-year concession of the Freeport Terminal (signed in: 2004). The existing port hierarchy in the Mediterranean port region might be altered by the presence of actors involved in core operations, investments in dredging, hinterland connections and infrastructure, logistics and market-oriented strategies (cf. Zohil & Prijon, 1999).

Finally, there is the rise of logistics and distribution centres in European ports. Companies opt for a hybrid structure of both centralised and local distribution facilities; hence there is some degree of decentralisation in European distribution structures. The recent enlargement means that the new EU covers a much larger geographic region, making more likely the potential of a two-tiered European distribution structure (Notteboom & Winkelmans, 2004). The latter will consist of central European distribution centres together with regional distribution centres established in Southern Europe and the Eastern Mediterranean port region. Which ports will manage to attract these is still questionable and the port hierarchy might be altered substantially. In Greece, a law under preparation in 2005 addresses the establishment and operation of intermodal distribution centres, acknowledging the necessity to adjust to an environment of greater complexity and dynamism.

6. CONCLUSIONS

The inefficiencies of the existing traditions of port organisation, the adjustment pressures of the economic environment and the fiscal constraints exposed the misalignment of strategy and structure to Greece's environment (following Baltazar & Brooks, 2001) and contributed to the reconsideration of the Greek port policy in the late 1990s. A vastly changing and complex environment was combined with a failed efficiency-oriented strategy that focused on the delivery of basic services, and a mechanistic structure of centralised decision-making characterised by procedural standardisation (Table 7.3).

Until now, policy reforms are marked by the devolution of port governance, via the corporatisation of 10 ports and the public listing of the two major trans-European ports of Piraeus and Thessaloniki, with the State retaining a 75% stake. Whether these developments constitute a permanent pattern, or the prelude to the imminent privatisation of some or all these ports, remains to be seen. Nonetheless, this suggests that port governance in Greece is not aligned to a configuration that might advance port competitiveness.

The economic and institutional environment is characterised by high levels of uncertainty, complexity and dynamism. Both successes and failures of ports in the Mediterranean region are common, especially in the case of dynamic market niches. Stakeholders' relations are volatile and loyalties are undermined as port users seek to place themselves in the best conditions to benefit from globalisation and trade restructuring. The progress of European integration expands the potential of changes in the port hierarchy at local, regional and, not least, continental level.

The current Greek ports strategy is based on the need to overcome long-term inefficiencies. Greek ports have experienced a model of Fordist

Table 7.4. Greek Ports Governance and the Matching Framework.

		Configuration 1	Configuration 2
Environment			High uncertainty High complexity and dynamism
Strategy	Current	Efficiency-oriented focus on delivery of the basic product and service	
	Target		Effectiveness-oriented strategy focused on delivery of peripheral product and services
Structure		Mechanistic Centralized decision-making characterized by procedural standardization (subject to bureaucratic interferences)	

Source: Based on Baltazar and Brooks (2001).

impersonal structures, inflexible labour regimes and the lack of capacity to provide several value-added services. They are currently seeking to build relationships with port users, potential investors and/or other ports, aiming at efficiently delivering core and additional services. Yet, another strategy is in the agenda. All ports seek ways to be engaged in an effectiveness-oriented strategy that would enhance their capacity to integrate in intermodal logistics chains, and not least the relationship with port users. This strategy would match the requirements of the contemporary economic context (Table 7.4).

However, both the current and the desired strategies are undermined by a centralised decision-making process, characterised by procedural standardisation, and are subject to bureaucratic interferences. Governmental involvement is not limited to general regulatory, planning or financial issues. An interventionist state interferes in strategic decisions and daily port life. This is the case even in the ports that are listed on the stock exchange (Piraeus and Thessaloniki) and thus are subject to published business plans and stricter corporate responsibility rules. Ministerial interventions in the summer of 2005 resulted in successive port workers' strikes that further undermined the competitive position of the port. The lack of labour reform, and the delays in readjusting the balance among the various actors (i.e. by establishing intra-port competition) is not irrelevant to this practice; such adjustments imply significant political costs.

The absence of a matching configuration contributes to the contemporary difficulties that Greek ports experience in enhancing competitiveness. Nevertheless, as national port reforms began only five years ago, the governance of Greek ports is still in a state of flux. Infant autonomous corporate entities have yet to finalise their business plans, organisation structures and strategies. Moreover, both the national and the international institutional settings are in transition. The rules that would allow direct private sector involvement in the provision of port services are under consideration, at both national and EU levels. Given the dynamics of the economic and policy environment and the challenges and opportunities they pose to the port hierarchy, further legislative amendments and a reconsideration of political practices might further modify the structures of the sector and thus enable Greek ports to adjust their strategies.

REFERENCES

Baltazar, R., & Brooks, M. R. (2001). The governance of port devolution. A tale of two countries. *Proceedings of the 9th World Conference on Transport Research* (CD-ROM), Seoul.

Barros, C. P., & Athanassiou, M. (2004). Efficiency in European seaports with DEA: Evidence from Greece and Portugal. *Maritime Economics and Logistics, 6*(2), 122–140.

Brooks, M. R. (2004). The governance structure of ports. *Review of Network Economics, 3*(2), 168–183.

CEU. (1997). *Green Paper on seaports and maritime infrastructure*. Brussels: Commission of the European Union (Com (97)678, final. 10.12.1997).

CEU. (2004). *Proposal for a directive of the European Council and the European Parliament on market access to port services*. Brussels: Commission of the European Union (Com (2004)654, final. 13.10.2004).

Chlomoudis, C. I., Karalis, V. A., & Pallis, A. A. (2003). Port reorganisation and the worlds of production theory. *European Journal of Transport and Infrastructure Research, 3*(1), 77–94.

Chlomoudis, C. I., & Pallis, A. A. (1997). Investments in transport infrastructure in Greece: Have the EU initiatives promoted their balanced and rational distribution? *World Transport Policy and Practice, 3*(4), 23–29.

Chlomoudis, C. I., & Pallis, A. A. (2002). *European port policy: Towards a long-term strategy*. Cheltenham, UK: Edward Elgar.

Cullinane, K. P. B., & Song, D. W. (2002). Port privatisation policy and practice. *Transport Reviews, 22*(1), 55–75.

ESPO. (2005). *Annual report 2004*. Brussels: European Sea Ports Organisation.

Goulielmos, A. M. (1999). Deregulation in major Greek ports: The way it has to be done. *International Journal of Transport Economics, XXVI*(1), 121–148.

Hellenic Ports Association (HEPA). (2003). *Greek ports 2003*. Thessaloniki, Greece: HEPA.

INU. (1995). *Greek ports (in Greek)*. Piraeus, Greece: International Nautical Union.

Juhel, M. H. (2001). Globalisation, privatisation and restructuring of ports. *International Journal of Maritime Economics, 3*(2), 128–138.

MMM. (2002). *Strategy for a national ports policy.* Piraeus, Greece: Ministry of Mercantile Marine, http://www.uen.gr (accessed: May 2005).

Notteboom, T. E., & Winkelmans, W. (2001). Reassessing public sector involvement in European Seaports. *International Journal of Maritime Economics, 3*(2), 242–259.

Notteboom, T. E., & Winkelmans, W. (2002). Stakeholder relations management in ports: Dealing with the interplay of forces among stakeholders in a changing competitive environment. *Proceedings of the IAME 2002 conference* (CD-ROM), Panama City, Panama.

Notteboom, T. E., & Winkelmans, W. (2004). Overall market dynamics and their influence on the port sector. In: ESPO (2004), *Factual report on the European port sector* (pp. 4–72). Brussels: European Sea Ports Organisation.

Pallis, A. A. (1997). Towards a common ports policy? EU-proposals and the industry's perceptions. *Maritime Policy and Management, 24*(4), 365–380.

Pallis, A. A. (2002). *The common EU maritime transport policy: Policy Europeanisation in the 1990s.* Aldershot, UK: Ashgate.

Pallis, A. A., & Vaggelas, G. K. (2005). Port competitiveness and the EU port services directive: The case of Greek ports. *Maritime Economics and Logistics, 7*(2), 116–140.

Psaraftis, H. N. (2005). Tariff reform in the Port of Piraeus: A practical approach. *Maritime Economics and Logistics, 7*(4), 356–381.

Thessaloniki Port Authority (ThPA). (2005). Strategic-investment planning of Thessaloniki Port Authority. Retrieved July 2005 from http://www.thpa.gr

Wang, J. J., Koi-Yu Ng, A., & Olivier, A. D. (2004). Port governance in China: A review of policies in an era of internationalizing port management practices. *Transport Policy, 11*(1), 237–250.

World Bank. (2000). *Port reform toolkit.* Washington, DC: World Bank.

Zohil, J., & Prijon, M. (1999). The MED rule: The interdependence of container throughput and transhipment volumes in the Mediterranean ports. *Maritime Policy and Management, 26*(2), 175–194.

CHAPTER 8

PORT GOVERNANCE IN TURKEY

Ersel Zafer Oral, Hakki Kisi, A. Güldem Cerit,
Okan Tuna and Soner Esmer

ABSTRACT

Turkey is located between Asia and Europe and attracts attention with its economic development. Turkey's landbridge position in north–south and east–west transportation means that ports are of vital importance to the efficiency of logistics activities of the country. Although Turkey has a strategic position in terms of logistics and shipping, its approximately 160 ports do not enjoy the usual benefits of ports. The ports and piers can be classified in terms of whether they are operated by public sector, affiliated sector, regional municipalities or the private sector. This paper focuses on the Turkish port management and administration system and the possibilities for applying good governance in Turkish ports during the current process of privatization of public ports. It concludes that the privatization process has not been completed yet and there are many legal and practical issues to be resolved.

1. INTRODUCTION

Turkey is located at the crossroads of the trade between Asia and Europe and encircled by the Black Sea, the Marmara Sea, the Aegean Sea and the Mediterranean Sea. It has borders with Georgia, Armenia, Azerbaijan

Devolution, Port Governance and Port Performance
Research in Transportation Economics, Volume 17, 171–184
ISSN: 0739-8859/doi:10.1016/S0739-8859(06)17008-0

(Nakhichvan) and Iran to the east, Bulgaria and Greece to the west, Syria and Iraq to the south and Russia, Ukraine and Romania to the north. The area of the country is 814,578 km^2 and the population approximately 70 million. There are about 160 ports along its 8,300 km of coastline.

The Mediterranean Sea is the main transportation corridor between Far-East Asia and Europe. Gioia Tauro and Algeciras are the most important hub ports in the Mediterranean Sea (General Director of Railways, Ports and Airports Constructions (CRPA), 2005a). Haifa, Damietta and Piraeus are the main ports in the region close to Turkey. The most important Mediterranean gateway ports to Europe are Genoa, Barcelona and Valencia. In the Eastern Mediterranean and Black Sea regions there are no significant gateway ports.

At present, the weight of the demand and the privileged location of ports such as Gioia Tauro, Algeciras or Cagliari mean that the majority of hub ports are located in the Western Mediterranean. However, the principal hubs are likely to change, given that the Eastern Mediterranean is experiencing far greater growth than the Western Mediterranean. It is foreseeable that this tendency will be consolidated as the development of Turkey and the Balkan countries advances and commercial exchanges between the countries of the Near East are deregulated (CRPA, 2005b).

This chapter attempts to define port governance system in Turkey during the current process of privatization of public ports.

2. AN ANALYSIS OF TURKISH PORTS FROM THE PERSPECTIVE OF PORT COMPETITION

With globalization, good port governance has gained importance due to strong competition among the ports. Port governance should be taken as a key issue for the advancement of port competitiveness (Chlomoudis & Pallis, 2004). From a historical point of view, port governance in Turkey can be classified into three main periods: a nationalization period, a period of both public and private port operations and, most recently, a privatization period during which the government is withdrawing from port operations. However, from the administrative point of view, the existing ports in Turkey are classified into four groups: public, municipal, affiliated and privately owned ports.

Turkey's current port policy is supported by legislation such as Law 618 (ports, dated 1925), Law 815 (cabotage, dated 1926), Law 3621 (coasts, dated 1990) and Law 4046 (privatization). The Undersecretariat for

Fig. 8.1. Location of the Main Ports in Turkey.

Maritime Affairs (UMA) oversees ports in its role as the national maritime administration.

Municipal ports in Turkey can be considered as negligible. They do not play an important role in the overall marine transportation of Turkey because of their low share of cargo throughput. Fig. 8.1 shows the major ports in Turkey.

Turkey has influence in the region that includes the Middle East, Eastern Mediterranean, Black Sea, the Balkans and Central Asian countries, namely Turcic countries. Turkey has strong economical and cultural relations within the region, so it has a vision of ports as part of its role in global trade. In this context, Turkey plans that the North Aegean port in Candarli, in the north of Izmir, will be a regional container hub port. Also, some foreign and national private companies are constructing high-capacity container terminals close to Istanbul.

The final report of a study detailing a nationwide port development master plan for Turkey analyses the existing container ports as follows (Overseas Coastal Area Development Institute of Japan (OCADIJ), 2000):

Istanbul (container ports in the Marmara Sea): Container ports in the Marmara Sea cannot be major competitors in Eastern Mediterranean transport. They can certainly act as feeder ports and if a container port in the Marmara Sea can collect a certain amount of local container cargo, the port can possibly attract large container vessels to call at the port directly (CRPA, 2005a).

Izmir (container ports in the Aegean Sea): Izmir port is in a good location between the Eastern Mediterranean and the Black Sea and can be a major competitor in both. Izmir is the third largest city in Turkey and an important business centre. The port has a vast agricultural and industrial hinterland. It is the port for the Aegean Region's industry and agriculture and plays a vital function in the country's exports. All different types of commodities and cargo groups are handled in the port and port expansion studies continue. The port is also connected to the national railway and highway network.

Mersin Port (container ports in the Mediterranean Sea): Mersin Port can be a major competitor in Eastern Mediterranean transport (CRPA, 2005a). It is the main port for the Eastern Mediterranean Region's industry and agriculture and a gateway for the Turkish provinces of Gaziantep and Adana as well as for Iraq and Iran. The port's rail link and its easy access to the international highway make it an ideal transit port for trade to the Middle East. With its modern infrastructure and equipment, efficient cargo handling, vast storage areas and its proximity to the free trade zone, Mersin is one of the most important ports in the Eastern Mediterranean. The facilities at the port handle general cargo, containers, dry and liquid bulk and Ro–Ro.

Turkish ports serve the national economy, but they have insufficient capacity in terms of infrastructure, superstructure, equipment, etc. for transit cargoes, so they hardly compete with the regional ports. The major ports in Turkey used to be operated within a monopolistic regime until the privatization of some of them. Currently, privatization administration of Turkey is very keen on fair competition (Table 8.1).

Municipal ports, such as Ayvalik, serve the tourism and passenger market, although they also handle some bulk and general cargo. Municipal ports are comparatively small-scale and are operated by the municipalities where the ports are located. Affiliated ports are owned and operated by either large state-owned or private industrial companies and these ports usually serve the tramp and bulk market. The last group is made up of privately owned ports; most of these primarily handle their own cargoes but do serve other customers.

Private ports have specialized terminal operations that usually serve the bulk and tramp market. Some are occupied with the liner market by serving containerized cargo. The private ports/terminals are in a rather competitive position compared with publicly operated ports. Generally, they prefer to specialize and operate more efficiently. The decision-making process in private ports is much quicker and more efficient at investing, especially in cargo

Table 8.1. Major Ports in Turkey.

Ports	Length (m)	Depth (m)	Handling Capacity (Ton/Year)	Ships (Ship/Year)	Storage Capacity (Ton/Year)		Container Capacity (TEU/Year)
					Open	Closed	
Haydarpasa	2,765	(−6,−12)	6,488,300	2,651	471,360	362,384	264,000
Derince	1,132	(−4,5–15)	1,910,900	567	2,951,760	200,000	–
Samsun	1,756	(−6,−12)	2,284,100	1,130	8,556,720	192,304	–
Mersin[a]	3,180	(−6,−14,5)	5,510,800	3,052	8,109,024	562,992	203,376
Iskenderun[a]	1,427	(−10,−12)	3,223,600	640	8,991,120	294,320	–
Bandirma	2,788	(−10,12)	2,636,100	4,277	1,868,280	144,000	–
Izmir[b]	2,959	(−4,−12)	4,931,600	3,635	565,920	377,648	265,728
Marport[c]	2,000	(−14,5)			409,000		900,000

Source: Turkish Chamber of Shipping (2004), www.arkas.com.tr

[a]Privatized.
[b]Under process of privatization.
[c]Private.

handling equipment. The management style of the private port sector has minimized internal bureaucracy.

While some private ports adopt customer relationship management techniques in their relations with port users, this does not seem to be a practice of public ports. The main concerns of public ports are social and economic issues. Their principal aims are to increase the economic benefits of the port for the nation or region and to cooperate with labour unions in seeking to achieve this. While private multipurpose and container ports are much more focused on value-added services and a non-union labour force to maximize their profits, the public port enterprises have operated ports by involving strong labour unions in the issues. After privatization, the labour unions are rather weakened or have been eliminated.

The Overseas Coastal Area Development Institute of Japan (OCADIJ, 2000) analysed the relationship between port administration, port management and the institutional framework and identified the following four policies:

1. Autonomous port administration
2. Flexible port management
3. Financial independence
4. Port development in liaison with regional development

Although the fourth development policy has always been implemented in liaison with regional and national authorities, the first policy is not exercised in Turkey at all. On the other hand, the second and third policies are exercised only by the private sector.

The OCADIJ (2000) made the following two recommendations with respect to the importance of Turkish ports increasing the productivity of their cargo handling and service levels: (1) effectively use existing facilities and (2) be more competitive.

As a result of this analysis, it is observed that there is a fast transition and unpredictable result for the port users after the privatization process, and Turkey has not set a specific model to fit its nature and needs. For instance, the privatization contracts of Port of Mersin and Port of Antalya have different content and logic, in terms of articles concerning operation periods, investment obligations, increasing cargo throughputs, etc. Since public ports cannot be flexible, do not have financial independence, and are not focused on clear goals on delivering services, they have become inefficient. Inefficiency has also been caused by political interference in employment issues and the high turnover rate in top managerial positions.

3. PORT-RELATED BODIES

The main port-related governmental organizations can be summarized as follows: Prime Ministry, State Planning Organization (SPO), Undersecretariat for Maritime Affairs (UMA), Ministry of Transport (MOT), Ministry of Health, Ministry of Public Finance, Ministry of Interior, Ministry of Public Works and Settlement (MPWS), Ministry of Industry (MOI), Ministry of Agriculture, Ministry of Environment, General Directorate for Construction of Railways, Seaports and Airports (CRPA), State Economic Enterprises such as Turkish State Railways (TSR) and Turkish Maritime Organization (TMO), municipalities, customs, immigration police, etc.

The relationship between governmental organizations is shown in Fig. 8.2, and the functions and responsibilities of the main port-related organizations are described below.

The Prime Ministry and the SPO consider the total balance of investment in Turkey and judge the feasibility of specific projects. The MOT coordinates all the development of ports in Turkey and has responsibility for

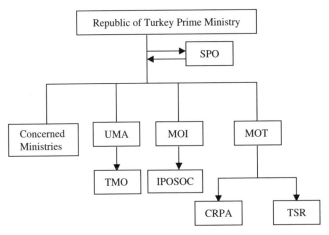

Fig. 8.2. Relationships between Governmental Organizations. Abbreviations: SPO, Prime Ministry and the State Planning Organization; UMA, Prime Ministry and the Undersecretariat for Maritime Affairs; MOI, Ministry of Industry; MOT, Ministry of Transport; TMO, Turkish Maritime Organization; IPOSOC, Industrial Ports of State-Owned Companies; CRPA, General Directorate for the Construction of Railways, Ports and Airports; TSR, Turkish State Railways.

setting the port tariff of TSR ports. The MOI controls and coordinates the industrial ports of state-owned companies (IPOSOC). The Ministry of Public Finance funds port investments and collects taxes. The Ministry of Health is responsible for the control of, and measures related to, public health (quarantine, patent, etc.). The Ministry of Agriculture is responsible for fisheries and approves new port investments and development. The Ministry of Environment approves environmental impact assessment studies of ports. The Ministry of the Interior is responsible for the police, immigration, etc. The CRPA undertakes all planning, research, construction and maintenance work on ports belonging to the public and affiliated sectors. The Prime Ministry and the UMA coordinate political, economic and legal aspects concerned with international maritime issues according to national policy. The UMA used to set the port tariff of TMO ports. Such public ports are not operated by TMO any longer since they have been privatized. It is the UMA that acts as the maritime authority in Turkey and administers the maritime vessel traffic system. This relates to aspects such as the entry and exit of ships into and out of port and the regulation of shipping and navigation, etc. MPWS prepares territorial application plans and defines the land use principles for all industrial sectors, including the transportation sector. State Economic Enterprises, such as the TSR and TMO, operate, develop and maintain owned ports. The TSR also undertakes miscellaneous transportation by providing connections between railways and ships and establishing and operating the required superstructure, such as warehouses, silos, fuel facilities, etc. TMO's additional responsibilities were to provide and undertake loading and discharging operations by constructing and establishing the required facilities, to provide some services for ships, such as fresh-water, fuel oil, etc., and to construct and operate the required superstructure facilities at its owned ports. Municipalities are concerned with city–port relations and environmental impacts and they provide some services to ports such as fresh-water, garbage collection, etc.

Weak coordination and conflicts of authority may happen from time to time among these related bodies. For example, deepening the draft in the Port of Izmir has always been a matter of conflict and some infrastructure investment has also experienced similar problems. There is a complex system of involvement in the investment, operation and administration process. A single issue might be a concern of at least two or three bodies affiliated with different ministries. This situation causes chaos in the decision-making process and coordination. Needless to say, inefficiency arises in the port governance process in Turkey.

4. PORT INVESTMENT, DEVELOPMENT AND MANAGEMENT

The coastline of Turkey is legally public property owned by the state, and its use has to be such that it contributes to the public interest. So long as the coastline is open for public use, the construction of port facilities such as wharves, piers, breakwaters, etc. can only be constructed by permission from central government authorities.

The CRPA undertakes the development of ports. It formulates development plans, constructs port infrastructure for public and municipal ports and performs some maintenance. All of the port development projects, including those in affiliated and private ports, are controlled by the following three departments of CRPA: the Planning and Project Department, the Harbour Survey and Design Department, and the Harbour Construction Department.

The Ministry of Transport proposes port development projects prepared by the CRPA to the SPO. The SPO considers the total balance of investment in Turkey by prioritizing projects, as well as judging the feasibility of specific construction projects. When they are approved by the SPO in the Five-Year Development Plan, the Ministry of Finance finances the investment. In the case of private ports and industrial enterprises, they are constructed and financed by affiliated institutes. The CRPA only approves their projects and controls all construction.

In general, prospective investors, including all private and public institutions or individuals who construct marine structures such as berths, docks, shelters, mooring facilities, piers and breakwaters must get approval from the relevant government organizations, including the CRPA. First, the governor of the province reviews the investment plans containing all the details and properties of the structures, as well as reports, maps and other related documents, on the background and local information of the area. Then, the MPWS examines the proposal in terms of the general and regional planning principles, and general settlement plan decisions. The Prime Ministry UMA, MOT and Ministry of Environment are also involved in this process. After obtaining approval from the MPWS, the CRPA examines the implementation projects from a technical point of view. The investors go ahead with construction with full responsibility and control still lying with the CRPA. Following the completion of construction, the Prime Ministry UMA gives the investors permission to operate the new facility.

The ports of Turkey are classified into four groups according to an operational point of view: public ports, municipal ports, affiliated ports and

Table 8.2. Administrative Classification of the Turkish Ports.

Operators		Classification	Total Length of Ports and Pier (m)
TSR ports	7	Public	16,007
TMO ports	7	Public	2,623
Industrial ports of state-owned companies	37	Affiliated	30,662
Municipal ports	45	Regional municipalities	8,875
Private sector ports	51	Private sector	22,094
Privatized TMO ports	13	Private sector	9,481
Total	160		89,742

privately owned ports. The main public ports are operated by TSR and the TMO. Administrative classification of the Turkish ports is shown in Table 8.2. The TSR manages ports connected to the railway system. These major public ports are general-purpose ports under the control of the MOT, which approves the budgets and annual programmes and plans of both TSR and TMO. The TMO was privatized as TMO Inc. Co. as a state enterprise in 1995.

TSR ports are managed by the Ports Department from its headquarters in Ankara, the capital of Turkey. The Ports Department is responsible for the management, the overall planning and functioning of the ports, and their coordination. The CRPA coordinates with the TSR headquarters in formulating port development plans. Each individual port has a Port Manager who is mainly in charge of operations, including all services to ships and cargoes, using their own labour and equipment. Each port also consults and advises headquarters in planning its port development. The ports operated by TSR are shown in Table 8.3.

TSR ports procure superstructure (warehouses, CFS, cranes, and cargo handling equipment, etc.) and engage in operations in their own right. However, cargo handling and marine services are also provided by private companies under the control of port management bodies. TSR ports provide marine services (pilotage, tugs, mooring, fresh-water supply and bunkering). For example, in the Port of Izmir, pilotage, tugs and mooring services are provided by TMO, which also carries out the pilotage service in many of Turkey's ports. Table 8.4 shows the privatized ports of TMO.

Affiliated ports are special industrial ports and industrial enterprises, which are state-owned or private companies. These ports fall under the control of the MOI and are mostly confined in purpose to the particular

Table 8.3. Distribution of the Main Cargo Groups Handled at the TSR Ports (Tons).

Ports	Year	General Cargo	Container	Dry Bulk	Liquid Bulk	Total
Haydarpasa[a]	2001	2,540,788	2,202,474	45,750	0	4,789,012
	2002	2,769,618	2,338,146	5,121	2,073	5,114,958
	2003	3,203,175	2,503,643	16,085	65	5,722,968
	2004	3,320,515	3,128,689	8,421	306	6,457,931
Mersin[b]	2001	1,609,420	2,982,231	2,977,333	6,059,698	13,628,682
	2002	1,004,885	3,858,623	3,073,544	5,825,813	13,762,865
	2003	1,102,223	5,128,919	3,128,805	6,116,431	15,476,378
	2004	1,064,862	5,924,054	2,679,438	7,514,939	17,183,293
Iskenderun[b]	2001	262,989	317	465,350	968,369	1,697,025
	2002	223,185	303	492,317	886,200	1,602,005
	2003	468,848	3,646	557,331	1,226,892	2,256,717
	2004	173,214	8,282	630,519	1,421,583	2,233,598
Samsun[a]	2001	1,005,852	11,941	1,523,142	3,000	2,543,935
	2002	838,999	7,880	1,750,028	13,988	2,610,895
	2003	619,153	0	2,127,347	22,236	2,768,736
	2004	756,183	0	2,318,711	37,340	3,112,234
Derince	2001	406,840	4,988	127,913	8,479	548,220
	2002	692,529	5,232	400,042	51,172	1,148,975
	2003	765,798	15,667	603,025	67,149	1,451,639
	2004	1,090,900	11,184	799,990	65,377	1,967,451
Bandirma[a]	2001	183,437	14,176	2,477,019	304,458	2,979,090
	2002	133,702	8	2,080,954	145,026	2,359,690
	2003	160,529	0	2,399,140	164,207	2,723,876
	2004	768,797	284	2,290,429	183,750	3,243,260
Izmir[c]	2001	496,005	4,671,425	2,986,219	272,420	8,426,069
	2002	567,725	5,439,787	3,457,351	187,851	9,652,714
	2003	614,348	6,478,213	3,765,593	251,445	11,109,599
	2004	673,254	7,659,365	3,947,449	220,197	12,500,265
Total	2001	6,505,331	9,887,552	10,602,726	7,616,424	34,612,033
	2002	6,230,643	11,649,979	11,259,357	7,112,123	36,252,102
	2003	6,934,074	14,130,088	12,597,326	7,848,425	41,509,913
	2004	7,847,725	16,731,858	12,674,957	9,443,492	46,698,032
	2005[d]	3,533,691	7,298,559	4,492,409	3,348,198	18,672,857

Source: Turkish State Railways (2005).
[a]Public ports.
[b]Private ports.
[c]Under process of privatization.
[d]As of the end of May 2005.

Table 8.4.   The Privatized Ports of Turkish Maritime Organization
(TMO).

Ports	Length	Depth	Ton/Year	Ships	Storage (Ton/Year)
Alanya	239.00	(−6,−10)	–	240	–
Antalya	1,900.00	(−4,−10)	3,338	2,975	4,714
Marmaris	462.00	(−12)	–	1,460	–
Gulluk	358.90	(−10,−12)	3,750	240	–
Kusadasi	920.12	(−11)	–	2,400	–
Ceşme	480.00	(−7.5,−10)	–	1,060	–
Dikili	178	(−6,−8)	500–550	200–220	–
Gokceada	500.00	(−5,−6)	312	104	–
Darica	25.00	(4–5)	–	–	–
Canakkale	100.00	(−6,−6.5)	300	104	–
Lapseki	200.00	(−6,−6.5)	100	100	–
Tekirdag	1,014.00	(−4,−9)	2,900	1,050	361
Istanbul	1,120.00	(−6.5,−10)	–	5,250	–
Kabatepe	320.00	(−6,−8)	–	–	–
Sinop	197.20	(6.4,11.95)	–	250	400
Ordu	269.00	(−8,−9)	865	350	1,300
Giresun	1,022.00	(−8,−10)	1,394	1,575	1,375
Trabzon	1,525.00	(2.5,10)	3,839	2,839	3,193
Rize	130.00	(−5)	529	140	–
Hopa	1,145.00	(4.5,10)	1,394	1,425	1,228
Total	12,124.63		14,788	20,700	12,571

Source: Turkish Maritime Organization Inc (2005), Turkish Chamber of Shipping (2004).

needs of industrial concerns. These ports are generally considered to be quasi-entrepreneurial administrations, independent and free from political pressure and functioning on an entrepreneurial basis, having independent budgets. Some state industrial enterprises are under the control of the Privatization Administration, as privatization has been progressing recently. The construction of infrastructure at public ports is met from the national budget, while the maintenance of these structures is undertaken by the respective port management bodies at their own expense. Superstructure and cargo handling facilities are established by port management bodies and/or provided by operating companies.

Municipalities manage municipal ports. These public ports are comparatively small and limited to relatively small volumes of coastal traffic serving the local needs of provincial towns. Some municipalities manage ports with their own port management division, but this does not involve large-scale development.

Privately owned ports are constructed and managed by the private sector after obtaining the permission of central government. Small-scale private sector port development has taken place in the region around the Marmara Sea. These small-scale port developments are based on private capital. Some of these ports were privatized in the last 10 years, having previously belonged to the TMO.

The ports operated by the TSR, namely Bandirma, Izmir, Samsun, Derince, Mersin and Iskenderun, are scheduled to be privatized. However, privatization will be carried out by a method that does not involve the transfer of the title of the ports, such as the transfer of operational rights, lease or other management method. The decision of the Supreme Privatization Council sets forth that the privatization will be finalized within 12 months (Privatization Administration of the Republic of Turkey, 2005).

With the amendment made to the Privatization Law by Law 5189, foreign entities (through a Turkish subsidiary) can also acquire the operational rights of a port. For example, in 2005, PSA gained the operational rights to one of the major ports in Turkey. Pursuant to Law 815 (on Domestic Shipping), pilotage services can only be carried out by Turkish citizens.

5. DISCUSSION

An overwhelming amount of cargo handling and cargo transfer in seaborne trade in Turkey is carried out at public ports. However, major public ports with the highest port throughput in Turkey appear not to be operated efficiently. Politicians and bureaucrats interfere in the port industry in order to meet both self-serving political objectives and industrial objectives. Bureaucracy is always a barrier. From the point of view of the port business, this is neither a flexible nor a workable system, as centralization creates a great deal of difficulty with respect to decision-making. The central planning of the ports means that some specific and special needs are missed. Employment has always been exposed to political interference, but authority and responsibility are not well defined.

The private ports in Turkey have physical deficiencies in cargo handling equipment and storage yards. They have inadequate financial resources and difficulties in investing in port development. Most private ports located in the Marmara region fall outside the coordination provided by national port policy and are in destructive competition with each other. This is reflected in their very low port tariffs and in the fact that they have no idea what other ports are doing or investing in.

Industrial enterprise ports are, of course, very important for their plants' industrial activities, but they are not a significant presence in the port sector. It is useless to criticize their way of administration and operation, however, even though their capacity utilization rates have always been questionable.

The Turkish Competition Authority, Ministry of Transport, Ministry of Finance and the Privatization Administration of Turkey are among the related bodies for privatization. However, there has never been a single supreme organization in Turkey to coordinate port investments, port development and port competition, especially for the port privatization period. As far as the authors are concerned, there should be an integrated supreme body to coordinate all the ports according to a national port policy that is compatible with EU transport policy. Such a coordination entity can comprise representatives of port operators, port users, municipalities, related government agencies like CRPA, Customs, Prime Ministry Undersecretariat for Maritime Affairs, NGOs like the Turkish Chamber of Shipping and universities. Another area of coordination is required for participation of local authorities and NGOs in the port administration for good governance. Unless the privatization practices consider this vital concern, the main drawback of the privatization process will emerge, resulting in serious local conflicts. It is too soon to assess the outcome of Turkey's privatization programme but, for these reasons, it is too early to consider the future as promising.

REFERENCES

Chlomoudis, C. I., & Pallis, A. (2004). *Port governance and the smart port authority: Key issues for the reinforcement of quality port services.* Istanbul: WCTR.

General Director of Railways, Ports and Airports Constructions (CRPA). (2005a). *North Aegean (Candarli).* Port – Interim report. CRPA and the Ministry of Transport and Communication, SENER-EUROESTUDIOS-DOLSAR Consortium, Vol. 1, April.

General Director of Railways, Ports and Airports Constructions (CRPA). (2005b). *Mersin new container terminal feasibility – Progress report.* CRPA and the Ministry of Transport and Communication, with the collaboration of TCT, ALATEC and IDOM. Vol. 1, April.

Overseas Coastal Area Development Institute of Japan. (OCADIJ). (2000). *Final report for the study on the nationwide port development master plan in the Republic of Turkey (ULI-MAP).* Japan International Cooperation Agency, CRPA and Ministry of Transport and Communication, August.

Privatization Administration of the Republic of Turkey. (2005). Preliminary information on the privatization project of TSR ports. April, Ankara.

Turkish Chamber of Shipping. (2004). *Deniz Sektörü Raporu 2003.* Istanbul: DTO Yayınları.

Turkish Maritime Organization Inc. (2005). www.tdi.com.tr/

Turkish State Railways. (2005). www.tcdd.gov.tr/liman/esya.htm

www.arkas.com.tr

CHAPTER 9

THE RIVER PLATE BASIN – A COMPARISON OF PORT DEVOLUTION PROCESSES ON THE EAST COAST OF SOUTH AMERICA

Ricardo J. Sánchez and Gordon Wilmsmeier

ABSTRACT

The ports of Buenos Aires and Montevideo operate in the same market, which stretches along the River Plate Basin. The congruent core hinterland includes the metropolitan areas of Buenos Aires and Montevideo. The extended hinterland embraces Paraguay, Bolivia and several interior regions of Argentina.

This chapter analyses the performance of the two ports at two points in time – 1999/2000 and 2003/2004. In 2001/2002, Argentina and Uruguay suffered from a deep economic crisis and developed various strategies and environments for port governance in order to deal with these changing conditions. The Matching Framework is applied to understand the different outcomes of the port devolution processes in each country.

1. INTRODUCTION

Port activity is a competitive industry whose environment determines the conditions for competition and success. Organizations in the port industry

Devolution, Port Governance and Port Performance
Research in Transportation Economics, Volume 17, 185–205
Copyright © 2007 by Elsevier Ltd.
ISSN: 0739-8859/doi:10.1016/S0739-8859(06)17009-2

tend to be complex, due both to the participation of diverse players and interests, and the recent port devolution process, which has altered environmental conditions and services over the last few decades. Such a process is far from being stable in its start-up and subsequent evolution, and organizations need to be almost constantly adjusting to changes in the environment.

This chapter analyses the evolution of the two main container ports in the River Plate port range, Buenos Aires and Montevideo, by applying the theoretical approach of the Matching Framework (Baltazar & Brooks, 2001). The two ports are considered to operate in the same market, which stretches along the River Plate port range.[1] The congruent hinterland extends throughout regions of Argentina and Uruguay situated along the River Plate, and the metropolitan areas of Buenos Aires and Montevideo (the core hinterland).[2] The location in the River Plate delta extends the hinterland along the river system of the Uruguay and Paraguay-Paraná waterways, and therefore reaches as far as Paraguay, Bolivia and northeast Argentina (the extended hinterland). These geographical settings create a competitive environment.

The recent rapid expansion of the Brazilian Rio Grande Port intensifies the competitive situation between the Ports of Buenos Aires and Montevideo in the River Plate port range, offering ample conditions to compete for cargo destined for the extended hinterland.

On the basis of these conditions, this chapter analyses the performance of these two ports at two points in time (1999/2000 and 2003/2004). During this period, both countries suffered from a deep economic crisis and developed different strategies of port governance to cope with the changing conditions. The Matching Framework is applied to understand the different outcomes of the port devolution processes in each country in this competitive environment.

The first section focuses on the River Plate Basin as a single competitive market for containerized trade. The development and key issues of the port devolution process in each port provide insight into the performance of both ports in the period from 1995/1996 to 2003/2004. The chapter outlines the political, economic and institutional environment, including a brief description of the economic and social crisis in 2001/2002.

The second section analyses the ports' performance by applying the Matching Framework at two points in time: 1999/2000 and 2003/2004.[3] Emphasis is placed on the changes in the ports' strategies and structure over time and on the comparison of performance in similar economic environments.

2. THE CASE OF THE RIVER PLATE BASIN

The River Plate Basin is a central gateway to the southeastern and central markets of South America. The river and the connected Paraguay-Paraná river system are the main arteries for the movement of agricultural products from the highly productive interior regions of Argentina, Uruguay, Paraguay and Brazil. In recent years, increasing effort has been put into improving navigation on the Paraguay-Paraná river system. Buenos Aires and Montevideo, as the biggest deepwater ports in the region, have to provide efficient and effective operations to serve as competitive interfaces for their agricultural hinterlands.

2.1. Economic Crisis: New Dynamics

The state of the economy in Uruguay depends to a large extent on the economic performance of neighbouring countries, Argentina and Brazil. Uruguay's geographical situation between two much larger economies and its location at the head of the River Plate places the country in an advantageous position as a logistics centre.

However, the deep economic crisis at the turn of the century revealed the negative sides of Uruguay's interdependency with the adjacent economies. Beginning in late 1999, foreign investors became worried about Argentina's ability to pay its large public sector debt, especially in the wake of Brazil's January 1999 currency devaluation. Argentina's and Uruguay's economies are closely linked to Brazil's, which generated concern about the impact of the Brazilian currency devaluation on the Argentine and Uruguayan economies.

The crisis brought an abrupt change in the economic environment, creating uncertainty and therefore putting new pressures and challenges on the whole economic system. The recession of 1999–2000 led into the 2001–2002 crises, whose effects included a drop in GDP of about 17 per cent in Argentina and Uruguay, default on debt payment, a sharp increase in poverty, etc. The GDP per capita fell by around 20 per cent, bottoming out in 2002 at about 10 per cent below its level in 1995.

The crisis had a more direct impact on Argentina, because the exchange rate of the Argentine peso was fixed to the United States dollar at that time. This made Argentine exports relatively expensive. Exports fell over 10 per cent between 1998 and 1999, but recovered rapidly after the fixed exchange rate was eliminated. Imports to Argentina declined about 40 per cent between 1998 and 2002, but recently imports have recovered sharply and are expected to pass the 1998 level in 2005.

For a long time, the Uruguayan Government and the private sector tried to combat the inferior position in trade through a variety of policies and investment strategies. Since 1923, the country has adopted national policies to attract industry and trade to free-trade zones. Federal legislation guarantees free-trade zone participants exemption from all corporate and national taxes, value-added taxes, customs duties, sales taxes and social security taxes for foreign workers. It also places no time restrictions on goods stored in warehouses. The free-port system is a key factor in the country's competitiveness in the regional market, as the other countries do not have this system.

In the last four years, Uruguay has conducted studies on different transport corridors to the neighbouring countries, especially in relation to port access. The required physical improvements and the institutional and financial strategies needed to make these improvements happen have been recognized by the national institutions as necessary for the health of Uruguay's economy.

Ports handle around 90 per cent of external trade in terms of volumes.[4] Dependence on maritime trade was a key driver in the processes of reforming port structures.

2.2. *Port Devolution under Competitive Market Conditions*

Until the early 1990s, investment in the infrastructure and maintenance of the public ports of Argentina and Uruguay was almost fully undertaken by the national governments under a tool port model. The government was in charge of managing the ports, making infrastructure investments and planning national port policies. In Argentina and Uruguay, private stevedoring companies conducted port operations under a permit system, in general at a low business risk level and in a purposively competitive environment. However, the management of port services did not result in efficient and effective terminal operations. Stevedoring personnel were grouped in public entities. In Uruguay, they were part of the Administración Nacional de Servicios de Estiba (ANSE),[5] and in Argentina they were managed by the Ente Nacional de Contrataciones Navales (ENCON)[6] and the Ente Nacional de Contratación y Garantización de Jornales (ENGONAR).[7] Moreover, the national governments were responsible for the condition of the navigation channels in both countries.

The direct government provision of port services resulted in inefficiency, excessive delays and waste of financial resources, resulting in a drop in the competitiveness of national production activities relative to the rest of the world. The main task of investment was conducted inefficiently and inadequately. In Argentina, public investment decreased from US$69 million (1980) to oscillate around US$4 million annually between 1984 and 1991 (Sgut, 2000). In both countries, short-term expenses consumed the greatest share of revenues, while the financial resources for investment provided by the central governments decreased steadily. Nevertheless, insufficient improvements were implemented in both ports.[8]

2.3. Changing Structures: A Response to Inefficiency

2.3.1. Argentina

Before the reform, Argentine's port system structures had disparate organizational structures with restricted interaction, unclear and confused lines of responsibility, and functional overlap resulting in failures in decision-making. A first response to the precarious economic situation in the 1980s was a series of private initiatives, led by the agricultural exporting industries and directed at building more efficient terminals. Since Argentina was striving towards a more export-oriented economy with growing export flows, the National Congress passed a series of laws to push towards an opening of the economy in the beginning of the 1990s, driven by various intentions including the process of democratization and recovering the deficits of outstanding investments. The liberalization process included the deregulation of certain economic activities and the introduction of a convertibility law.[9] In this context, Argentina was one of the first South American countries at the beginning of the 1990s to allow privatization of transport infrastructure (see Table 9.1).

Prior to the reform, the state port authority (AGP) administered the ports, but lacked autonomy in decision-making to guarantee an economically efficient operation of the ports. The port sector worked inefficiently, struggling under innumerable regulations, bureaucratic traps and sectoral privileges, especially in the port labour sector. The excess of rules and regulations stalled the potential of private investment in the ports. Secondly, Argentina did not have a national port development framework, which also limited necessary investment.

In general, the problems prior to the reform were

- excess of work force and regulation, in the stevedoring as in the responsible government bodies,
- inefficient use of the port area,

Table 9.1.   Coexisting Institutions Influencing the Port Sector Prior to
the Reform.

- Administradora Nacional de Aduanas (ANA) (National Customs Administration) – fiscal
  control, tariffs and authorization of exports
- Prefectura Naval Argentina (PNA) – pilotage services, monitoring of the port and access
  channels for security reasons, registration of stevedores
- Local port authorities – coordination of port activities and resolution of labour conflicts
- Port Labour Police
- Dirección Nacional de Construcciones Portuarias y Vías Navegables [National Service for
  Port Construction and Inland Waterways] – dredging of access channels and docks
- Maritime agencies organized loading and discharge operation for stevedoring companies.
  These companies provided their own equipment and Operated with storage areas owned by
  the AGP (State Port Authority)

- deficits in the provision and maintenance of infrastructure and port superstructure in the port sector from the state,
- insufficient facilities for the increasing container operation in ports, and
- security problems.

After 1991, several decrees were passed deregulating activities in the port sector[10] in order to establish free-market mechanisms for service provision, reduce tariffs and achieve a higher degree of international competitiveness. This reform included

(a) Creation of administrative and management agencies for several ports,[11] and the transfer of other national ports to the jurisdiction of sub-national government.
(b) Liberalization of contracting towing, stowage, etc. services.
(c) Free-crew selection by ship owners.
(d) Free pricing for tariffs and freight.
(e) Port service provision to be allowed 24 h a day.
(f) Greater legal safety for pre-existing and future private ports.

In addition, flexible start-up rules were established.

These measures included the deregulation of waterborne transportation and its related services, a loosening of specific operating conditions and the elimination of some freight taxes. However, the port of Buenos Aires operated as a public port until 1994. During this time, 30 private stevedoring companies were active throughout the ports. This situation did not allow any investment, either in the form of private investment because of the dispersed and almost atomized market of port activities, or by the public authorities, which had no intention of investing. Sgut (2000) describes this situation as chaotic, with unclear responsibilities and insufficient and ineffective services at very high prices.

The port reform was part of a general change in the nation's economic policy, based on private initiative and operation of key components of infrastructure. One of the most important aspects of this initiative was labour reform, which eliminated restrictive labour practices, promoted stable relations between labour and terminal operators, and reduced the number of labourers at the ports. These efforts, driven by the government and the Congress, started in 1989 as a result of the general political direction of opening, decentralization, privatization and deregulation, and not from specific measures (Sgut, 2000). In this framework, the restructuring of ports and the transport sector was intended to

- improve the efficiency and investment in the transport system,
- strengthen foreign trade,
- increase port productivity,
- make Argentine ports competitive in a global environment,
- enhance the planning and control functions of the state,
- promote free competition, and
- eliminate the deficits of state-owned enterprises, especially railway sector.

Port operation in Puerto Nuevo (Buenos Aires) was transferred to private participation in 1994. The port was reorganized into six terminals, with separate calls for bids to promote intra-port competition.[12] Concessions for terminals 1–5 were granted to globally operating port companies. Terminal 6 was closed a few months after the concession was granted to a local group (Sánchez & Sgut, 2002).

After the Puerto Nuevo concessions were granted by the national government, a new container terminal was built at a distance of a few miles and under a different jurisdiction. This terminal has had an outstanding performance record since its inauguration in 1996 and today is handling the largest share of all containers in Buenos Aires. Other ports in the region started to compete. Zárate operated with two gantry cranes and an increasing throughput until 2004. These developments contributed to a growing inter- and intra-port competitive environment between Buenos Aires and other neighbouring Argentine ports.[13] The pro-intra-port competition structure of the port reform led to a situation in which Argentina's ports spend more time competing with each other than with foreign competitors.

In July 2000, the authorities allowed the merger of private cargo terminals operating in Buenos Aires port, an option that had been prohibited in the mid-1990s.

After the reform, the port authority in Buenos Aires was supposed to terminate the previous bureaucratic agency (AGP) and replace it with a new, smaller one with different functions. However, AGP was not closed, nor was a new agency established, which created institutional uncertainty until 2004. In that year, with the introduction of a new decree on the Buenos Aires port, it was decided that the AGP was to continue with its prior characteristics. This situation did not send a clear signal of institutional strength to the port industry and shipping lines.

A key aspect of the new structure is that the administrative agencies have to contribute to the transfer and/or privatization efforts of port services and terminals. At the same time, they are responsible for dredging, signalling, buoy placement, etc. These units are still located at different jurisdictional levels – national, provincial or municipal. In some cases, they are port management consortia in which economic organizations participate; in other cases, it is only the government and its officials in a bureaucratic entity. Each organization has different decision-making and financing mechanisms. In short, the post-reform structure leans towards decentralization and prevents the coordination of common port policies and strategies, contradicting key principles for the success of port policies such as close collaboration between ports, port authorities and national authorities (Cullinane & Song, 2002).

The lack of coordination makes port access landside and waterside in Buenos Aires a major issue. The terminals are located in central urban locations, and both rail and highway access must compete with the regular city traffic when serving the port. This is particularly a concern for truck traffic, which carries 85 per cent of the port-related cargo.[14] Dredging is a critical and expensive task for the port. However, the access channel has not been dredged at the new required drafts, which carries the risk that the port may become inappropriate for direct deep-sea calls.

Additionally, until now, Argentine's transport policy has not been integrated with national economic or trade policies. Over the past 10 years, the Government's responsibility for transportation has passed through many departments in the executive branch. This continual relocation appears to reflect a lack of commitment to an integrated transportation system and a weak position for the transport secretary.

The government has never provided sufficient funding for the development of infrastructure, and presently lacks money to invest in the necessary transport infrastructure; required maintenance on important parts of the country's transportation infrastructure is being deferred. Border crossings, especially within MERCOSUR, are an additional problem.

2.3.2. Uruguay

The 1992 Port Law and its associated decrees[15] were landmarks in the process of enhancing the competitiveness of Uruguayan ports. The goals of the Uruguayan port policy are

(a) continuity of services, "service started, service completed" principle;
(b) provision of safety;
(c) provision of regularity;
(d) maximum productivity and efficiency;
(e) mandatory coordination and collaboration for the best service;
(f) free competition;
(g) equal rights; and
(h) freedom of choice for consumers, etc.

The legislation reflects the idea that the efficiency, safety and reliability of a port system are its guiding principles, and defines free-port contexts and their operational arena, port customs areas, the basic merchandise system and foreign trade zones. In the port devolution process, the main mechanism was to discontinue direct port operation by the state, and call on the private sector to join a new type of public–private partnership (PPP) for better port management.

The Uruguayan port reform process also involved a clear definition of port policy,[16] and the role of the Administración Nacional de Puertos (ANP),[17] thereby adding efficiency to the provision and competitiveness of the ports and foreign trade. In turn, the new ANP rules include maintaining the capacity of rendered services, granting and controlling permits and/or concessions for operating services, contracting dredging services and ensuring free merchandise circulation and competition as well as advising the government on transportation policies.

The passenger terminal, refrigerated storage facilities and warehouses at the Montevideo port were transferred smoothly to the private sector by granting permits (for about two years) through a bidding system, or concessions (for terms of 10 or more years). This transfer was accompanied by a staff reallocation process, reducing the headcount from 3,370 in 1992 to 860 at the end of 2004. Most retired employees still of working age were relocated to privately run service companies.

However, the creation of a specialized container terminal faced several obstacles. Four calls for bids were necessary. The first three ended in failure, creating uncertainty for the carriers. In 1999/2000, Montevideo started to lose direct calls because of the lack of equipment and good management and investment, creating a new urgency for successful privatization of the terminal. Finally, in 2001, a transparent and quick auction led to the transfer to a Uruguayan international business group; operations began in 2002.

2.4. The Path Towards Success: Strategies

2.4.1. Uruguay

The Port of Montevideo is better positioned on the River Plate than the Port of Buenos Aires because the river's currents deposit silt on the south side of the river along the Argentine coast and, in particular, at Buenos Aires. Dredging the harbour and its access channels is thus significantly easier for Montevideo. Nevertheless, the differences in the scale of operation between the two ports should allow Buenos Aires to recover the cost associated with its geographical disadvantage. Currently, Montevideo has a minimum water depth of 10.5 m, while Buenos Aires can only provide a water depth of 9.5–9.75 m.

After winning the Montevideo port concession in 2001, the Katoen group's strategy was to invest in this terminal so that it could act as a regional hub for container transhipment. Key aspects of the port's attractiveness for the group were its central location in the southeastern South American market, its free-trade zone status, more competitive transit times to major destinations than from Buenos Aires, low port costs[18] and the river connection Praguay–Paraná to the inland production areas. Operating hours, port stays and waiting times for ships improved notably.

Since the start of the port devolution process in 1992, container movements have multiplied from 61,800 TEUs and 13,300 transhipments in 1992 to 194,400 and 230,400, respectively in 2004; movement of reefers grew significantly from 5,200 in 1992 to 82,800 in 2004. Operating costs decreased by 50 per cent for imports and 75 per cent for other operations between 1992 and 2004. However, these figures also show the success of the container terminal after the economic crisis in 2001–2002, as it outperformed its competitor ports in terms of transhipments (see Fig. 9.1) and container movements (see Fig. 9.2) in general.

In short, the privatized port terminal quickly added significant capabilities for taking advantage of competitive market conditions. Montevideo had the strongest and steadiest growth rates in the period in the River Plate Basin from 1996 to 2004. The Port of Rio Grande benefited the most, leading to

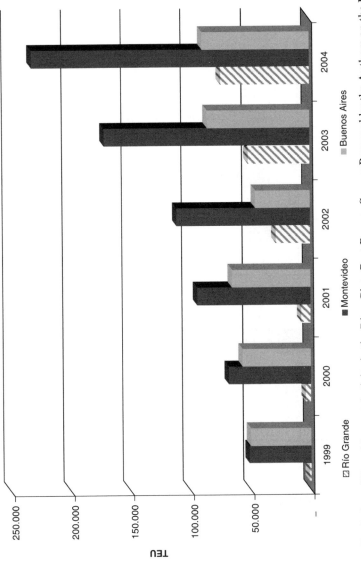

Fig. 9.1. Evolution of Transhipment Activity in the River Plate Port Range. *Source:* Prepared by the Authors on the Basis of Data from Port Authorities.

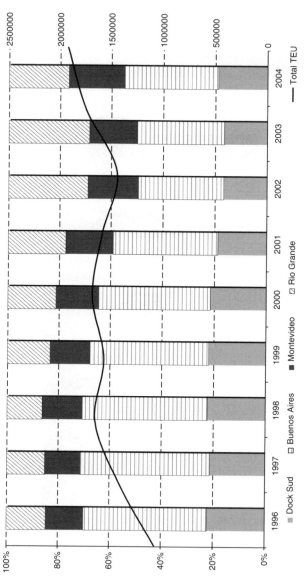

Fig. 9.2. Evolution of the Container Market in the River Plate Port Range. *Source:* Prepared by the Authors on the Basis of ECLAC (2005).

almost a doubling of container movements from 1999 to 2003. Buenos Aires ports were affected most by the economic crisis. However, a strong recovering process can be observed for Buenos Aires since 2002 (see Fig. 9.3).

2.4.2. Argentina
Even though the port devolution process led Buenos Aires to be the busiest port in Latin America in the year 2000, the port lost a significant share of the direct competitive container market between 1997 and 2002 (see Fig. 9.2).

Uncertainty in Argentine's port environment grew when shipping lines started to change their port of call from Buenos Aires to Montevideo or even to Rio Grande (Brazil), especially in 2002/2003. A main reason for these shifts has been the River Plate toll, which increased 45 per cent in 2003, and higher port costs in general in Buenos Aires. A further advantage of Montevideo over Buenos Aires is the existence of the free-trade zone there. Moreover, in reaction to an increasing number of ship calls, Montevideo started to offer new transhipment services to Buenos Aires (Ward, 2004).

Only in 2004, Buenos Aires was able to improve its overall share significantly by recapturing activities from Rio Grande rather than Montevideo. During the same period, Montevideo was able to steadily increase its market share in an almost stagnant container market in the River Plate Basin.

The average size of vessels handled in the Buenos Aires area grew from 7,000 NRT (1990) to 12,000 NRT (2002). Meanwhile, the average discharge and loadings per berthing increased from 500 TEUs in 1990 to 843 TEUs in 2000, and 2004 volume was almost double that of 2000. Operating times were reduced and the average vessel port stay time decreased from over six days prior to the reform (in particular, in 1991) to a little more than two days after the reform. Unit costs (per TEU) for loading and discharging decreased from US$300 in 1990 to US$90 in 2000–2004.

Another important factor in port organization strategies for both countries is their difference in foreign trade power. Argentina's foreign trade increased from US$40 billion (1995) to US$50.2 billion (2000), and Uruguay's from US$4.9 billion to US$5.7 billion. Throughout the period, Argentina's trade was eight to nine times higher than that of Uruguay (ECLAC, 2004).

Argentine port strategies have concentrated on national activities partly for reasons of scale, whereas the growth potential of Uruguayan ports is tied to capturing cargoes from neighbouring countries. Argentina's basic strategy is primarily to handle cargoes for the internal market in an intra-port competitive framework. Meanwhile, Uruguay's restricted internal market leads the port to seek success in inter-port competition at the operator and port

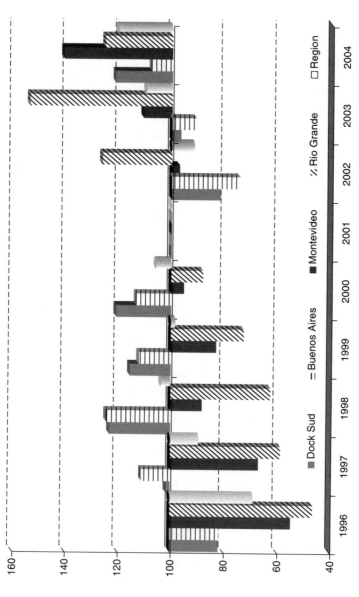

Fig. 9.3. Evolution of Port Movements (TEU), 1996–2004, Base 2001. *Note:* The Figure Shows an Index of the Growth of TEU Movements (Base: 2001 = 100). *Source:* Prepared by the Authors on the Basis of the ECLAC (2005).

authority level,[19] focusing on capturing transhipments and offering value-added services in addition to its own cargoes.

These differences in national strategies are not new, and have a historical background as illustrated by the following extract of a debate in the Uruguayan Senate (1991):

> We should not forget that our country was born as an independent nation mainly due to the conflict of interests between the Buenos Aires and Montevideo ports. Ultimately, this was the only actual reason why Uruguay came to be, not just the 'Eastern Bank' but a Republic. It was precisely the interests of Buenos Aires, favouring that city, and the comparative advantages of Montevideo which have always been behind the conflict that gave rise to our nation. Also, it shows what Great Britain ultimately pointed out: that the River Plate should not be a port of only one nation.[20]

3. MATCHING FRAMEWORK ANALYSIS

Adapting the theoretical framework of Baltazar and Brooks (2001) to the case of the River Plate Basin, we identified the configurations shown in Table 9.2 for the two ports at both points in time under analysis.

The configurations cannot be clearly categorized as implying a superior fit, as defined by Baltazar and Brooks (2001). We identify hybrid configurations in both cases. As regards the environment, we introduced a slight variation to Baltazar and Brooks, changing the "high" and "low" uncertainty concepts to a relative uncertainty. Thus, we refer to "more" or "less" uncertainty, understanding that this better describes the situation of the environment in the different time periods.

The environment in Argentina has changed from less uncertainty to more uncertainty between 1999 and 2004, mainly because of the institutional framework. In 1999, the uncertainty was apparently less than after the crisis. The level of uncertainty then increased, mainly due to the general situation and the different approaches and views of the port market players and the authorities, which had become apparent in the lack of port modernization after the honeymoon phase at the beginning of the reform. The freezing of tariffs as an answer to the devaluation of the Argentine currency did not include the port sector, which continued to operate at free tariffs and use the US dollar as its working currency. However, the port sector was unable to turn this situation into a factor to regenerate trust and improve the infrastructure environment (i.e., investment in port infrastructure) in order to enhance its competitiveness.

In Montevideo, the environment was one of high uncertainty during the economic crisis. The port devolution process ended successfully with

Table 9.2. The Modified Matching Framework.

	Buenos Aires		Montevideo	
	1999/2000	2003/2004	1999/2000	2003/2004
Environment	Less uncertainty Less complex and dynamic	More uncertainty High complexity and dynamism	More uncertainty More complex and dynamic	Less uncertainty High complexity and dynamism
Strategy	Efficiency oriented Focus on delivery of basic services	Effectiveness oriented Focus on delivery of peripheral services	Effectiveness oriented Focus on delivery of peripheral services	Effectiveness oriented Focus on delivery of peripheral services
Structure	Mechanistic/organic Centralized decision-making characterized by procedural standardization	Mechanistic Decentralized decision-making characterized by procedural standardization	Mechanistic Centralized decision-making characterized by procedural standardization	Organic/mechanistic Decentralized decision-making characterized by cross adjustment of standardized procedures

Source: Prepared by the authors.

the construction of the specialized container terminal. After the awarding of a concession for this terminal, port development gained even stronger impetus. The stable institutional framework conditions (tax relief, free-port status, greater stability of policies, etc.) contributed significantly to the observed reduction in uncertainty, at a time when the economic crisis was as serious for Uruguay as for Argentina, if not even more serious.

The important institutional aspect of its free-port status has brought advantages in terms of inter-port competition. Montevideo can benefit in several ways from its foreign trade zone. Peter Voss Hall (2002) argues that these benefits can be defined as static and dynamic. He refers to the tax situation as a static advantage, while the dynamic benefits arise from the possibility to use the policy as a regulating tool to cope with arising uncertatinties in demand levels, product mix, trade regulations, etc. Further, he sees it is a tool that allows institutionalized relationships between the port authority and trade firms in regard to trade promotion, real estate development, and so on. Thus, the free trade zone is an important building block contributing to an organic structure.

A comparison of the perception of the port and overall infrastructure quality and the institutional environment ex ante and ex post the economic crisis in Argentina and Uruguay in the fields of transparency of government policy-making, effectiveness of law-making bodies and hidden trade barriers supports the above mentioned observations. While the infrastructure in general and that of ports in particular is perceived to be almost equal in both countries, Uruguay was able to provide a safer institutional environment[21] than Argentina and therefore create a basis and supporting framework for more successful port performance. Uruguay shows significant advantages in terms of the transparency of the government's policy-making and the effectiveness of law-making bodies. The burden of regulation and hidden trade barriers also seems to be significantly less.

However, interesting changes can be observed in the strategies. Montevideo has maintained the strategy of inter-port competition and peripheral services over time, including transhipment capture goals (see Fig. 9.1), value-added services and free merchandise systems, with corporate and historical precedents. Indeed, the authors interviewed different players in the River Plate port market, and they all agreed with this observation. In Montevideo, the different interest groups (port operators, maritime agents, national authorities, etc.) were practically unanimous in their views.

While Montevideo continued its historical strategy, Buenos Aires shifted its focus from the delivery of basic to the delivery of peripheral services.[22]

This concept was already familiar but had never been put into practice; operators and management focused mainly on capturing cargoes to/from the Argentine market,[23] allowing Buenos Aires to become the busiest port in Latin America and the Caribbean in 2000.[24] After the crisis, a more noticeable willingness to capture transhipments and provide value-added services can be observed as a result of the drop in its regional market share especially in 2004 (see Figs. 9.2 and 9.3). The institutional framework, with the contradictory viewpoints of operators, port and customs authorities, has set up barriers to that strategy, limiting the options for better port development. This resulted in a constant level of transhipment activity in 2003/2004, while Montevideo was able to more than double its transhipment activities in 2004, compared to 2002.

Buenos Aires' strategy arose in response to the economic crisis, and until now it has been implemented for too short a time to allow assessment. Thus, when the change of economic environment finally forced Buenos Aires to change its strategy towards regional inter-port competition, the port faced two drawbacks: firstly, its competitor has been on this path for a longer time, and secondly, its institutional environment is less stable and more uncertain.

The structure changed in both ports during the study period. In the case of Buenos Aires, it changed from partly organic to mechanistic, mainly driven by competition, which included the decision to allow for a horizontal merger of terminals to reach higher levels of centralization and greater cost efficiency. However, Montevideo developed in the opposite direction from a mechanistic to a hybrid mechanistic/organic structure, because the port put emphasis on the provision of ancillary services (free-trade activities, logistics services, local transit to the competing Buenos Aires port). This trend towards more organic structures allowed Montevideo to better adjust its services to the increasing numbers of customers and a growing diversified market.

A further difference between Buenos Aires and Montevideo is the decentralized decision-making at the political level. This difference constrains port performance in Buenos Aires, because here it is not characterized by mutual cross-adjustment. Mutual cross-adjustment and an organic structure, however, are regarded as necessary for a good fit of the Matching Framework configuration, where decentralized decision-making exists.

4. CONCLUSION

Neither of the ports can clearly be associated with either of the two Matching Framework configurations. The analysis shows that a change in one of the three Matching Framework spheres does not necessarily imply changes in the other spheres. Moreover, changes seem not to be unidirectional. As seen in the Matching Framework comparison table, the main differences between the two ports and historical periods are related to changes in the environment. The environment is important because a satisfactory performance of the infrastructure market (including port devolution aspects) requires a smooth balance between political stability and the flexibility needed to adapt industry conditions to changing economic environments. A successful balance depends on the form of the relationships between stakeholders, which can create an environment that promotes port development. This institutional characteristic requires skill in balancing the demands of the different groups involved, in order to establish a fine equilibrium between flexibility and certainty. Institutional weakness, however, may cause imbalances that allow for different kinds of opportunistic behaviour, promoting decisions that favour short-run interests at the expense of the interest of the whole society.[25]

Theoretically, the environment, stability, risk level or certainty, and the inter-sectoral equilibrium are limited by the institutional endowment of the country, where the standards and rules of the game between bureaucratic decision-makers, private firms and social sectors are crucial.

In the port sector, the standards and forms of relationships between customs, port authorities, port industry, operators, workers, shippers and others create different environments for competitive ports. This framework explains port performance on the basis of institutional endowments, which include conflicts between different players, the capacity of lobbying groups, the efficacy of policy-makers and the mechanisms of understanding conflict and cooperation that lead to higher or lower overall performance. Also relevant are the design of regulatory instruments and incentives and the scope of a port devolution process. In this regard, a high level of pro-competitive behaviour from customs is decisive.

While the Matching Framework has explained the importance of the fit between environment, strategy and structure, this paper shows, using the examples of Buenos Aires and Montevideo, that institutional endowment plays a decisive role in reaching the equilibrium of maximum port performance. This is due to the fact that the institutional framework determines the matching of capacities in the relationships between environment, strategy and structure as well as each institution's performance.

The Port of Buenos Aires suffers from an unclear institutional structure, which imposes decentralized decision-making, but mutual cross-adjustment between the institutions does not exist. The clearer structure in Uruguay reveals the advantages, as implied in the Matching Framework, over the mixture in the neighbouring country, leading to greater success in port performance and a greater ability to capture cargo in the competitive market of the River Plate Basin.

Uruguay benefited somehow from the differences in strategy between the Ports of Buenos Aires and Montevideo. While the economic environment put both ports to a performance test, the inter-port competition strategy made Montevideo more efficient. Furthermore, Uruguay was not suffering from fixed exchange rates, which made port operations less costly.

The case study of the River Plate Basin illustrates how the Uruguayan port industry performed better than that of Buenos Aires over the same period, creating a more pro-development environment based on a firm institutional framework.

NOTES

1. The word "range" refers to a geographically defined area encompassing a number of ports with a largely overlapping hinterland that thus serve much the same customers (Van de, Voorde, & Winkelmann, 2002, p. 6)

2. Ports and terminals considered Montevideo, specialized container terminal, and Buenos Aires: Puerto Nuevo, T1 to T5 and Dock Sud, container terminal.

3. In comparison to Baltazar and Brooks (2001), the matching framework is applied in an extended sense, including not only the primary port sector, but also the countries' general institutional situations.

4. ECLAC (Economic Commission for Latin America and the Caribbean, 2004), International Transport Database, for the fiscal years 2000–2003.

5. National Administration of Stevedoring Services.

6. National Agency for Port Recruitment.

7. National Agency for Recruitment and Wage Regulation.

8. A gantry crane was built in Montevideo and a tugboat was bought prior to the reform.

9. The law fixed the value of the Argentine peso to the United States dollar.

10. Main instruments: Port Law No. 24093 and decrees 1740/91, 2284/91 and 817/92.

11. Buenos Aires, Rosario, Quequén, Bahía Blanca, Santa Fe and Ushuaia.

12. The terminals were numbered 1–6, and the first two were part of a single concession.

13. Ports such as Dock Sud, Zárate and La Plata.

14. Eight per cent is moved by rail and the remaining share by waterborne transport.

15. Main instruments: Ley de Puertos (Port Act) No. 16246 and Decretos (decrees) 412/1992 and 413/1992.

16. See Section 3 of decree 412/1992.

17. National Port Administration.
18. Up to three times lower than in Buenos Aires in 2003.
19. For details on the concept of port competition, see Van de Voorde and Winkelmann (2002).
20. Senator F. Bouza, Session Log of the Uruguayan House of Senators, quoted in Cámara de Senadores de la República Oriental del Uruguay (1992, p. 115).
21. On the basis of International Institute for Management and Development (IMD), World Competitiveness Reports, 2000, 2004 and 2005, Lausanne.
22. Peripheral services consist mainly of transhipment and value-added services.
23. Macroeconomic conditions lent a hand, as the Convertibility Law resulted in high figures for imports.
24. For more details, see ECLAC (2005).
25. Private opportunism affects the investment rationale, efficiency and performance (overinvestment, for instance), and government opportunism, even if seemingly beneficial to current consumers (i.e., forcing low prices), is likely to reduce investment level and quality, which ultimately is harmful to consumers.

REFERENCES

Baltazar, R., & Brooks, M. R. (2001). The devolution of port management: A tale of two countries. Presented at the World Conference on Transport Research, Seoul, July.

Cámara de Senadores de la República Oriental del Uruguay. (1992). *Diario de Sesiones de la Cámara de Senadores de la República Oriental del Uruguay*. Tomo 345. Montevideo, Uruguay.

Cullinane, K., & Song, D.-W. (2002). Port privatization policy and practice. *Transport Reviews*, *22*(1), 55–75.

ECLAC (Economic Commission for Latin America and the Caribbean). (2004). Preliminary overview of the economies of Latin America and the Caribbean 2004 (LC/G.2265-P/I). Santiago, Chile, December. United Nations publication, Sales No. E.04.11.G.147.

ECLAC. (2005). *Maritime profile*. Online, available at www.eclac.org/transporte/perfil

Sánchez, R., & Sgut, M. (2002). Conditions of settings and strategies in the Argentine port reform. Presented at the Port Performance Research Network (PPRN) meeting, Panama, 12 November.

Sgut, M. (2000). Estudio sobre Reestructuración Portuaria – Impacto Social – Puerto de Buenos Aires (Argentina). Working Paper, ILO, Geneva.

Van de Voorde, E., & Winkelmann, E. (2002). A general introduction to port competition and management. In: M. Huybrechts, et al. (Eds), *Port competitiveness: An economic and legal analysis of the factors determining the competitiveness of seaports*. Antwerp: De Boeck.

Voss Hall, P. (2002). *The institution of infrastructure and the development of port-regions*. Ph. D. dissertation, University of California at Berkeley.

Ward, R. (2004). Turnaround time for BA. *Containerisation International*, *37*(3), 47–50.

CHAPTER 10

PORT GOVERNANCE AND PRIVATIZATION IN THE UNITED STATES: PUBLIC OWNERSHIP AND PRIVATE OPERATION

James A. Fawcett

ABSTRACT

Owing to its history as a nation fashioned from a federation of relatively autonomous states, port governance in the United States is largely in the hands of those 50 states rather than the federal government. Evolving from accustomed private ownership in the 18th and 19th centuries, seaports in the US are now commonly owned by public agencies, governed by public boards and yet often operated by tenants who are lessees of those agencies. The following discussion unravels the complex ties of governance and privatization in US seaports.

1. BACKGROUND

The approach of the United States toward seaport ownership and devolution is complex, evolving from a multitude of factors, the more important of which are historical attitudes of the public to commerce, juridical

Devolution, Port Governance and Port Performance
Research in Transportation Economics, Volume 17, 207–235
Copyright © 2007 by Elsevier Ltd.
ISSN: 0739-8859/doi:10.1016/S0739-8859(06)17010-9

interpretation of the public trust doctrine over submerged lands, a tradition of public agencies operating as quasi-private enterprises, a longstanding interest of the federal government in maintaining navigable waterways and to accidents of history such as wars. Perhaps surprisingly in a country as devoted to capitalism as the US, its large general cargo seaports have remained largely in public hands while smaller, special-purpose seaports, often handling primarily bulk cargoes, are more commonly found in private hands. What we shall see in the following discussion is a complex array of seaport ownership and operation divided further by various public–private arrangements often using the capability of public agencies to fund infrastructure. There are powerful historical antecedents to the philosophy by which seaports are operated in the United States, and a brief historical review is essential to our appreciation of current styles of port governance and ownership in the US.

In that regard, this discussion will begin by considering briefly how the history of the country, its geography and the predilections of its people influenced attitudes toward governance in general. With that brief history as a foundation, we will explore the intertwined role of the states and federal government in port governance, with special attention to that role in the 20th century, followed by a similar discussion of privatization as found in the US. Although the 50 states are essentially autonomous in decisions regarding port governance, we will see that the schemes that emerge fall into a few standard patterns depending upon state or local custom, history or competitive advantage. In a large country with over 350 commercial seaports, there is broad opportunity for variety (Newman & Walder, 2003).

America is a seagoing nation not solely out of choice but also out of necessity. She is surrounded on her two flanks and to her south by vast oceans and joined to Canada on the north and Mexico on her southwest side. Despite the fraternity of a common language and compatible, if dissimilar, political institutions, Canada was even more sparsely populated in its early days than the United States. While today Canada may be a major trading partner, the population base did not support the needs of the US at that time. The US relationship with Mexico was even more challenging. In addition to being separated both by language and culture, Mexico was physically separated by a substantial wilderness beyond the western boundaries of the US colonies. Thus, the US had to look seaward rather than north or south for trading partners, and its seaport infrastructure became essential to its economic survival.

Further, early US residents were of a seagoing nature. They had emigrated from Europe by sea and often had strong attachments to the sea in their countries of origin – England, the Netherlands, Scotland and France. Early settlements in the US relied upon coastal locations for communication, trade and sustenance from the former European homelands as well as Africa and the Caribbean. Early settlements at Boston, Jamestown in Virginia, and New York all had their substantial foundations laid by seafaring folk.

We may conceive of the United States in its formative years as an island nation surrounded on four sides by water: the East Coast, the Gulf of Mexico and the Mississippi River and on the fourth side by the Great Lakes. Moreover, her early settlers had little choice but to utilize that connection to the lakes and sea as a transportation network, a source of sustenance from fishing and as the means to develop trade with the states of Europe and elsewhere.

One of the struggles of the new nation revolved around the role of a central government as opposed to the role of the states. Remembering that the colonies were at times enclaves of persons of similar religious beliefs yet distinct from the religious traditions of other – often adjacent – colonies, the attraction of a country composed of mutually dependent but still individualistic states held great attraction to a people fleeing religious persecution. Fearing loss of identity and accustomed religious traditions, the form of governance of a new nation was of singular importance. Such was its importance that three of the nation's founders, James Madison, Alexander Hamilton and John Jay, penned a series of 85 articles, *The Federalist* (Hamilton, Madison, & Jay, n.d.), developing arguments around a strong federation of states that were autonomous in many matters but yet collectively united.

The words of both the Declaration of Independence and Constitution of the United States (1987) reflect that notion in the linguistic construction of a union of states. While we might argue that the notion of relatively autonomous states was grounded in a history of fear of centralized power in the English crown and its persecution of religious minorities, the result over the past 216 years has been diversity of styles of state governance, as well as a continuous and robust debate over the role of the states as opposed to the central government (a central government which we, interestingly enough, now commonly refer to as the "federal" government. Thus the notion of a federation of states continues in our common terms of reference).

2. PORT GOVERNANCE

If political repression experienced by colonial immigrants in their countries of origin forged a sense of caution over powerful central governments, and if that caution influenced their approach to forming a union of sovereign states – the *United* States – then a further devolution of central government power in the form of the public trust doctrine further influenced the operation of seaports in the incipient nation. Eichenberg, Fletcher, and Kehoe (2005) describe the public trust doctrine as having roots in ancient Roman law, later embodied in English common law where submerged lands (i.e. ports, lands with port potential and their approaches) were owned by the crown. But, as they note, "[u]nder the constitutional principle that all new states join the Union on an equal footing with the original thirteen states, state ownership of tidelands was extended to all new states regardless of whether ownership as expressly conferred to the state" (p. 4). Thus, the states can define the application of the public trust doctrine according to their own needs, and the US Constitution supports that devolution of authority over these lands. Nevertheless, the submerged lands must remain in public ownership.

California is a case in point: title to the underlying water resources within the port are retained not by local government but by the State of California under the "Tidelands Trust," (California State Lands Commission, n.d.) and this is common practice throughout the country after the decision of the US Supreme Court more than 110 years ago in *Illinois Central R.R. Co. v Illinois* 146 U.S. 387, 452 (1892), where the court held that a state's title to tide and submerged lands, "is a title held in trust for the people of the State that they may enjoy the navigation of the waters, carry on commerce over them and have liberty of fishing," free from obstruction or interference from private parties. The court decreed that these principles apply uniquely to submerged lands as opposed to dry land where the state *did* have the power to transfer ownership to private hands if such a transfer were in the best interests of the public. Of course the concept becomes more complex when lands that were formerly submerged are converted by landfill into dry land, often in seaports.

Subsequent to the Illinois Central case federal legislation, notably the Submerged Lands Act of 1953 (43 U.S.C. §1301 et seq.) specified that "the right and power to manage, administer, lease, develop, and use the said lands and natural resources all in accordance with applicable State law be, and they are, subject to the provisions hereof, recognized, confirmed, established, and vested in and assigned to the respective States ..." (43 U.S.C. §1311(a)(2)). And yet, the federal government retains rights of commerce, navigation and national defense, if not ownership, in the three-mile territorial sea of the states under section 1314(a) of the Act. That provision specifies,

The United States retains all its navigational servitude and rights in and powers of regulation and control of said lands and navigable waters for the constitutional purposes of commerce, navigation, national defense, and international affairs, all of which shall be paramount to, but shall not be deemed to include, proprietary rights of ownership, or the rights of management, administration, leasing, use, and development of the lands and natural resources which are specifically recognized, confirmed, established, and vested in and assigned to the respective States and others by section 1311 of this title.

Thus, states had the power to regulate uses of these lands, in the public interest, despite limitations of sale. Historically, they often sought to make use of these lands as seaports. In the 19th century, for example, they frequently allocated dry port land to railroads that had the capital and incentive to develop port facilities. If we conceive of the seaport as a modal shift point for cargo, the interest of railroads in seaports is quite logical. Important contemporary seaports such as the Port of Los Angeles had their genesis as freight entrepôts (Queenan, 1983) and were originally creations of railroads (Sherman, 1995). However, their influence at the waterfront – not to mention their ability to act monopolistically in cargo movement – eventually brought public condemnation and a call for reform. Hershman (1988, p. 11) cites three reasons for public ownership and operation of the waterfront in the 20th century, "the progressive movement in local politics, dissatisfaction with railroad control of the waterfront, and opportunities to assume ownership of surplus federal government facilities." Citing earlier work by Olson, Hershman (1988, p. 40) notes,

Resentment toward the railroad companies was very similar in both coastal and inland cities. Over a period of years, waterfront areas had developed into dirty and congested sites that imposed intense social and economic costs on communities ... [w]hen civic leaders realized that the public interest, as they conceived it, could not be served with continued private control of the port, many communities, at varying times and rates, began a process of shifting harbor ownership and control from the private sector to the public sector. This signified the beginning of the public port entity, created to oversee harbor development and effectively manage port operations.

Two other factors evolved in the US influencing public port ownership, the two world wars and the influence of the US Army Corps of Engineers. Entering World War I in April 1917, the US sent thousands of troops by ship to Europe to join in the war effort, necessitating new port infrastructure to accommodate the troop, logistic and naval forces that would support the Allies in defeating Germany. However important the effort, it paled in comparison to the seaport buildup as the US entered the Second World War in 1941. Taking the war overseas to the enemy meant that ships, both merchant and military, needed to be constructed hurriedly and in vast numbers. Accordingly, shipyards, both naval and private, were rapidly founded, expanded and operated at full speed to support the war. Moreover, since much of the war in the Pacific was fought in sea battles or on islands, the need for shipping was intense, especially on the West Coast, and the nation responded by enhancing its public port facilities there. After the war, vast numbers of the facilities were declared surplus and transferred to local governments. (Nevertheless, some shipyards and naval facilities continued to operate into the 1990s.) Later when they were declared surplus, the existing public ports in which they were located absorbed them as the Port of Long Beach absorbed the US Naval Shipyard and Naval Station at Long Beach, California, in 1997.

The influence of the US Army Corps of Engineers has also had a critical impact on public ownership of US seaport facilities (Allen, 1996). Beginning in the 19th century, as the cargo fleet grew both in numbers and size of vessels, dredging became essential to the viability of seaports. The federal government, under provisions of its navigation servitude referenced in the Submerged Lands Act but with a foundation in the commerce clause (US Constitution, Article 1, Sec. 8), maintained those shipping channels that were designated a national interest. In 1824 in the landmark case of Gibbons v. Ogden (22 U.S. 1), the US Supreme Court determined that interstate commerce free of state regulation was in the interest of the nation at large. The decision permitted the fledgling US Army Corps of Engineers to take action to facilitate that interstate commerce through construction and maintenance of navigation channels by construction of locks and dams and, of particular importance to coastal seaports, dredging. That process continues to this day where Congress allocates funds for the Civil Works Program of the Corps of Engineers, a mission separate and distinct from its role as combat engineers.

Empowered under Section 10 of the Rivers and Harbors Act of 1899 (2003), the US Army Corps of Engineers is responsible for maintaining the navigability of "federal navigation channels" in the US. The Code of Federal Regulations further defines these channels as "navigable waters of the US" and specifies that they are critical to maintaining waterborne commerce.

Section 329.4 – General definition. Navigable waters of the United States are those waters that are subject to the ebb and flow of the tide and/or are presently used, or have been used in the past, or may be susceptible for use to transport interstate or foreign commerce. A determination of navigability, once made, applies laterally over the entire surface of the waterbody, and is not extinguished by later actions or events which impede or destroy navigable capacity.

In practice, based on its expertise, the Corps of Engineers annually identifies for Congress those dredging or lock and dam maintenance projects needing federal funding either for maintenance or new construction. They are aggregated into the annual (usually) Water Resources Development Act providing the federal share of funds for the maintenance or construction. The federal government does not bear the full burden of this construction under the theory that the locality derives significant benefits along with the nation at large. Thus, virtually all of the projects are cost-shared either with local government or the states, often both. The cost burden shared by these so-called "local sponsors" varies depending upon the stage of the project (reconnaissance, feasibility or actual construction) and the nature of the project (e.g. beach renourishment, dredging, lock and dam repair or construction). In the case of coastal seaports, the contribution of the federal government to the financial feasibility of port growth as well as modernization (channel deepening) hinges on the ability of the port to secure a commitment from Congress to assist local agencies, through the Corps of Engineers, a critical role in seaport development.

As federal funds are utilized to maintain these channels, all shippers pay fees on cargo movement to the Harbor Maintenance Trust Fund whose resources are held in trust to be utilized for maintenance of the federal navigation channels.[1]

Although dredging is critical to the functioning of both public and private ports, and federal support of dredging through the civil works program of the Army Corps of Engineers is essential for many – if not most – port dredging projects, historically there are always more claims from seaports for dredging and other Corps of Engineers projects than Congress will fund. Therefore, essential to this process is local political influence in Congress, which often determines which projects are selected from among the many seeking federal funding. Indeed, seaport agencies spend a great deal of time and political capital ensuring that their projects are successful in obtaining Congressional funding, thus assistance from the Corps. Delays in that process often result in business foregone, as carriers may seek alternate ports at which to call. Since business lost in this way may take years to recover, there is understandably intense competition in the halls of Congress for federal support.

Clearly, the most important measure that the federal government can take in US coastal ports is to provide dredging such that new, deep draft vessels can make port calls (Sherman, 1995). As Erie (2004) observes with respect to the Port of Los Angeles, which had been limited to a main channel depth of 35 feet for almost 50 years, involvement of the local congressman was essential to drawing the attention of Congress to funding further harbor deepening. Determined and tenacious work by the congressman was necessary to appropriate the fund for the project.

> More than one-third of the world's new container fleet was unable to enter the main channel; and an even greater percentage of petroleum and chemical carriers was turned away. It made little sense for the L.A. port to construct container cranes and terminals if ships could not reach them. Working closely with port and city officials, San Pedro Congressman Glenn Anderson led a prolonged campaign for federal funding, eventually securing $36 million to dredge the harbor to forty-five feet. In 1981, after ten years of funding, permit and approval delays, dredging finally began. In 1983, a deepened L.A. harbor was opened to hundreds of huge new container ships. (Erie, 2004, p. 90)

Through this process the federal government retains a vital role not only in maintaining the viability of US seaports, but also in maintaining a close political relationship to all major ports in the country. Federal contributions support not only maintenance dredging, but also channel deepening projects that allow them to compete for the newest ships as vessels continue to grow in displacement and draft. In many ways, this dependence of US ports on federal support for both maintenance and new construction dredging has inhibited greatly the willingness of entrepreneurs to privatize them.

With this background as a foundation, we now turn to a discussion of the forms of seaport governance within the framework of federal, state and municipal institutional arrangements.

3. FEDERAL INVOLVEMENT IN US SEAPORT GOVERNANCE

While the history of the nation has granted considerable autonomy to the states and recognized their sovereignty in many ways, at least in the case of US navigation infrastructure, federal involvement remains both facilitative of their aspirations and determinative through both regulation and funding (Dunfee, 1977–1978). In effect, the federal government sees itself as the overseer of US seaports through implementation of its navigational role.

For example, as noted, the Rivers and Harbors Act of 1899 gives the US Army Corps of Engineers responsibility for maintaining navigable waterways, locks, dams and other facilities that bear on navigation and therefore upon trade. The US Department of Transportation provides funding for terrestrial roadway networks to every seaport in the US. While not directly maritime in nature, the roadways serve to move goods to and from US seaports. The United States Coast Guard, operating through the new US

Department of Homeland Security, provides vessel and maritime law enforcement security in every major US seaport. The Department of Homeland Security also hosts bureaus of Immigration and Customs Enforcement, Customs and Border Protection, the Transportation Security Agency, the Federal Emergency Management Agency, the US Secret Service, and the Citizenship and Immigration Service. In multipurpose port districts, where the district also operates an airport (e.g. in cities such as New York, Portland and Oakland), the Federal Aviation Administration may also provide federal services to the port agency. So, there is overlapping governmental influence in US ports with ownership of the capital infrastructure in state hands and their subdivisions; yet, there are so many regulatory and funding strings attached to that ownership through funding and regulation that, in effect, these seaports are de facto partnerships between the states or localities and the federal government. Despite the apparent autonomy of states and local governments with respect to their seaports, in practical fact the federal government retains extensive and yet, to the casual observer, subtle powers over the nation's maritime infrastructure. Indeed, those powers are fundamental and pervasive. While the commercial operations of seaports seek revenues and cargo throughput as well as promote economic development in their regions, the federal government retains vitally important roles in funding development.

4. PORT GOVERNANCE AND PRIVATIZATION

Governance in the US is primarily concerned with who exercises oversight over ports and how that oversight is legitimized. If ports are public agencies, governance clearly demands a different approach than if they are owned by public companies whose primary obligation is to shareholders or private companies whose responsibility is to its owners. Port managers retain an obligation to their constituent-owners, and public ownership is the norm. Where seaports are privately owned, their obligations differ from public agencies; also, their opportunities for financing and use of publicly provided services are limited.

Hershman (1988, p. 338) distinguishes among public ports in the US, separating the "landlord" and "operating" characteristics of seaports and sheds light on the subtle distinctions that exist between the so-called "public" and "public–private" port operations. (See Table 10.1 for a listing of landlord, operating and "limited operating" ports in the United States). Hershman draws a distinction between the landlord and operating port

Table 10.1. U.S. Public Port Operating Status.

Landlord	Operating	Limited Operating
Anchorage, Port of	Alabama State Port Authority	Albany, Port of
Bridgeport Port Authority	Beaumont, Port of	Corpus Christi Authority, Port
Brownsville, Port of	Bellingham, Port of	of
Canaveral Port Authority	Everett, Port of	Houston Authority, Port of
Cleveland–Cuyahoga County	Freeport, Port of	Maryland Port Administration
Port Authority	Georgia Ports Authority	Port Everglades
Coos Bay, Port of	Grays Harbor, Port of	Portland (OR), Port of
Detroit/Wayne County Port	Guam, Port Authority of	Seattle, Port of
Authority	Hawaii DOT, Harbors Division	Shreveport-Bossier, Port of
Duluth Seaway Port Authority	Lake Charles Harbor and	Stockton, Port of
Galveston, Port of	Terminal District	Tacoma, Port of
Greater Baton Rouge Port	Longview, Port of	Vancouver (WA), Port of
Commission	Massachusetts Port Authority	
Greater LaFourche Port	Milwaukee, Port of	
Commission	North Carolina State Ports	
Green Bay, Port of	Authority	
Humboldt Bay Harbor District	Olympia, Port of	
Iberia, Port of	Orange, Port of	
Indiana Port Commission	Oswego Authority, Port of	
Jackson County Port Authority/	Oxnard Harbor District/Port	
Port of Pascagoula	Hueneme	
Jacksonville Port Authority	Panama City Port authority	
Kalama, Port of	Port Angeles, Port of	
Long Beach, Port of	Port Arthur, Port of	
Los Angeles, Port of	Port Lavaca–Point Comfort,	
Manatee County Port Authority	Port of	
Miami, Port of	Puerto Rico Ports Authority	
Mississippi State Port	Sacramento, Port of	
Authority, Gulfport	San Diego Unified Port District	
New Orleans, Port of	South Carolina Ports Authority	
New York/New Jersey, Port	South Jersey Port Corporation	
Authority of	South Louisiana, Port of	
Oakland, Port of	St Bernard Port, Harbor and	
Palm Beach District, Port of	Terminal District	
Pensacola, Port of	Tampa Port Authority	
Philadelphia Regional Port	Virginia Port Authority	
Authority	Wilmington (DE), Port of	
Plaquemines Port, Harbor and		
Terminal District		
Redwood City, Port of		
Richmond (VA), Port of		
San Francisco, Port of		
Toledo-Lucas County Port		
Authority		
34 Ports	32 Ports	11 Ports

Source: Rexford Sherman, American Association of Port Authorities (2005).

agencies as two public ports, one of which is operated such that the "majority of its facilities" are leased to others for operations (the landlord port) and the other (the operating port) that "manages the day-to-day activities on its terminal by scheduling vessel calls, arranging stevedoring services, employing longshore labor, and other similar functions" (Hershman, 1988, p. 339). Both types of ports are regarded as public agencies but the operating port is "more public" than the landlord port in his descriptions.

There are a few large "operating ports" in the country, but the landlord port is more or less the standard model in the US. In terms of the style of operation, both compete aggressively with neighboring ports to secure market share, market their services like private companies, borrow capital, fund major infrastructure development and manage themselves in a manner resembling private companies as well. For this reason, Boschken (1988) has described them as "public enterprises." For purposes of our discussion, and based on the extensive role we have described for the federal government, the ports of this nation are not private but, then again, they are not altogether public either. They are essentially public–private partnerships.

In the US, the "public enterprise" duality characterizes seaports perhaps more clearly than other public agencies. That is, the port is responsible for facilitating economic development via private enterprise but is also often a public agency responsible for its actions as it manages the port in the public interest. Port managers routinely face this dilemma in scope of activities, as they seek to effect private sector activities that promote community economic development, while at the same time being cognizant of the needs of the public for environmental quality, recreation and the general needs of the public in that enterprise.

As Boschken (1988) notes, ports retain the dual characteristics of public agencies, nevertheless existing in an environment where they must routinely act as quasi-private companies all the while retaining their responsibility – both fiscally and politically – to their constituent public. This duality has great attraction for the public as it correctly interprets the public responsibilities that the seaport must exercise to protect the public interest when the same port has prodigious powers to negatively affect the quality of life for the public both in the region that it serves and for those who may live in close proximity.

In both landlord and operator ports, much of the responsibility for capital improvements rests with the public agency. Naturally, this would be so in an operating seaport but it is also essentially true for landlord ports as well. Tenant leases on port property are typically of a sufficiently short duration that it would be effectively impossible to obtain credit in the commercial markets for the short duration of most leases (10–20 years) given the high cost of fixed shoreside cargo handling equipment. Container cranes, bulk loaders, rail infrastructure and other capital equipment are simply beyond the realm of financial feasibility for most terminal lessees to amortize over the relatively short duration of those leases. However, the public financing capability of port authorities gives them unique access to capital markets and allows them, in some cases, to float bond issues that are tax-free to the buyer. (As opposed to common practice in developing nations such as Mexico where private access to capital is an incentive to governments to privatize seaports; Eaton, 1997). Indeed, one of the more compelling reasons for public ownership of these vital facilities is in the ports' capability to borrow sufficient funds for their own capital expansion.

4.1. State Management of Seaports

We have established that the federal government retains substantial powers in US seaports, but given the history of the nation we have also shown that the 50 states reserve significant autonomous powers as well. At one extreme are those states that retain governance over their seaport at the state level by virtue of state seaport management agencies. States may also charter so-called "special districts" whose jurisdiction is less extensive than the entire state and may encompass a region extending beyond the reach of municipal or county boundaries (Washington Public Port Districts, 2005). These agencies are often endowed by the state with substantial powers common to state agencies and responsive often to a board of commissioners and ultimately to the state itself.

4.2. State Port Authorities: Single and Multipurpose

As we have noted, the disaggregated port management model in a country as large as the US has created a variety of styles for port organization and management. Because of the power of a seaport to create economic activity in a region, where US states are large, individual municipalities tend to create and manage port authorities (e.g. California, Texas and Florida). However, in geographically smaller states or those with only one primary port or a primary and one or more smaller seaports, for reasons of administrative efficiency, these states often establish a single state commercial seaport agency to manage all operations. Table 10.2 gives examples of states that have established these authorities.

Table 10.2. State Port Authorities.

State	Port Agency	Ports Included
Delaware	Diamond State Port Corporation	Wilmington
Maryland	Maryland Port Administration	All
Virginia	Virginia Port Authority	All
North Carolina	North Carolina State Ports Authority	All
South Carolina	South Carolina State Ports Authority	All
Georgia	Georgia Ports Authority	All
Alabama	Alabama State Port Authority	All
Mississippi	Mississippi State Port Authority	All
Illinois	Illinois International Port District	Chicago
Indiana	Ports of Indiana	All
Hawaii	Hawaii Department of Transportation	All
Pennsylvania and New Jersey	Delaware River Port Authority	Philadelphia (Cruise Terminal @ Pier 1) and Camden (AmeriPort Intermodal Rail Center)
Louisiana	Louisiana Department of Transportation and Development	Louisiana Offshore Oil Port (LOOP)

Source: Derived from various sources including port websites.

A study by the American Association of Port Authorities (2000) found that state or commonwealth port management agencies were relatively common (about 17% of the total reporting jurisdictions) but that there was quite a variety in governance schemes, owing largely to the autonomy of states to best determine the governance scheme for its ports. Table 10.3 presents a summary of that study.

As governing bodies for seaports, state agencies are chosen about as often as municipal corporations. Here again, public–private management of the assets of the port is the common mode of operation. In some cases, these agencies have grown as large and comprehensive as the bi-state Port Authority of New York and New Jersey managing tunnels, bridges, airports and seaports (Port Authority of New York and New Jersey, n.d.; Rodrigue, 2004). New York and New Jersey are not the only jurisdictions to develop these "multipurpose" port districts. Table 10.4 depicts the 11 seaport agencies in the US that follow this comprehensive model. There is a rich public–private interaction within these seaports. For example, virtually all are involved in leasing land or facilities within their jurisdiction from basic

Table 10.3. Port Governance by Type.

Type of Governance	Number of Ports
Bi-state port authorities	2
State administrative departments or agencies	5
State or commonwealth port authorities	15
County port departments	2
County port authorities	3
Municipal agencies	16
Special purpose navigation districts	57

Note: Table 10.3 was developed in 2002 and Table 10.1 in 2005. The difference in total number of ports is due to the three-year gap between the two, the fact that the charts are noting different characteristics of ports, governance (Table 10.3) and management type (Table 10.1), and the data perhaps reflect changes in membership in the American Association of Port Authorities by the reporting ports.
Source: American Association of Port Authorities (2002). *Port activities, functions and types.* (http://www.aapa-ports.org/industryinfo/port_activities.htm).

Table 10.4. Comprehensive "Multipurpose" Port Districts.

Port District	Size of Seaport (Volume of Traffic)	Activities Managed
Port Authority of New York and New Jersey	Large	Seaport, airport, tunnels, bridges, real estate, commuter rail
Massachusetts Port Authority (MassPort)	Small	Seaport, airports, Tobin Bridge
Toledo-Lucas County Port Authority	Small	Seaport, airport, real estate, rail terminal
Puerto Rico Ports Authority	Small	Seaport, airport
Virgin Islands Port Authority	Small	Seaport, airport, real estate
Port of Bellingham	Small	Seaport, airport, marina, rail station, real estate
Port of Seattle	Large	Seaport, airport, real estate marinas
Port of Olympia (Washington)	Small	Seaport, airport, real estate, marina
Port of Grays Harbor (Washington)	Small	Seaport, airport, real estate, marina
Port of Portland (Oregon)	Small	Seaport, airports
Port of Oakland	Medium	Seaport, airport, real estate, marina

Source: Various sources including port district websites.

warehouses to retail facilities. Many of these comprehensive agencies provide public facilities such as the Elliot Bay Fishing Pier and Park at Pier 86 and the Smith Cove Park and bike path at Terminal 91 in Seattle. They may also provide marinas such as those owned by the ports at Bellingham, Seattle, Olympia, Grays Harbor and Oakland. Additionally, some provide foreign trade zones in which warehouse space can be leased.

These comprehensive ports offer a unique model of port management with a focus substantially beyond marine cargo movement into air cargo and passengers. For example, their airports range from international facilities at Seattle, Boston, New York, San Juan, St. Thomas, Portland and Oakland to small general aviation airports. The larger of these provide myriad opportunities for private enterprise, such as airline offices, terminal facilities, hangars, retail facilities, at times hotels, offices and hangars for fixed base operators at general aviation airports. By virtue of the wide range of their activities, these comprehensive or multipurpose port districts have a greater impact on the local communities than if their operations were limited solely to the movement of intermodal freight and bulk cargoes.

State chartered port districts are often established in mid-sized cities where scale economies can be achieved by integrating both airport and seaport activities. At times, the two developed concurrently. However, co-operation and integration of facilities can also be a peremptory act when neighboring jurisdictions could develop, especially, airports that would serve a region, thus capturing the revenue from their neighboring competitor jurisdictions. Seaports have a place-dependent quality but airports, by their nature, can serve a region and be located wherever land is available. So, merging a port district and an airport, both placing similar demands on public transportation managers, is in some cases an essential method of securing economic development for a jurisdiction by foreclosing the option of a competing facility developed by a neighboring jurisdiction.

4.3. Regional Port Management Special Districts

Work by the American Association of Port Authorities (AAPA, 2000) established a census of port agencies by type for their membership. That summary is presented in Table 10.3. Although the governance taxonomy is not completely consistent between the two studies, it is clear that the "special-purpose navigation district," conforming more or less to Olson's "area-wide special district" or "special municipal district" has become the preferred model as seaports developed in the United States. Olson (1992) breaks down port governance structures into seven categories (see Table 10.5) that encompass the range of structure found.

These agencies, often extending over broad regional boundaries, encompass areas that might not have the fiscal capacity to develop a port on their own but nevertheless where a port is needed to serve the region (Fawcett & Marcus, 1991). In Washington State, for example, port districts may serve entire counties or, in other cases, a portion of a county (State of Washington, 2005; Washington Public Port Districts, 2005). Established under state laws, these agencies may have various motivations for existence including a means of foreclosing competition between communities seeking port facilities where there is insufficient demand to support multiple ports. In those cases, the state may intervene and establish regional port authorities to meet the needs of the region.

Cost sharing among small communities is one motivation but another more powerful one is the capacity to float bond issues. Like municipal governments, special districts often have the capability under state laws to issue revenue bonds to fund infrastructure improvements. Unlike general obligation bonds that often require super-majority voter approval, revenue bonds can be issued by the commissioners of the special district, without voter approval, on the basis of expected earnings of the special district. Since special districts also encompass jurisdictions that are not coterminous with

Table 10.5.   Taxonomy of Port Governance Structures in the United States (After Olson, 1992).

Bi-state jurisdictions	Created by the federal government, joint management by two or more states
State port management agency	States create the governing agency for ports – usually statewide
Area-wide special district	Special districts that are independent of other local government political boundaries
County port districts	Port districts established by a county
Special municipal district	Special district established by local government (often established by counties)
City port agency	A port management department of a municipality
Territorial	Ports created within territories or commonwealths of the United States

Note: Multipurpose port districts incorporating a seaport in combination with one or more other function (airports, tunnels, bridges, parks or other activities) may appear in any one of the governance structures shown. The term "multipurpose" is a functional appellation rather than one referring to governance.
Source: Chart modified after Olson (1992, p. 16).

existing political boundaries, and with the capability to float revenue bond issues, they often have policy latitude not ordinarily available to municipal government agencies. Not surprisingly, as Table 10.3 shows, they are the preferred method of governing port agencies in the US because of their flexibility.

4.4. Municipal Port Management Agencies

In contrast, the municipal agency is less popular as a model for port management than the special district. I have explored some of the reasons for establishment of special districts and they are often compelling in a region where there is competition for economic development opportunities. The municipal option is more common when a port has a long history of development and the municipality of which it is a part does not have strong competition from other nearby governments. Apparently contradicting this model, the two adjacent seaports that handle the largest volume of container cargo in the US are located adjacent to one another, appear in an aerial photo to be a single seaport and are owned and operated by two competing governments: the Port of Los Angeles and the Port of Long Beach, California. Each port is owned by its city, managed as a municipal agency and is in vigorous competition with its neighbor, to the benefit of port customers, it might be argued.

Nevertheless, in each of these cases there are strong historical reasons for the governing regimes. Los Angeles was strongly influenced in its development by the railroads and grew out of a vision by a local entrepreneur, Phineas T. Banning (Queenan, 1983). Long Beach, on the other hand, was blessed with abundant oil resources and had the financial wherewithal to pursue its own interests in port development independent of the competitive juggernaut posed by Collis P. Huntington and the Southern Pacific Railroad that largely controlled early development of the Port of Los Angeles at San Pedro and Wilmington.

Thus, the two busiest cargo ports in the US are creatures of their municipal governments but are anomalies in terms of the model of governance presented here, largely for local historical reasons; one of these is that the railroads played such a large and, at times, pernicious role in seaport development in the region. This history and the reform movement in California politics around the turn of the 20th century caused the public and local governments to be very suspicious of another (Queenan, 1983). Moreover, at that time, the State of California had assumed management of the state's most prominent port, San Francisco, and left the cities to the south to develop their own nascent seaport operations, which they did (Olson, 1992). The transition is instructive in that we see here the influences governing the movement *from* private *to* public in the early part of the last century. That process was by no means unique to California. Olson reminds us that the Port of San Francisco was merely the first publicly owned seaport in the US, having been chartered by the state in 1853. Subsequent years have seen that process continue to the point where, as we have noted, virtually all general cargo seaports in the country are now publicly owned.

5. PORT PRIVATIZATION

The forces of federalism, reform of abuse by monopolistic railroads, world wars and the need for public funding to operate and expand increasingly large seaports has meant that almost all *general cargo* ports in the US are now publicly owned. However, there remain substantial facilities under private ownership, generally those ports handling bulk freight of all kinds: sand, gravel, forest products, minerals, project cargoes, petroleum, chemicals and metals. Recalling the history of some of those that remain gives us some insight into how our modern general cargo ports have evolved from the days when private companies built facilities for their exclusive use. It also provides a backdrop against which we can gain insight into the differences between them and today's modern port agencies.

One of the earliest still in private hands is the port at Marcus Hook, Pennsylvania, where Joseph Pew and the Sun Oil Company brought in crude oil from the Spindletop oil field in Texas by ship, refined it on site and transported the refined products via railcar to Sun Oil Company's market (Sunoco, Inc., n.d.; Philadelphia Regional Port Authority, n.d.). Pew selected the location because it was close to his other Pennsylvania operations, had good water at the berth and was adjacent to an existing major rail line, facilitating the distribution of his refined product. In 2002, it celebrated its 100th anniversary and is typical of private seaports found in the US: single purpose, often owned over a long history, utilized mainly for bulk commodities and often for liquid bulk products. While many early private ports were owned by railroads, nowadays refiners or importers and exporters of bulk (e.g. petrochemicals, agricultural products or minerals) and neo-bulk products (e.g. structural steel and project cargoes) often hold those ports in private hands. In San Francisco Bay there are two private seaports, one at Pittsburg owned by the Posco Division of US Steel that imports the so-called "hot bands" (coil steel) from Korea and a second private terminal at the Port of Richmond, a liquid bulk terminal owned by the Standard Oil Company of California (Chevron) although the balance of the port is owned by the City of Richmond. The private ports remain, often because they have long ago been amortized on the books of the owner or owner–user, they are often consumptive of valuable alongside space that the owner could not afford to lease in a modern cargo port and remain useful as long as the owner moves or hosts shippers that move these high-volume, low-value cargoes.

Since purely private seaports are relatively unusual in the US outside of the movement of bulk and neo-bulk cargoes, we need a means of discerning a distinction between public and private operations. Fortunately, Baltazar and Brooks (2001) have helped us conceptualize seaport functions and how they might be allocated in a modern seaport in a general model of port governance and operations (Table 10.6).

The authors note that the allocation of functions to Regulators will depend upon the regulatory regime of the country and that the allocation between Operator and Landlord may be redistributed depending upon local custom. Nevertheless, Table 10.6 provides an elegant method of visualizing many of the important port services, leaving to us to specify how they are

Table 10.6.   Baltazar and Brooks' (2001) Port Devolution Matrix.

Governance	Regulator Functions	Port Functions	
		Landlord	Operator
Public	• Licensing, permitting • Vessel traffic safety • Customs and immigration	• Waterside maintenance (e.g. dredging)	• Cargo and passenger handling • Pilotage and towage
Mixed public/ private	• Port monitoring • Emergency services • Protection of public interest on behalf of the community	• Marketing of location, development strategies, planning • Maintenance of port access	• Line handling • Facilities security, maintenance and repair • Marketing of operations • Waste disposal
Private	• Determining port policy and environmental policies applicable	• Port security • Land acquisition, disposal	• Landside and berth capital investment

Source: Baltazar and Brooks (2001).

divided in practice. Using their matrix as a guide, let us first examine examples of the types of port operations that have become privatized in landlord seaports. Based on our earlier characterization of public seaport ownership in the US, this is the environment where we are most likely to observe activities that have become privatized.

5.1. Private Lessees in Public Ports

Whether state ports are managed by a single state agency or by a special district, municipal or county port authority, the principles of privatization are similar. Irrespective of the port operating agency, whether state, special district or local government, public–private operation has essentially the same genesis in practice: the seaport makes a decision that one or more of its functions can be best operated by a private contractor. The motivations for privatization are legion. On its most prosaic level, the port may be unable to hire adequately trained staff or adequately manage the task contemplated. More likely, the port agency finds that it will be more economically efficient to outsource the function. Perhaps the port seeks to avoid the administrative burden of hiring additional staff or of incurring the administrative overhead of that staff. In some cases it cannot effectively provide the capital necessary to adequately equip a given port function and must outsource the work. In other cases, port managers want to deflect responsibility for a port operation that they consider particularly risky or likely to incur public criticism. In other words, functions may be outsourced because of lack of an adequate labor force, for fiscal reasons, for bureaucratic reasons within the agency or simply for political reasons. The end result is that some of the functions of the port are shifted to the private sector while the ownership of the port remains in public hands.

The outsourcing process is common to many public agencies and relatively simple: the port issues a request for proposals (RFP) or request for qualifications (RFQ) for a particular aspect of port operations that it seeks to privatize and widely publishes it. (An RFQ is often a preliminary step to issuing an RFP; the RFQ sets standards for who is actually qualified and has the financial wherewithal to bid on the matter in question.) A potential vendor responds to the RFP and presents a proposal responding to the terms of the RFP along with other bidders. The agency requesting bids is subject to public contracting requirements of the jurisdiction (usually state standards). After a vetting process supervised by the public agency, a time-limited contract is let for private performance of what is usually a service.

The client here may be either the port governing body or the state or local contracting authority for the port. In most cases, these offers are made by a port management agency itself. In the US, the fundamental characteristic is that, where public funds have been used to develop a port facility, the land and water area of the seaport is almost never privatized. Rather, services performed on port land are what are most commonly privatized since, in most states, the water area itself must remain in public ownership under principles of the public trust doctrine.

5.2. Landlord Seaports and Privatization

The bulk of privatization, as defined here, occurs in landlord seaports and in the functions that most readily lend themselves to shifting from public to private operation. It is somewhat remarkable, however, to examine Table 10.1 and note that in this survey by the American Association of Port Authorities, there are virtually as many operating seaports (fully publicly managed and operated) in the survey as there are landlord seaports. It is only when we begin to break down the categories that we see that most of the large cargo ports in the nation are, indeed, landlord seaports (e.g. Los Angeles, Long Beach, Miami, New Orleans, New York and New Jersey, Oakland and Philadelphia) where there is a mix of public ownership and private operation (Brinson, 1995). Baltazar and Brooks (2001) have again provided a template for the types of services that we may see privatized in these ports (e.g. waterside maintenance [dredging], land acquisition and disposal, marketing of location, development strategies and planning, maintenance of port access, port security, land acquisition and disposal, cargo and passenger handling, pilotage and towing, line handling, facilities security, maintenance and repair, marketing of operations, waste disposal, landside and berth capital management). Using their model as a template, let us consider how some of these functions lend themselves to private operation in the publicly owned general cargo ports of the US.

5.3. Waterside Maintenance

The main channels of most major US seaports are considered (declared) to be "navigable waters of the United States" and, as such, the responsibility to maintain the navigational integrity of the channels has rested with the civil works program of the US Army Corps of Engineers. Theoretically, the

Harbor Maintenance Tax and Trust Fund, for which monies are collected from imported cargo and passengers, provides funding for Corps of Engineers' channel deepening and maintenance projects. The "local sponsor" referred to earlier is fully responsible for those areas of the port outside the "navigable waters of the US." So, where alongside depth must be increased or maintained, either the port or its tenants must bear the cost of such dredging. Where the alongside facility is publicly operated, the port bears the cost of this maintenance dredging; where it is a privately operated facility, the owner/tenant bears the cost. The nature of the land use adjacent to the marine berth usually determines who shall pay, thus the nature of landside ownership is critical to our understanding of the issue.

5.4. Landside Acquisition and Disposal

The fundamental characteristic of a landlord seaport is its management of the land adjacent to the amenity resource, the shipping channel (Fawcett, 2004). Public contributions of capital have made the channel navigable and created great value for the landside resources. Moreover, since the adjacent land resources are so valuable, ports often seek to expand those resources by using powers of eminent domain, condemning private land for public use and recycling it into port-related uses, outright purchase from willing landowners, or more exotic techniques such as creating new land by discharging dredged material from channel deepening projects behind dikes built in shallow water areas that sequester the material permitting it to de-water and consolidate. The Port of Los Angeles has created hundreds of new acres of land over the past 20 years through this technique.

Nevertheless, ports generally do not seek to operate their own facilities on these valuable lands. Instead, they lease the property to carriers, terminal operators of all kinds for the reasons noted above. Ports then outsource the operation of these facilities to private firms for a period of years, often 10–15. The tenant on these terminals complies with port regulations and pays ground rent on the terminal as well as the tariff for goods moved through the terminal. However, tenant-operated terminals in landlord ports are *leased* rather than *owned*, with ownership continuing to reside in the hands of the port.

This arrangement facilitates all manner of public oversight of private operations. For example, in creating the terms of a lease, the port retains the power to demand specific performance from the tenant over the term of the lease. For example, as air pollution has become an issue in many large seaports, the port can demand that all vessels using a given leased terminal be equipped to discharge cargo at "cold iron" (engines shut down and all power for operation of ship's equipment derived from shore power). On container terminals, the port may specify queuing arrangements for trucks waiting to load cargo. Hours of operation may be specified in the lease and environmental regulations, likewise, may be specified. Thus, while the port retains ownership of the terminal, the terms of the lease can dictate the manner in which the private operator of that terminal conducts his business. The terminal has become a public–private partnership, common in the US and encompassing a wide diversity of contractual arrangements between ports and their tenants. In this manner, landlord seaports become property managers while simultaneously acting as intermediaries between the larger public sector of environmental, commercial, transportation and law enforcement regulators and the port's tenants. It is clearly a complex relationship.

5.5. Development Strategies and Planning

Again, there is a shared interest between port tenants and port authorities in planning for new development, while overall port planning is a bedrock function of port authorities, and cannot be delegated by the port whether a consultant acts as the port's agent or the port staff does the planning. Planning by a tenant in most cases must be approved by the seaport for consistency with its overall development plans. It must also be approved by a suite of local, state and federal environmental regulatory agencies, not least of which is often the state's coastal management agency that implements its own state coastal planning process subject to the authority granted by the federal Coastal Zone Management Act of 1972 (2004). Where ports are publicly owned, and precisely because the public owns them, port managers have a political and ethical responsibility to engage the public in consultation over port modifications and improvements, above and beyond the strict dictates of regulatory requirements.

5.6. Maintenance of Port Access

Seaward access to a port is clearly a function of the port as a public asset. The port exists by virtue of the investment (public in the United States) that has created the seaway or channel that potential carriers can use to access landside port facilities. Without that investment, there would be no seaport. As we have seen, there are private ports in the country but most of them benefit from the investment made by the public sector in dredging and maintaining ship channels. However, there can be a private component to this as well because the landside approaches to port facilities are not always under the exclusive authority or control of the port agency. In some cases, landside access is created or enhanced as a condition of a terminal lease or the lease itself is discounted because access is limited. There are many small non-maritime seaports along the Gulf Coast where the terminal operator, rather than a government agency, has built infrastructure to make a small seaport on land alongside the Gulf Intracoastal Waterway. Absent the capital of the terminal operator, small local governments might not be able to afford such a project. Where seaports are large and the public benefit of access to the port is more visible to the public and politicians, landside infrastructure tends to be provided more readily by the public sector.

5.7. Cargo and Passenger Handling

Commonly, services such as cargo movement and passenger management are outsourced even in operating seaports in the US. Because of the diversity between large and small ports and between operating and landlord, there are many possible permutations where these services are provided, so I can speak here in only the most general of terms.

The temporal nature of the work and uneven demand for their services dictates that most often the stevedore and longshore labor will be provided by a contractor either to the port agency or to the terminal operator who is a lessee of the port. For the operating port, contractor services relieve the agency of the demands of providing public agency personnel benefits for employees for whose services there may be limited demand. For lessees who are terminal operators, the same principle applies even though they may operate in an arena where the requirements for personnel benefits are perhaps less demanding. Nevertheless, it is uncommon to find stevedore and longshore services provided by port employees.

5.8. Pilotage and Towing

Harbor and river pilots are generally private contractors in the United States (American Pilots' Association, n.d.). They are quite often organized into pilot associations for convenience of administration and deployment in a port or along a river but they commonly are independent contractors. Nevertheless, the individual states regulate the industry, set fees and license the associations and the individuals to ensure that there is adequate protection for vessels entering and leaving state ports and that rates are sufficient to provide capable pilot service. The purpose behind this regulation is to provide adequate pilotage much as a public utility serves the greater public interest. In a rather odd confluence of regulation, the federal government licenses pilots through the US Coast Guard to ensure professional competence, in contrast to the essentially fiscal and business regulation imposed by states. In a very few cases like the Port of Los Angeles, the harbor pilots are port employees, but that is the rare exception to the rule.

Private contractors also provide barge, tug and towing services within ports for essentially the same reasons of intermittent demand as with stevedoring services. In a riverine environment, barge and towing service is provided in the same manner as other marine carriers operate with the exception that the service is domestic and regulated exclusively by US law.

5.9. Facilities Security, Maintenance and Repair

There is shared responsibility for security as well as maintenance and repair regardless of whether the US port is an operating or landlord port. The overall responsibility for security in US ports, with the passage of the *Maritime Transportation Security Act of 2002* (2002),[2] rests with a range of actors including the port itself, terminal operators, vessel operators and contractors. In that complex arrangement, the role of security is at times a public and other times a private responsibility. At the level of the port, the agency is responsible for overall safety in cooperation with various federal agencies, most notably the US Coast Guard. Each seaport in the US is now responsible for developing a safety protocol for its facilities in accordance with the MTSA.

At the terminal level in a landlord port, the MTSA merely memorializes some of the best practices and procedures that terminal operators ideally have been following and which insurance companies have encouraged for many years. Since the MTSA implements in the United States the provisions of the International Ship and Port Facility Security Code (International Maritime Organisation, n.d.), its provisions are familiar to vessel operators and terminal operators who may be based outside the US.

Similarly, the ports and their tenants share responsibility for maintenance and repair, depending upon which owns or is charged in a lease with primary authority for the infrastructure in question. That joint charge may be further shared within a leased container or bulk terminal when the port retains ownership of large equipment such as container cranes or bulk loaders. In those circumstances, the lease document will specify the responsible party for maintenance and repair.

6. CONCLUSION

In a nation with more than 350 commercial seaports and an economic system and culture that values entrepreneurship and individualism, one could be forgiven for assuming that a good portion of the seaport resources of that nation would be in private hands. Remarkably, in the United States we see that governments – of various types – own virtually all the nation's major general cargo seaports. The form of governance of the nation's seaports, thus, has its foundation in the template established for the nation itself as it evolved through the 18th and 19th centuries, namely, respect for local or regional autonomy but also a skepticism of outright private ownership of such economically critical infrastructure. States have autonomy in this nation and they have either built seaports of their own accord or empowered local governments to create the maritime infrastructure. In most cases, we have also observed that a local government or a special district of government is the port manager with boards of directors drawn from the region or community.

In terms of operation, despite the governance regimes that we observe, most of the nation's seaports share a good portion of their management with the private sector. At once the ownership regime appears biased toward government and yet, in practice, vast portions of many of the nation's largest seaports are leased to private firms to operate and manage. Those firms operate all manner of marine terminals – container, bulk, liquid bulk, auto, breakbulk and neo-bulk – and do so on land leased from the ports. When firms lease terminals, they also contract with other private firms to provide pilotage, stevedoring, marketing, towing and a host of other services essential to efficient port operation.

Two basic themes drive this dichotomous system. The public seeks to retain some level of control over major public infrastructure since memories of monopolistic ownership and abuse of these facilities, essential to the nation's foreign trade, are still relatively fresh in the public mind. The robber barons of the 19th century demonstrated that private ownership of what were, and are, fundamentally public utilities could not be countenanced by the American public. Thus, the reformist approach to sequestering seaports in public ownership remains a powerful theme. But there is a practical side to public ownership as well. The ability of public agencies to develop prodigious capital resources sufficient to build the modern seaport trumps that of virtually any private firm to do the same. In such an environment, wise management of publicly critical infrastructure dictates that they will remain in public hands. Thus, we find a mix of public and private interests in American seaports and, for this country at least, the system seems to function well.

Disclaimer: This research was developed with support from the National Sea Grant College Program, National Oceanic and Atmospheric Administration (NOAA), US Department of Commerce, under grant number NA 16RG2256. The views expressed are those of the author alone and do not necessarily reflect those of the US Government, NOAA or any of its sub-agencies.

NOTES

1. For additional information, see: Institute for Water Resources, US Army Corps of Engineers (2003).
2. See http://www.uscg.mil/hq/g-cp/comrel/factfile/Factcards/MTSA2002.htm.

REFERENCES

Allen, R. (1996). *A history of the USACE civil works program*. Remarks presented by Ron Allen, Assistant Chief Counsel for Legislation and General Law at the Headquarters, U.S. Army Corps of Engineers. Abstract retrieved on 30th September 2005 from http://www.hq.usace.army.mil/cecc/pubs/cwhist.htm

American Association of Port Authorities. (2000). *Port activities, functions, types: Facts*. Alexandria, VA: American Association of Port Authorities. [Also http://www.aapa-ports.org/industryinfo/port_activities.htm].

American Association of Port Authorities. (2005). *U.S. port ranking by cargo volume 2003, short tons*. Alexandria, VA: American Association of Port Authorities (AAPA Advisory, February 21).

American Pilots' Association. (n.d.). *Pilotage in the United States*. Retrieved on 4th August 2005 from http://www.americanpilots.org/pilotage.htm

Baltazar, R., & Brooks, M. R. (2001). *The governance of port devolution: A tale of two countries*. World Conference on Transport Research, Seoul, July.

Boschken, H. L. (1988). *Strategic change and organizational design in Pacific Rim ports in transition.* Tuscaloosa: University of Alabama Press.

Brinson, J. R. (1995). *Public port authorities and private sector service providers.* Research paper presented to the AAPA Professional Port Manager Program. Alexandria, VA: American Association of Port Authorities.

California State Lands Commission. (n.d.). *The public trust doctrine.* Retrieved on 5th August 2005 from http://www.slc.ca.gov/Policy%20Statements/Public_Trust/Public_Trust_Doctrine.doc

Coastal Zone Management Act of 1972, 16 U.S.C. 33, §1451 et seq (2004).

Constitution of the United States. (1987). Washington, DC: Library of Congress.

Dunfee, G. E. (1977–1978). *Territorial status of deepwater ports.* San Diego: San Diego Law Review (15 San Diego L. Rev. 603).

Eaton, D. W. (1997). *Transformation of the maquiladora industry: The driving force behind the creation of a NAFTA regional economy.* Tucson: Arizona Journal of International and Comparative Law (14 Ariz. J. Int'l. & Comp. L. 747).

Eichenberg, T., Fletcher, K., & Kehoe, K. (2005). The legal context of submerged lands leasing & ownership. In: M. W. Beck, K. M. Fletcher & L. Z. Hale (Eds), *Towards conservation of submerged lands: The law and policy of conservation leasing and ownership* (pp. 4–16). Narragansett, RI: Rhode Island Sea Grant.

Erie, S. P. (2004). *Globalizing L.A.: Trade, infrastructure, and regional development.* Stanford, CA: Stanford University Press.

Fawcett, J., & Marcus, H. (1991). Are port growth and coastal management compatible? *Coastal Management, 19*(3), 275–295.

Fawcett, J. A. (2004). Short-sea shipping: Reducing vessel traffic impacts to the bay and delta. Paper presented to the CalFed Bay-Delta Science Conference, Sacramento, CA, September.

Hamilton, A., Madison, J., & Jay, J. (n.d.). *The federalist.* New York: Modern Library College Editions.

Hershman, M. (Ed.) (1988). *Urban ports and harbor management: Responding to change along U.S. waterfronts.* New York: Taylor and Francis.

Institute for Water Resources, US Army Corps of Engineers. (2003). *Annual report to Congress: Status of the Harbor Maintenance Trust Fund, fiscal years 2000, 2001 and 2002,* http://www.iwr.usace.army.mil/iwr/pdf/hmtfreport00-02.pdf

International Maritime Organisation. (n.d.). *International Maritime Organisation: Marine environment.* Retrieved on 1st August 2005 from http://www.imo.org/home.asp

Maritime Transportation Security Act of 2002. (2002). Pub. L. 107–295, November 25, 2002, 116 Stat. 2064.

Newman, D., & Walder, J. H. (2003). Federal ports policy. *Maritime Policy and Management, 30*(2), 151–163.

Olson, D. J. (1992). Governance of U.S. public ports: A preliminary survey of key issues. Paper prepared for delivery at the Marine Board Port Governance Roundtable, 10 November 1992. Washington, DC: National Research Council (Marine Board).

Philadelphia Regional Port Authority. (n.d.). *History.* Retrieved on 5th August 2005 from http://www.philaport.com/history.htm

Port Authority of New York and New Jersey. (n.d.). *Port Authority of New York and New Jersey: Seaport.* Retrieved on 2nd August 2005 from http://www.panynj.com

Queenan, C. P. (1983). *The port of Los Angeles: From wilderness to world port.* Los Angeles: Los Angeles Harbor Department.

Rivers and Harbors Act of 1899, §10, 33 U.S.C. § 403 (2003).

Rodrigue, J.-P. (2004). Appropriate models of port governance: Lessons from the port authority of New York and New Jersey. In: D. Pinder & B. Slack (Eds), *Shipping and ports in the 21st century.* London: Routledge.

Sherman, R. B. (1995). *Privatization and its implications for U.S. public seaport agencies.* Washington, DC: American Association of Port Authorities.

State of Washington. (2005). Revised Code of Washington. *Port districts authorized – Purposes – Powers – Public hearing.* [RCW 53.04.010]. Olympia, WA: State of Washington (revised as of January 20).

Submerged Lands Act of 1953, 43 U.S.C. §1301 et seq., as amended.

Sunoco, Inc. (n.d.). *About Sunoco: Sunoco history.* Retrieved on 1st August 2005 from http://www.sunocoinc.com

Washington Public Port Districts. (2005). Retrieved on 31st October 2005 from http://www.washingtonports.org/port_information/histories/historyofports.htm

CHAPTER 11

PORT DEVOLUTION AND GOVERNANCE IN CANADA

Mary R. Brooks

ABSTRACT

The devolution of port authorities in Canada has not been without debate over the past 70 years. This paper provides a brief introduction to the role of ports in Canada and then examines the history of port policy and devolution, concluding that past policies were considered to have failed due to their inability to respond to changing circumstances. The paper then introduces the current port policy, and discusses its evaluation by two Review Panels. It draws conclusions about the three-pronged approach to port governance in Canada and the difficulties faced by ports under the existing policy, setting the stage for later chapters in the book.

1. INTRODUCTION

This paper is intended to provide an appreciation of port policy in Canada, its history and its progression from a centralized model of port governance to a decentralized and commercialized one. To understand port policy in Canada, it is important to begin with an understanding of the role ports play in the country's trade-based economy.

Devolution, Port Governance and Port Performance
Research in Transportation Economics, Volume 17, 237–257
Copyright © 2007 by Elsevier Ltd.
All rights of reproduction in any form reserved
ISSN: 0739-8859/doi:10.1016/S0739-8859(06)17011-0

Canada is a small country in terms of population but large in terms of land mass and resource wealth. A politically stable member of the G-8, its GDP per capita ranks sixth in the world according to the *OECD Factbook 2005* (OECD, 2005), and the country is highly export-dependent. Since 1998, between 85 and 86 percent of Canadian exports by value have been destined for the United States, and this has been a stable relationship (Transport Canada, 2005, p. 5). Trade with the US moves predominantly by truck, and marine trade with the US accounts for a mere three percent of the value of Canada's US trade. Canada's trade with the rest of the world is marine- and air-dependent with the marine mode carrying more than two-thirds of Canada's trade by value to its non-US trading partners. Like Australia, Canada is a significant supplier of raw materials to the world's manufacturing industries. For these goods, ports provide an essential economic service.

The top five ports in Canada by tonnage account for 41.3 percent of Canada's marine traffic, and the top 10 account for 61.5 percent. Domestic traffic constitutes only 31 percent of total marine traffic (Table 11.1). Only three of Canada's top 10 ports (Vancouver, Montreal and Halifax) have substantial container terminal operations. A map illustrating the location of Canada's top 10 ports is found in Fig. 11.1.

In Canada, the regulation of shipping is federally controlled, and ports fall under federal regulation. The provision of channels and their approaches, their operation and maintenance are considered the purview of the national government. Dredging for a new berth, however, is not considered maintenance and therefore is not a public good. Although terminals may be provided by the private sector and built on either public or private land, they are subject to national regulation. There are some rights on the east coast that pre-date Canada's independence from Great Britain but these are small in number.[1] Gratwick and Elliott (1992) provide a detailed listing of the statutes and laws of relevance to port governance and port operations in Canada and so these will not be discussed in this paper.

This paper begins by providing a history of port policy and devolution in Canada. It builds a picture of how port policy evolved before introducing the current port policy, first proposed by the *National Marine Policy* and ultimately implemented in a revised form via the *Canada Marine Act, 1998.*[2] The paper discusses the evaluation of the policy by two Review Panels and identifies their concerns as well as those of others at the time. The paper draws conclusions about the three-pronged approach to the governance of ports in Canada and the difficulties faced by ports under the existing policy.

Table 11.1. Top 10 Ports in Canada in 2003 and Percent of Domestic Traffic.

Port	Cargo Tonnage (000 tonnes)	Percent Domestic	Category (1)
Vancouver	67,948	2.8	Canada Port Authority
Come-by-Chance	43,694	44.3	Local/regional port still to be devolved by Transport Canada
Saint John	25,880	12.8	Canada Port Authority
Port Hawkesbury	22,927	3.2	Local/regional port (2)
Sept-Iles	22,682	15.8	Canada Port Authority
Quebec	20,349	20.0	Canada Port Authority
Montreal	20,291	21.9	Canada Port Authority
Port-Cartier	17,439	24.2	Canada Port Authority
Newfoundland Offshore (3)	17,129	98.7	Private
Halifax	14,214	21.3	Canada Port Authority
All other ports	170,459	43.9	
Total traffic	443,012	30.8	

Notes:
(1) The type of port is that indicated in the transfer inventory at the Port Programs and Divestiture web site of Transport Canada as of 30 June 2005.
(2) This port has a public wharf and a number of private terminals, some of which service offshore oil and gas and oil storage.
(3) This is defined by Statistics Canada as shipments from the Terra Nova and Hibernia offshore oil and gas platforms, for the most part direct to foreign markets.
Source: Calculated from data provided by Statistics Canada (2005, Table 13).

2. THE HISTORY OF PORT LEGISLATION AND CONTROL IN CANADA

In 1867, the *British North America Act* (now the *Constitution Act, 1867*) placed matters relating to shipping and navigation solely under federal jurisdiction. The inefficiencies of administrative control became evident during the Depression years and, following a survey of the national ports (Gibb, 1932), the Federal Government implemented Canada's first ports policy in 1936 with the creation of the National Harbours Board (NHB), a Crown Corporation under the *National Harbours Board Act*.[3] The initial five ports, plus two in the Province of Quebec, had their harbour commissions disbanded and control replaced by the Ottawa-based NHB; these seven were

Fig. 11.1. Top 10 ports in Canada with Seattle and New York. *Source:* Created for the author.

later joined by another eight prior to the 1980s (Gratwick & Elliott, 1992).[4] Based on centrally determined command and control principles, the system provided a standard set of port charges to be applied across the country; there was no local marketing expertise (Goss, 1983). By the early 1980s, the 15 NHB ports handled approximately half of Canada's seaborne trade.

While the reform of the 1930s centralized control and management of Canada's essential ports, nine ports remained under their local harbour commissions.[5] *The Harbour Commissions Act of 1964* standardized the incorporation process and their activities with each commission having a Board of federal and municipal appointees and significant local autonomy. Gratwick and Elliott (1992) conclude that under the 1964 legislation, harbour commissions had greater discretion in management and an easier approval process than the NHB ports, even though they had to have their major borrowing and large capital projects approved by the Minister of Transport.

In addition to the NHB ports and the harbour commissions, more than 500 smaller ports (and government wharves) were directly administered by Transport Canada, with responsibility for major repairs and investment in the hands of the Minister of Public Works. In 1973, more than 2,000 public harbours and government wharves were transferred to the Department of the Environment.

While NHB ports remained firmly under the central control exercised by the NHB, a structure criticized for its inflexibility to adapt to a changing competitive environment, smaller ports and harbour commissions had greater flexibility and local responsiveness capability. Responsibility for governance of ports in Canada was fragmented.

While moves towards devolution began in the mid-1970s, the government's first reform efforts finally found a footing in the 1980s with the release in May 1981 of the Liberal government's White Paper (Gratwick & Elliott, 1992; McCalla, 1982). The resulting legislation amended the *National Harbours Board Act* and created the Canada Ports Corporation. With the passage of the *Canada Ports Corporation Act*, the way was paved for a parent Board of Directors in Ottawa and each former NHB port to become a Local Port Corporation (LPC). The Minister of Transport appointed Directors of LPC Boards. As is common in publicly traded companies, they in turn appointed the Chief Executive Officer. However, the Boards did not control financing decisions; budgets for the LPCs were sent to the Federal Government in Ottawa for approval and, if investment was required, it was financed through the borrowing power of the Government of Canada. The ports continued to be eligible for financial support from government, and were also able to borrow from private financial institutions as well as from the federal coffers. The financing capacity for major capital works under this model of governance were dependent in some measure on the relationship between individual Board Members and the Minister as much as they were on the persuasiveness of the business case for investment. The control of the capital investment was in line with the vision of the government of the day and not with that of the port's managers or its LPC Board.

The implementation of the *Canada Ports Corporation Act* failed to fully overhaul the port administrative structure. At the time, McCalla (1982, p. 291) noted that the changes "will bring all ports under a national ports policy, which is something that Canada has lacked, but creating a policy without a strategy of how it is to be implemented does not guarantee the policy's success."

While this early devolution effort moved a small step away from the government-dominated command and control structure ports were under previously, the model did not make significant progress along the devolution continuum.

[B]ureaucratic rules, regulations and financial strictures of the 1984 Financial Administration Act prevented major ports from operating as true commercial entities. ... [T]he legislation did not create commercialized and market-responsive institutions required to

meet the turbulent pressures of continental and global economic competition of the
1990s. (Ircha, 1997, p. 125)

With the election of a Conservative government in Canada in 1984, Canada's
policy environment headed in a direction closer to that of the Reagan government in the US. There was a philosophical redirection towards greater
devolution with the privatization of Air Canada and the beginnings of a plan
to transfer airports to local control. By the time the Conservative party was
defeated in the federal election of 1993, four of Canada's largest airports had
been "commercialized" by leasing them to locally controlled not-for-profit
entities. The devolution programme initiated by the Conservatives had
momentum, continued under the supervision of the Chrétien Liberal government, and was reinforced by the privatization of two large transport
entities: Canadian National (one of Canada's two Class I railways) and
NavCanada (the air navigation system).

The Chrétien Liberal government set its own agenda for continuing
transport devolution activities with two speeches made in the spring of 1994
by the Honourable Doug Young, then Minister of Transport:

> What this nation must have is an integrated and affordable national transportation
> system. One that emphasizes safety and reliability. One that is efficient. And one that
> builds strong, viable companies in all modes. (Transport Canada, 1994a)

> Transport Canada will not abandon its responsibility to ensure safe and secure trans-
> portation standards, rules and regulations. (Transport Canada, 1994b)

This was followed by the release of the *National Marine Policy 1995*
(Transport Canada, 1995). In it, the Federal Government set the following
objectives for port reform:

- ensure affordable, effective and safe marine transportation services;
- encourage fair competition based on transparent rules applied consistently across the marine transport system;
- shift the financial burden for marine transportation from the Canadian
  taxpayer to the user;
- reduce infrastructure and service levels where appropriate, based on user
  needs; and
- continue the Government of Canada's commitment to safe transportation, a clean environment and service to designated remote communities.
  The government will also maintain its commitment to meeting all constitutional obligations (Transport Canada, 1995, p. 3).

The strategy proposed was a clear one: to build a port system that would
be financially self-sustaining, autonomous and meet these objectives. The

driving force behind the continuation of the devolution programme was a recognition that Canada's fiscal situation was in deficit and unsustainable, and that the government's contribution to the financing of transportation in Canada needed radical change. Port devolution was only part of a larger programme of reform.

As the legislation implementing reform proceeded through Parliament, there was a lack of political will to go the distance gone with airport devolution. After a Cabinet shuffle, the new Minister of Transport was not as enamoured as his predecessor with the creation of fully independent port authorities (similar to Canadian airport authorities). As a result, while the objectives of the devolution policy did not change, different governance structures and management systems than those used for airports were promulgated. National Port System ports became federal agencies (e.g. not devolved any further than before but now without access to federal funds; borrowing limits were still controlled by the Minister). The Liberal government did not fully adopt the not-for-profit corporation model of airports when it came to ports. Here the politics of devolution were more intense and, although airports were generally larger revenue generators, port governance in Canada proved more difficult to change.

3. GOVERNANCE STRUCTURES PROPOSED IN 1995

The *National Marine Policy* (NMP) proposed three categories of ports in Canada, as noted in Table 11.2. Each had different ownership models, organizational structures and management processes, with different reporting mechanisms and accountability to the public. As is evident from Table 11.1, size and volume of traffic are not perfectly correlated with the model chosen by government; the largest are not necessarily categorized as Canada Port Authorities.

The categories proposed by the NMP were eventually implemented; as of the end of 2004, only one operating harbour commission remained in Canada – the Port of Oshawa (Transport Canada, 2005, p. 67). Most planned devolution has taken place (Table 11.3). However, the divestiture of local/ regional ports did not proceed at the pace anticipated and the timeline for completion of divestiture has been extended by Cabinet an additional four years to 31 March 2006 (Transport Canada, 2005, p. 68).

Responsibility for providing port activities also varied by port management model, and is documented in Appendices A–C. Based on the Baltazar and Brooks' (2001) port devolution matrix, the allocation of functions between public, private and mixed governance models illustrates this

Table 11.2.   Port Governance in Canada.

Type	Criteria (1) Under NMP (1995)	Outcome (June 2005)
Canada Port Authorities (managed by Board made up of nominated representatives of user groups and various levels of government)	Be financially self-sufficient Is vital to domestic and international trade Is part of the network of 40 ports accounting for 80 percent of Canada's marine traffic Serves a large market area Has links to major land transport infrastructure 8 ports were required to apply (Fraser River, Halifax, Montreal, Prince Rupert, Quebec City, Saint John, St. John's and Vancouver). Others to apply and be dealt with on a case-by-case basis	19 ports applied for and were granted CPA status
Local/regional ports	Can apply for CPA status (must be financially self-sufficient in this case) or be transferred to provincial governments, municipal authorities, community organizations, or private interests (in this order) Transfers to be completed in 6 years	All but 62 have been devolved
Remote ports	Is a port in an isolated community, reliant on marine transportation and a government wharf	26 remote ports remain under Transport Canada
	To remain with the Federal Government 60 ports listed in the NMP	The others were transferred to provincial or private interests

Note: These criteria were stated in the policy and the policy listed all ports meeting the criteria in each group in the appendices of the policy.
Source: National Marine Policy 1995 (Transport Canada, 1995) for the first two columns; the third column detail is based on ports listed at the Transport Canada Port Programs and Divestiture web site (http://www.tc.gc.ca/programs/ports).

Table 11.3. Implementation of Devolution in Canada.

As of 30 June, 2005, 461 of the 549 Transport Canada facilities earmarked for devolution had been transferred, demolished or had their status as a public harbour terminated	
Deproclaimed (between June 1996 and March 1999)	211
Transferred to provincial governments	40
Transferred to other federal government departments	65
Divested to local interests	120
Demolished or Transport Canada's interest terminated	25
Local/regional ports remaining under Transport Canada purview (still to be devolved)	62
Remote ports	26
Total	549
Canadian Port Authorities	19
Public Ports in Canada prior to the *National Marine Policy*	568

Source: Transport Canada Port Programs and Divestiture web site (http://www.tc.gc.ca/programs/ports).

variance between approaches. It is clear that the Canadian government retains a strong government regulator function in two of the three models (CPAs and Remote), those deemed as essential infrastructure to the national posts system (CPAs) and those where service is deemed a public good (Remote ports). Furthermore, the government retains a strong regulatory role, albeit a smaller one, in those ports it determined to be of local/regional significance. While Canada's commercialization programme did not go as far as the privatization programmes seen in other countries, the commercialized nature of both CPAs and local/regional ports is a model that can deliver many of the "benefits" of privatization. What the appendices do not illustrate is that implementation of the governance model was not uniform in each port class (as not all activities were available in each port) but the mix of activities in these appendices approximates the pattern of responsibility for activities in each port class and confirms Baird's (2000) position that activities in the regulator, landlord and utility (operator) will vary by port.

Canada Port Authorities and the approach taken to their commercialization will be discussed later, as these ports attracted most of the media and academic attention. Very little has been written about port policy and its implementation in the other two groups of public ports.

As a transition arrangement, Harbour Commissions that did not apply to be CPAs were to be allowed to continue as commissions for up to two years and, if they were not CPAs at the end of the period, would be devolved or divested, in essence being treated as local/regional ports.

With local/regional ports, Transport Canada followed a "privatization" programme that transferred land and chattels. The approach was not technically privatization (but more decentralization) as there was and is an implied hierarchy of who may "acquire" the port and its assets. First priority was given to other federal departments, then provincial agencies, next municipal or other public bodies and finally, as a last resort, private entities. The port must operate as a not-for-profit without recourse to the federal government. These transfers were negotiated under the following principles:

- no offer that leaves the Crown financially worse off as a result of divestiture will be accepted;
- the Crown receives best value for port land and other assets;
- a new port owner will not enjoy any windfall profits from the subsequent sale of lands, assets and/or chattels; and
- Transport Canada fully upholds its fiduciary responsibility with respect to First Nations (Transport Canada, 2004).

As part of the divestiture negotiation, Transport Canada recognized that some ports might not meet the standards required of private sector entities in Canada; for example, the government (as operator) cannot be sued by the government (as regulator) for environmental, safety or fiduciary violations. The transfer of ownership to the private sector may, therefore, be accompanied by legal liability. As a result, Transport Canada established the Port Divestiture Fund to "cover a portion of the costs incurred by the new owner or operator to achieve compliance with regulatory or insurance requirements, to fund feasibility studies, or to reduce potential liability" (Transport Canada, 2004).

The fund could also be used to make lump-sum payments to facilitate the takeover of a port or to assist local interests to take over a group of ports and reduce costs by rationalizing the accompanying infrastructure. The requirements for performance monitoring post-transfer are limited to the filing of verification statements (to ensure that new port owners are meeting their obligations under the transfer agreement) and audits initiated by Transport Canada.

For those ports remaining under Transport Canada administration that have yet to be divested, the port policy objectives are:

- contributes to the achievement of Canada's international trade objectives as well as national, regional and local economic and social objectives;
- functions efficiently;
- provides port users with accessible and equitable transportation services; and
- works in coordination with other marine activities and surface and air transportation systems (Transport Canada, 2004).

According to Transport Canada (2005, p. 69), these ports as a group are not financially solvent; revenues for fiscal year 2003–2004 were C$12.4 million while expenses totalled C$21.8 million. However, this should not be seen as an indication of failure of the policy. When reporting the financial state of ports, Transport Canada does not separate Remote ports, which provide services to Canada's remote and northern communities (a public good), from those that have yet to be devolved and for which there may not be an eventual "buyer." For example, revenues from Come-by-Chance, a port that services a major east coast refinery, offset some public obligations elsewhere but the extent of cross-subsidization is not reported.

One measure of success might be the completion of the planned devolution programme in the time that has passed since the announcement of the policy or the passage of the *Canada Marine Act, 1998*. Yet, as of 30 June 2005, 62 of the ports for which "privatization" was planned remain under Transport Canada control in spite of the existence of the Port Divestiture Fund. The process has clearly run into difficulty and the reasons for the government's failure to complete the programme have not been investigated to the author's knowledge. While it is possible that government underestimated the attractiveness of the opportunity to local interests, it is also possible that the bureaucrats responsible for divestiture have had insufficient incentive to complete the process. These two possibilities are highly speculative and are only presented to underline the point that the costs of the programme and its performance need to be investigated and evaluated.

Transport Canada's primary measure of programme success is departmental spending. According to the annual report for 2003–2004 (Transport Canada, 2004), prior to any port divestiture, the department spent C$22 million annually for maintenance and, in 2003–2004, only C$5.4 million was budgeted for this activity. If the purpose of devolution was only to download costs in a time of financial exigency (the mid-to-late 1990s), it can be concluded that the programme has been an unqualified success.

4. PORT MANAGEMENT POLICIES

The management structure proposed in 1995 was modelled on earlier devolution practice and experience. As specified in the *National Marine Policy* (NMP),

> Canada Port Authorities will be federally incorporated as not-for-profit corporations with powers and responsibilities similar to those of corporations established under the Canada Business Corporations Act. Although they won't issue shares, they will be private-sector organizations with a mandate to operate with full commercial discipline. (Transport Canada, 1995, p. 13)

> The nominating groups for each corporation, which will be sanctioned for each port by the Government of Canada prior to passage of the new Act, will then appoint the members of the new board who will assume operational control. (Transport Canada, 1995, p. 18)

This policy was eventually promulgated in Section 4(e) of the *Canada Marine Act, 1998,* which noted that one of the objectives of the legislation is to "provide a high degree of autonomy for local and regional management of components of the system of services and facilities and be responsive to local needs and priorities." Furthermore, Section 4(f) sets, as an objective, management that "encourages, and takes into account, input from users and the community in which a port or harbour is located." The management structure for Canada Port Authorities was rigidly specified in the *Canada Marine Act, 1998.*

A major criticism of the new CPA management model was that the Minister of Transport controlled appointments to the Boards of Directors of the various CPAs (Brooks, Prentice, & Flood, 2000). While Community input was sought, the Minister decided on his choice of candidates and was able to reject candidates put forward by local interests. With this method of appointment, Board decisions could reflect ministerial priorities if the appointees chose loyalty to the Minister over fiduciary responsibility to the entity, as required under the governing legislation. This problem did not apply to the vast majority of ports devolved.

The transition from a central approach to port management to a decentralized model focusing on business-like elements of competition proved to be a challenging one in Canada. In the end, airports had greater flexibility through the Letters Patent process than ports achieved and so, rather than leapfrog the governance model closer to the privatization end of the Devolution Continuum presented in Chapter 1, the government implemented a less commercialized devolution option. Progress in the overall programme of devolution of transport entities towards more fully commercialized models had more than stalled; it had reversed direction towards greater government influence.

5. EVALUATION OF THE NMP IMPLEMENTATION FOR PORTS

While the initial philosophy of port devolution in Canada appeared to follow in the footsteps of airport devolution policy, the path diverged for political reasons (noted previously). In part, this had to do with the governance structures chosen. The author agrees with Ircha (1997, p. 134) who, in his study of Canadian port reform, found that the airport model of Board appointments was not favoured by Canadian port managers. He concluded that "the port managers' perspectives may reflect imprinted institutional values developed from decades of strong federal presence ... in Canada's ports." In other words, it was difficult to change organization culture in ports when the government was not prepared to substantially alter the governance structure as it had done with airports. As already noted, ports remained government agencies, not independent for-profit or not-for-profit entities.

The operation of a commercially oriented entity usually requires the appointment of an *independent* Board of Directors to set the policy for the entity and provide advice and guidance to the management team. In a publicly traded private sector company, new Board Members are generally chosen by existing Board Members as replacements are required or on a scheduled determined by the entity's by-laws. Board Members are usually appointed by the relevant government minister, in cases where the entity is owned and controlled by the government and where there is a management Board. These two reflect the extremes usually seen in Canadian governance. The process of Board member selection in the Port of Halifax provides a useful illustration of the dissatisfaction felt by the community about the Federal Government's reversal of direction away from greater devolution. According to Brooks et al. (2000, pp. 138–139), "the dissatisfaction [about Board appointments in the case of CPAs] stemmed from (1) the perceived lack of accountability to the user groups, and (2) the appointment rather than nomination of user directors." The airport model had incorporated a Board structure that resulted in the majority of Board Members being appointed by the Board (following the private sector approach) rather than by the Minister (who appointed only two). The actual Board Members first chosen in the Port of Halifax under the new model reflected the Minister's reward (patronage) system rather than the skill mix needed and desired by the community. As was noted by Cayo (1998, p. A17), the result was not much of a change: "the decisions will still be made by the same old crowd." How often this was replicated across the country has not been examined.

The outcome of port devolution efforts in the late 1990s incited debate long after the enactment of the *Canada Marine Act, 1998*. The *National Transportation Act, 1987* and its subsequent revising legislation, the *Canada Transportation Act, 1995*, both envisaged a regular review process and made the process a mandatory requirement of the Minister. (Such mandated regular reviews are a clear sign of good governance accountability.) The review process for the *Canada Transportation Act, 1995* began in 2000 and the Review Panel (CTAR) determined that its mandate was very broad and recommended in its report (Public Works, 2001) an early review of the *Canada Marine Act, 1998* among other things (Recommendation 9.15). The majority of the CTAR's port policy recommendations (Table 11.4) focused on governance and reporting issues, providing confirmation that devolution was not perceived to have gone as well as it might have.

Based on the CTAR report, the Minister of Transport did indeed begin an earlier-than-mandated review of the *Canada Marine Act, 1998*. This Review Panel (CMAR) was concerned about the inability of CPAs to finance large-scale capital infrastructure in particular and devoted an entire chapter (Chapter 5) of its report (Public Works and Government Services Canada, 2001) to the issue. Recommendations 5.1–5.4 addressed that concern:

> Recommendation 5.1: The Government of Canada should make investments in infra-structure for CPAs where the amount of capital needed is beyond the ability of the CPA to finance from its cash stream as is currently provided for and where the business case for such investment has been approved by the appropriate government department. (Transport Canada, 2002, p. 26)

Recommendations 5.2 and 5.3 reflected concern about access to financing and borrowing limits, again related to the ability of ports to finance infra-structure investment, while Recommendation 5.4 focused on making access to capital more like that of competing US ports, including the use of tax-exempt bonds. Ports in the US are, in some cases (Seattle being an example), able to raise taxes from local citizens, in other cases able to draw on the financial largesse of their state governments and, in still others, able to issue tax-exempt, junk or municipal bonds. Canadian CPAs have only the ability under the *Canada Marine Act, 1998* and their Letters Patent to raise financing by pledging their revenue stream from operations. They cannot

Table 11.4. Recommendations of the Canada Transportation Act
Review Panel (2001).

Recommendation 9.9
The Panel recommends that the provisions of the Canada Marine Act making the Crown
responsible for liabilities of Canada Port Authorities be removed

Recommendation 9.10
The Panel recommends that borrowing limits in the letters patent of Canada Port Authorities be
removed

Recommendation 9.11
The Panel recommends that the number of directors on Canada Port Authority boards
appointed directly by the Minister of Transport be reduced to two

Recommendation 9.12
The Panel recommends that well defined limits be placed on Canada Port Authorities' use of
for-profit subsidiaries

Recommendation 9.13
The Panel recommends that
• a for-profit Canada Port Authority subsidiary be allowed to provide a service to the port only
  if it is the successful bidder in a fair and open competitive tendering process; and
• if a port in its own right (rather than through a subsidiary) undertakes activities that compete
  with commercial firms, it be required to demonstrate that the decision is in the port's financial
  interest

Recommendation 9.14
The Panel recommends that Canada Port Authorities be required to develop comprehensive
performance measurement systems and to make the resulting information publicly available

Recommendation 9.15
The Panel recommends that a review of the Canada Marine Act be initiated by the beginning of
2002

Source: Public Works and Government Services Canada (2001).

subsidize operations through engaging in consulting or service subsidiaries
or transfer pricing from non-port activities. Their activities and their revenue sources are limited by the Letters Patent. This puts them at a distinct
disadvantage in comparison with ports in the US.

The governance model imposed by the Federal Government via the *Canada Marine Act, 1998* left ports in a situation worse than airports when it
came to financing needed investment. The Government may have been
concerned about the extensive debt being incurred by Canadian airports in
replacing aging infrastructure, but at least airports had access to the financing mechanisms (such as bonds and profit-making service subsidiaries)
denied to ports, which had neither access to the bond market nor the ability

to engage in non-port revenue-generating activities. Furthermore, by making payments to the Federal Government based on a percentage of gross revenue (the airports paid a predetermined rent), the capital reserves of the CPA ports were further "taxed" even if the funds were clearly required to finance investment for substantial growth in demand. While a small percentage of gross revenue may be seen to be better than a fixed charge, the fixed charge has two benefits: (1) it is negotiated as part of the lease negotiation and if the necessity of capital investment is obvious, it may be set at zero; and (2) a fixed charge provides an incentive to grow revenue for the port without additional "tax" on that revenue.

There were several other recommendations in CMAR intended to fine-tune the CPA model from a financing, investment and governance perspective.

Clearly the concerns of both Review Panels were not unfounded. Since the promulgation of Canada's port policy through the *Canada Marine Act, 1998,* several CPAs have sought increased borrowing limits from the Minister of Transport to address inadequacies in infrastructure funding. Given the severe increase in demand from Asia for west coast port facility development, the Port of Vancouver was finally successful in 2004 in increasing its limit to C$510 million from the previous limit of C$225 million (Transport Canada, 2005, p. 69). Port capital investment is generally lumpy and not readily financed based on incremental cash flows; by the time banks are interested in providing debt financing, the port has already passed the time when construction should have begun and capacity is stressed.

In sum, the CMAR Panel questioned the appropriateness of the underlying management structures for implementing the CPA governance model. Recommendations 5.16 and 5.17, for example, focused on the appointment of Board Members (Directors) and made an effort to resolve outstanding issues associated with the perception of ministerial interference. However, it did not question the departmental nature of the model at all, or, because the *Canada Marine Act, 1998* did not legislate ports other than CPAs, the effectiveness of the non-CPA governance models in devolving ports in Canada.

6. CONCLUSIONS

The process of port devolution in Canada was highly political and hotly debated. The most important element of transport devolution policy in Canada was its use of non-share capital, not-for-profit entities. While some entities (particularly Air Canada and Canadian National, both of which competed with private sector businesses) were privatized, those seen as part of a national infrastructure network were devolved using a non-share capital type of structure. While the UK opted for a privatization route to deliver the benefits of productivity and financial responsiveness, Canada decided to invoke community responsiveness through the creation of not-for-profit, stakeholder-focused entities and is one of the few countries (if not the only one) to take this approach.

The key issues for Canadian ports today centre on the ability of CPAs to finance capital investment. The restrictions imposed on CPAs by the initial legislation were clearly seen by all stakeholders to be too severe when compared with financing restrictions in other countries, in particular the US. By making recommendations for new legislation, the CMAR Panel hoped to (a) encourage the government to more closely examine its governance of ports through a five-year mandated review process, and (b) provide ports with the tools they needed to compete in a commercial environment. Notably, the CMAR Panel did not see that further devolution was necessary as it was remarkably silent on this issue. This is particularly surprising as the chattering classes now talk of making airports, which had been devolved to not-for-profit corporations, even more commercial by going the rest of the way along the devolution continuum to complete privatization. This inconsistency in treatment between ports and airports and continuing divergence in policy is a concern to Canadians following devolution implementation.

Neither Review Panel dealt with other port devolution structures. This is particularly important as some of the non-CPA ports account for a significant volume of traffic. While the CPA ports responsible for the largest portion of marine traffic have come under the microscope, non-CPA ports have not received the scrutiny that should take place if the policy proposed more than a decade ago is to be fully and properly assessed.

NOTES

1. Pre-Confederation (1867) water lots exist and have grandfathered rights that must be taken into account in port development planning.
2. The legislation was passed in June 1998 with effect from 1 January 1999.
3. Ports were considered to be in dire straits and merited earlier attention than did Canada's shipping policy. The latter, which was modeled on the British approach, was not examined or evaluated until the 1960s (Letalik & Gold, 1992).
4. The essential five were Halifax, Montreal, Quebec City, Saint John and Vancouver; the two in Quebec were Chicoutimi and Trois Rivières; the additional eight

were Baie des Ha! Ha!, Belledune, Churchill, Port Colborne, Prescott, Prince Rupert, Sept îles and St. John's.

5. The nine were Hamilton, Oshawa, Port Hope, Thunder Bay, Toronto and Windsor on the Great Lakes and Fraser, Nanaimo and Port Alberni on the west coast. Each harbour commission operated under its own private act.

REFERENCES

Baird, A. (2000). Port privatisation: Objectives, extent, process and the U.K. experience. *International Journal of Maritime Economics*, 2(3), 177–194.

Baltazar, R., & Brooks, M. R. (2001). The governance of port devolution: A tale of two countries. *World conference on transport research*, Seoul, Korea, July.

Brooks, M. R. (2004). The governance structure of ports. *Review of Network Economics: Special Issue on the Industrial Organization of Shipping and Ports*, 2(2), 169–184. Available at http://www.rnejournal.com

Brooks, M. R., Prentice, B., & Flood, T. (2000). Governance and commercialization: Delivering the vision. *Proceedings of the Canadian transportation research forum*, June, 1 (pp. 129–143).

Cayo, D. (1998). The politics of a super-port. *The Globe and Mail*, 2 November, p. A17.

Gibb, A. (1932). *National ports survey*. Ottawa: F.A. Auckland, Kings Printer.

Goss, R. (1983). *Policies for Canadian seaports*. Ottawa: Canadian Transport Commission.

Gratwick, J., & Elliott, W. (1992). Canadian ports: Evolving policy and practice. In: D. VanderZwaag (Ed.), *Canadian ocean law and policy* (pp. 237–260). Toronto: Butterworths.

Ircha, M. C. (1997). Reforming Canadian ports. *Maritime Policy and Management*, 24(2), 123–144.

Letalik, N., & Gold, E. (1992). Shipping law in Canada: From imperial beginnings to national policy. In: D. VanderZwaag (Ed.), *Canadian ocean law and policy* (pp. 261–288). Toronto: Butterworths.

McCalla, R. J. (1982). Canadian port administration: Its future structure. *Maritime Policy and Management*, 9(4), 279–293.

OECD. (2005). *OECD factbook 2005: Economic, environmental and social statistics*. Paris: Organisation for Economic Co-operation and Development.

Public Works and Government Services Canada. (2001). *Vision and balance: Report of the Canada Transportation Act Review Panel*. Ottawa: Public Works and Government Services Canada, June.

Statistics Canada. (2005). *Shipping in Canada 2003 (54-205)*. Ottawa: Statistics Canada.

Transport Canada. (1994a). Transport Minister Douglas Young speech: National Transportation Day, June 3.

Transport Canada. (1994b). 2001: A transportation Odyssey. Address by Transport Minister Douglas Young to the Toronto '94 Conference, Toronto, June 13.

Transport Canada. (1995). *National marine policy*, December.

Transport Canada. (2002). *The Canada Marine Act – beyond tomorrow: Report of the review panel to the minister of transport (TP-1407B)*. Ottawa: Transport Canada.

Transport Canada. (2004). *Annual report on port divestiture and operations 2003–2004*. Available at http://www.tc.gc.ca/programs/ports/annualreport0304.htm

Transport Canada. (2005). *Transportation in Canada 2004: Annual report (TP-13198E)*. Ottawa: Transport Canada.

APPENDIX A. ALLOCATION OF FUNCTIONS FOR CANADA PORT AUTHORITIES

Governance	Regulator Functions	Port Functions	
		Landlord	Operator
Public	• Licensing, permitting • Vessel traffic safety • Customs and immigration	• Waterside maintenance (e.g. dredging)	• **Cargo and passenger handling** • Pilotage and towage
Mixed Public/ Private	• Port monitoring • Emergency services • Protection of public interest on behalf of the community	• Marketing of location, development strategies, planning • Maintenance of port access	• Line handling • **Facilities security, maintenance and repair** • **Marketing of operations**
Private	• Determining port policy and environmental policies applicable	• Port security • Land acquisition, disposal	• **Waste disposal** • **Landside and berth capital investment**

Note: The font typeface indicates allocation. **Bold** indicates private sector ownership and provision while *italics* indicates mixed public–private ownership and provision. If the government retains ownership via a government corporation, the function remains regular in style.
Source: Brooks (2004; Table 4). With permission from *Review of Network Economics*.

APPENDIX B. ALLOCATION OF FUNCTIONS FOR LOCAL/REGIONAL PORTS IN CANADA

Governance	Regulator Functions	Port Functions	
		Landlord	Operator
Public	• *Licensing, permitting* • *Vessel traffic safety* • Customs and immigration	• **Waterside maintenance (e.g. dredging)**	• **Cargo and passenger handling** • <u>**Pilotage and towage**</u>
Mixed Public/ Private	• Port monitoring • Emergency services • Protection of public interest on behalf of the community	• **Marketing of location, development strategies, planning** • **Maintenance of port access**	• <u>**Line handling**</u> • <u>**Facilities security,**</u> **maintenance and repair** • **Marketing of operations**
Private	• Determining port policy and *environmental policies* applicable	• **Port security** • **Land acquisition, disposal**	• **Waste disposal** • **Landside and berth capital investment**

Note: The font typeface indicates allocation. **Bold** indicates private sector ownership and provision while *italics* indicate location-dependent mixed ownership and provision. If the government retains ownership via a government corporation, the function remains regular typeface. <u>Underline</u> means may not be provided.

Source: Brooks (2004; Table 5). With permission from *Review of Network Economics*.

APPENDIX C. ALLOCATION OF FUNCTIONS FOR REMOTE PORTS IN CANADA

Governance	Regulator Functions	Port Functions	
		Landlord	Operator
Public	• Licensing, permitting • Vessel traffic safety • Customs and immigration	• Waterside maintenance (e.g. dredging)	• *Cargo* <u>and</u> <u>passenger</u> *handling* • <u>Pilotage</u> *and* **towage**
Mixed Public/ Private	• Port monitoring • Emergency services • Protection of public interest on behalf of the community	• Marketing of location, development strategies, planning • Maintenance of port access	• <u>Line handling</u> • **Facilities** **security**, maintenance, and repair • Marketing of operations
Private	• Determining port policy and environmental policies applicable	• Port security • Land acquisition, disposal	• **<u>Waste disposal</u>** • *Landside and berth capital investment*

Note: The font typeface indicates allocation. **Bold** indicates private sector ownership and provision while *italics* indicate location-dependent mixed ownership and provision. If the government retains ownership via a government corporation, the function remains regular typeface. <u>Underline</u> means may not be provided.

Source: Brooks (2004; Table 6). With permission from *Review of Network Economics.*

CHAPTER 12

PORT REFORM: THE AUSTRALIAN EXPERIENCE

Sophia Everett and Ross Robinson

ABSTRACT

A decade after port reform was initiated in Australia it is becoming clear that objectives of corporatisation are not being realised. For example, political intervention persists, which thwarts commercial benefits being realised. The chapter suggests that this is not a problem of political interference per se. Rather it is a product of the model of corporatisation set in place and is imbedded in legislation.

1. INTRODUCTION

Australia has undergone a decade of port reform, which has led to market restructure, ownership change and the roles and responsibilities of the former port authorities being much reduced. These changes, which were translated into privatisation and corporatisation strategies, emerged from a number of directions and for a number of reasons, including the perceived urgency to achieve significantly improved efficiency in the nation's ports, and the need for governments, particularly state governments that were the owners of ports, to rationalise structures and budgets. Furthermore, it was consistent with the broader strategy of microeconomic reform adopted by state and federal governments alike.

Devolution, Port Governance and Port Performance
Research in Transportation Economics, Volume 17, 259–284
Copyright © 2007 by Elsevier Ltd.
All rights of reproduction in any form reserved
ISSN: 0739-8859/doi:10.1016/S0739-8859(06)17012-2

Government business enterprises (GBEs) came under particular scrutiny in Commissions of Audit in a number of states and became the focus of continuous monitoring (Report of Victorian Commission of Audit, 1993). Ports, and particularly port authorities, were caught in the crossfire. First, as the nation's cargo handling agencies they were expected to deliver operational efficiencies to their key players or stakeholders and particularly to shippers/cargo owners and shipowners. Second, as statutory authorities and GBEs they were expected to conform to the newly emerging Treasury, Ministerial and Cabinet demands for financial viability.

These demands for change were a result of governments at both federal and state levels being under increasing pressure from the 1980s onwards to perform. State governments, in particular, came under severe economic pressure and drastic reduction in government spending was called for. An ethos subsequently emerged, accompanied by a strong push for deregulation and by demands made on governments to either make these operations profitable or withdraw from them altogether. Within this market-oriented environment, ports that had previously been seen as providing a public utility function as statutory authorities now were expected to perform as efficient and competitive businesses.

2. BACKGROUND TO REFORM

Understanding port reform cannot be divorced from the prevailing economic/political/ideological environment that framed these policies. This background is briefly discussed below.

Australia from the middle of the 1970s saw some significant philosophical and policy changes that would change the face of transport. This would have a dramatic impact on ownership of transport infrastructure following the deregulation of government monopolies. The subsequent opening up of the market to competition and the phenomenon that followed enabled modal integration, indeed ownership of supply chains by a single owner or alliance (Robinson, 2003). This policy, generally referred to as microeconomic reform, shifted government focus from the macroeconomy away from the welfare state of the Keynesian paradigm to productivity and efficiency improvements at the firm or industry level and trending towards free market philosophies.

The paradigmatic changes in economic policy that formulated these changes did not develop initially in Australia. Rather, they were adopted by the Australian governments, federal and state alike, following overseas developments particularly in the United States of America (USA), the United Kingdom (UK) and in New Zealand (NZ).

While initially the focus of reform was on the labour market, microeconomic reform, particularly following the formal enactment of National Competition Policy (NCP) in 1995, led to an intensive period of corporate restructure and ownership change as well as deregulation. Initially, reform focused on the private sector – the iron and steel, mining and automobile industries, for example – but following the enactment of NCP, public sector operations also came under scrutiny. Consequently, the port and rail sectors underwent considerable restructure.

Reform in Australia was not merely following overseas trends, however. There existed sound economic reasons why a paradigm change was required. The benefits of the Keynesian paradigm, while yielding substantial achievements in the initial period such as the virtual disappearance of unemployment for a number of decades (Quiggin, 1996) did not persist. Problems began to emerge from the 1960s onwards associated with growing inflation followed by the oil shocks and subsequent economic crises of the 1970s. This period saw the emergence of a spate of different problems – the incidence of stagflation, for example – that is, the incidence of high unemployment levels occurring simultaneously with rising inflation. These created anomalies for the Keynesian economists who found them not readily explicable in terms of classical economic theory.

The recession of the late 1970s led to a switch in focus from macroeconomic issues to the microeconomy. Particular problems emerged in Australia, for the recession meant not only a significant decline in economic growth but also an increase in government spending on social policies created by the recession itself – unemployment relief, welfare assistance and assistance to industry, for example. These escalated to the extent that by the mid-1980s the government spending was exceeding government revenues. The impact was such that by 1985/1986, Australia's foreign debt represented 30 percent of the GDP compared with some 7 percent at the beginning of the 1980s (Beresford, 2000).

In response to this dilemma, a new orthodoxy emerged based on a revival of classical ideals of liberalism and theories of neo-classical economics. Initially led by economists such as Von Hayek and Friedman, the new orthodoxy was rapidly adopted by countries such as the UK under Thatcher and the USA under Reagan. Australia followed this trend with the New Right, as it came to be known in Australia, and proponents argued that Australia was over-governed, over-taxed and over-regulated and that deregulation and drastic reductions in government spending were called for. The argument was based on the premise that the private sector could carry out most functions more efficiently than government. Commercial operations should, therefore, be undertaken by the private sector, wherever possible.

The push for reform was formalised in 1995 with the introduction of NCP. As noted above, reform had initially been focused on the private sector, but following the release of the Hilmer Report (1993) the public sector also came under scrutiny. Particular attention was devoted to public sector monopolies providing commercial services. In the transport sector this had particular relevance for ports and railways.

To introduce competition into these sectors, Hilmer recommended their restructure. Some problems required particular attention, for although privatisation was an option, as in the UK, he warned that where the incumbent firm had developed into an integrated monopoly during its protection from competition, structural reforms were required to dismantle excessive market power and increase the contestability in the market. He cautioned against issues relating to privatisation per se for while this avenue could provide efficiency benefits, there was a risk that, without appropriate restructuring, the anti-competitive structure of the former public sector monopoly could be transformed into a private sector monopoly.

3. REFORM OF AUSTRALIAN PORTS

Within the context of the New Right ideology and the recommendations in the Hilmer Report, public sector reform was initiated. Many areas of public sector transport were chronic loss-makers requiring ongoing subsidy. Sayers (1989) indicates that the NSW railway, for example, had deteriorated to the point where the cost of subsidy was up to 9 percent of the state's recurrent budget and the loss was estimated to be in the vicinity of A$1 billion annually. In the port sector, reform was called for with users demanding productivity and efficiency improvements and equitable and transparent accounting systems; coal exporters, for example, argued that they cross-subsidised other loss-making services in the port (Everett, 1988).

State governments, under severe economic pressure and having jurisdiction over ports, came under pressure from both users and shippers to either make the ports profitable and efficient or withdraw from those commercial activities. In this new market-oriented context, ports that had previously provided a public utility function would be restructured into efficient, profitable and competitive businesses. The mechanism for restructure was privatisation and corporatisation.

All state governments in Australia, having jurisdiction over ports, have reformed the port sector as indicated in Table 12.1. Prior to reform, ports had been established as statutory authorities with responsibility for basic port functions – pilotage, pollution control and marine safety issues. In some instances, commercial operations such as coal- and grain-loading facilities were the responsibility of the port authorities, and until the 1980s, ownership and management of container terminals were in some instances also undertaken by the port authorities.

As Table 12.1 suggests, some ports were privatised although the dominant model adopted in Australia was that of corporatisation. Each state government, however, enacted legislation implementing its own corporatisation

Table 12.1. Reform Models in Australian States.

State	Corporatisation	Privatisation
Victoria		
Melbourne	×	
Geelong		×
Portland		×
Hastings	×	
NSW		
Sydney	×	
Newcastle	×	
Port Kembla	×	
Queensland		
Brisbane	×	
Gladstone	×	
Bundaberg	×	
Rockhampton	×	
Mackay	×	
Townsville	×	
Cairns	×	
South Australia		
All ports		×
Western Australia		
All ports	×	
Northern Territory		
Darwin	×	
Tasmania		
All ports	×	

model. In addition, while some ports were privatised, corporatisation also entailed some elements of privatisation as initially corporatised ports became landlords and commercial operations, with few exceptions, were privatised. The port of Gladstone was one of these exceptions in which the corporatised port under Queensland government ownership was, and remains, responsible for coal-loading operations.

Australian state governments, as noted above, have embraced different but closely related strategies – privatisation and corporatisation. Some confusion persists with the definition of these models and, as the terms are imprecise, note the fundamental differences between these rather broadly defined strategies.

- Privatisation is the transfer of public assets to the private sector. This can be accomplished by outright sale, lease or contracting out. It can refer to the sale of the entire port, such as the port of Geelong, or part of a port operation, a terminal for example, a tug or pilotage operation.
- Corporatisation constitutes a corporate restructure and relates exclusively to the port authority. A public presence in the port is maintained, albeit much reduced. A government-owned corporation is established by legislation replacing the statutory authority, essentially to undertake landlord functions. Private sector business principles are adopted and, in most instances, commercial operations in the port are transferred to the private sector. While numerous models of corporatisation have been put in place, fundamentally they are either Government-Owned Companies (GOCs) or statutory state-owned corporations (SSOCs). The differences between these two models and their implications are discussed in later sections.

As noted above, state governments adopted different strategies and were driven by different objectives. Victoria, for instance, initially adopted privatisation strategies. This state government was in dire financial straits at the time and the sale of high cost, loss-making ports relieved government of an ongoing burden. The sale also reduced the public debt. States adopting the privatisation model were generally driven by two fundamental issues – first, high cost and poor productivity and, second, reduction of the public debt.

The following section will discuss briefly the different models set in place in each state.

3.1. Victoria

The state government of Victoria adopted a number of different models. The ports of Geelong and Portland were privatised. It was government's initial intention to also privatise the port of Melbourne. In response to intense opposition from port users, this strategy was abandoned and the port was subsequently corporatised. The port of Hastings, the second commercial Melbourne port, was also offered to the private sector. The bid from the owners of the port of Geelong, Toll Ports, was rejected by the Australian Competition and Consumer Commission (ACCC) on the basis that a single owner of both the Geelong and Hastings ports would inhibit competition. The model ultimately adopted by the port of Hastings is somewhat of a hybrid, with the port itself being corporatised but the management contracted out.

3.2. New South Wales

All three commercial ports in New South Wales (NSW) were corporatised as SSOC. When corporatisation was initially introduced in NSW the preferred model proposed by government was that of a GOC (State Owned Corporations Act, 1989). Under this model commercial ports would be created in an identical manner as a private sector company, subject to the same regulatory regime as any other company. Following a state government election and a change of government from a Liberal to a Labor government in 1995, the corporatisation legislation, which had already been tabled in parliament, was amended as the newly elected government and was ideologically opposed to the creation of a port company model in which government was distanced from day-to-day control. Consequently the model in place in NSW is that of a SSOC (State Owned Corporations Amendment Act 1996).

3.3. Queensland

Queensland has adopted the corporatisation model establishing its ports as government-owned corporations. This is frequently interpreted as a GOC when, in fact, the model set in place is that of a SSOC. The Queensland government, as in the case of the NSW Labor Government, was also reluctant to reform its ports in a hands-off fashion. Rather, it was argued that government entities should be reformed with a structure that enabled them to

(a) operate, as far as practicable, on a commercial basis and in a competitive environment,
(b) retain public ownership of the entities, and
(c) allow the State, as owner on behalf of the people of Queensland, to provide strategic direction to the entities by setting financial and non-financial performance targets and community service obligations.

The objectives of corporatisation in Queensland were to improve overall economic performance, and the ability of the government to achieve social objectives by improving the efficiency and effectiveness as well as account-ability (Government Owned Companies Act, 1993).

3.4. South Australia

South Australia (SA) is the only state in which corporatisation has been a precursor to privatisation. All SA ports were corporatised in 1994 under a single corporation – the South Australian Ports Corporation (PortsCorp) (Public Corporations Act, 1993). In 2000, however, the government of SA announced its intention to sell off the state's ports and privatise PortsCorp. This meant that all South Australian ports were sold in 2002 under a bid-ding process and were sold to a single buyer – Flinders Ports Pty Ltd. This purchase included the acquisition of the port infrastructure, a 99-year lease and operating licence for the Port of Adelaide and the six regional ports of Port Lincoln, Port Pirie, Port Giles, Klein Port, Thevenard and Wallaroo.

3.5. Western Australia

The government of Western Australia set in place yet another model. In that state, government initially rejected both the privatisation and corporatisat-ion models and introduced a strategy of commercialisation. Under this model, ports would retain statutory authority status with trade facilitation being a major function. While landlord responsibilities were to be retained, a customer focus strategy was implemented. At the port of Fremantle, the state's major container port, broad objectives were

(a) to ensure that port services and facilities are reliable, competitive and meet customer needs;
(b) to work with customers to facilitate trade opportunities; and
(c) to continue to improve capability to provide value for customers and pro-vide for long-term business sustainability (Public Corporations Act, 1993).

The Western Australia government subsequently corporatised its ports as SSOCs.

3.6. Tasmania

The government of Tasmania has been the only Australian state to implement a port structure similar to that of the private sector although public ownership is retained. The Tasmanian government opted for a model that differs significantly from the SSOC model implemented elsewhere. Indeed the ports in Tasmania are the only ports to be corporatised as GOCs with the enactment of the Port Companies Act, 1997 (Tasmania). When the ports were corporatised they were formed as companies "limited by shares and incorporated under the Corporations Law" (Port Companies Act, 1997, Tasmania). The significance of GOCs and SSOCs will be discussed in some detail later in the chapter.

4. THE SUCCESS OR OTHERWISE OF CORPORATISATION

Considerable debate has occurred regarding the benefits of corporatisation generally (Bottomley, 1997; Wettenhall, 1995) and particularly in relation to ports (Hirst, 2000). Evaluation of port performance pre- and post-corporatisation is, for a number of reasons, fraught with difficulties. First, it is an exercise with apparently dubious benefits as, in most instances, there is no basis for comparison. With the exception of the port of Gladstone, which has retained responsibility for coal-loading operations, all other ports have generally privatised commercial operations. This means that the entity existing at present is significantly different from that existing before corporatisation occurred.

Second, a further challenge is the actual measuring of performance, and a major issue in performance monitoring of all government businesses is determining what the appropriate measures are. All GBEs report partial financial performance measures, such as earning before interest and taxation. These, the Productivity Commission (2005) suggests, are relatively simple to calculate, intuitive and easy to understand. The Commission points out, however, that any single indicator represents only part of the financial performance and might be misleading about the overall performance of the corporation.

Third, a further difficulty in the monitoring of performance relates to the fact that not all ports undertake the same functions. This difficulty is exacerbated by the variance in size, the range of services they provide and, as indicated in Table 12.2, the activities they perform. Furthermore, a number of ports have other business interests and responsibilities, such as airports. The Port of Brisbane Corporation and the Hobart Ports Corporation each have a substantial interest in their local airports, for example. Indeed, the port of Brisbane in 2003 owned 38 percent of the Brisbane Airport Corporation while the port of Hobart owned 98 percent of Hobart International Airport. If these activities are measured as part of the overall port performance, a distortion in final results will occur.

Table 12.2. Activities – Port GTEs (2003–2004).

Port GTE	Jurisdiction	Activities				
		Port Facilities Management	Pilotage	Stevedoring	Cold storage	Airport operations
Newcastle Port Corporation	NSW	✓	✓	×	×	×
Port Kembla Port Corporation	NSW	✓	✓	×	×	×
Sydney Port Corporation	NSW	✓	×	×	×	×
Port of melbourne Corporation	Victoria	✓	×	×	×	×
Victorian Channels Authority[a]	Victoria	×	✓	×	×	×
Victorian Regional Channels Authority[b]	Victoria	×	×	×	×	×
Gladstone Port Authority	Queensland	✓	×	×	×	×
Port of Brisbane Corporation	Queensland	✓	×	×	×	×[c]
Cairns Port Authority	Queensland	✓	✓	×	×	✓
Townsville Port Authority	Queensland	✓	×	×	×	×
Port Corporation of Queensland	Queensland	✓	✓	×	×	✓
Mackay Port Authority	Queensland	✓	✓	×	×	×
Fremantle Port Authority	WA	✓	×	×	×	×
Bunbury Port Authority	WA	✓	✓	×	×	×
Port Hedland Port Authority	WA	✓	✓	×	×	×
Dampier Port Authority	WA	✓	✓	×	×	×
Geraldton Port Authority	WA	✓	✓	×	×	×
Albany Port Authority	WA	✓	✓	×	×	×
Burnie Port Corporation	Tasmania	✓	✓	×	×	×
Hobart Port Corporation	Tasmania	✓	✓	✓[d]	✓	×[c]
Port of Devonport Corporation	Tasmania	✓	✓	×	✓	✓
Port of Launceston Pty Ltd	Tasmania	✓	×	×	×	×
Darwin Port Corporation	NT	✓	✓	×	×	×

Source: The following table is from www.ccnco.gov.au/research/crp/gte0304/gte0304.pdf
[a] The Victorian Channels Authority(VCA) was transferred to the Port of Melbourne Corporation in November 2003 and the VCA was abolished pursuant to section 181 of the PSA on 31 March 2004.
[b] The Victorian Regional Channels Authority (VRCA) began operations on 1 April.
[c] Investment only – not direct operation
[d] Subsidiaries of the Hobart Ports Corporation provides stevedoring services in several SA ports and in Tasmania.

Despite these limitations, some performance measures are listed in Table 12.3 and a 'whole of sector performance' is used with aggregate measures such as total revenue, operating profits, return on assets and return on equity. This table indicates that the performance in all ports improved in the period 1999/2000 to 2003/2004. In this instance there is a valid basis for comparison as the entire period is post-corporatisation. Table 12.4 indicates that specific indicators such as profit and revenue for particular ports have increased in most of the major Australian ports for that period. Note, however, that this covers the corporatisation period but does not include pre-corporatisation data.

The return on equity ratio increased from 3.1 to 5.9 percent in 2003/2004 for the port sector as a whole (Productivity Commission, 2005, p. 267). In Victoria, the government regulator, the Essential Services Commission, proposed

Table 12.3.   Whole of Sector Performance Indicators.

	Units	1999–2000	2000–2001	2001–2002	2002–2003	2003–2004
Size						
Total assets	$m	3,149	3,356	4,521	4,997	5,722
Total revenue	$m	605	606	821	890	1,024
Profitability						
Operating profit before tax	$'000	128,949	140,715	130,548	163,659	315,874
Operating sales margin	%	29.5	31.2	23.6	23.9	35.9
Cost recovery	%	171.3	146.7	130.8	131.3	156.0
Return on assets	%	6.1	6.1	4.7	4.8	7.2
Return on equlity	%	3.1	3.9	2.3	3.1	5.9
Financial management						
Debt to equity	%	32.4	32.3	30.1	32.8	30.2
Debt to total assets	%	23.2	23.3	21.5	23.8	22.5
Total liabilities to equity	%	44.6	43.0	42.7	44.9	42.3
Interest cover	Times	3.2	3.5	2.7	3.6	5.5
Current ratio	%	105.1	128.2	136.8	130.4	157.1
Leverage ratio	%	144.6	143.0	142.7	144.9	142.3
Payments to and from government						
Dividends	$'000	65,346	68,533	100,838	91,136	99,165
Dividends to equity ratio	%	3.1	3.0	3.2	2.8	2.6
Dividends payout ratio	%	102.4	77.1	141.3	88.8	44.4
Income tax expense	$'000	65,149	51,869	59,163	61,075	92,629
CSO funding	$'000	9,926	12,184	13,994	15,354	7,778

Source: Productivity Commission (2005). Financial Performance of Government Trading Enterprises (1999–2000 to 2003–2004), Melbourne, July.

Table 12.4.   Selected Performance Indicators for Major Ports (1999–
2000 to 2003–2004).

	Revenue ($m)		Profit before Tax ($'000)	
	1999–2000	2003–2004	1999–2000	2003–2004
Newcastle	38	40	16	13
Port Kembla	32	47	8	28
Sydney	108	138	45	51
Gladstone	86	188	25	83
Brisbane	84	117	26	41
Fremantle	60	77	14	17
Hobart	17	36	76	6
Melbourne	–	101	–	14

Source: Productivity Commission (2005). Financial Performance of Government Trading Enterprises (1999–2000 to 2003—2004), Melbourne, July.

a benchmark return on equity of 7.3 percent for the Melbourne Port Corporation (MPC) and 6.7 percent for the Victorian Channel Authority (VCA).

In 2003/2004, the Productivity Commission determined that only 9 of the
23 ports had a return on equity ratio of above 7 percent, with the Port of
Melbourne Corporation (PoMC) achieving 1.1 percent and the VCA achieving 0.2 percent. The Commission points out, however, that the return on
equity for the newly restructured PoMC was for the first year of operations
and was substantially affected by restructuring (*ibid.*).

5. CORPORATISATION OR MICROECONOMIC REFORM?

A further reason for the difficulty in assessing with any degree of accuracy the
benefits of corporatisation is the fact that port authority restructure followed
a period of intensive microeconomic reform. This had been underway for
more than a decade before corporatisation was introduced. This was significant for ports, particularly in the high-cost labour market. Labour market
reform was such that the Sydney Ports Authority reduced its workforce from
3000 in 1989 to less than 600 immediately prior to corporatisation in 1995
(Hayes, 1995). This clearly had a significant impact on the Authority's bottom
line and, it must be noted, preceded the corporatisation of the port.

Furthermore, in that period, the Maritime Services Board of NSW, the
parent body of NSW ports, had embarked on an intensive program of

privatisation and rationalisation of services. The massive reduction in the labour force and the transfer to the private sector or closure of antiquated commercial facilities that had been replaced with new privately owned terminals, clearly had a significant impact on the overall performance. The coal-loading facility in Port Kembla, for example, was leased for a 20-year term to a consortium of coal companies. In addition, grain-loading facilities in Port Jackson were closed when a new grain-loading terminal was built in Port Kembla; subsequently, this was also privatised and sold to the grain industry.

The difficulty in evaluating the success of corporatisation is that there is little basis for comparison. The structure of the port market was significantly different in the 1980s from that in the 1990s when corporatisation was introduced and attempting to evaluate benefits pre- and post-corporatisation is akin to comparing apples with oranges. Furthermore, it is very difficult to evaluate performance because of the difficulties in assessing what benefit was a product of corporatisation per se and what was the result of the broader microeconomic reform program.

6. HOW EFFECTIVE IS THE CORPORATISATION MODEL?

Apart from difficulties associated with accurately measuring benefits of reform, how effective have these policy changes been, or are likely to be? Have they achieved the goals that might be expected from such strategies, or the goals as set out in the policy statements, the legislation and Statements of Corporate Intent? This section will focus on these issues.

When assessing the likely benefits of corporatisation and privatisation, it is important to fully recognise the limitations of the policy frameworks and that the options for better outcomes are likely to be circumscribed. In the case of corporatisation, and particularly under a SSOC model, not only is public sector ownership maintained, but also there is direct ministerial involvement. Under the SSOC models, the potential for ongoing political interference remains and, indeed, political input and direction remains of central importance.

It has been noted that port reform was to address the inefficiencies existing in Australian ports. A distinction must be made, however, first between capital city ports and regional bulk ports – the latter are highly efficient and comparable with or exceeding 'best practice' globally. Secondly, it is important to differentiate between the port corporations and the operators within the port. It is too frequently the case that the term 'port' is used to denote some homogeneous entity. This is certainly not the case, particularly with a corporatised port in which the commercial operations, such as the terminals, have been privatised and relate to the port corporation only in so far as tenancy is concerned.

The port corporation in this instance is likely to have little impact on the direct performance of its tenant although, as the landlord, it can indirectly affect the operator by introducing competitors into the port.

7. OWNERSHIP – PUBLIC OR PRIVATE?

Is a corporatised or privatised port more effective than the former statutory authority? This again is difficult to evaluate as privatized ports are unlikely to release confidential information to the public and particularly to competing ports. The rationale for private ownership in the first place was driven by the perception that private owners operate a port more efficiently because economic survival depends on long-run profitability. Private ports, consequently, are perceived to be more responsive to user needs, more aware of costs and more innovative than public sector ports where operating losses had, in the past, been absorbed by governments. This factor, it was believed, had created complacency, a factor exacerbated by the role of the 'job for life' attitude of the waterfront labour.

But the debate about whether publicly or privately owned ports are preferable has by no means been resolved. Although the view that private sector ownership was inherently more efficient was argued by proponents of the New Right, this certainly did not have universal endorsement among governments and their respective 'think tanks' and advisory bodies.

Arguments for the retention of ports as public sector operations had a long history in Australia insofar as it was widely recognized that governments traditionally played a developmental role in ports and provided facilities and services that may not realize full cost recovery except in the long term. In this philosophical context, it was seen that governments had the responsibility for the provision of essential port facilities and services in order to generate and sustain trade and associated economic activities. In some circles it was argued that this function was more appropriately provided by the public sector (Industry Commission, 1993). Indeed, when state governments were debating whether or not to privatise their ports the Commission argued that

Seaports are integral parts of the nation's transport system. Their location has had a major influence on the development of Australia's capital infrastructure. Their operation then and development have implications that extend beyond the interest of any single entity or group of firms and as such there is a public interest in the development and maintenance of the major ports and facilities. This public interest in ports is the overriding rationale for the establishment and/or maintenance of public port authorities. (Industry Commission, 1993, p. 39)

The concepts of privatisation and corporatisation, however, are not new. Despite the rhetoric of the 1980s and 1990s, privatisation, in its many forms and guises, has occurred for much of the 20th century – the sale of the Australian Commonwealth Lines of Steamers in 1928, for example (Everett & Robinson, 1993). What is new about the current trend is the extent of government asset sales and the ideology that drives this philosophy.

Nor is the concept of corporatisation new, having existed in some form or another since the 19th century. It can be traced back to the 1880s when state railway enterprises, which had traditionally operated as ministerial departments, were vested in a new kind of administrative body described as the statutory authority or public corporation. They were incorporated under its statutory charter as a body corporate and the action was a clear move to distance GBEs from the rigidities and hindrances of standard public service and governmental processes and to allow them the freedom and flexibility of private businesses. A century later, however, we are still pursuing this ideal as we have been unable to effectively distance government from day-to-day operations.

We have, in fact, generally failed to transform ports into commercial businesses along the lines of private sector operations. While we have emulated, partially at least, private sector business practices we have been unable to operate like private sector counterparts because we have been unable to draft and enact appropriate legislation. Effective corporatisation demands effective legislation, an objective that, to date, has been elusive. Some reasons for this will be discussed in the following section but there is little doubt, in the opinion of the authors, that the issue lies not with the concept of corporatisation per se but with the legislation.

8. A PARADIGMATIC PROBLEM?

The transformation of Australian ports into corporatised entities was driven by the necessity to improve efficiency in the port environment and commercialization of public sector ports more generally. The strategy, in particular, was about distancing government from day-to-day operations – the political element that was recognised as the cause of sub-optimal performance.

There is widespread agreement that corporatisation has yielded significant productivity improvements and raised efficiency levels (Hirst, 2000).

Dissatisfaction persists, however, with the fact that government influence has not been removed and that ongoing political interference continues to impede commercial objectives. Hirst argued, for example, that while the corporatised ports were far more responsive to customer needs and were prepared to work more closely with their customers to obtain better transport solutions, many continuing disadvantages with the corporatised structure persist, which portrays on-going bureaucratic and political interference in port issues that negatively impact port operations and efficiency (Hirst, 2000).

This problem of persistent interference is not restricted to Australia, however. Notteboom and Winkelmans (2001) suggest that political management structures in Europe impeded many public port organizations from developing flexibility and versatility to cope with a lack of productivity and innovation, and to respond adequately to structural changes in the world economy. Brooks (2001) points to problems in port performance and suggests that it may be an issue of port governance and questions whether commercially oriented governance structures are capable of delivering the expectations of both government and the community.

Everett (2005), on the other hand, has adopted a different perspective and argues that in the case of many Australian ports, the issue is not one of political interference per se or whether government and community expectations can be satisfied simultaneously. Rather, that political interference may be the cause of sub-optimal performance but is itself the effect of the legislation and the model implemented. Treating the effects of political interference and resulting inefficiency neglects the fundamental cause of the problem – the legislation.

The case study in the following section illustrates this issue and discusses the port of Melbourne, corporatised by legislation as a SSOC in 1995. This study exemplifies problems associated with the attempts to set in place a port business while being unable to remove government from day-to-day operations. In addition, it exemplifies a policy consistent with legislated National Competition Policy where the objective was a 'level playing field' and the removal of perceived advantages associated with government ownership rather than the creation of an efficient port business per se. The case study further demonstrates a weakness in effective port planning when a plethora of government instrumentalities have input into the planning and policy processes relating to the port.

Legislation in 1995 corporatising the port of Melbourne put in place a business model which, as a result of emerging anomalies, required amendments in 2003. A problem resulted from the drafting of the policy with little input from the port operator. The study represents a government-owned entity – the port – with a charter to operate as a commercially viable business which has its strategy developed by a government department and ministers whose responsibility was establishing government policy to meet the political and philosophical agenda. Meeting these objectives is not a problem for a government department as such, but the case study suggests that it is a problem for a government operation with the responsibility for running a commercially focused business.

9. CORPORATISING THE PORT OF MELBOURNE: A CASE STUDY

Prior to reform in 1995, Victorian ports, like those elsewhere in Australia, were structured as statutory authorities. The ports of Melbourne and Hastings were operated by the Port of Melbourne Authority (PMA), which had the management responsibilities for these ports as well as the outer ports of the eastern half of the state and all foreshore and marine assets. The two other commercial ports in Victoria – Geelong and Portland – had their own authorities.

In 1992, the PMA and the port authorities of Geelong and Portland were declared reorganising bodies under the *Victorian State Owned Enterprises Act, 1992*. This legislation, Russell (2001) points out, closely mirrored that of New Zealand's (State-Owned Enterprises Act, 1986) with the primary goal of requiring government-owned businesses to become successful, prudent, commercial businesses, with social responsibility equivalent to private companies. Under this model, non-commercial activities became subject to defined Community Service Obligation regimes.

In 1993, the Victoria State Government, in dire financial straits, embarked on a privatisation program. The ports of Geelong and Portland were sold and although the port of Melbourne was also targeted for privatisation, in response to intense opposition from port users, it was subsequently corporatised.

Under the Port Services Act, 1995, some major changes were introduced in relation to the functions and responsibilities of the Melbourne port. It separated, for example, the responsibility for the land and water sides. The 1995 Act created the MPC and the VCA, dividing the responsibilities of the former Melbourne Port Authority between them. The MPC became a landlord managing landside assets of the former PMA and the VCA had jurisdiction for maintaining navigation aids, channels and harbour control. Regulation of dangerous goods, pollution and land use planning were transferred to other government entities including the Department of Infrastructure and the

Department of State and Regional Development. In addition, regulatory control was transferred to the Office of the Regulator-General (ORG), which also had the responsibility for regulating prices.

Under the Guiding Principles for Victoria Port Reform an environment was created where

- competition and conditions existed that encouraged the most cost-effective service provider;
- asset ownership rested with the party best able to make use of the asset;
- the private sector would have the predominant role in commercial service provision and port investment; and
- non-commercial activities would be separated from commercial activities.

The role of the newly created MPC was "to plan and coordinate the development of, and manage, land within the Melbourne port area and construct infrastructure within that area and to make land and infrastructure within the Melbourne port area available to port service providers" (Port Services Act, 1995).

The restructure meant a significant refocus for the Corporation from a previously held wide range of tasks to a narrow role focused almost exclusively on property management. The MPC, as noted above, lost its responsibility for navigation and channels as well as its regulatory responsibilities. It could no longer set prices and undertake other essential port functions such as trade development, strategic land use and transport planning, environmental management, safety, marketing and community relations – all became the responsibility of other government organisations. The Department of Infrastructure assumed the port planning role and the Department of State and Regional Development assumed responsibility for trade facilitation (Department of Infrastructure, 2001). Other changes in responsibilities included the coordination and clean up of spills in Victoria waters, which was taken over by the Marine Board of Victoria, while responsibility for dangerous goods and marine pollution investigations and prosecutions was transferred to the Environment Protection Authority. Port policy, previously the responsibility of the Minister for Transport, now was transferred to the shareholding minister, the Treasurer.

In due course, the redistribution of responsibilities and the institutional arrangements under the 1995 legislation were seen to have contributed to a neglect of strategic issues and to constrained funding for public investment. Not only was there a neglect of strategic issues, but also strategies essential for the port's further business growth were ignored or not implemented. This was due, in part, to the uncertainty relating to the responsibility for particular tasks. Furthermore, organizations vested with the charter to perform particular tasks did not have the resources and the organization that did have the resources did not have the charter. Channel deepening, for example, which was recognised by the MPC to be essential for the future growth of the port if it was to accommodate the next generation container ships, was one of these anomalies. The MPC had recognised the need for channel deepening to cater for post-Panamax vessels, but the dilemma was caused because while the MPC had the resources to deliver this and the ability to recoup the expenditure through cargo growth, it did not have the charter to deliver channel deepening. The VCA, on the other hand, the corporation with the charter to deliver channel deepening, did not have the resources to implement the strategy nor the means of recouping the capital expenditure. Inaction in this matter was the result!

9.1. Reform Post-2003

Difficulties associated with the creation of a 'split' port led to further reform and the enactment of the Port Services (Port of Melbourne Reform) Bill, 2003. The amended legislation attempted to address these anomalies, setting in place a more holistic strategy for the port. The MPC and the VCA were disbanded and replaced with the PoMC. The creation of the PoMC meant reuniting the land and water side of the port into one port entity. Its new objective was to deliver a "broader charter defined for the Corporation by the Government" and a key element of this entailed the "synergistic benefits from managing the land and sea components of the port as one to provide *inter alia*:

- An enhanced ability to drive efficiencies across the land and water logistics chain
- Integrated whole of port infrastructure planning and development, in particular to drive and optimise benefits from channel deepening
- A 'one stop shop' (single agency) for customers
- A whole of port approach to the management of risk
- A greater ability to market the port as a global player and to undertake joint marketing exercises with other agencies such as the City of Melbourne
- Integrated environmental planning and control
- A single interface for the management of community relationships
- Increased efficiencies associated with reduced management overheads, allowing the freeing up of resources for delivery of the new strategic objectives" (Port of Melbourne Corporation, 2003, p. 7).

The 1995 legislation had created the MPC strictly as a landlord and, as noted above, had separated the responsibility for the land and water. Under the 2003 legislation these two essential assets were reintegrated and the newly created PoMC would now have responsibility for the land and water side (PoMC, 2003, p. 11). Furthermore, its marketing and trade facilitation role was restored to the extent that the PoMC had the responsibility "to promote and market the port of Melbourne". In addition, under the new framework the port was not seen as an isolated precinct but an integral part of a supply chain with the responsibility "to facilitate the integration of infrastructure and logistics systems in the port of Melbourne with relevant systems outside the port" (PoMC, 2003, p. 11).

The new legislation also introduced policy and accountability changes. The issue of accountability had been a rather peculiar one and had fluctuated in the past between the Shareholding Minister (the Treasurer) and the Portfolio Minister (Minister for Transport). This had led to some interesting outworkings as these ministers had been driven by different objectives. The Treasurer's focus had been return on investment, a not unreasonable objective given the initial position of the Corporation as a landlord. The Minister for Transport, on the other hand, while being aware of the delicate financial situation existing in Victoria in the early 1990s, was concerned with the efficiency of the port and meeting customer needs.

Under the 2003 amended legislation the responsibility was once again delegated to the Minister for Transport with "each port corporation subject to the general direction and control of the minister." There was a proviso, however, and an assurance that the "Treasurer would not be excluded as each port corporation was subject to any specific direction given to it by the portfolio minister but with the approval of the Treasurer" (*ibid.*, p. 20).

10. AN EFFECTIVE BUSINESS MODEL?

This raises the question, has the restructure and ongoing reform in the port created an effective business model? Is this latest restructure the panacea for ongoing port ills or an endeavour that will hopefully yield improved results? The objective and the broader charter of the Corporation continue to be determined by government – the owner of the port and the single shareholder.

Companies which are market leaders, irrespective of ownership, are those with the ability to be innovative (Robinson, 2003). This, Moss Kanter (1992) suggests, is not simply a matter of hardware or technology. These are necessary but not sufficient conditions for success and the latest restructure of the port of Melbourne is primarily about hardware and who has responsibility for what. While the 'who' has changed, the 'what' remains unchanged.

Success in an organisation requires an effective long-term strategy – a crucial element that has to date evaded the Corporation. The current restructure of the port is not simply a result of problems associated with the 1995 corporatisation model but reflects the failure to implement a long-term effective business model – a situation that has been ongoing for decades. Indeed, since the 1960s, the port has undergone restructure every three or four years and has had no less than five CEOs or acting CEOs in less than a decade.

Arguably this is not a problem of infrastructure or hardware per se, or of operational efficiency. Rather it is a product of the structure of the organisation itself and about the role of management and shareholders in the organisation. Arguably, the need for revision in 2003 and the recommended strategy was not determined by the operator – the port – but by politicians and bureaucrats, ones not necessarily with the requisite skills or operational and business expertise. Furthermore, decisions are likely to be made to meet political and ideological, rather than commercial, needs.

This raises the question, Do politicians and bureaucrats have the requisite business acumen to determine business strategy for the port or indeed for any government-owned business? This is particularly relevant as politicians and governments operate in a three- to four-year electoral cycle. This issue of tenure and being seen to implement policies successfully is exacerbated as senior bureaucrats in Australia are now also employed on a non-tenured contract basis. This loss of tenure might mean that the solutions sought focus on visible achievements occurring during the cycle rather than on the longer-term benefits of the port. The outcome of this may be a strategy that serves short-term cyclical expectations and political demands of the strategist, rather than longer term business needs of the port and its users.

Corporatisation and attempts to create effective business models that emulate private sector operations fall short of meeting objectives. This, it must be noted, is not a product of inappropriate or inadequate infrastructure, nor of poor calibre of port managers, but reflects ineffective business models and government legislation. This calls into question whether governments, or government instrumentalities, with the responsibility of delivering the 'public good' can reconcile this with that of a commercially oriented business. It raises a further important issue, can a model be created in which government is the owner of a port in which the operator has independence from its owner and from political control? Governments have attempted to address this issue for a century or more and corporatisation constitutes one more attempt to commercialise government operations with variable success because direct political control has not been removed.

This is not a port management issue as such, nor is it one of infrastructure. Rather it is one of legislation and organisational structure that ensures ongoing political input on a day-to-day basis. This problem, arguably, will not be resolved until a more effective corporatisation model is developed and more effective legislation is enacted. The following section will examine an alternative model for the creation of a government-owned business. It will also discuss some of the problems associated with this in the existing Westminster system of government.

11. CREATING AN EFFECTIVE GOVERNMENT-OWNED BUSINESS

How then do we create a company emulating a private sector model while maintaining government control? Some ports, such as those in Tasmania, have been corporatised as GOCs. Other government entities, particularly at the federal level, have also been created as GOCs – Australian Rail Track Corporation, for example, the company created by the Commonwealth Government as the access provider for the national rail track and network.

11.1. Government-Owned Companies

The creation of a company, government-owned or otherwise, under the Corporations Act requires a constitution that includes a company's corporate governance rules, a statement on the structure of the company, its management, responsibilities of the board, etc. This constitution comprises a Memorandum and Articles of Association and is registered with the Australian Securities and Investment Commission (ASIC). Under this model, the port company becomes subject to the same regulatory regime as private sector organisations – a regime that will ensure that conditions in the constitution are not breached. In the event that a breach does occur, the port company, or in some instances the Minister who represents the owner of the port, becomes personally liable under the *Corporations Act.*

Under the GOC model, the port company, while being government-owned, is not accountable to Parliament, which is the requirement under the Westminster system. Rather, the company is subject to the provisions of the Corporations Law and accountable to ASIC. Under this model, ASIC and not Parliament is the regulator and ASIC must be satisfied in relation to the company's performance and protocol.

11.2. Statutory State-Owned Corporation

Corporatised ports in mainland Australia, on the other hand, have been created as SSOCs. This means that while the structure of the organization and its stated objective may emulate that of a private company, its legislation and regulatory mechanism is significantly different. SSOCs are subject to a regulatory regime embedded in the legislation and under a Westminster system accountability is by way of a direct line to the minister who reports to the cabinet, which is ultimately accountable to Parliament. ASIC does not have a role in the oversight of this model.

The adoption of a SSOC model in most Australian states has ensured that direct ministerial control and lines of accountability as per the tenets of the Westminster system are maintained. Indeed, with the SSOC model, political and ministerial input is pivotal in the day-to-day operation of the ports. In some states, Ministers determine prices and/or infrastructure investment and divestment. Furthermore, policy and strategy are determined by the minister, who is advised by his/her department and who may or may not be driven by commercial objectives.

12. A MORE EFFECTIVE MODEL?

Arguably, the GOC is a preferable model for the pursuit of commercial objectives. However, some problems associated with it have been raised by politicians and in legal circles, particularly with regard to the constitution and consistency with the Westminster system of government.

The SSOC is created by its own, rather than generic, statute and is subject to the provisions of that statute. In all instances of corporatised port legislation, ministerial input on a day-to-day basis has not only been maintained, but is inherent in the legislation. This means that the fundamental cause of sub-optimal performance – political interference – has not only been maintained but is embedded in the legislation.

It has created a dilemma because, on the one hand, as corporatised entities ports are expected to operate as private sector businesses. And on the other, government ownership brings with it ministerial, political and policy responsibilities that may or may not be commercially oriented but might be justified on the basis of the 'public interest'. The potential for conflict under this model has been difficult to resolve.

The Productivity Commission has found that this tension can be addressed by establishing sound government procedures and by clearly defining the non-commercial objectives (*Op cit,* p. 49). The Commission argues further that private businesses are subject to laws and regulations designed to ensure that public interest is protected. Similarly, under the SSOC model, shareholding ministers may also impose public interest requirements through ownership controls.

Identifying and understanding these requirements may be difficult for a number of reasons – the values and policy platforms of the incumbent government, for example. Ministers, as elected representatives, have the responsibility for resolving the inevitable trade-offs between conflicting commercial and possibly non-commercial but public interest objectives.

The problem is also occurring at the inter-government level, particularly as a result of the plethora of regulators involved. Although the ports are state government-owned, the stevedoring companies traditionally have been under the regulation initially of the Prices Surveillance Authority, which has subsequently become part of the ACCC, a federal government instrumentality. The particular problem arises as state government instrumentalities are the regulators for the port corporation and a federal government entity regulates particular businesses within that port – the stevedoring company, for example.

Researchers have attempted to analyse port privatisation and corporatisation within a framework that suggests that an organisation's performance is a function of the match among the characteristics of the organisation's environment, strategies and structures (Baltazar & Brooks, 2001). This appears a rather simplistic conceptual framework and does little to identify the reasons why a mismatch may occur. This chapter has illustrated that a crucial element of effective corporatisation is effective legislation. It can be argued that an 'organisation's environment' includes the political environment, which determines, drafts and enacts legislation. Only in the vaguest and most superficial manner is the notion of an organisation's environment relevant, however.

13. CONCLUDING COMMENTS

A decade after port corporatisation was introduced by Australian state governments, it is becoming questionable whether the model provides an appropriate business model. While productivity levels have improved (Hirst,

2000), it raises the question whether we have created an effective or optimum business model.

Clearly, there are some unresolved issues that have been raised and a dilemma has emerged. Retaining public ownership has ensured ongoing political input, and freedom in day-to-day operations has not been realised. This is a product of legislation, for effective corporatisation demands effective legislation – legislation that means that government can retain ownership but with a model that has the shareholder at a distance. While the current situation persists in which the minister determines policy, sets prices, determines investment strategy and can intervene to suit other than commercial demands, it is unlikely that port business can be likened to a private sector business model.

The model in place ostensibly emulates the private sector and while commercial objectives are articulated, the reality is that political objectives frequently override commercial ones. If the cause of poor port performance is the direct result of political interference, and if we are unable to put in place appropriate legislation that will eliminate political interference, would the more effective model be privatisation?

REFERENCES

Baltazar, R., & Brooks, M. R. (2001). The governance of port devolution: A tale of two countries. *World conference on transport research*, Seoul, Korea, July.

Beresford, Q. (2000). *Governments, markets and globalisation*. St. Leonard's, NSW: Allen & Unwin.

Bottomley, S. (1997). Corporations and accountability: The case of Commonwealth government companies. *Australian Journal of Corporate Law*, 7(2), 156–158.

Brooks, M. R. (2001). Good governance and ports as tools of economic development: Are they compatible? Paper presented at the international association of maritime economists annual conference, Hong Kong.

Department of Infrastructure. (2001). *The next wave of port reform in Victoria: An independent report to the Minister for Ports*.

Everett, S. (1988). *The location of transport infrastructure and the policy-making process: Port terminals and modal networks for NSW export coal in the post 1970s*. Unpublished PhD thesis. The University of Wollongong.

Everett, S. (2005). Developing a government owned business: A more effective model for corporatised ports. *Proceedings of the international association of maritime economists annual conference*, Cyprus.

Everett, S., & Robinson, R. (1993). Making the Australian flag fleet efficient: Dysfunctional policy processes and the 'play of power'. *Maritime Policy and Management*, 25(3), 269–286.

Government Owned Companies Act. (1993). Government of Queensland.

Hayes, J. (1995). Benefits of port reform in NSW. Paper presented at the Chartered Institute of Transport. Transport 1995 Seminar, 8 August.

Hirst, J. (2000). Future management trends in Australian ports. Paper presented at AAPMA conference Pan Pacific.

Industry Commission. (1993). *Port authority services and activities.* Canberra: AGPS.

Moss Kanter, R. (1992). *The change masters.* London: Routledge.

Notteboom, T. E., & Winkelmans, W. (2001). Reassessing public sector involvement in European seaport. *Maritime Policy and Management, 28*(1), 71–89.

Port Companies Act. (1997). Government of Tasmania.

Port of Melbourne Corporation. (2003). *Draft Strategic Plan.*

Port Services Act. (1995). Government of Victoria.

Port Services (Port of Melbourne Reform) Bill. (2003). Government of Victoria.

Productivity Commission. (2005). Financial performance of government trading enterprises 1999-00 to 2003-04, Melbourne, July.

Public Corporations Act. (1993). Government of South Australia.

Public Corporations Act. (1993). Government of Western Australia.

Quiggin, J. (1996). *Great expectations: Microeconomic reform in Australia.* St. Leonard's, NSW: Allen & Unwin.

Report by the Independent Committee of Inquiry (Hilmer Report). (1993). National competition policy, Canberra, AGPS.

Report of the Victorian Commission of Audit. (1993).

Robinson, R. (2003). Port authorities: Defining functionality within a value-driven chain paradigm. *Proceedings of the international association of maritime economists annual conference*, Pusan.

Russell, W. (2001). *The next wave of port reform in Australia.* Melbourne: DOI.

Sayers, R. (1989). In transition: NSW and the corporatisation agenda. *Proceedings of seminar MGSM*, Sydney.

State-Owned Enterprises Act. (1986). Government of New Zealand.

State Owned Corporations Act. (1989). Government of New South Wales.

State Owned Enterprises Act. (1992). Government of Victoria.

State Owned Corporations Amendment Act. (1996). Government of New South Wales.

Wettenhall, R. (1995). Corporations and corporatisation: An administrative history perspective. *Public Law Review, 6*(1), 7–23.

CHAPTER 13

THE PORT OF SINGAPORE AND ITS GOVERNANCE STRUCTURE[☆]

Kevin Cullinane, Wei Yim Yap and
Jasmine S. L. Lam

ABSTRACT

This chapter provides an overview of the port of Singapore and then focuses on the governance structure within which it operates. An analysis is undertaken of the main sources of cargoes that pass through the port, either to and from its hinterland region or as transhipments. The overall performance of the port is assessed across a range of criteria and the different forms of operation that take place within the port are explained. In considering the governance structure within which all this takes place, particular attention is paid to the role of the Maritime and Port Authority (MPA) of Singapore as the statutory regulatory authority and to the ownership structure of the container-handling sector and the potential for privatisation within it. The chapter concludes with an exposition of the major challenges facing the port and an analysis of the potential future of the port itself and its governance structure.

[☆]The authors would like to stress that the views presented in this work do not represent those of their respective organisations.

Devolution, Port Governance and Port Performance
Research in Transportation Economics, Volume 17, 285–310
ISSN: 0739-8859/doi:10.1016/S0739-8859(06)17013-4

1. INTRODUCTION

The British East India Company founded the modern port of Singapore in 1819 as a centre for entrepôt trade. The port served mainly as a regional distribution centre for cargo traffic originating from and destined for the Malayan hinterland (Huff, 1994). From the late 1960s, however, Singapore began to focus on positioning itself as a transhipment centre for international cargo. Hence, the throughput of the port now consists not only of local cargo from Malaysia and neighbouring Indonesian islands, but also includes cargoes that are transhipped through the port to and from Europe, East/North-East Asia, Australasia and the Indian subcontinent. Over the years, Singapore has become recognised as a focal point for some 200 shipping lines with links to more than 600 ports in over 120 countries worldwide.

The aim of this chapter is to provide an overview of the port of Singapore. Section 2 provides a background to the port, analysing the sources of its hinterland and transhipment cargoes, its overall performance across a number of parameters and the nature of the operations that take place within it. This is followed in Section 3 by a focus on the governance structure for the port, focussing in particular on the role of the relevant regulatory agency, the ownership structure of the container handling sector and the potential for privatisation within this sector. Section 4 analyses the major challenges facing the port, while Section 5 concludes.

2. BACKGROUND

2.1. The Maritime Industry Context

From the perspective of calculating its contribution to the national economy, the government of Singapore classifies the maritime industry as comprising the following individual sectors and activities: the port, shipping lines, shipping agencies, shipbuilding and repair, offshore, marine equipment, bunkering, ship chandlery, cruise, inland water transport, shipbroking and chartering, classification societies, ship management, maritime logistics, maritime-related legal services, maritime-related insurance, maritime-related finance, maritime-related education and training, maritime-related research and development and maritime-related information technology. In terms of economic contribution, Table 13.1 shows that in 2002 the maritime industry in Singapore (thus defined and as a whole) generated US$4.7 billion worth of direct value added and accounted for 4.9 percent of GDP.

Table 13.1. Contribution of Singapore Maritime Industry to the
Economy in 2002.

Indicator	Value
Number of establishments	4,400
Number of employees	86,500
Direct value added	US$4.7 billion
Direct value added as % of GDP	4.9
Direct and indirect value added	US$7 billion
Direct and indirect value added as % of GDP	7.3

Source: Singapore Maritime Cluster Study (2004).

In 2002, the maritime industry in Singapore was estimated to comprise over 4,400 establishments, employing a total of around 86,500 people. This represents about 4.3 percent of the total workforce nationwide. Including the indirect value added component would raise the GDP contribution of the maritime industry to US$7 billion, or 7.3 percent of national GDP. To put these figures into some sort of perspective, the size of the maritime industry in terms of the absolute value added that it generates is, therefore, comparable to that of Norway, Hong Kong and the Netherlands. However, the contribution of maritime activities to GDP is even more significant in Singapore, where maritime value added accounts for approximately 5 percent of GDP, whereas the contributions of maritime value added to the economies of Norway, Hong Kong and the Netherlands is below 3 percent.

The top five activities in the maritime industry accounted for almost 80 percent of direct value added, 70 percent of employment and 50 percent of revenue in 2002. These activities consisted of the port, shipbroking and chartering, shipbuilding and repair, shipping lines and agencies and logistics services. The contribution of the port[1] was particularly significant in accounting for over 20 percent of direct value added. This was followed by shipbroking and chartering and shipyard activities, with each accounting for about 15 percent of direct value added. In terms of employment, the shipyards alone accounted for more than one-third of people employed in the maritime industry. The yards were followed by the logistics and port sectors, which accounted, respectively, for 14 and 11 percent of people employed in the maritime industry. The major revenue generators in 2002 were bunkering and shipping lines, with both sectors together accounting for 58.4 percent of total revenue generated by the maritime industry. Shipbuilding and repair activities followed in a distant third position, accounting for 8.5 percent of revenue generated by the maritime industry.

2.2. Port Performance

The port of Singapore has gained significant advantage from its strategic location on one of the world's busiest shipping lanes, the Malacca Straits. Geography alone, however, does not guarantee success. The port owes much of its success to the resident port and maritime-related community that strives to provide world-class and competitive products and service standards to meet the stringent requirements of port users. This has enabled the port of Singapore to remain one of the world's busiest in terms of vessel arrivals, bunker sales, cargo tonnage handled and container throughput.

Table 13.2 shows a summary of port performance between 1999 and 2004. The total tonnage of shipping calling at the port grew to a new world record of 1.0 billion gross tons (GT) in 2004. Average vessel size increased by 26.3 percent compared to five years previous in 1999. The most significant increases in the average size of vessel calling at the port over this period were in LPG/LNG carriers and in container vessels.

Strong growth was also achieved for total cargo handled; this grew by 13.1 percent from 2003 to reach 393.4 million freight tonnes (FT) in 2004. Containerised cargo, which totalled 223.5 million FT or 21.3 million TEUs, continued to account for the largest share of the total tonnage handled at

Table 13.2. Summary of Performance (1999–2004).

Indicator	1999	2004	% Change from 2003	1999–2004 CAGR
Containerised	176,569	223,500	16.6	4.8
Conventional	11,985	17,376	17.7	7.7
Oil	124,386	129,328	4.8	0.8
Non-oil bulk	12,963	23,183	29.7	12.3
Total cargo volume ('000 FT)	325,902	393,388	13.1	3.8
Container throughput ('000 TEUs)	15,945	21,329	15.9	6.0
Bunker Sales ('000 tonnes)	18,891	23,567	13.3	4.5
Container vessels	332.0	397.1	10.0	3.6
Oil tankers	244.7	276.5	6.6	2.5
Chemical tankers	7.7	24.5	24.4	26.0
LPG/LNG tankers	20.9	30.7	−5.5	8.0
Regional ferry	10.4	9.9	3.1	−1.1
Total vessel arrivals (million GT)	877.1	1042.4	5.7	3.5

Source: MPA (2005).

56.8 percent. This was followed by bulk oil cargo, which accounted for 32.9 percent. Together, these two cargo categories accounted for almost 90 percent of the total tonnage of cargo handled in the port of Singapore.

Bunker sales also grew to a new world record of 23.6 million tonnes in 2004. This is mainly attributed to various measures that have been introduced to strengthen the integrity of the bunker trade – a strategy prompted by the prospect of greater competition in the region, mainly from new Malaysian and newly privatised Sri Lankan bunker suppliers (Anonymous, 2002a). These measures include more stringent bunker checks for 'ship-to-ship' transfers and bunker tankers loading at terminals. Constituting one of Asia's largest flag fleets, the Singapore Registry of Ships (SRS) also grew from the 11th largest merchant fleet in 1999 to become the world's 6th largest fleet in 2004, totalling 27.7 million gross tons.

2.3. Sources of Cargo Throughput

Local cargo handled by the port of Singapore comes mainly from what is referred to as the SIJORI region (SIngapore-JOhor-RIau; refer to Fig. 13.1). This Indonesia–Malaysia–Singapore growth triangle, officially formed in 1994, is based on the concept of trilateral cooperation that emphasises sub-regional development within the wider regional context of the Association of Southeast Asian Nations (ASEAN). Through strategies that engender mutual cooperation, the main aim of establishing SIJORI is to facilitate economic development by enhancing the attractiveness of the region for foreign investments. The SIJORI triangle now constitutes one of Southeast Asia's most vibrant economic zones, with Singapore as its focal point. Hinterland cargo from the SIJORI region comprises mainly container and liquid bulk cargo and, as its name implies, the structure of hinterland transportation can be split into three smaller geographical areas, namely Johor–Melaka, the Riau Islands and internal to Singapore.

Cargo originating in the Johor–Melaka area is mainly containerised. Despite the presence of a railway line running north–south from Kuala Lumpur to Singapore, the primary form of hinterland transportation from this area is road haulage. Most hauliers prefer to use the North–South Expressway and Federal Highway Route 2 that run parallel to the railway line. In recent years, the Malaysian government has achieved some success in encouraging industries in this area to use Port Klang as their gateway port in preference to the port of Singapore (Robinson, 2005; Robinson & Everett, 1997). The motivation for this is obviously national self-interest, but the comparative success of this strategy can be largely attributed to continuing

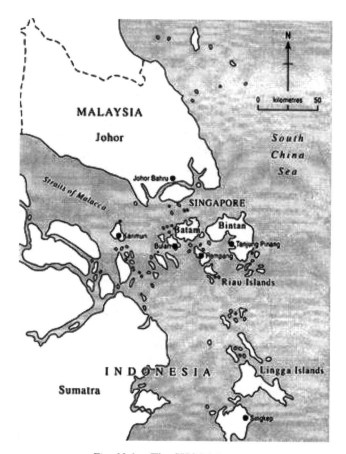

Fig. 13.1.   The SIJORI Region.

improvements in the handling efficiency of Malaysia's national load centre (Mak & Tai, 2001). As for seaborne feeder movements, most of the cargo from the Johor–Melaka area is usually transported via ships and barges from the ports of Pasir Gudang, Muar, Melaka, Mersing and Batu Pahat.

Cargo originating from the Riau Islands consists of manufactures transported entirely in containers, as well as conventional cargo and minerals transported either in containers or in loose form. Ports connected to the area include Belawan, Bintan, Tanjung Pinang and Panjang.

Cargo originating from within Singapore itself consists mainly of containers that are transported by road haulage. However, a very significant amount of liquid bulk cargo is also transported via both road haulage and pipeline.

Transhipment cargo is sourced mainly from Singapore's traditional hint-erlands in Southeast Asia. However, as illustrated in Fig. 13.2, the geographical position of Singapore on the mainline Europe–Far East shipping route has meant that the port is also able to serve as the transhipment point for cargo originating from, or destined for, the European, East/Northeast Asian and Australasian markets. In the case of container traffic, Singapore has benefited greatly from the significant transhipment potential it offers for container traffic movements between Europe and East/Northeast Asia.

2.4. Port Operations

As can be seen in Fig. 13.3, in general, port facilities in Singapore are mostly concentrated in the southwestern part of the country. Containers are mainly handled at five locations in the southern part of the island. Through its Singapore-based subsidiary, PSA Corporation, PSA International operates the Tanjong Pagar, Brani, Keppel and Pasir Panjang terminals, while Jurong Port Pte Ltd runs a comparatively small container terminal at Damar Laut. Conventional cargo is handled at the multi-purpose terminals located at Pasir Panjang Wharves, Sembawang Wharves and Jurong Port, while petrochemicals are handled at Jurong Port and Jurong Island. The major shipyards are located at Tuas View and cruise facilities are located at the southern tip of the country at the Singapore Cruise Centre.

2.4.1. Container Operations

In 2004, 33 percent of vessels arriving at the port of Singapore (representing 599.8 million gross tonnes) called for the purpose of engaging in cargo operations (MPA, 2005). As can be seen in Fig. 13.4, containerised cargo formed the largest category of cargo handled by the port, totalling 223.5 million FT or 21.3 million TEUs, in 2004. PSA International accounted for about 98 percent of the market share in container handling, while Jurong Port Pte. Ltd handled most of the remaining 2 percent. This dominance of PSA International in Singapore's container handling market is largely attributable to the incumbent monopoly position it enjoyed dating back to the early 1970s when Singapore began to handle its first containers.

A high level of terminal efficiency (Cullinane, Song, & Gray, 2002), coupled with the absence of any credible competition within the region, allowed PSA International to double its container throughput roughly every four

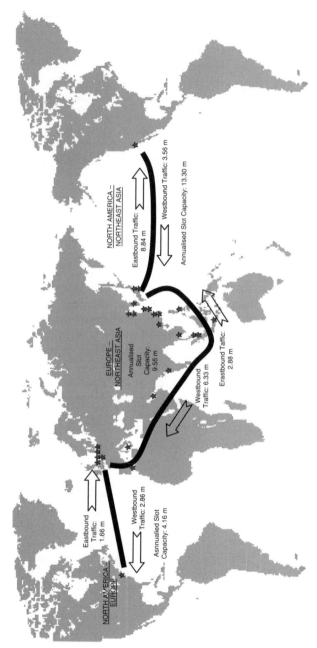

Fig. 13.2. Flow of Container Traffic in 2003 (in TEUs).

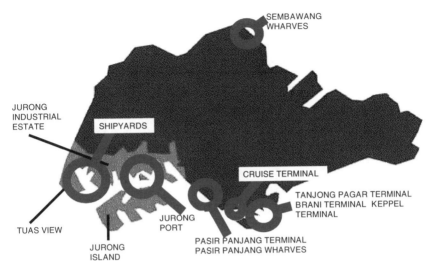

Fig. 13.3. Port of Singapore.

years up to 1994, after which the next doubling occurred only in 2004. It should be acknowledged that the early rapid expansion in container throughput (pre-1994) at the port of Singapore was facilitated by the rapid adoption of the containerisation concept within the main economies (Thailand, Malaysia and Indonesia) whose trade feeds into the hub of Singapore.

Although Singapore is widely acknowledged as offering one of the highest service standards in the region, the recent past has seen both Maersk–Sealand and Evergreen relocating their Southeast Asian regional transhipment hubs to the Port of Tanjung Pelepas (PTP); a brand new port development located on what was once a greenfield site lying just across the border in Malaysia.

Maersk–Sealand announced its decision to relocate its hub from Singapore to PTP in August 2000. This decision had the immediate effect of diverting about 2 million TEUs of container throughput, or about 10 percent of PSA International's volumes handled in Singapore. In pursuit of its global strategy of vertical integration into the ports sector, Maersk–Sealand simultaneously took a 30 percent stake in PTP, which was estimated to be worth US$165 million (*The Business Times*, 2000). Nevertheless, the strong connectivity offered by Singapore has ensured that Maersk–Sealand continues to tranship a significant share of its container traffic through the port of Singapore (Yap & Lam, 2004).

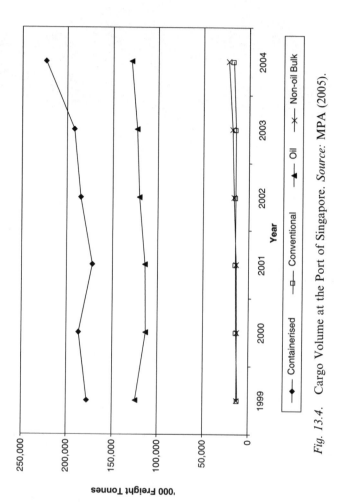

Fig. 13.4. Cargo Volume at the Port of Singapore. *Source:* MPA (2005).

The Taiwanese container shipping line, Evergreen, announced its decision to relocate its regional hub from Singapore to PTP in January 2002 (*Maritime Asia Today*, 2002). In terms of the subsequent loss of container throughput in Singapore, the impact of this decision was considerably less than in the case of Maersk–Sealand. However, most significantly, PTP had shown that its presence in the competitive regional container handling market had already become sufficiently strong to be able to attract the largest and third largest container shipping lines in the world to call at its facilities as regional hub, rather than at those of Singapore. In the case of both companies, the lower costs of PTP were cited as the primary motivator for their decisions (*Agence Presse France*, 2002). However, many commentators have since suggested that the major motivator behind the move of Maersk–Sealand to PTP was its desire to exert control over its own destiny through its investment in PTP (Anonymous, 2002b). These same commentators pointed to the port of Singapore's reversal of policy on the issue of granting licenses for dedicated terminals as a clear indication that this was the case (Anonymous, 2000).

By virtue of this experience, both the port and country of Singapore have experienced firsthand and most acutely the manifest effects of the threat posed by regional port competition. In response to these decisions and in anticipation of similar actions being taken by other shipping lines within its customer base, PSA International has moved to lock in its customers with new long-term contracts. These agreements usually involve guarantees on terminal service levels and include such aspects as berthing priorities, berthing flexibility and rate discounts among other benefits. In return, shipping lines will offer guarantees on minimum throughput (Hand, 2002). Most significantly, it also initiated a new policy of offering existing and prospective customers the possibility of operating dedicated or joint venture terminals. December 2003 witnessed the signing of the first deal for a joint venture terminal in the port of Singapore; this was with the Chinese state-controlled container shipping line COSCO and was worth US\$94.34 million (Anonymous, 2003). Concurrently, in a bid to become more cost-competitive, PSA International instigated a wage restructuring exercise where salaries were reduced by up to 14 percent (PSA International, 2003a) and staff headcount by 496 (PSA International, 2003b).

2.4.2. Oil Cargo Operations

Bulk oil cargo (including chemicals) accounts for the second largest category of cargo handled in Singapore, totalling 129.3 million FT or 32.9 percent of the total cargo handled by weight in 2004. Singapore is one of the major petroleum refining centres in the world, with total crude oil refining capacity of 1.3 million barrels per day. The three main refineries include those operated by

- Exxon Mobil with a capacity of 580,000 bbl/d
- Royal Dutch/Shell with a capacity of 430,000 bbl/d
- The Singapore Refining Corporation with a capacity of 285,000 bbl/d

To maintain Singapore's long-term status as the energy hub in the region, Jurong Island was created in the 1990s by reclaiming land to conglomerate seven smaller islands to produce a cost-competitive location for highly integrated, world-scale petrochemical plants. Representatives of the petroleum and petrochemicals industry that have operations on Jurong Island include Exxon Mobil, Shell, Chevron Texaco, BASF, Sumitomo Chemical and Mitsui Chemical. The aim of the government is to develop Jurong Island into a self-sufficient petrochemical complex that incorporates all of the key accompanying support facilities. Reclamation work is scheduled for completion by the end of 2005, when the size of Jurong Island will be 3,200 ha. By 2010, the island will be able to accommodate five refineries and 150 companies with an estimated fixed asset investment of US$24 billion. One problem on the horizon, however, is that the demand for oil products in Singapore's traditional export markets has gradually been met by import substitutions from more local production facilities, such as the refineries that have been established and developed in India, Malaysia and Thailand.

2.4.3. Bunkering Operations

Singapore is the largest bunkering centre in the world. A record volume of 23.6 million tonnes of bunker fuel was sold in 2004, generating revenues in excess of US$8.5 billion. The strategic location of the port on one of the world's busiest shipping routes ensures a steady source of potential demand. Singapore's bunker market is characterised by intense competition and speculation between buyers and sellers and the Maritime and Port Authority of Singapore (MPA) has always placed a heavy emphasis on ensuring bunker quality, the adherence to international standards, and maintaining the cost-competitiveness of the industry.

Some recent measures aimed at protecting the quality of bunkers sold in Singapore include the Accreditation Scheme for Bunker Suppliers and the Code on Quality Management for the Bunker Supply Chain. These measures are also complemented by restrictions on Outside Port Limit bunkering and on supplies from unknown sources and, therefore, of unknown quality. The MPA advocates that adhering to international norms will help create an assurance of high product quality and instil greater confidence in its bunker customers. In an effort to achieve this objective, revisions have been made to the Singapore Standard for Bunkering (SSCP60) that includes new provisions on health, safety and the environment. In the attempt to boost business volumes while keeping bunkering costs low, the Special Bunkering Anchorage scheme has been extended to include both the Eastern and Western sectors of the port. In these locations, vessels of 20,000 gross tons and above stand to enjoy more than a 50 percent concession in port dues. With such measures in place, the authorities intend to maintain Singapore's lead as the world's busiest bunkering port.

3. GOVERNANCE STRUCTURE

3.1. Regulatory Administration

The main organisation currently involved in the regulation of the port is the MPA. Until 1996, cargo operations and port regulation were handled by three government agencies. These were the National Maritime Board (a statutory board in the charge of the Ministry of Transport that overlooked matters relating to the training of seafarers), the Port of Singapore Authority (a statutory board in the charge of the Ministry of Transport that overlooked matters relating to cargo operations and port regulation) and the Marine Department (a department in the Ministry of Transport overseeing the ship registry).

In February 1996 the *MPA Act* (1996) passed into law. The MPA was established by combining the regulatory functions of the three entities, while the commercial and marine activities of the original Port of Singapore Authority were separated to form the PSA Corporation, a wholly owned entity of Temasek Holdings, the investment company of the Singapore government. The changes are represented in Fig. 13.5 below.

As the port authority, the MPA manages and administers the port of Singapore through the regulation of essential port and marine services and facilities. However, it is not involved in any operational aspects of the terminal businesses. In fact, the MPA performs several functions, including:

(a) *Regulating and licensing port and marine services and facilities.* This includes discharging the MPA's responsibilities with respect to Singapore's role as both a flag and port State, ensuring that regulations are kept up to date and complied with as well as adhering to various

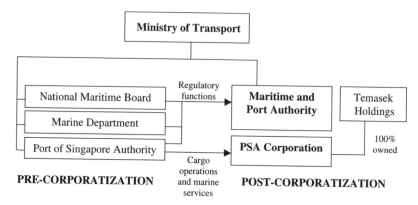

Fig. 13.5.   Structural Changes to the Governance of the Port of Singapore.

International Maritime Organisation (IMO), and other international conventions, to which Singapore is a party. In terms of regulating Singapore's container handling sector, the MPA is responsible for issuing licenses to operators (currently just two) to handle cargo. In the case of the PSA, the license extends to the provision of pilotage and towage services. It is the practice in Singapore not to make public the terms and conditions under which these licenses are issued.

(b) *Protecting the marine environment and ensuring navigational safety and maritime security.* This includes, inter alia, the prevention of marine incidents by managing vessel movements, the protection of the marine environment through proactive measures such as those aimed at oil spill prevention and working with other Singapore government agencies and the private sector in the implementation of various security measures. As discussed in Lam and Bhattacharyya (2005), the latter is especially the case in relation to the number of threats to maritime security that are faced in the Malacca Strait.

(c) *Managing Singapore's merchant fleet.* This includes the promotion and marketing of the Singapore Registry of Ships (SRS). The emphasis in this activity has been placed on promoting Singapore as a high-quality register. This has been done by targeting high-quality owners and ships to register with the SRS, ensuring the appropriate qualification and welfare of seamen employed on board Singapore registered ships, and providing high-quality support services to the shipping community.

(d) *Working with various government agencies and industry partners to de-velop and promote Singapore as a leading hub port and International*

Maritime Centre (IMC). This includes assessing industry trends and developing strategies which, for example, are aimed at identifying and nurturing new business areas, applying appropriate port planning so as to maintain Singapore's status as a premier hub port and, by encouraging both local participation and foreign investment, enhancing the breadth and depth of services that Singapore currently offers. Thus, the MPA has been proactive in attempting to reinforce the competitiveness of the port of Singapore against regional competition. For example, in 2004 it announced a package of cuts in port dues worth US$1.5 million for vessels bunkering in the port, car carriers, regular passenger ship calls and major ship repair work with the intention of boosting these sectors (Hand, 2004a). Indeed, it is precisely this sort of measure, taken in support of its port (especially when, in the past, they have related specifically to container shipping), which has prompted Hong Kong to complain within a World Trade Organisation forum that Singapore's allocation of assistance to ships to ensure that they continue to use the Port of Singapore constitutes a form of services subsidy (Zarocostas, 2004).

(e) *Safeguarding Singapore's maritime interests in the international arena.* This responsibility extends to being the *de facto* advisor on maritime-related matters to the government of Singapore. This role also includes managing the MPA's foreign relations in order to enhance Singapore's profile in the international maritime community – an aspect that includes building up links with important maritime countries and international organisations such as the IMO.

The MPA is headed by a chief executive who oversees the day-to-day operations of the organisation. The chief executive in turn reports to the Board of Directors, which is chaired by the current Permanent Secretary of the Ministry of Trade and Industry (MTI). The MPA falls under the purview of the Ministry of Transport (MOT) and major policy decisions are made at the ministerial level, usually in close consultation with members of the maritime community; in particular, the Singapore Shipping Association and the Singapore Maritime Foundation. The benefit of this arrangement is that policies are formulated with a strong focus on their relevance and importance to the commercial needs of the maritime community. This system also facilitates direct and speedy feedback from the commercial maritime community to the minister on various issues where improvements are felt to be necessary. The potential drawback of this system is that it does require a capable administration in order for there to be effective management of communication and information channels. Further details on the workings of the MPA can be obtained from the organisation's website at www.mpa.gov.sg.

Apart from the MPA and MOT, other government agencies with relevant domain knowledge are also involved in the management, regulation and development of selected maritime industries. For instance, the Economic Development Board of Singapore oversees the shipbuilding, ship repair, offshore and marine equipment sectors while the Singapore Tourism Board oversees the cruise sector. Both these statutory boards report to the MTI. Port security arrangements are provided by the Police Coast Guard and the Republic of Singapore Navy. The former reports to the Ministry of Home Affairs, while the latter reports to the Ministry of Defence.

Hence, the regulation and management of the port relies on the concerted effort of various government agencies and private sector bodies to come together on specific issues to bolster the competitive advantage of maritime Singapore vis-à-vis current and upcoming competitors. The guiding principle that unites these organisations is the common avowed objective of creating a 'pro-business environment' where the maritime business community can thrive and prosper in an internationally competitive context.

3.2. Ownership and Control of Container Operations

Although the government has announced that it has "set aside land for 20 additional berths in Pasir Panjang Terminal for whoever can best run the berths, be it PSA, Jurong Port, other port operators or shipping lines" (Government of Singapore, 2002a), PSA International is by far the most important container handling company operating in Singapore.

Before 1997, the Port of Singapore Authority was a public port authority under the Government of Singapore. The Authority not only owned the terminal facilities, but also managed and controlled the operational aspects of its container terminals. On 1 October 1997, however, the PSA Corporation was formed by the corporatisation of the former Port of Singapore Authority. This corporatisation involved the transformation of the port of Singapore from its previous status as a government body to one where it is independent of government. It has commercial objectives and takes decisions on a commercial basis, akin to a private sector company. At the same time, it does remain an entirely government-owned entity, with a wholly owned subsidiary of the Government (i.e. Temasek Holdings (Private) Limited) holding a 100 percent share of the PSA Corporation's shares (Cullinane & Song, 2001). The corporatisation also marked the separation of terminal operations from the port authority function, which came to be respectively performed by the PSA Corporation and the MPA.

The government felt that these changes to the governance structure of the port of Singapore were required in order to enhance the commercial flexibility of PSA to operate and to invest effectively in the fiercely competitive environment that was emerging in the regional port landscape. It was felt that this, together with the perceived greater discipline and market orientation of a private sector entity, would sharpen PSA's competitive edge in this environment. In addition, the changes would provide PSA with the ability to better respond to the requirements of the port's bigger customers. Finally, and most importantly, it was suggested that the new arrangements would facilitate PSA's strategy to "go global" (Government of Singapore, 1996, 1997).

In December 2003, the PSA Corporation was restructured so that it now covers only Singapore's domestic container terminal operations. At the same time, this downsized entity became a 100 percent subsidiary of a new holding company, PSA International. All PSA Corporation's international business at that time transferred to PSA International, which again, although run as an independent concern on a strictly commercial basis, ultimately remains a government-owned entity since Temasek Holdings remains a 100 percent government-owned investment holding company (PSA International, 2003c). Included in the reorganisation was an earlier move made in March 2003 to transfer PSA Corporation's airport handling, cruise terminal and exhibitions businesses, as well as its interest in cable car operator Singapore Cable Car (Pte) Ltd, to Hazeltree Holdings Pte Ltd, a 100 percent owned subsidiary of Temasek Holdings. The transfer was aimed at sharpening PSA's business focus on its core competence in port development, management and operations in anticipation of the restructuring that was to follow in December.

Under the new structure, there are four business regions (Singapore, Europe, China, and Asia and the Middle East), each headed by a regional chief executive officer responsible for business performance. The reorganised structure moved PSA International from a Singapore-centric company to one that had become globally oriented in its business focus. Subsequent to the restructuring in 2004, Moody's raised its rating of PSA Corporation to Triple A in recognition of the Port of Singapore's response to regional price competition and the strategic and economic importance of the port to the government and nation (Hand, 2004b).

Over the years and especially since corporatisation, what is now PSA International has adopted a strategy of diversifying its portfolio of container terminal operations on a global scale. At the time of writing, PSA International operates 17 terminal facilities in 11 countries worldwide (www. internationalpsa.com), including

- Singapore port (through its 100 percent ownership of PSA Corporation)
- Delwaide and Churchill Docks, Europa and Noordzee Terminals and Deurganckdok West in Antwerp, Belgium and OCHZ terminal in Zeebrugge, Belgium
- Holland Terminals in Rotterdam
- Voltri Terminal Europa and Venice Container Terminals in Italy
- Sines Container Terminal in Portugal
- Tuticorin Container Terminal in India
- Dalian, Fuzhou and Guangzhou Container Terminals in China
- Incheon Container Terminal in Korea
- Muara Container Terminal in Brunei
- Eastern Sea Laem Chabang Terminal in Thailand
- Hibiki Container Terminal in Japan

The booming Chinese economy and an expansion of container volumes emanating from the rest of Asia has helped PSA International's profits reach US$544 million in 2004, an increase of 29 percent over the previous year. At only 5 percent, however, revenue growth was notably slower than global growth in container throughput, which was up 15.5 percent year-on-year to 33.11 million TEUs. PSA International itself attributes this partially to the sale of non-port businesses (Hand, 2005).

Throughput at the port of Singapore was up 14.1 percent to 20.62 million TEU in 2004. However, although its home base port still accounted for 62.2 percent of total container throughput handled by PSA International in 2004, Singapore port operations accounted for only 45 percent of group revenues. Despite this, it did remain the most important source of group profit, with PSA International indicating that the port of Singapore accounted for over 50 percent of its profit, which itself grew by 17.2 percent compared to 2003 figures. Interestingly, the net profit reported in 2004 (but relating to 2003) for the newly established Singapore-based company PSA Corporation actually exceeded total group net profit (Hand, 2005).

In terms of future plans for further expansion and/or diversification, COSCO is keen to expand upon its cooperation with PSA International and the two companies already have links not only in Singapore but also in China and the European port of Genoa. The seemingly close working relationship is cemented and complicated by the fact that Temasek Holdings owns a stake in the Chinese company's Singapore-listed arm, Cosco Corp (Anonymous, 2003).

At the time when Maersk-Sealand took a 30 percent stake in PTP, reports suggested that PSA was also keen to buy into PTP. Similarly, in 2003, there were rumours that PSA's parent, Temasek Holdings, was involved in negotiations for closer cooperation or even a merger with PTP's main shareholder, the Malaysian Mining Corp. Ultimately, this had to be formally denied by the latter in a statement delivered to the Kuala Lumpur stock exchange (Hand, 2003). More recently, informed opinion would appear to suggest that PTP might include its main competitor as an investor in its second phase expansion (Hand, 2004c). The rationale for such a rapprochement might lie with the fact that Singapore is now in a much stronger competitive position than it has been previously and also with the fact that this may overcome the limitations on space for the further expansion of port facilities that currently exist within Singapore.

At the time of writing, speculation again abounds. This time that Temasek Holdings is seeking to acquire a controlling stake in P&O, the UK ports and ferries group (Gray & Hand, 2005). If true and if successful, this could lead to the creation of the world's largest container terminal operator. In addition, if last year's takeover of Neptune Orient Lines (NOL) by Temasek is also added into the equation, it might be the case that the holding company is seeking to establish itself in a leading position in both container terminals and liner shipping and perhaps even decide to merge PSA with NOL. Such a strategy would pitch the newly merged conglomerate into the same competitive arena as the A.P. Møller–Maersk Group.

Yet more speculation in the press (Hand, 2005) suggests that PSA International is expected to revive its long-awaited initial public offering (IPO) on the Singapore stock exchange. An IPO of PSA International shares was originally planned for 2002 but was postponed following the company's loss of the Singapore-based regional transhipment business of Maersk–Sealand and Evergreen to PTP. At that time, it was expected that US$1.22 billion would be raised from the sale of a 20–25 percent stake in the company (Hand, 2004d). It is interesting to consider that the form of privatisation chosen at that time reflected the fact that the port of Singapore favoured a common user system, rather than dedicated berths, for its terminals (Cullinane & Song, 2001), possibly in the belief that they are inherently more efficient (Lim, 1996). Given the volte-face on this policy that has taken place since then, it remains to be seen whether the IPO option will actually remain the preferred method of securing private sector involvement in the share ownership of PSA International, should it desire to do so.

With the recent sharp upturn in PSA International's business fortunes and the continuing prospect of a generally healthy shipping industry, the time seems ripe for such a move. Should an IPO of PSA International actually come to fruition in the foreseeable future, it would represent Singapore's biggest in over a decade. It would also help silence some of those WTO members, including the US, that have called on Singapore to reduce the level of government-linked corporations within the economy which, according to some estimates, account for as much as 40 percent of the total capitalisation on the Singapore stock exchange (Zarocostas, 2004).

It has already been mentioned that the second container terminal operator in Singapore is Jurong Port Pte Ltd. Handling approximately 0.7 million TEUs of containers in 2004, Jurong Port was awarded the second port licence by the Maritime and Port Authority of Singapore in January 2001 after having been a division within Jurong Town Corporation (JTC), a statutory board under the purview of the MTI, for over 30 years. Jurong Port remains 100 percent owned by JTC. According to the government, this represented a move (however small) to encourage greater intra-port competition (Government of Singapore, 2002b).

4. FUTURE CHALLENGES

High cargo volumes and shipping tonnage generally translate into more business opportunities for the port. However, the port of Singapore faces increasing competition from neighbouring competitors who are able to offer ever-improving service quality at competitive prices.

The major challenges the port of Singapore faces as a consequence of the increasingly competitive environment in which it has to operate can be summarised as three main areas:

1. Port competition in Southeast Asia is likely to centre on transhipment containers, with price being the key determinant of port choice decisions. Singapore's ability to overcome the competition it faces from other ports in the region depends crucially on how well it can leverage the competitive and comparative advantages of each ASEAN member in order to sustain its position as the region's premier transhipment hub.

   Other than fulfilling customer demand for the delivery of high-quality services at competitive prices, there is also the need to strengthen feeder connections within the region and, more importantly, to generate greater intra-regional and external trade volumes. The current strategy being pursued by Singapore for securing this expansion of trade mainly involves the establishment of bilateral Free Trade Agreements and multilateral agreements within ASEAN and between ASEAN and other regional groupings or countries (ASEAN, 2005; Government of Singapore, 2005). Although these agreements do not relate directly to cargo handling, the greater trade volumes that are expected to be generated from these initiatives will yield greater opportunities in an expanded regional transhipment market not just for Singapore, but for all the region's major port players. This could mitigate, at least to some extent, the deleterious impact of price competition across the region's port sector as a whole. This proposed cooperative approach is also likely to lead to better management of economic resources that are currently devoted to intense port competition and enable the region's ports to engage in competition through more economic and sustainable means.

2. As an alternative approach to enhancing the relative competitiveness of the port of Singapore and, thereby, overcoming the intensifying port competition in the region, the government is actively pursuing a policy of developing the country to achieve the status of IMC. The rationale for this policy is to move up the value chain by focusing on knowledge-intensive activities that are harder for competitors to emulate and, in consequence, propel Singapore's maritime industry onto the next level of economic competence. The same higher-level activities also have the potential to generate stronger and wider positive multiplier effects for the maritime cluster and the national economy overall. However, a world-recognised IMC usually possesses a wide range of ancillary services that exhibit no particular dependence on a single or narrow group of activities.

As we have already seen, the Singapore maritime industry is dominated by the port, shipbroking and chartering, shipbuilding and repair, shipping lines and agencies and logistics services. Hence, the government has been focusing much effort on increasing the breadth and depth of other maritime ancillary services by capitalising on its leading status in cargo-handling, bunkering and ship repair. By so doing, it is attempting to firmly anchor these sectors within the IMC-development strategy. This will involve identifying key maritime activities for specific or joint development in order to generate greater economic spin-offs for the whole of Singapore.

3. On the landward side, the constraint on available space for development translates into high land values that dictate the need to optimise the management of land-based facilities such as those involved in cargo-handling activities. On the seaward side, a similar lack of appropriate

cheaply navigable space demands the better management of anchorages and channels. Although the local port community has made extensive use of technology-based solutions to alleviate the concerns over lack of space both ashore and on the seaward side of operations, congestion on either side of the berth will inevitably transform into higher generalised costs and lower profit margins for the port user.

This set of circumstances should, and when it arises will, ultimately lead to the undermining of the long-term competitiveness of the port of Singapore and, therefore, the growth in its long-term volume of business and the revenue that accrues from this. Although port development plans are in place (most notably with respect to the expansion of the Pasir Panjang terminal) and even taking into account the greater competition that Singapore faces within the region, demand is fast outpacing planned supply of capacity and shortfalls are likely to occur sooner than originally expected. On the seaward side of the port, given the narrowness of the Straits of Singapore, there are obvious physical constraints on the size of navigable channels and anchorage capacity. This limiting factor is likely to become increasingly influential as ever-larger containerships are deployed by shipping lines.

5. CONCLUSIONS

Although Singapore currently enjoys a clear lead in the region in terms of both cargo volumes handled and shipping volumes received, the port community recognises that it must improve on its competitiveness by offering higher standards of service and productivity at competitive prices. If it fails to do so, it is fully cognisant of the risk of being replaced by upcoming regional competitors.

This risk is especially prevalent since port competition in Southeast Asia is expected to centre on the business of container transhipment. As a port where transhipment represents 80 percent of its container handling activity (Hong, 2002), Singapore is especially vulnerable to competition. As can be deduced from Table 13.3, this is despite having the highest level of connectivity by shipping lines and their services, the importance of which in winning transhipment business has been identified by UNCTAD (1993).

Hutchison Port Holdings took a 31.5 percent stake in Westport Terminal in Port Klang in August 2001 (World Cargo News, 2001). Together with the share in PTP taken by APM Terminals that has already been discussed,

Table 13.3. Index of Connectivity by Container Shipping Services for Selected Ports[a].

Number of:	Shipping Lines Calling		Shipping Services Calling	
	1999	2003	1999	2003
Singapore	100	100	100	100
Port Klang	52	63	35	43
Tanjung Pelepas	0	13	0	14
Laem Chabang	28	34	13	18
Tanjung Priok	40	41	17	19
Manila	31	32	23	23

Source: Containerisation International Yearbooks (2001–2005).
[a]Singapore = 100.

these investments have enabled the ports in question to compete head-on with PSA International in Singapore. Besides providing good service, both PTP and Westport, like Singapore, will be able to handle the mega container vessels that are planned for the future. Ocean Shipping Consultants (2003) forecast that transhipment volumes in Southeast Asia would double from 18.1 million TEUs in 2002 to 36.5 million TEUs by 2010. It further forecast that while Singapore's transhipment volumes, which are mainly handled by PSA International, would grow by 77.9 percent in the same period to hit 24.2 million TEUs, its market share would decrease from 75.0 to 66.3 percent.

Given the strategic importance of the port sector to the economy, it seems clear that the government of Singapore will continue to have a significant involvement in the commercial decisions of the sector. Thus, devolution of powers and/or even privatisation are unlikely to reach anywhere near the lengths attained in the UK for example. The reason why a comprehensive public–private split in the governance structure is unlikely is that it could result in the government losing control of its ability to meet its objectives for long-term economic development. However, at the same time, the government of Singapore does recognise that it has to foster efficiency in the commercial operation of the nation's port sector and, as such, has been and continues to be motivated to devolve commercial decision-making to those entities that are best equipped to focus on these commercial, rather than wider economic, perspectives. Treading the fine line of achieving an appropriate balance in devolving decision-making authority will be one of the critical challenges faced by the government in the further development and success of the port in future years.

One could argue that the changes that have thus far been made to the governance of the port of Singapore have enabled the port to continue to perform well. Apart from the growth in cargo-handling volumes mentioned earlier, the port has also been voted the "Best Seaport in Asia" from 1989 to 2004 in the Hong Kong-based Asian Freight Industry Awards and terminal operator PSA International has also been voted the world's "Best Container Terminal Operator" for the past 15 years.

The most important impact of the changes in governance has been that it has allowed PSA International and Jurong Port to focus on port-related business, while MPA addresses the broader issue of ensuring that the entire maritime cluster continues to remain competitive and an important contributor to Singapore's economic performance. Following the change in its responsibilities, the MPA has acted on behalf of the government of Singapore in actively pursuing a policy of developing the country to become an IMC. The rationale for this is to move up the value chain by focusing on knowledge-intensive activities that are harder for competitors to emulate and, in consequence, to propel Singapore's maritime industry to the next level of economic competence.

These same higher-level activities also have the potential to generate stronger and wider positive multiplier effects for the maritime cluster and the national economy overall. However, a world-recognised IMC usually possesses a wide range of ancillary services that exhibit no particular dependence on a single or narrow group of activities. In fact, as has been shown earlier, Singapore's maritime industry is dominated by its port, shipbroking and chartering, shipbuilding and repair, shipping lines and agencies and logistics services. Hence, the government has been focusing much effort on increasing the breadth and depth of other maritime ancillary services by leveraging or capitalising on its leading status in the port-related activities of cargo-handling, bunkering and ship repair. By so doing, it is attempting to firmly anchor these sectors as the core of the IMC development strategy.

In conclusion, Singapore's rich maritime heritage has allowed its maritime industry to remain dynamic and competitive until now. That same industry has also witnessed the nurturing of many companies that have progressed to become world-renowned international players. However, there are major challenges that must be faced and while the port will obviously continue to be one of the main lynchpins of the maritime cluster, it is clearly important that the cluster's core is expanded to include other maritime activities that are knowledge intensive and that generate wide-ranging and extensive economic spin-offs. The ultimate goal is therefore to create an environment in

Singapore that is conducive to the future prosperity and success of maritime-related businesses.

NOTE

1. The port sector is defined by the Singapore Department of Statistics to include exclusively port operation services such as stevedoring and lightering activities.

REFERENCES

Agence Presse France. (2002). Singapore port operator loses Taiwan's Evergreen to Malaysia, 3 April, Singapore.

Anonymous. (2000). Maersk takes PTP stake. *World Cargo News*, September, p. 6.

Anonymous. (2002a). Singapore status faces a double challenge. *Lloyds List*, 16 September.

Anonymous. (2002b). No regrets as Maersk settles in at Tanjung Pelepas. *Lloyds List*, 24 May.

Anonymous. (2003). COSCO plans more links with PSA. *Lloyds List*, 5 December.

ASEAN. (2005). ASEAN free trade agreements, ASEAN Secretariat, www.aseansec.org/economic/afta/afta_agr.htm

Containerisation International Yearbooks. (2001–2005). London: Informa Maritime & Transport.

Cullinane, K. P. B., & Song, D-W. (2001). The administrative and ownership structure of Asian container ports. *International Journal of Maritime Economics*, *3*(2), 175–197.

Cullinane, K. P. B., Song, D-W., & Gray, R. (2002). A stochastic frontier model of the efficiency of major container terminals in Asia: Assessing the influence of administrative and ownership structures. *Transportation Research A*, *36*(8), 743–762.

Government of Singapore. (1996). Maritime and Port Authority of Singapore Bill, Second Reading, 18 January.

Government of Singapore. (1997). Port of Singapore Authority (Dissolution) Bill, Second Reading, 25 August.

Government of Singapore. (2002a). *National Day Rally Speech*, 18 August, www.gov.sg/nd/ND02.htm

Government of Singapore. (2002b). Speech by Prime Minister Goh Chok Tong at PSA's gala dinner to celebrate 30 years of containerisation in Singapore, Ritz-Carlton Millenia, 28 June, http://app.mfa.gov.sg/pr/read_script.asp? View,1448.

Government of Singapore. (2005). Singapore's free trade agreements, Ministry of Trade and Industry, http://app.fta.gov.sg/asp/overview/overview.asp

Gray, T., & Hand, M. (2005). P&O shares rise on Singapore bid talk. *Lloyds List*, 20 May.

Hand, M. (2002). Hanjin Shipping extends agreement with PSA Corp. *Lloyds List*, 24 July.

Hand, M. (2003). Tanjung Pelapas shareholder denies merger talk. *Lloyds List*, 27 August.

Hand, M. (2004a). Singapore slashes up to 50 percent from port dues. *Lloyds List*, 22 December.

Hand, M. (2004b). Restructured PSA gains ratings upgrade from Moody's Company. *Lloyds List*, 16 February.

Hand, M. (2004c). Tanjung Pelapas linked to PSA. *Lloyds List*, 24 August.

Hand, M. (2004d). PSA remains coy on possible 2005 initial public offering. *Lloyds List*, 22 December.

Hand, M. (2005). Asian exports push PSA profits up to $544 m. *Lloyds List*, 11 March.

Hong, Y. N. (2002). Chairman of PSA Corporation's speech at the gala dinner to celebrate 30 years of containerisation in Singapore, Ritz-Carlton Millenia, 28 June. www. psamuara.com.bn/news/articles%20on%20psa's%2030th%20anniversary/speech_by_ dr_yeoninghong_28_june_2002.htm

Huff, W. G. (1994). *The economic growth of Singapore: Trade and development in the twentieth century*. Cambridge: Cambridge University Press.

Lam, J. S. L., & Bhattacharyya, A. (2005). *Analysis of the threats to maritime security*. Paper presented at the International Association of Maritime Economists Annual Conference, Limassol, 23–25 June.

Lim, D. (1996). Global ports of the 21st century. *SingaPort'96 and SibCon'96 Conferences*, Singapore.

Mak, J., & Tai, B. K. (2001). Comment: Port development within the framework of Malaysia's transport policy: Some considerations. *Maritime Policy and Management, 28*(2), 199–206.

Maritime Asia Today. (2002). Evergreen/PTP switch a done deal – YF Chang, 15 January.

MPA. (2005). Summary of port performance 1999–2004, www.mpa.gov.sg/

MPA Act. (1996). Government of Singapore, http://statutes.agc.gov.sg

Ocean Shipping Consultants. (2003). *The World Containerport Outlook to 2015*. Surrey, UK: Ocean Shipping Consultants Limited.

PSA International. (2003a). *News release*, 24 July, www.singaporepsa.com/html/news/ nr030222.pdf

PSA International. (2003b). *News release*, 22 February, www.singaporepsa.com/html/news/ nr030724.pdf

PSA International. (2003c). New corporate structure to position the PSA Group for global growth and expansion, 1 December, www.internationalpsa.com/news/nr031201.pdf

Robinson, R. (2005). Liner shipping strategy, network structuring and competitive advantage: A chain systems perspective. In: K. P. B. Cullinane (Ed.), *Shipping economics* (Vol. XII). Research in Transportation Economics. Amsterdam: Elsevier.

Robinson, R., & Everett, S. (1997). *Malaysian cargo through Malaysian ports: Defining appropriate policies*. Research Report, Maritime Institute of Malaysia – Institute of Transport Studies, University of Sydney, pp. 22–24.

Singapore Maritime Cluster Study. (2004). Contribution of Singapore maritime industry to the economy in 2002. Presented at the 3rd Maritime Forum, Singapore.

The Business Times. (2000). Pelepas port will not end up only a Maersk hub: Ipsen, 23 August.

UNCTAD. (1993). Strategic planning for port authorities. Monograph UNCTAD/SHIP/646. Geneva: United Nations.

World Cargo News. (2001). Westport traffic growing rapidly. April.

Yap, W. Y., & Lam, J. S. L. (2004). An interpretation of inter-container port relationships from the demand perspective. *Maritime Policy and Management, 31*(4), 337–355.

Zarocostas, J. (2004). HK outrage over Singapore port services support. *Lloyds List*, 18 June.

CHAPTER 14

PORT GOVERNANCE IN HONG KONG

Dong-Wook Song and Kevin Cullinane

ABSTRACT

This chapter on the port of Hong Kong begins by describing its cargo throughput and infrastructure. It moves on to review the development and current state of the port governance structure and the economic and port policies that have been pursued by the government of Hong Kong in recent years. It is established that the port of Hong Kong and the wider economy are critically dependent on container flows to and from the Pearl River Delta across the border in the Chinese mainland. It is shown that Hong Kong's economic and port policies have largely been formulated as a response to the newly emergent geopolitical context that prevails in South China and the trade flows that move into and out of the area. This is also evident in a port governance structure that has recently been revamped. The nature of the recent changes to this structure and their motivation is explained and the emergent relationship between government (including the Marine Department as the port authority), terminal operators and the shipping industry described. A three-tier hierarchy is posited in terms of the roles taken by central government, the Marine Department as port authority and the two industry advisory councils that have been created. The new structure is criticised for distancing the port and shipping sectors from the wider logistics industry by limiting their involvement in fora that influence logistics policy in Hong Kong and the whole South China region.

Devolution, Port Governance and Port Performance
Research in Transportation Economics, Volume 17, 311–329
ISSN: 0739-8859/doi:10.1016/S0739-8859(06)17014-6

The chapter finishes by highlighting the importance and potential impact on the port of impending competition legislation and imminent port development planning.

1. INTRODUCTION

The shift of the global economy towards the newly industrialised countries in Asia and the consequently better trade links that now exist between Asian nations and the rest of the world have resulted in rapid economic development and growth in the Asian region. Since international trade is carried predominantly by sea transport, the major container ports in the region have played a crucial role in contributing to Asia's economic development.

The Chinese economy is currently widely regarded as the world's most vibrant. As the entrepôt for the Chinese mainland, Hong Kong enjoyed a particularly rapid pace of economic development up until the Asian financial crisis that occurred towards the end of 1997. At the time when Hong Kong commenced its dramatic economic growth in the 1960s, it developed its container terminals in order to accommodate regional container traffic. Its counterparts in the Chinese mainland did not keep pace with this rate of development and, as a consequence, there was no serious competition for the port of Hong Kong from Chinese mainland ports in the 20 years preceding the mid-to-late 1990s. The role of Hong Kong as the regional logistics hub emerged and progressed accordingly. However, since China's economy moved into double-digit growth in the early 1990s, the port of Hong Kong has increasingly faced a substantive challenge from ports in the Chinese mainland, particularly those located in South China (see Fig. 14.1).

The latest available statistics show that, in terms of annual container throughput, 10 Asian ports are ranked among the top 20 container ports in the world. With the exception of just one of the past 10 years, the port of Hong Kong has topped the world's league of busiest container ports.

In 2001, in terms of bulk cargoes, 16.6 million tonnes of wet bulk and 20.3 million tonnes of dry bulk were handled in Hong Kong, figures that compare to 122.0 million tonnes of containerised cargo – a not insignificant proportion (Cullinane & Cullinane, 2003). Coal provides the main source of energy for Hong Kong's three main power stations. Cement, stone, sand and gravel are vital import cargoes to support Hong Kong's prodigious building industry. Most of the major international oil companies have terminals in Hong Kong and import petroleum and petroleum products.

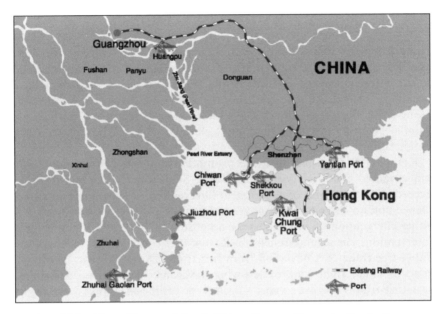

Fig. 14.1. Hong Kong and South China. *Source:* Drawn by the Authors.

Despite the fairly significant volume of bulk cargoes imported into Hong Kong, their value is comparatively small, as is their contribution to the economy and domestic employment. Together with the fact that the terminals which handle bulk cargoes are all owned and controlled by private industrial concerns that have imported these cargoes for their own use, this has meant that bulk movements through the port of Hong Kong barely register on the political horizon. This situation is reflected not only in the remit of the Hong Kong Port Development Council, where the avowed focus is solely and exclusively on container handling, but also in the design and implementation of different port governance structures over time. In recognition of this understandable perspective, this chapter mirrors this approach by concentrating only on those elements of the port of Hong Kong's activities (i.e. the container port) that relate directly to the formal design and evolution of port governance.

This chapter aims to provide a vision of the current status of Hong Kong port by reviewing the development of its governance structure and the economic and port policies it has pursued. It will be shown that these have largely been formulated as a response to the newly emergent geopolitical context that prevails in South China and the trade flows that move into and out of the area.

2. TRADING PATTERNS

Hong Kong has always been open to China, and Hong Kong businessmen currently receive no special treatment from the Chinese mainland compared to other foreign business people (Fung, 1996). Owing to its geographical proximity, extended family relationships and linguistic closeness, however, Hong Kong businesses often receive informal extra incentives, particularly from local authorities in Guangdong (Yeh & Mak, 1995; Chow & Ng, 2004; Gold, Guthrie, Wank, & Li, 2004).

Krause (1981) classifies the trade promotion strategies of the Asian newly industrialised countries (viz. Hong Kong, Singapore, Korea and Taiwan) according to the degree of government intervention, as shown in Table 14.1. Depending on the pattern of government trade promotion, the strategy is either interventionism or *laissez-faire* according to the degree of government intervention. The second-dimension Krause uses concerns the trade pattern: either free trade or a deviation from free trade. Deviations can be categorised as either right-wing or left-wing deviations. Right-wing deviations, resulting from excessive export subsidies or tariffs on imports, are trade-promoting. Left-wing deviations, however, entail excessive restrictions on exports or imports and are trade-reducing.

In the context of the Krause framework, Hong Kong follows a non-deviationist and *laissez-faire* approach, which has impacted every aspect of the Hong Kong economy, including the port industry. In other words, the HKSAR Government minimises its involvement in the economy and in trade, adopting what is very close to a free trade position. This philosophy is applicable to almost all industry sectors.

To a large extent, the world continues to view Hong Kong as the main conduit for accessing the Chinese mainland. A large volume of trade moves through Hong Kong to and from the Chinese mainland, particularly the Pearl River Delta (PRD) region of Guangdong province just across the border. This has sometimes been referred to as entrepôt trade. In other

Table 14.1.   The Trade Promotion Strategy of the Four Asian Tiger Economies.

Type	Free Trade	Deviation
Interventionism	Singapore	Taiwan, Korea
Laissez-faire	*Hong Kong*	—

Source: Krause (1981).

words, the role of Hong Kong in China's trade is by and large in the role of an intermediary. In this respect, two notable forms of trade between the Chinese mainland and Hong Kong are worth discussing: that is, re-exports and outward processing (for detailed information, see Sung, 1998).

In brief, *re-exports* take place when imports to Hong Kong are consigned to a buyer in Hong Kong, who takes legal possession of these cargoes. In this process, the buyer in Hong Kong carries out a value-added economic activity (e.g. grading, packaging, bottling and assembling), then re-exports the goods elsewhere. On the other hand, *outward processing* arrangements are made if companies subcontract all or part of their production processes. These trading patterns often occur between Hong Kong companies and manufacturing entities in China. Raw materials or semi-manufactures are exported to China for further processing. The Chinese mainland entities engaged in this process can be local enterprises, joint ventures or some other form of business involving foreign investment.

These two trading patterns are an important characteristic of the trade between Hong Kong and China. It is speculated that Hong Kong's trade pattern with China will continue to grow over the next few years. Given its geographical location and its know-how in banking and finance, insurance, telecommunications and transportation, Hong Kong is expected to continue to serve as the gateway to and from the Chinese mainland, benefiting each side via trade, so as to create and sustain economic synergy in the region (Wang, 1997).

The importance and development of the role played by the port of Hong Kong in facilitating the inward and outward movement of the economy's freight trade can be gleaned from Fig. 14.2. This reveals the primary positions of both seaborne and river trade and the fact that, in comparison to these modes, those of rail and air are largely insignificant. This is despite the fact that Hong Kong is the largest commercial airport for the handling of freight cargo in the world. The overwhelming significance and growth in importance of containerised cargoes carried on seaborne trips (even in simply volume terms, let alone value terms) can be seen in Fig. 14.3.

3. THE PORT OF HONG KONG

With the exception of 1998, when it was relegated to second position by Singapore, the port of Hong Kong has been the world's busiest container port for the whole of the last decade. It has experienced a remarkable growth in throughput over time, reaching approximately 22 million TEUs

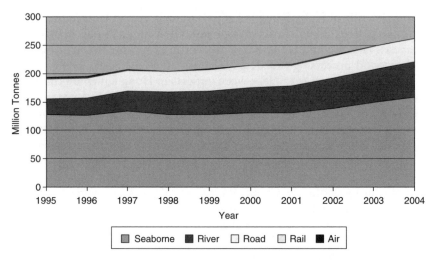

Fig. 14.2. Hong Kong's Total (Inward and Outward) Freight Movements by Mode. *Source:* Hong Kong Port Development Council (2005).

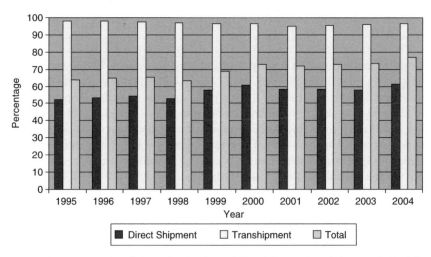

Fig. 14.3. Percentage of Containerisation of Total (Inward and Outward) Freight Movements. *Source:* Hong Kong Port Development Council (2005).

in 2004 (Containerisation International, 2005). The port is served by about 80 international shipping lines, providing over 400 container liner services per week that connect to over 500 destinations worldwide. The layout of the container terminals within the port of Hong Kong is shown in Fig. 14.4.

The port has nine fully operational container terminals in Kwai Chung with a combined capacity of approximately 18 million TEUs per annum. They occupy 285 ha of land, providing 24 berths and 8,530 m of deep-water frontage. The water depth is 15.5 m. In 2004, container terminal throughput amounted to 13.4 million TEUs, representing 61% of the port's total container throughput. The remaining 39% was handled at mid-stream sites, the River Trade Terminal (RTT), Public Cargo Working Areas (PCWAs), buoys and anchorages, and other wharves.

Historically, Hong Kong's port has expanded in tandem with the rapid economic development of Southeast Asian countries and China. This growth has been characterised by greater international trade connections between these areas and the rest of the world. The role of Hong Kong in global, as well

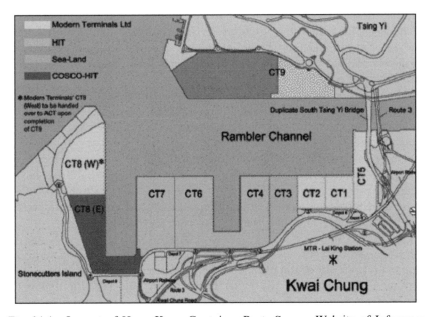

Fig. 14.4. Layout of Hong Kong Container Port. *Source:* Website of Informare (http://www.informare.it/news/gennews/2000/20001052.asp). On February 23rd 2005, Dubai Ports International (DPI) announced that it had completed the acquisition of CSX World Terminals (Sea–Land in this diagram).

as regional, container transport is well recognised and the concept of Hong Kong as a regional hub port or load centre cannot be separated from the regional manufacturing hubs of Southeast Asia and South China in particular.

In the current climate, any planning in Hong Kong cannot be implemented without taking into account the 'China factor'. While Hong Kong tries to retain its position as global and regional hub port for South China in the future, it is vitally important that Hong Kong recognises the need for regular contact and cooperation with the authorities just across the border in Shenzhen. This is facilitated by information exchange on port development strategies and by understanding what facilities are required in order to optimally meet the needs of its customers in the region. At the same time, the port of Hong Kong faces more severe competition, particularly in the form of the challenges posed by ports such as Singapore in the region and Yantian in the local South China context. As a consequence, Hong Kong's position as a leading load centre and transhipment centre for cargoes destined for, or originating from, China is under threat from a number of developments. These include: its reversion to China in 1997 and the concomitant reduction in autonomy that has since been perceived; port development in China itself; the potential for the opening of direct shipping links between Taiwan and the Chinese mainland, and; the easing of restrictions on access to, and investment in, Chinese ports for overseas shipping lines and container handling companies.

Almost every major international container shipping company is now engaged in trade with China and, in consequence, the number of direct line services to an increasing number of Chinese ports has grown exponentially. This occurrence is a phenomenon of just the last few years. In the past, the large proportion of container trade with China was transhipped via Hong Kong, Korea and even Taiwan. In spite of the fact that a significant amount of container trade bound for China is still transhipped through these three territories, mainly Hong Kong, the Chinese government operates a deliberate policy to reduce the ratio of cargoes that do.

About 90% of South China's cargo is handled through Hong Kong. Of the total container throughput in Hong Kong, about 70% relates to goods shipped into or out of the Mainland. Hong Kong is, therefore, the gateway to southern China, and the Pearl River Delta now has about 480,000 factories, employing over 5 million workers and accounting for 40% of the total exports of the Chinese mainland. A large part of this output is, in fact, in no small part due to the foreign direct investment of Hong Kong entrepreneurs, who have been investing in adjoining Guangdong province, especially the Pearl River Delta, for more than two decades. The extent of that investment amounts to about US$45 billion, with Hong Kong-owned enterprises employing an estimated 10 million people (Federation of Hong Kong Industries, 2003). The combined container throughput of Shenzhen ports in 2003 amounted to a total of more than 10 million TEUs. As is graphically illustrated in Fig. 14.5, this constitutes an obvious threat to Hong Kong's port and associated businesses.

With respect to the development of port facilities, the first purpose-built RTT was completed at the end of 1999. Designed as a consolidation point for containers and bulk cargoes shipped between Hong Kong and the PRD ports, it provides a cheaper and environment-friendly alternative to road transport. Similarly, Container Terminal 9 is currently nearing completion. It occupies an area of 70 ha and has six berths with a design capacity to handle a total of 2.6 million TEUs a year. This will make it Hong Kong's largest container terminal. When it is completely operational (due at some time in 2005) the total handling capacity of Hong Kong's container

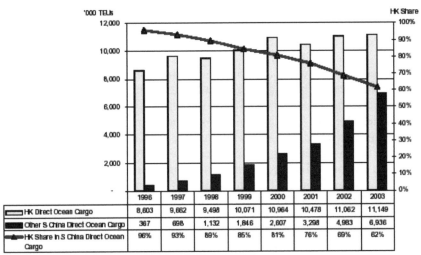

Notes: South China Direct Ocean Cargo includes South China direct ocean cargo handled by Hong Kong, Shenzhen and Guangzhou (no ocean to ocean transhipment). TEU means Twenty-Foot Equivalent Unit, which is the standard unit for counting containers and for describing the capacities of container ships or terminals. One 20 Foot ISO container equals 1 TEU. *Shenzhen ports account for the vast majority of throughput, as compared with Guangzhou.

Fig. 14.5. South China Direct Ocean Cargo: Hong Kong versus Shenzhen (and Guangzhou) Ports.*

terminals will exceed 15 million TEUs a year. Coupled with continuous efficiency improvements and the existence and further development of other container handling facilities (such as the RTT and mid-stream operations), the government of Hong Kong expects to be able to cope with forecast cargo throughput of about 30 million TEUs by 2010 (Ip, 2003; Hong Kong Economic Development and Labour Bureau, 2004).

4. PORT GOVERNANCE AND ADMINISTRATION

4.1. Context

The port of Hong Kong operates on the basis of what the Hong Kong government advertises as a unique public and private sector model. In fact, this model is one that is employed, to a lesser or greater extent, in many parts of the world. The port governance model that prevails in Hong Kong is one where the private sector is left to finance, develop and operate terminal facilities, while government concentrates on providing the back-up infrastructure needed to service the port, as well as strategic planning for port development. Thus, the port of Hong Kong is owned and managed by private companies, while the government provides merely the land, navigation channels, infrastructure and utilities. In fact, the Hong Kong government is the lessor of land sites to the private terminal operating companies. Neither the government nor the Marine Department (the de facto 'port authority' with respect to many aspects of port operations) owns or operates container terminal facilities. These are all privately owned and operated by four private companies, either independently or in partnership with each other: Modern Terminals Ltd., DPI Terminals, Hongkong International Terminals (HIT) Ltd. and COSCO-HIT Terminals (HK) Ltd.

The high level of operational efficiency has in the past enabled the terminals to handle throughput at levels that are higher than design capacity and there is no reason to believe that this will not again be possible, or indeed necessary, in the future. In addition, there are also over 20 large and small 'midstream' operators that handle containers from ships anchored in the harbour and numerous companies operating in what is referred to as 'the (Pearl) river trade'.

4.2. Government Involvement

Three government departments are directly and obviously involved in the governance of transport in Hong Kong. They constitute the official organisations that monitor, supervise and coordinate daily routine affairs, as well as strategic development plans. The *Transport Department* is responsible for road and railway administration, the *Civil Aviation Department* for airport administration, and the *Marine Department* for administration of port and maritime affairs. The *Transport Department* is directly responsible to the *Secretary for the Environment, Transport and Works*, while the Civil Aviation and Marine Departments report directly to the *Secretary for Economic Development and Labour*.

The *Secretary for Economic Development and Labour* heads the Economic Development and Labour Bureau (EDLB – formerly, the Economic Services Bureau) and assumes policy responsibility for ensuring that Hong Kong has the economic infrastructure to maintain its position as a major international and regional centre for business, aviation, port, maritime and logistics services and for tourism. The EDLB and the Marine Department are the principal bodies within the government responsible for Hong Kong's maritime matters, although various other government bureaux and departments do become involved with various aspects of maritime affairs on an ad hoc basis in areas such as insurance, tax, environment, work safety and practices, etc.

4.3. Industry Participation in Strategy Formulation

As shown in Fig. 14.6, the *Secretary for Economic Development and Labour* is responsible for the overall formulation of port and maritime policy under the advice of what was formerly the *Hong Kong Port and Maritime Board* (PMB) and what is now two separate bodies: the *Hong Kong Maritime Industry Council* (MIC) and the *Hong Kong Port Development Council* (PDC).

The PMB was created on 1 June 1998 to advise the government on Hong Kong's development as both a leading port and a premier international shipping and maritime services centre. In May 2000, it was restructured to include responsibilities for promoting logistics services development. It was originally formed from the Port Development Board, which itself was established in April 1990, to advise the government solely on port planning and port strategic development. The original rationale for reorganising and renaming the PDB came from the political and economic desire to promote the synergy between the port and shipping sectors.

The PMB brought together key players from the private sector and government to determine and promote solutions to problems. Two committees were set up in order to effectively discharge its functions: the *Port*

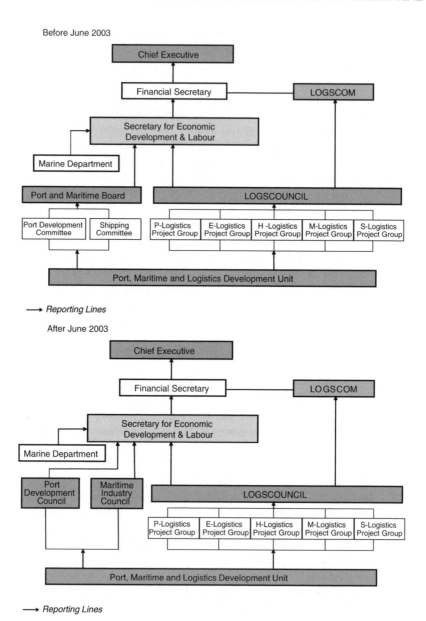

Fig. 14.6. Logistics Administration and Organisation in Hong Kong. *Source:* Updated from Cullinane and Song (2001).

Development Committee focusing on strategic port-related issues and the *Shipping Committee* focusing on developing Hong Kong as an International Maritime Centre.

The PDC and MIC were created to replace the PMB in June 2003. The decision to re-organise the PMB into these two new and separate advisory bodies was justified on the basis of a recommendation from the PMB itself and based on the findings of a major consultancy project on strengthening Hong Kong's role as an international maritime centre (Hong Kong Port and Maritime Board, 2003). This report suggested that the then existing arrangements, involving a role for the PMB, had strengthened the perception that the government is absorbed with the port and that it considers the maritime industries to be exclusively concerned with container shipping. The increasing importance attached to 'logistics' in recent years had only served to reinforce that impression still further. The existing Shipping Committee of the PMB was deemed to lack authority and there was a need, it was felt, to develop the perception of the maritime industries in their own right and to show that the port and logistics, while important, constitute just part of the range of Hong Kong's maritime industries.

Given the focus of the report on enhancing Hong Kong's position as an international maritime centre, most of its policy recommendations related to the strengthening of what had formerly been the role of the Shipping Committee within the PMB. When addressing the potential future for what had been the Port Development Committee within the PMB, the report was far less prescriptive, suggesting only that it might be kept separate from the Logistics Council, put underneath it or merged with it.

The MIC focus is on the promotion of Hong Kong as an international maritime centre and the development of the maritime industry. It provides advice to the government on the formulation of measures and initiatives to further develop Hong Kong's maritime industry. As an integral part of this remit, it assists the government to promote the Hong Kong Shipping Register, the shipowning and ship management services, and the development of human resources for the maritime cluster.

The PDC, on the other hand, provides advice to the government on port development strategy and port planning, including the development of new cargo handling and container terminal facilities to meet future demand. It will also assist the government in the promotion of Hong Kong as a regional hub port and a leading world container port.

The PMB previously, and now the PDC and MIC, strive to produce an improved infrastructure and business environment for port operators as well as for users of the port. One example of their work has been to facilitate the movement of containers to and from the Pearl River Delta. To this end, steps have been taken to streamline the flow of vehicles across the land boundary between Hong Kong and mainland China, to expand the control point handling capacity at Lok Ma Chau (one of the three border crossings for trucks), to establish empty truck lanes with fast customs checks, and to extend customs clearance hours at night at Lok Ma Chau/Huanggang. The major port development project that is underway is the building of the new six-berth Container Terminal 9, but a 'Competitive Strategy and Master Plan' for future port development (including an assessment of the need for Container Terminal 10 in the future) is currently being prepared by independent consultants.

The *Secretary for Economic Development and Labour* is also responsible for more general logistics policy under advice from the *Steering Committee on Logistics Development* (LOGSCOM), which is headed by the more senior *Financial Secretary*, to whom the *Secretary of Economic Development and Labour* is directly responsible. LOGSCOM provides the policy steer and accelerates measures to take forward the concept of 'Logistics Hong Kong'. The *Hong Kong Logistics Development Council* (LOGSCOUNCIL) is chaired by the *Secretary for Economic Development and Labour* and is tasked with implementing the directives it receives from the LOGSCOM. It also provides a forum for public and private sectors to discuss and coordinate matters of concern to, or affecting, the logistics industry and to carry out joint projects.

The *Secretary for Economic Development and Labour* has been appointed chairman of both the MIC and the PDC, in addition to also being the chairman of the Hong Kong Logistics Development Council (LOGSCOUNCIL).

Having previously provided the same for the PMB, the *Port, Maritime and Logistics Development Unit* (PMLDU) provides secretariat support to both the MIC and the PDC, as well as to their various committees, on all matters pertaining to the development of Hong Kong as a world leading port and a premier international shipping and logistics centre. The Secretariat also plays an active role in the promotion of the port of Hong Kong, its shipping industry and logistics services, both locally and overseas.

4.4. Operational Administration: The Marine Department

The Port of Hong Kong is unusual in that it does not have a port authority that provides the port infrastructure and controls it. Most of the port facilities are privately owned and operated, with minimal interference from government. The government's role is limited to undertaking long-term strategic planning for port facilities and to provide the necessary supporting infrastructure, such as roads and the dredging of access channels to the terminal.

The Marine Department is headed by the *Director of Marine* and is responsible for all navigational matters in Hong Kong, as well as the safety standards of all classes and types of vessels. Its stated mission is 'to promote excellence in marine services', which includes:

- facilitating the safe and expeditious movement of ships, cargoes and passengers within Hong Kong waters;
- ensuring compliance with international and local safety and marine environmental protection standards with respect to ships registered and licensed in Hong Kong and using Hong Kong waters;
- administering the Hong Kong Shipping Register and developing policies, standards and legislation in line with international conventions;
- ensuring compliance with international and local requirements on the competency of seafarers for ships registered and licensed in Hong Kong and using its waters, and regulating the registration and employment of Hong Kong seafarers;
- coordinating maritime search and rescue operations within Hong Kong's international area of responsibility and ensuring compliance with international conventions;
- combating oil pollution in Hong Kong waters, collecting vessel-generated refuse and scavenging floating refuse in specified areas of Hong Kong waters;
- providing and maintaining in the most cost-effective manner the number of government vessels that departments need to conduct their business.

The Marine Department provides a number of major maritime services:

- *Vessel Traffic*: The Vessel Traffic Centre helps ensure a safe and efficient marine traffic flow in the busy waters of Hong Kong. The service is supplemented by the department's patrolling launches, radar stations scattered around Hong Kong waters and by two local traffic control stations.
- *Pilotage Services*: The department regulates and monitors the pilotage service, which is operated by the pilots themselves as a private company. Pilot numbers and the quality of their services are kept under constant review and are closely monitored by the Pilotage Advisory Committee.

- *Hydrographic Office*: This carries out hydrographic surveys and charting services, including the issuance of Notices to Mariners for updating of charts, to fulfil international obligations and to safeguard Hong Kong's maritime trade.
- *Dangerous Goods*: The Dangerous Goods Management Unit carries out random inspections of vessels, with a view to ensuring the safe carriage of dangerous goods by sea in compliance with international and local requirements so as to minimise potential hazards to people and the environment.
- *Maritime Search and Rescue*: The Maritime Rescue Co-Ordination Centre, which is manned 24 h a day by professionally trained staff, is responsible for coordinating all maritime search and rescue operations within Hong Kong waters and the international waters of the South China Sea that are bounded by latitude 10° N and longitude 120° E.
- *Mooring Buoys*: The department provides and maintains 56 buoys within the port at which ships may perform mid-stream operations. Most are typhoon moorings, where vessels may remain secured during tropical cyclones.

5. CONCLUDING REMARKS

As shown in Fig. 14.7, the administration system for the port of Hong Kong can be depicted as a three-tiered hierarchy. Since it maintains ownership over the land upon which the container terminals are built, the government constitutes the highest tier in the administrative structure. Under the HKSAR Government, the Marine Department acts in the capacity of port authority and deals with all navigational matters in the port and surrounding area. It has responsibility for vessel traffic management, the safety standards of all classes and types of vessels and other regulatory matters, and is involved in the strategic planning of port developments. The PDC and MIC (formerly the PMB) are also involved in the planning of new port developments, but are merely advisory, rather than governmental, bodies.

Given the organisational structure that prevails and the various entities involved in the administration of the port of Hong Kong, it is clear that the government and people of Hong Kong perceive its port and maritime industries as critical to the success of its wider logistics policies and development. Since its port and the maritime transport associated with it constitute the mainstay of Hong Kong's transport logistics, it was natural that prior to June 2003 the PMB was the organisation coordinating all port and logistics-related

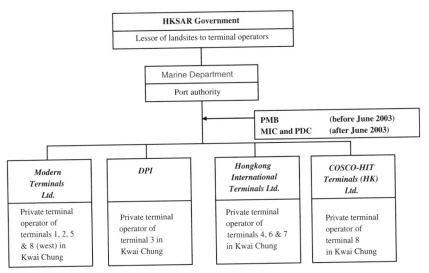

Fig. 14.7. Port Administration in Hong Kong. *Source:* Updated from Cullinane and Song (2001).

issues and policies. With the replacement of the PMB in June 2003 by the PDC and MIC, however, mainstream responsibility for logistics matters seems to have been diverted even more to the LOGSCOUNCIL and the LOGSCOM.

Back in 1998, the original reorganising and renaming of the PDB emerged as a response to the demand for the greater promotion of any existing synergy between port and shipping sectors. This demand emanated most strongly from that part of Hong Kong's maritime community that were extant customers of the port. In effect, the Hong Kong-based shipping industry was given, for the first time, a formally recognised and substantive advisory role in the formulation of government policy not only towards the maritime industries, but more specifically towards the port itself.

By implication, this attempt to promote any existing synergy between shipping and port sectors would appear to have been judged a failure. Only five years later, the shipping industry lobby in Hong Kong has gained an even stronger political voice. By emerging from under the wings of the PMB, it now merits its own advisory group, having equal status with that relating to port issues.

Although there are structures in place to allow each of the Councils to make each other aware of current thinking and discussions on policy recommendations, these are much less formal than was previously the case. In addition, both the PDC and the MIC appear to have lost any responsibility for, or even linkages to, the wider discussion and policy advice relating to the general logistics context. This might emerge as having important political ramifications.

Under previous structures, the port and shipping sectors were correctly recognised as the lynchpin of Hong Kong's logistics sector. In the absence of deliberate measures to prevent it from happening, the new structure has potentially undermined this. The currently high profile of 'logistics' in Hong Kong, especially as it relates to the intermediation of mainland China trade, extends to the 'man on the street' and even to popular youth culture. The divorcing of port and shipping industry representation from that of the wider logistics debate may be replicated in the minds of the people of Hong Kong and, in the longer run, have adverse consequences for the whole maritime sector. Of course, any deleterious effect of this type that might materialise will have to be netted against the gains that may accrue from a greatly strengthened maritime sector in Hong Kong and the emergence of Hong Kong as a true International Maritime Centre. Giving Hong Kong's shipping and wider maritime industry greater ownership and control over its own fortunes was, after all, the intention behind the recommendations that were initially made and subsequently implemented on splitting up the PMB into the two more focused advisory groups.

Another issue that is of more immediate concern to the future of the port of Hong Kong is the ongoing discussion on the introduction of a new competition policy that would apply *prima facie* to all economic sectors. By potentially impacting directly the operation of liner shipping services that call at the port, there is a danger that the imposition of such a policy may have the adverse effect of inducing switches in port choice among the users of Hong Kong's container port.

Also in the immediate short term, the publication of a new report on the port of Hong Kong's future is imminent at the time of writing. 'Hong Kong Port: Master Plan 2020' has been commissioned so that a competitive strategy and master plan for port development over the next two decades can be formulated. It is intended that this report will not only resolve the ongoing debate as to the preferred location for the site of new container port facilities, but will also make recommendations on how best to enhance competitiveness in the region by identifying new improvement initiatives. The latter are bound to have as their focus the improvement of landside logistics infrastructure and the consequent shortening of access times to and from the main source and destination of cargoes in the Pearl River Delta. In the absence of measures that prove successful in achieving this, the question of the location and form of new container handling facilities may well prove to be an irrelevance.

REFERENCES

Chow, I. H. S., & Ng, I. (2004). The characteristics of Chinese personal ties (Guanxi): Evidence from Hong Kong. *Organization Studies, 25*(7), 1075–1093.

Containerisation International. (2005). *Yearbook*. London: Informa.

Cullinane, K. P. B., & Cullinane, S. L. (2003). *Working paper 1: Maritime and shipping industry profile*. Study to Strengthen Hong Kong's Role as an International Maritime Centre, report for the Port & Maritime Board, Government of the Hong Kong Special Administrative Region.

Cullinane, K. P. B., & Song, D.-W. (2001). The administrative and ownership structure of Asian container ports. *International Journal of Maritime Economics, 3*(2), 175–197.

Federation of Hong Kong Industries. (2003). Made in PRD: The changing face of HK manufacturers. Retrieved from http://www.industryhk.org/english/fp/fp_res/files/prde.pdf

Fung, K. C. (1996). Mainland Chinese investment in Hong Kong: How much, why and so what? *Journal of Asian Business, 12*(2), 21–39.

Gold, T., Guthrie, D., Wank, D., & Li, B. B. (2004). Social connections in China: Institutions, culture, and the changing nature of "Guanxi". *American Journal of Sociology, 109*(6), 1533–1534.

Hong Kong Economic Development and Labour Bureau. (2004). Study on Hong Kong Port – master plan 2020: Draft executive summary. Retrieved from http://www.pdc.gov.hk/eng/plan2020/pdf/annex.pdf

Hong Kong Port and Maritime Board. (2003). *Study to strengthen Hong Kong's role as an International Maritime Centre*. Retrieved from http://www.mic.gov.hk/eng/bulletin/docs/Final_Report-10_Mar_03.pdf

Hong Kong Port Development Council. (2005). *Summary statistics of port traffic of Hong Kong*. Retrieved from http://www.pdc.gov.hk/eng/statistics/docs/Sep2005.pdf

Ip, S. (2003). Speech given by the Secretary for Economic Development and Labour at the opening ceremony of Container Terminal 9. Retrieved from http://www.pdc.gov.hk/eng/bulletin/speech22072003.htm

Krause, L. (1981). Summary of the Eleventh Pacific Trade and Development Conference on Trade and Growth of the Advanced Developing Countries. In: W. Hong & L. Krause (Eds), *Trade and growth of the advanced developing countries in the Pacific Basin*. Seoul: Korea Development Institute.

Sung, Y. W. (1998). *Hong Kong and South China: The economic synergy*. Hong Kong: City University of Hong Kong Press.

Wang, J. J. (1997). Hong Kong container port: The South China load centre under threat. *Journal of the Eastern Asia Society for Transportation Studies, 2*(1), 101–114.

Yeh, G. O., & Mak, C. K. (Eds). (1995). *Chinese cities and China's development: A preview of the future role of Hong Kong*. Hong Kong: Centre of Urban Planning and Environmental Management, University of Hong Kong.

CHAPTER 15

PORT GOVERNANCE IN CHINA

Kevin Cullinane and Teng-Fei Wang

ABSTRACT

This chapter begins by describing China's policies of economic reform since the inauguration of its open door policy in 1978. This provides the historical context for the country's concurrent reform of its port industry. The evolution and gradual decentralisation of the port governance system is analysed within three distinct phases of development that have taken the sector from one where ownership and decision-making were highly centralised to one where policies of corporatisation and privatisation have been recently reinforced by China's new Port Law of 2004. The chapter concludes by asserting that it is still too early to tell whether the latest phase of reforms will prove to be successful in solving China's port problems – particularly the capacity issue – and points to possible implications of the reforms for overseas investment and future levels of concentration within the market.

1. INTRODUCTION

China's economy has experienced phenomenal growth since the introduction of its open door policy and accompanying economic reforms in 1978. As shown in Fig. 15.1, this can be seen quite starkly in the development of its GDP and international trade figures.

Devolution, Port Governance and Port Performance
Research in Transportation Economics, Volume 17, 331–356
Copyright © 2007 by Elsevier Ltd.
All rights of reproduction in any form reserved
ISSN: 0739-8859/doi:10.1016/S0739-8859(06)17015-8

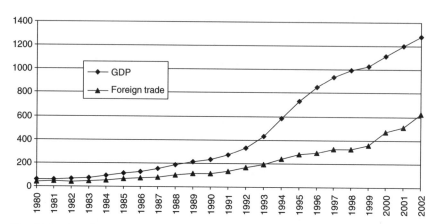

Fig. 15.1. Foreign Trade and GDP of China (in Billion US$). *Source:* Department of Water Transport (1998, 2002).

As an integral element of its economic development strategy, the economic structure of China has also changed dramatically over the last few decades. It has moved from an economy that revolved around central government planning to one where decision-making is more widely distributed within what is increasingly emerging as a market-based economy. As such, there has been a significant shift from a high degree of centralisation to a situation where there is much greater decentralization – a trend that has been characterised by the devolution of decision-making authority. New governance structures have either been put in place or naturally evolved in both governmental and commercial organisations in order to facilitate this.

As the conduit for the nation's international trade, the port industry in China has also had to develop rapidly in order to keep pace with an ever-expanding economy and cargo flows. Table 15.1 lists the most important ports in China and the growth over time of cargo throughput at these ports. This shows that the fastest growing ports are mainly those (such as Ningbo, Tianjin and, most notably, Shenzhen) that are competing directly for cargoes with already established ports; in this case, Shanghai, Dalian and Hong Kong, respectively. Also, Fuzhou and Xiamen have benefited greatly from favourable policy changes with respect to inward investment from, and shipping links with, Taiwan. Fig. 15.2 shows the locations of China's major ports.

Table 15.2 provides a wider perspective on this growth in throughput figures by categorising the various phases of port development that China

Table 15.1. Growth in Cargo Throughput at the Top Ports in China (in Million Tons).

	1990	1995	2000	2002
Shanghai	139.59	165.67	204.40	263.84
Ningbo	25.54	68.53	115.47	153.98
Guangzhou	41.64	72.99	111.28	153.24
Tianjin	20.63	57.87	95.66	129.06
Qingdao	30.34	51.30	86.36	122.13
Qinhuangdao	69.45	83.82	97.43	111.67
Dalian	49.52	64.17	90.84	108.51
Shenzhen	4.80	11.74	56.97	86.67
Fuzhou	5.61	10.32	24.26	39.07
Lianyungang	11.37	17.16	27.08	33.16
Rizhao	9.25	14.52	26.74	31.36
Yingkou	2.37	11.56	22.68	31.27
Xiamen	5.29	13.14	19.65	27.35
Yantai	6.68	13.61	17.74	26.89
Zhanjiang	15.57	18.95	20.38	26.27
Shantou	2.79	7.16	12.84	13.80
Haikou	7.56	7.85	8.08	10.73
Total	483.20	801.66	1,256.03	1,666.28

Source: Department of Water Transport (2002).

has experienced since its genesis in 1949. The start of the 'fast development stage' in 1980 closely corresponds to the implementation of China's 'open door' economic policy, as characterised by the rapid expansion of international trade and cargo flows through China's ports. With the demand for cargo handling in ports growing rapidly, the success or otherwise of China's economic experiment depended upon its ability to respond by developing and supplying port capacity. This situation has prevailed over most of the past 25 years; consistent double-digit economic growth has meant massive increases in cargo throughput which, despite almost constant problems of port congestion, have ultimately been catered for by new port or terminal development and greater efficiency in cargo handling.

By the end of 2002 China had 33,600 berths, including a total of 835 berths having the capability of serving ships of over 10,000 dwt. Of these berths, 3,822 are distributed among the country's coastal ports, with 700 berths capable of serving ships of 10,000 dwt. The remaining 29,778 berths are distributed across numerous inland waterway ports, including 135 berths that can handle ships of 10,000 dwt (Department of Water Transport, 2002).

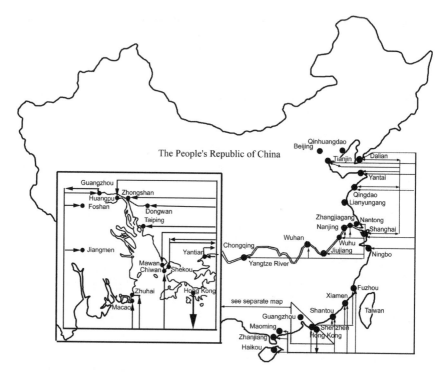

Fig. 15.2. The Three Major Geographical Port Groupings in Mainland China.

Table 15.2. Port Development in China[a].

	Starting Stage	Growing Stage	Fast Development Stage
Time period	1949–1972	1973–1980	After 1980
Number of existing deepwater berths	92 (as of 1972)	137 (as of 1980)	553 (as of 1998)
Number of new built deepwater berths	32	50	350

Source: Department of Water Transport (1998).
[a]The People's Republic of China was founded in 1949.

Most port development in recent years has focussed on the container sector. Similarly, most port policy has been formulated with container ports in mind, with the vast majority of port-related government legislation and policy aimed at expanding the capacity and enhancing the efficiency of the country's container terminals. The development of the container port sector within China has now reached a point where a new container port hierarchy has emerged that revolves around the role of many of China's major container ports as hubs for vast hinterland areas of the country (Cullinane, Cullinane, & Wang, 2005). The timescale and nature of this evolution is depicted in Fig. 15.3, as is the likely future scenario should current trends continue.

The focus of government port policy on the container sector is hardly surprising given the nation's push towards, and reliance upon, the export of manufactured goods. Indeed, the country already accounts for 40% of the container volume on the Asia/Europe trade and 50% on the transpacific (Containerisation International Yearbook, 2005). However, the role played by China's bulk ports and terminals can hardly be dismissed as an insignificant contribution to the country's economic miracle. The growth in China's manufacturing sector has provided an enormous stimulus to China's steel industry. Steel production in China increased by 73% between 2000 and 2004, at which time China's steel output accounted for around 23% of global steel production (Institute of Shipping Economics and Logistics, 2005).

In 2003, China's ports handled 575 million tonnes of coal, 306 million tonnes of ore and 384 million tonnes of crude oil and natural gas. In 2004, China imported 148.2 million tonnes of iron ore and is the world's biggest iron ore importer. The critical role of China's bulk ports and terminals is obvious, with Qingdao being the largest ore importing port in the world, with ore throughput of 41 million tonnes in 2004. Similarly, as China's main port for the export and import of coal, Qinhuangdao is by far the biggest dry bulk port in the world, with an annual throughput in 2004 of some 154 million tonnes (Department of Water Transport, 2005; Institute of Shipping Economics and Logistics, 2005).

It is widely agreed that the development of the port industry in China began in earnest in 1978, when national economic reform and open door policies were first introduced. Indeed, the evolution of port governance in China is consistent with, and can be largely explained by, the wider macroeconomic development of the nation. In line with the evolution of the whole economic system, China's port governance system has evolved from one characterised by a high degree of centralisation to one of decentralisation, and from a wholly planned economic system to one that is much more market-oriented.

This chapter analyses the evolution of the port governance system in China and characterises this process in terms of three distinct phases of

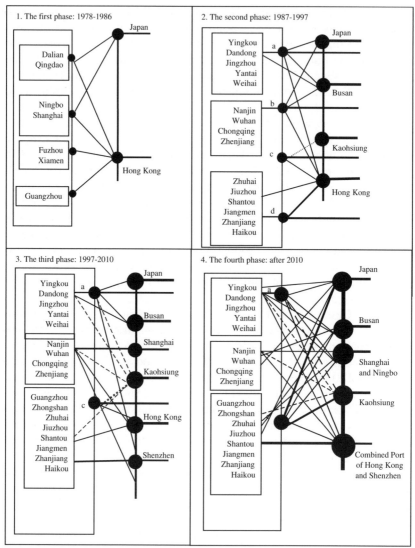

Fig. 15.3. Phases in the Development of Container Ports in China. *Source:* Cullinane et al. (2005).

development that mirror the changes in national economic policy. These phases are: 1979–1984, when the economy was centrally planned and a high level of centralised decision-making prevailed; 1984–2004, when authority and responsibility became increasingly decentralised and a process of partial privatisation was initiated in the form of minority shareholdings in joint ventures; and 2004–present, a period when decentralisation is being further reinforced through a legislative process that promotes both corporatisation and greater privatisation.

Section 2 of this chapter provides a context for the ensuing discussion by presenting the wider economic reform that has taken place in China over the past few decades. This wider tranche of reforms underpins the specific reform of the port industry and constitutes a large part of the rationale behind port reform in China. An important specific element of the general economic reforms that have been implemented within China has been the development of a system and codes of conduct for corporate governance. Section 3 provides a background to corporate governance in China and analyses its current status and implicit future evolution. The three phases in the development of port governance and administration in China that have already been summarised above are defined and analysed in detail in Section 4. In particular, the latest phase has been prompted by reforms that are embedded within a new piece of legislation, referred to as the *Port Law*, which came into effect on 1 January 2004. Section 5 summarises the nature of the changes in governance that China's programme of port reform has brought about and spells out the nature of the problems that these changes are seeking to address. The chapter concludes by asserting that it is still too early to tell whether the latest phase of reforms will prove to be successful in solving China's port problems, particularly the capacity issue, and points to possible implications of the reforms for overseas investment and future levels of concentration within the market.

2. ECONOMIC REFORM

2.1. Background

With the rapid development of China's economy and its increasing impact on the global economy, numerous efforts have been made to analyse the economic reforms that have been implemented in China over the past few decades. Just a few of the many examples include those of Nolan (1995), Li, Li, and Zhang (2000), Tian (2000) and Morris, Hassard, and Sheehan (2002).

The modern era of economic reform in China dates back to 1978 when, under the leadership of Deng Xiaoping, the open door policy was introduced with the intention to encourage trade and technology transfer and to reform state-owned enterprises (SOEs). The policy also involved the decentralisation and rationalisation of economic decision-making within the government.

Russia, other former socialist republics of the Soviet Union and some of the independent Eastern European countries opted for a 'big bang' approach to transformation of the market, concomitant with democratic political reform. In contrast, China embarked on what has subsequently been perceived to be a much more pragmatic and conservative longer-term evolutionary process based on experimentation with market mechanisms within a mixed economy, while at the same time resisting democratic political reform. Lin (2004, p. 2) pointed out that 'the [Chinese] approach is piecemeal, partial, incremental, often experimental, and especially without large-scale privatisation'. For instance, foreign-owned enterprises that were allowed to establish themselves in China were initially geographically confined to specially designated export processing zones, which in China are referred to as Special Economic Zones (SEZs). These have been created to promote investment and to provide a test-bed for experimenting with market economics on a controlled basis. There are specific incentives to promote investment associated with SEZs (ILO, 2006). Initially, however, the amount of foreign investment in any venture was limited to a 35% stake but this has since been raised, as has the actual number of such zones. There are now 15 Free Trade Zones and 39 SEZs within China (Customs General Administration of China, 2006). Since the introduction of SEZs, however, overseas investment has moved beyond their boundaries and spread across urban coastal China and, over time and remaining subject to government approval, the maximum foreign stake has now extended to majority ownership (Morris et al., 2002).

2.2. China's Accession to the World Trade Organisation

The conditions under which China acceded to the World Trade Organisation (WTO) mean that there are greater opportunities for foreign investors to compete with domestic investors on a more equal basis, with overseas investors enjoying better protection than previously through formal processes of judicial review and legal remedy (Li, Cullinane, & Cheng, 2003). The foreign ownership of assets in China will probably, therefore, become even more prevalent as time progresses and as agreed measures for the greater liberalisation of foreign direct investment are implemented in accordance with the stipulated timetable. It is important to recognise, however, that while foreign investors may benefit from receiving the same treatment as local enterprises, they will simultaneously lose the favourable treatment that in certain respects they previously enjoyed. The most poignant example of this will be the gradual removal of favourable tax rates that applied to foreign direct investment into China prior to WTO accession. Furthermore, where certain identified commercial or strategically sensitive activities are concerned, as is the case with shipping or logistics services (Li, Cullinane, Yan, & Cheng, 2005), restrictions will still be imposed on foreign investors. Realistically, however, it is probable that in the longer term these too will be lifted if China's experiment with even greater market-orientation proves successful.

In seeking to comply with the conditions of its accession to the WTO, the changes that China has already made to its legal system, and plans to make in the future, will no doubt give more confidence, reassurance and manifest legal protection to foreign investors. China's WTO accession should, however, be regarded as a double-edged sword, where both opportunities and risks exist for potential overseas investors in China.

2.3. Decentralisation and Privatisation

Morris et al. (2002) have highlighted the nature of decentralisation and privatisation within China's reform process as two of its major distinguishing features. As far as the former is concerned, the power to take strategic decisions shifted from central government to a more regional or local government locus, albeit remaining strictly under the control of the traditional Communist party organisation. In addition, there has been considerable decentralisation of decision-making, both strategic and operational, from central government to the management of the SOEs. Consequently, many SOEs under the direct supervision of central government were transferred to local governments and, in consequence, certain of the previous proliferation of central agencies have been gradually closed down. This is evidenced by a reduction in the number of central ministry-level agencies from 61 in 1982 to 41 by 1993 and 29 by 1998. In consequence, local governments have been given much more authority in administrative, economic and budgetary matters (Li et al., 2000).

Privatisation used to be regarded as a taboo concept in China, largely as the result of the political ideology that underpins the government and its institutions. Despite its increasing occurrence as a process that is influential in the formation of China's emerging economy, it is interesting that the term 'privatisation' is still frequently avoided within the arena of Chinese policy-making. The more politically neutral terms of corporatisation or marketisation are generally preferred and used more ubiquitously, despite the fact that these terms do relate to slightly different concepts that yield different implications (Everett & Robinson, 1998).

Li et al. (2000) have suggested that the implementation of privatisation processes per se was not the initial intention of the government when China embarked on its programme of reform. They identify, however, a number of important influential factors that have led to the implementation of privatisation processes within China. These are decentralisation policies, the emergence of competition, the need to establish legal and physical infrastructures and the diffusion of new ownership forms. Morris et al. (2002) argue that privatisation in China has occurred in two main areas. First, there has been privatisation at the macroeconomic level via the encouragement of private enterprises, particularly those that have received injections of foreign direct investment. Second, the remaining SOEs have been subject to a reform process, which has, to a certain extent, opened them up to market forces. It is important to emphasise, however, that a significant difference exists between the reform of SOEs and pure privatisation (Li et al., 2000; Lin, 2004). For example, Lin (2004) refers to the increasing presence of privatisation that does not relate to SOEs. Instead, a process of corporate privatisation has taken place within China as the result of the rapid development of newly established private enterprises that now account for a much greater proportion of productive output than was previously the case.

Lin (2004) provides a detailed picture of the reform of SOEs and summarises the four stages of this process as follows:

- *Stage 1 (1979–1983)*. The emphasis was placed on several important experimental initiatives that were intended to enlarge the enterprise economy by devolving enhanced commercial autonomy and expanding the role of financial incentives within the traditional economic system. These measures included the introduction of profit retention and performance-related bonuses and allowing the SOEs to produce at levels over and above the mandatory State plan. Enterprises that were involved in the export sector were also allowed to retain part of their foreign exchange earnings for use at their own discretion.
- *Stage 2 (1984–1986)*. The emphasis shifted to a formalisation of the financial obligations of the SOEs to the government and exposed

enterprises to market influences. From 1983, profit remittances to the government were replaced by a profit tax. In 1984, the government allowed SOEs to sell output in excess of quotas at negotiated prices and to plan their output accordingly, thus establishing the dual-track price system.

- *Stage 3 (1987–1992)*. An attempt was made to clarify the authority and responsibilities of SOE managers by formalising and widely adopting the contract responsibility system.
- *Stage 4 (1993–present)*. Attempts are being made to introduce modern corporate structures and systems to the SOEs.

In each stage of the reforms to date, government intervention gradually diminished, while the SOEs gained greater autonomy. Several studies have found that such reforms led directly to increased productivity in the SOEs, enhanced technical efficiency and to the movement of SOE production onto a location on the efficient production frontier (Gordon & Li, 1991; Dollar, 1990; Li, 1997). On the other hand, despite the improvement in productivity, the profitability of the SOEs declined and government subsidies increased (Lin, 2004).

3. CORPORATE GOVERNANCE

Sound and transparent corporate governance policies are widely recognised as an essential element in establishing an attractive investment environment that is characterised by a competitive and efficient financial market. It has also been suggested that good corporate governance facilitates better corporate performance, improved access to capital and informed entrepreneurial risk-taking, all of which contribute to the sound economic development of a nation (Hecklinger, 2004).

Following China's entry into the WTO, its implementation of a policy for the partial flotation of large-scale SOEs and the ensuing expansion of equity markets, the issue of corporate governance is viewed as pivotal to the success of China's enterprise reform and capital market development. In addition, appropriate corporate governance policies and codes play a crucial role in engendering greater investor confidence and allowing China to attract the international investors required to sustain its long-term economic growth.

China has undertaken crucial reforms and initiatives to enhance both corporate governance and the management of state-owned assets. The 'Code of Corporate Governance for Listed Companies in China' was issued by the China Securities Regulatory Commission (CSRC) and the State Economic and Trade Commission in January 2001.

This code is based on the Organisation for Economic Co-operation and Development (OECD) Principles of Corporate Governance (OECD, 2004) and the Chinese authorities have taken steps to enhance its enforcement through special inspections (Hecklinger, 2004). The code was established in accordance with the basic principles of China's Company Law, Securities Law and other relevant laws and regulations, as well as commonly accepted standards in international corporate governance. It was formulated to promote the establishment of a 'modern enterprise system' for China's listed companies, to standardise their method of operation and to foster the healthy development of the country's securities market. The Code sets forth, among other things, the basic principles for the corporate governance of listed companies in China, the available means of protection for investors' interests and rights and the basic rules of behaviour and moral standards for senior management (CSRC, 2001).

In late 2002, the 16th Communist Party Congress concluded that better management of state-owned assets would be one of the top priority areas for the current government and established the *State Assets Supervision and Administration Commission* (SASAC) in April 2003. The new commission represents a crucial step in separating the ownership function from the regulatory one within the Chinese administration.

A first report on corporate governance in China, highlighting the progress to date and remaining challenges, was published by the Shanghai Stock Exchange in 2004. The report indicates that China is still facing many challenges, some of which are commonly shared with other countries, while others are quite unique to China.

The second 'Policy Dialogue on Corporate Governance in China' was held in Beijing, China on 19 May 2005. The conference marked the first occasion where the OECD *Guidelines on Corporate Governance of State Owned Enterprises* (OECD, 2005) were presented after their endorsement by the OECD member countries at the end of April 2005. Issues discussed included: the role of the state as an owner; the composition and nomination of boards in SOEs; transparency and disclosure in SOEs; equitable treatment of (minority) shareholders; and the role and tasks of the different agencies involved in the regulation and supervision of SOEs.

Although progress has clearly been made, China's approach to corporate governance is not without its critics. Wang (2004) suggests that although China's governance codes appear to be well drafted, problems will inevitably arise in relation to over-regulation and under-enforcement, as is the case in any developing country. She highlights the continued propensity for abuse of the codes by the state and by majority shareholders (that may include the state). Wong (2005) goes so far as to argue that an Anglo-American model of corporate governance (such as that of the OECD) is wholly inappropriate for China, in that it fails to take seriously the prevailing institutional arrangements and practices which necessarily have an impact on and alter the logic of the model. He suggests that China should develop its own bespoke model that takes into account the specific economic and social needs of China, instead of simply adopting a corporate governance model developed for other countries. By so doing, China will then be affirming the explicit message of the OECD that 'there is no single model of good corporate governance' (Lin, 2001).

4. THE EVOLUTION OF PORT GOVERNANCE

4.1. 1979–1984: Planned Economy and a High Level of Centralisation

During this period, the port sector was extremely vertically integrated and centrally controlled. In its role as the representative of the Chinese central government, the Ministry of Communications was the owner of the ports and exerted total control over all port activities and decision-making, including the formulation of strategy and supporting policies, port planning, important infrastructure investment decisions and managing port operations. Local government at provincial or municipal level had no control over port authorities and all profits and losses from port operations were attributable to central government.

The advantage of this sort of system lay with the fact that the Chinese government could develop an overall national strategic plan across the whole of the country's port network and, given the limited available capital at this time, could concentrate on building several large ports. On the other hand, the main drawbacks of this system lay with the de facto lack of motivation or interest in improving operating efficiency on the part of the port management and local government. In other words, since they would not benefit or suffer from the economic performance of ports under their purview, neither operational management nor local government was motivated to further improve the quality of port production. This was a particular problem for the whole of China's port industry during this period.

Another serious drawback of this system was that insufficient funds were allocated for investment in most of the ports. Because of the limited amount of capital available from central government, only a few ports could invest sufficiently to ensure further growth and development. The result was serious cargo congestion in ports in 1981, 1983 and 1985 (Department of Water Transport, 1998).

4.2. 1984–2004: Towards Decentralisation

Recognising the critical drawbacks of a high degree of centralisation in port decision-making, especially the insufficient investment in port infrastructure and superstructure that resulted, a new system of governance was introduced in June 1984 in the port of Tianjin, one of the most important ports in China (Table 15.1). Under this system, the port of Tianjin was jointly managed by central government and the Tianjin municipal government. Subsequent to the implementation of this initial change in port governance structure for Tianjin only, almost all ports in China (except for Qinhuangdao) were placed under the joint purview of central and local governments by 1987. Local government obviously gained increasing control over the port sector during this time, as decision-making power and authority devolved from a highly centralised system to one of greater decentralisation. To summarise, during this whole period, there were three main types of port governance:

(1) Ports under the direct control of central government (i.e. through the Ministry of Communications). By 1987, the Port of Qinhuangdao was the only port that fell into this category.
(2) Ports under the joint control of both central and local governments, with most ports falling into this category.
(3) Ports under the control of local government. This form of governance applied mainly to ports that were small in production scale.

Along with the trend towards decentralisation of port governance, policies on port investment and management also changed dramatically during this period. Investment would be sourced not only from China's central government, but also from local government, foreign investment and through commercial bank loans.

One particularly important event to note was the establishment in 1987 of the Nanjing International Container Terminal Company Ltd., a joint venture between Nanjing Port Authority and US-based Encinal Terminals that was the first Sino-foreign port enterprise in China. This indicated a new stance in the policy of the Chinese government towards foreign investment in its port industry; from this point on, foreign investment in China's port industry was not simply allowed or merely tolerated, but actively encouraged. Such policies were detailed in the *Interim Regulations of the State Council of the PRC on Preferential Treatment to Sino-Foreign Joint Ventures on Harbour and Wharf Construction* and *Instructions on Reform and Further Development of Transportation System* promulgated and implemented by the State Council in 1985 and the Ministry of Communications in 1992, respectively.

One of the reasons behind the introduction and subsequent encouragement of foreign investment in China's port industry was the Chinese government's realisation that with the development of global transportation and the implementation of new transportation patterns, such as hub-and-spoke systems, its ports would face increasingly strong competition from counterparts in neighbouring countries. To increase the competitive edge of China's port sector, the most efficient approach was to introduce, or at least allow and encourage, the inculcation of sophisticated management from foreign investors (Cullinane, Wang, & Cullinane, 2004). During this period, the joint venture was the only way that foreign investors (including any potential investors from either Hong Kong or Macau) could enter the Chinese port market.

Paradoxically, the upper limit on stakes held by foreign investors in any single Chinese port was set at 49%. An official explanation or justification for this ceiling has been conspicuous by its absence. However, based on interviews with several port managers in China, the main explanations for this policy would seem to be the following:

(1) The ceiling allowed the Chinese government to retain the final say in matters relating to port planning and operation. Considering the different parties involved in port production and the variety of objectives they might possess, this is a particularly important objective for the Chinese government. An interesting example in this respect includes the investment of Hong Kong-based Hutchison Port Holdings (HPH) in the port of Ningbo. As a port investor, HPH attempted to increase container-handling charges in order to achieve its target rate of return on its investment. On the other hand, the port of Ningbo needed to maintain comparatively low container-handling charges in order to compete with its neighbouring rival port of Shanghai. By controlling the majority stake, Ningbo Port Authority retained the final say over port pricing (Hai, 2004).

(2) Given the booming cargo transportation market in China, maintaining the majority share in a port joint venture would ensure that Chinese parties to the agreement would accrue the largest share of the benefits from port operation.
(3) From the very early days of opening up the Chinese port sector to overseas investment, HPH had invested extensively in it. It was held by many that the ceiling on joint venture stakeholdings was deliberately aimed at limiting the influence of HPH in China (Asian Economic News, 2000).
(4) Finally, this sort of policy is consistent with the slow-paced conservative macroeconomic evolution in China.

Because of the existence of the 49% ceiling, foreign investors in China's port sector clearly did not have the right to control or decide upon important issues. To some extent, this dampened the enthusiasm of foreign investors. Nevertheless, by the end of this period the number of external investments in mainland China's port sector was quite considerable, although limited solely to container ports (see Table 15.3). Foreign direct investment in the port sector came especially from ethnic Chinese sources such as Hong Kong's HPH and Modern Terminals Limited and Singapore's PSA (Wang, Ng, & Olivier, 2004).

During this period, most strategic port activities were conducted by the local port authorities. These entities had two roles to play. On the one hand, as a government body, the port authority was the policy regulator, usually operating under the joint leadership of local and central government. As such, the local port authority was vested with the right to make final decisions on many important port development issues, such as selecting joint

Table 15.3. Major International Terminal Operators Investing in China's Main Container Ports.

Shanghai	Hutchinson Port Holdings, APM Ports, DP World
Shenzhen	Modern Terminals Ltd., Hutchinson Port Holdings, DP World, P&O Ports, APM Ports, Kerry Logistics
Qingdao	P&O Ports, APM Ports, DP World
Ningbo	Hutchinson Port Holdings
Tianjin	PSA International, DP World
Guangzhou	PSA International
Xiamen	APM Ports, Hutchinson Port Holdings
Dalian	PSA International, APM Ports
Fuzhou	PSA International

Source: Updated from Wang et al. (2004) and Cheng (2002).

venture partner investors from among those that had expressed a solid interest in investing. On the other hand, the local port authority was also an SOE that, in line with new government strictures, needed to operate in accordance with the market mechanism. This undoubtedly deterred port competition and impeded the development of the port industry. To overcome this problem, starting from the 1990s, the central government of China attempted to deprive port authorities of many autonomous functions so that the ports themselves could evolve towards purely business entities. Largely because of the complexity of the reforms themselves, however, this effort achieved only limited success (Wang et al., 2004).

Another problem with the dual role played by the local port authority lies in the overlapping power of the port authority and the local transportation department. Although the latter is less relevant to the daily operations of a port, it can still exert a significant influence on port planning and some other important issues. For instance, the local transportation department had the right during most of this period to plan the usage of land within a city or province. This could inevitably impinge upon the development of a port by influencing such factors as the amount of available land classified as port area or for expansion. As both port authority and local transportation department were government bodies and could not exert any authority over the other, disputes frequently arose.

4.3. 2004–Present: Decentralisation and Corporatisation

This phase began in earnest in 2004, although some changes of this type had been implemented towards the end of the previous phase. The *Port Law* and its complement, the *Rules on Port Operation and Management*,[1] have been in effect since 1 June 2004 and 1 January 2004, respectively. The *Port Law* is the first law in China dealing exclusively with the port industry and highlights the great importance attached to the port industry by the Chinese government.

The key issues in the *Port Law* and the *Rules on Port Operation and Management* rest with the introduction of a modern enterprise system into the port industry and the limitation of government intervention in port operations and management. As mentioned earlier, one serious drawback for port management in China was that the port authority was simultaneously both the regulator and a market player, and this seriously discouraged competition and impeded the further development of the port industry in China. Under the new *Port Law* and the *Rules on Port Operation and*

Management, the original port authority has been replaced with both a Port Administration Bureau and a port business enterprise. The former can be either an independent provincial or municipal Port Administration Bureau (the situation in Shanghai or Liaoning provinces) or vested in the provincial or municipal Transportation Administration Department (as in Guangdong or Zhejiang provinces). The latter form of organisation is simply a pure business entity, following and engaging in the open market, as well as in competition and cooperation with other business entities.

According to the *Port Law*, the Chinese central government will no longer retain any ownership of ports. The previous public ports owned or partly owned by the central government will be transferred to local provincial or municipal government. Under the framework of the *Port Law*, the central government and the relevant transportation department at the provincial level are responsible for strategic planning and formulating relevant policies and regulations for the development of the port network system at the level of the whole country and the province, respectively. The *Port Law* also implicitly defines the relationship between the strategic planning undertaken by central and local government. That is, strategic planning undertaken by local government needs to be approved by central government. To a great extent, this guarantees that the strategic planning conducted by local government will not contradict the overall strategic plan that is formulated by central government for the nation as a whole.

Another major characteristic of the *Port Law* lies with the fact that it regulates port investment and management. According to the *Port Law*, investors from both China and other countries are allowed to enter the port market. Also, according to the *Catalogue for the Guidance of Foreign Investment Industries* (revised version) approved by the State Council in 2004,[2] the ceiling on stakes in ports held by foreign investors has been abolished. Any qualified group or enterprise, whether from China or any other country, can apply to invest in port construction and/or operations. In other words, according to the *Port Law*, foreign investors can, on an independent basis (i.e. without the need for a Chinese joint venture partner), invest in and operate ports in China.

Apart from the obvious change in port ownership (i.e. the central government will no longer be involved in the ownership of ports), it is still too early to tell to what extent these new regulations are going to influence the port industry in China and its future development. Despite the usual scepticism over the effectiveness of law in China (Zheng, 1999), it is nevertheless fair to say that these laws at least provide the legal basis for the future evolution of the port sector in China.

5. SUMMARY AND CONCLUSIONS

The Chinese economy has undergone dramatic transformation during the past few decades. Concomitant with national economic development and reform, the port industry in China has experienced phenomenal growth, both in quantity and quality, over the same period.

Before 1984 ports were wholly owned by central government, with all authority and responsibility for port activities and their administration resting with the central government. In stark contrast, the *Port Law,* which came into effect on 1 January 2004, states that the Chinese central government will no longer own ports. The main responsibility of the Chinese central government now rests solely with the strategic planning of the port network for the whole country. In accordance with the *Port Law*, ports are now mainly owned and managed by different levels of local government, with private entities, including investors from other countries, permitted to invest in, own and manage Chinese ports.

Prior to 2001, the port authority in China was both the regulator and, in many senses, a market player. Following port reform, the former port authority was replaced with both a port administration bureau[3] and a corporatised business entity. While the former serve as the local government bodies responsible for the public administration of ports, the latter can participate fully in the open market and are willing and able to negotiate and conclude commercial deals with potential overseas joint venture partners. In essence, this has enabled the municipalities themselves to deal with foreign entities directly rather than, as previously, having to go through the relevant ministry in central government.

While port reform in China can be assessed as a rather passive and slow-moving process, this is similar to the pace and adventure exhibited on a larger scale in China's national macroeconomic reforms. During this long-running process, the governance and administrative structure of the port industry has evolved from one that was characterised by a high degree of centralisation towards a form where authority and management decision-making is much more decentralised. Figs. 15.4–15.6 summarise the transition in port governance that has occurred over the phases that have been analysed above.

When the Chinese government first opened up the nation's port sector to foreign direct investment, it quickly introduced a policy whereby no one overseas investor predominated. Despite the subsequent lifting of this maximum limit on overseas shareholdings in the port sector, there exists little likelihood that this line of thinking will disappear overnight, either in central

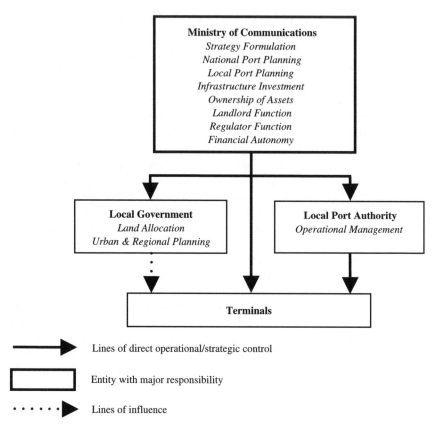

Fig. 15.4. China's Port Governance Model (Pre-1984).

government or at the local level. In fact, given HPH's existing position as the dominant market player in China, there have been quite well-supported suggestions that informal, but quite stringent, regulatory ceilings pertain in the case of applications from the company (Asian Economic News, 2000; Chadha, 2001).

The central idea or rationale underpinning port reform in China is the objective of freeing the port from the bureaucracy of government, and to increase the competitiveness and efficiency of the port industry, especially with respect to the provision of additional capacity. China's port infrastructure generally remains insufficient for satisfying future trade demand; capacity and efficiency upgrades have historically lagged behind the expansion of trade. Wang et al. (2004) attribute this not only to China's

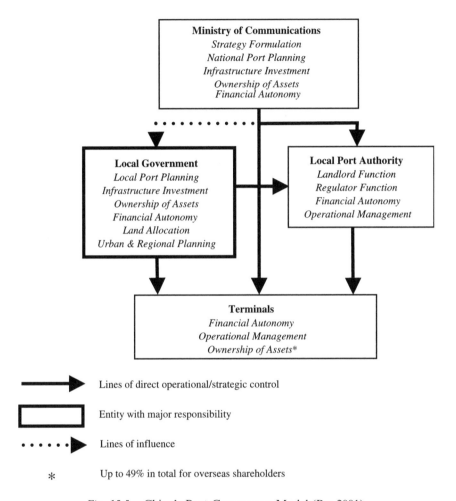

Fig. 15.5. China's Port Governance Model (Pre-2001).

phenomenal rate of economic growth, but also to the slow response of institutional processes and reforms. In consequence, the port sector in China has not reached the standards that might be expected elsewhere (UNESCAP, 2004).

Numerous plans are in place for expanding the country's bulk cargo-handling facilities and, for the first time, efforts are being made to attract foreign investors into this sector. Some success has already been achieved, with HPH having recently invested in an iron ore terminal in the port of

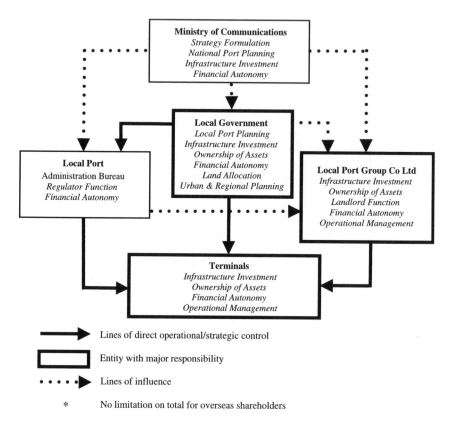

Fig. 15.6. China's Port Governance Model (Current).

Dalian. It is imperative that these initiatives do prove successful in increasing bulk port capacity; despite the phenomenal volumes that are currently handled, informed sources suggest that total available port capacity for moving bulk cargo is about 500 million tonnes less than the current demand. This suggests, therefore, that throughput in Chinese bulk ports could have been much higher in recent years if only the infrastructure could have handled the total demand and that supply bottlenecks in ports are effectively constraining China's economic growth rate (Park, 2004).

The situation with respect to bulk facilities contrasts sharply with that of container terminals. Planned annual investment in container port infrastructure is expected to be in the order of US$20 billion over the next 8–10 years (Baer, 2004). Irrespective of the expected future continuation of the very high growth rate in container throughput that has been experienced over recent years, when the scale of planned investment is of this magnitude, there is obviously considerable scope for timing and location errors. What should be avoided is the container port equivalent of the airport construction bubble that has recently occurred in Southern China. As the result of uncoordinated devolved municipal decision-making, five airports were opened in quick succession, leaving Zhuhai airport with the capacity to handle 100,000 aircraft a year while, in fact, handling only 25 a day while attempting to service its debt liability of $1.5 million per day (Baer, 2004). In terms of the currently planned expansion of container capacity, perhaps the most likely prospect of ill-advised investment and future overcapacity lies with Shanghai and Ningbo, where terminals with 52 and 30 berths, respectively are currently either under construction or planned (Cullinane, Teng, & Wang, 2005).

In an effort to manage the level of potential risk associated with port investment, particularly in the container sector, a trend of attempting to secure the unequivocal loyalty of shipping lines by particular ports has emerged; equity stakes in the most recent phases of terminal development to come on stream have increasingly been offered to shipping lines as 'dedicated' berths or terminals. If this approach continues and proves successful in attracting overseas investment, it will have the effect of opening up the container-handling market in China to a greater number of somewhat smaller shipowning operators of dedicated terminals. At the same time, such a trend will also have the dual benefit of rendering the market in China more competitive than it might otherwise become if the select club of global terminal operators – particularly those of ethnic Chinese origin – are left free to wield their undoubted power and influence over China's port industry and as global concentration of the container-handling sector continues unabated. If reports are to be believed, diversification in the ownership of port assets is something that, in any case, is a matter of some strategic priority for China's political leaders (Chadha, 2001; Airriess, 2001).

As well as the major problem of providing appropriate port capacity in the right location and at the right time, the sector also faces: a shortage of skilled labour; inadequate water depth of about 10 m instead of the 15 m required to accommodate fully loaded large container vessels; constraints in accessing the hinterland due to impoverished road, rail and waterway services; congestion in the port storage areas; unnecessary delays due to inefficient and slow customs and quarantine procedures and inefficient port services such as pilotage. In the past, it has been the system of port governance that has frequently been referred to as in need of improvement. It remains to be seen whether the latest port reforms will ameliorate some or any of the ills that currently beset port operations in China.

NOTES

1. The latter was formulated by the Ministry of Communications to accord with the forthcoming *Port Law*.
2. http://www.china.org.cn/chinese/PI-c/726125.htm, in Chinese.
3. Alternatively, in some port cities, certain of the port authority's former responsibilities have been embedded in the local transportation administration department or bureau.

REFERENCES

Airriess, C. A. (2001). The regionalization of Hutchison Port Holdings in Mainland China. *Journal of Transport Geography, 9*(4), 267–278.

Asian Economic News. (2000). China wants limits on Hutchison influence in ports, www. findarticles.com/p/articles/mi_m0WDP/is_2000_Oct_30/ai_66929817

Baer, M. (2004). Capitalising on the seaborne trade growth through strategic terminal investment in China. *China shipping conference*, Shanghai, 22–23 November.

Chadha, K. K. (2001). China limits Hutchinson on port stakes. *Asia Today* (August).

Cheng, C. (2002). A historical review of the development of containerized transport in China, Part III. *China Ports, 8*, 25–28 (in Chinese).

China Securities Regulatory Commission. (2001). Code of corporate governance for listed companies in China, Beijing, January 7, www.csrc.gov.cn/en/jsp/detail.jsp?infoid = 1061948026100&type = CMS.STD

Containerisation International Yearbook. (2005). London: Informa Publishing.

Cullinane, K. P. B., Cullinane, S. L., & Wang, T.-F. (2005). A hierarchical taxonomy of container ports in China and the implications for their development. In: T.-W. Lee & K. P. B. Cullinane (Eds), *World shipping and port development* (pp. 217–238). Basingstoke: Palgrave-Macmillan.

Cullinane, K. P. B., Teng, Y., & Wang, T.-F. (2005). Port competition between Shanghai and Ningbo. *Maritime Policy and Management, 32*(4), 331–346.

Cullinane, K. P. B., Wang, T.-F., & Cullinane, S. L. (2004). Container terminal development in mainland China and its impact on the competitiveness of the Port of Hong Kong. *Transport Reviews, 24*(1), 33–56.

Customs General Administration of China. (2006). www.customs.gov.cn (in Chinese).

Department of Water Transport. (1998). *China shipping development annual report*. People's Republic of China: Ministry of Communications.

Department of Water Transport. (2002). *China shipping development annual report*. People's Republic of China: Ministry of Communications.

Department of Water Transport. (2005). *China shipping development annual report*. People's Republic of China: Ministry of Communications.

Dollar, D. (1990). Economic reform and allocative efficiency in China's state-owned industry. *Economic Development and Cultural Change, 39*(1), 89–105.

Everett, S., & Robinson, R. (1998). Port reform in Australia: Issues in the ownership debate. *Maritime Policy and Management, 25*(1), 41–62.

Gordon, R., & Li, W. (1991). Chinese enterprise behavior under the reforms. *American Economic Review: Papers and Proceedings, 81*(2), 202–206.

Hai, B. (2004). Port investment in China, www.3rd56.com/wwwroot/list.asp?id=4733 (in Chinese).

Hecklinger, R. (2004). Welcome remarks by Deputy Secretary-General of the OECD, Policy Dialogue on Corporate Governance in China, Shanghai, 25 February.

ILO. (2006). Labour and social issues relating to export processing zones, Labour Law and Labour Relations Branch, International Labour Organisation, Geneva, www.ilo.org/public/english/dialogue/govlab/legrel/tc/epz/reports/epzrepor_w61/

Institute of Shipping Economics and Logistics. (2005). *Shipping Statistics and Market Review.* Bremen: ISL.

Li, K. X., Cullinane, K. P. B., & Cheng, J. (2003). The application of WTO rules in China and the implications for foreign direct investment. *Journal of World Investment, 4*(2), 343–361.

Li, K. X., Cullinane, K. P. B., Yan, H., & Cheng, J. (2005). Maritime policy in China after WTO: Impacts and implications for foreign investment. *Journal of Maritime Law and Commerce, 36*(1), 77–139.

Li, S. M., Li, S. H., & Zhang, W. Y. (2000). The road to capitalism: Competition and institutional change in China. *Journal of Comparative Economics, 28*(June), 269–292.

Li, W. (1997). The impact of economic reform on the performance of Chinese state enterprises, 1980–89. *Journal of Political Economy, 105*(5), 1080–1106.

Lin, C. (2001). *Private vices in public places: Challenges in corporate governance development in China,* Policy Dialogue Meeting on Corporate Governance in Developing Countries and Emerging Economies, OECD Development Centre and the European Bank for Reconstruction and Development, Paris, 23–24 April.

Lin, J. Y. (2004). *Lessons of China's transition from a planned economy to a market economy.* Peking University, Working Paper Series, www.ccer.edu.cn/download/2963-1.pdf

Morris, J., Hassard, J., & Sheehan, J. (2002). Privatization, Chinese-style: Economic reform and the state-owned enterprises. *Public Administration, 80*(2), 359–373.

Nolan, P. (1995). *China's rise, Russia's fall: Politics, economics and planning in the transition from Stalinism.* London: Macmillan.

Organisation for Economic Co-operation and Development. (2004). *OECD principles of corporate governance.* Paris: OECD Publications.

Organisation for Economic Co-operation and Development. (2005). *OECD guidelines on corporate governance of state-owned enterprises.* Paris: OECD Publications.

Park, K. (2004). China's port congestion delays deliveries, may hamper growth. *Bloomberg News* (May 17).

Tian, G. Q. (2000). Property rights and the nature of Chinese collective enterprises. *Journal of Comparative Economics, 28*(2), 247–268.

UNESCAP. (2004). *Development of shipping and ports in north-east Asia.* Bangkok: Transport and Tourism Division, United Nations Economic and Social Commission for Asia and the Pacific.

Wang, J. J., Ng, A. K. Y., & Olivier, D. (2004). Port governance in China: A review of policies in an era of internationalizing port management practices. *Transport Policy, 11,* 237–250.

Wang, M. (2004). Corporate governance in the People's Republic of China. *Proceedings of the 16th annual conference of the association for Chinese economics studies,* Australia (ACESA), Brisbane, 19–20 July.

Wong, L. (2005). Corporate governance in China: A lack of critical reflexivity? In: C. R. Lehman (Ed.), *Corporate governance: Does any size fit?* (pp. 117–144). Amsterdam: Elsevier.

Zheng, Y. (1999). From rule by law to rule of law? A realistic view of China's legal development. *China Perspectives, 25,* 31–43.

CHAPTER 16

PORT GOVERNANCE IN KOREA

Dong-Wook Song and Sung-Woo Lee

ABSTRACT

Until recently, Korea has been enjoying prosperous developments in container ports along with its national economy. Korean ports have passed through a variety of port governance stages. In the early period, only the government was a player in the industry, doing everything associated with management, operations and policy. However, the industry has gradually been opened to private entities to accommodate the changed environment. This evolutional movement results in a somewhat complex picture of governance structures in Korea. This chapter aims to discuss the evolution of such movements largely in terms of administrative and ownership structures since the 1960s.

1. INTRODUCTION

As shown in Fig. 16.1, the Republic of Korea (South Korea, hereinafter referred to as Korea) is located in North-East Asia. It is geographically connected to ocean economies such as Japan, the South-East Asian countries, North and South America, as well as to continental economies, such as China, Russia, Central Asian states and Eastern Europe. More specifically, Korea has geographical adjacency with the west coast of Japan and the north-east part of China.

Devolution, Port Governance and Port Performance
Research in Transportation Economics, Volume 17, 357–375
ISSN: 0739-8859/doi:10.1016/S0739-8859(06)17016-X

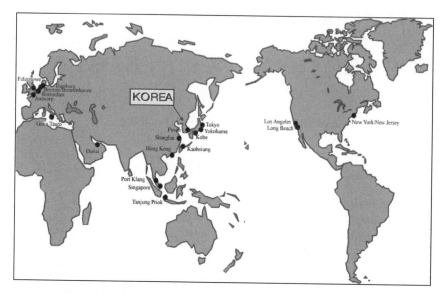

Fig. 16.1. Location of Korea. *Source:* Drawn by the Authors.

Thanks to this geo-locational benefit, Korea has enjoyed the prosperous development and growth of the national economy as a whole and container ports in particular. By connecting the currently divided North and South Korea and the Euro-Asian continent, the ongoing development of the Trans-Siberian Railway and Trans-China Railway will give Korea the further potential to become a world-class port or regional logistic hub.

This chapter aims to discuss the evolution of Korean port governance in terms of administrative and ownership structures over the years since Korea embarked upon its international trade in the early 1960s. In doing so, the rationale for and the benefits of the changes in governance structure that have occurred over the years will be analysed.

2. KOREAN PORTS IN GENERAL

Ports and associated facilities in Korea were largely established since the 1970s and a number of development and expansion plans are underway. This never-ending effort is understandable if we look at the way the Korean economy has achieved its remarkable progress in such a relatively short time. In principle, Korea can be regarded as an island country that is geo-politically separated from its northern counterpart due to their different ideological and political systems. In addition, Korea has adopted export-oriented economic development since the 1960s.

According to the classifications by Chenery and Syrquin (1986), Korea belongs to an outward- and industry-oriented economy. Song, Cullinane, and Roe (2001) noted that the industrialisation of this island-like country depends mainly upon the import of raw materials and the export of processed and finished products. Korea has adopted this economic strategy based on export expansion plans. The national economy, as a consequence, has developed remarkably with substantial increases in exports, which in turn has brought about a rapid increase in the volume of exports and imports. This development has a direct impact on the development of the port industry, as seaborne trade through ports accounts for almost 99 percent of the country's trade.

2.1. Port Classifications

Depending upon a managing entity, Korean ports are divided into designated *international trading* ports and *coastal* (or local) ports under both the Harbour Act (1967) and the New Port Construction Promotion Act (1996).[1] There are currently 28 international trading ports, 23 coastal ports and 9 new ports as shown in Table 16.1.

The 28 designated international trading ports are developed and managed by the Ministry of Maritime Affairs and Fisheries (MOMAF) and further subdivided into trade ports and coastal ports depending on the frequency of arrivals and departures of ocean-going ships. The 23 coastal or local ports are developed and managed by relevant regional and/or municipal governments. Apart from these 51 ports, new ports[2] are specially chosen and promulgated by MOMAF at existing trade ports or newly designated areas in order to ensure the smooth transportation of imports and exports.

Due to Korea's unique geopolitical situation, the major ports of the country (e.g. Busan and Gwangyang) are located on its southern coastline. As a result, the Ministry has recently paid great attention to the further development and enhancement of the Busan New Port and the Gwangyang Port, with the strategic intention to make them into container mega-ports that are strong candidates to become the regional hub for North-East Asia. Fig. 16.2 shows the key ports of the country.

In addition to Busan and Gwangyang ports, Incheon has supported the capital region, Seoul, as a gateway. Incheon has plans to start a big

Table 16.1. Port Classifications in Korea.

Classification	Construction and Operation	Location	Ports
International trading ports (28)	MOMAF[a]	Western coast (8)	*Incheon*[b], Pyeongtaek, Daesan, Tae-an, Boryeong, Janghang, Gunsan, Mokpo
		Southern coast (13)	Wando, Yeosu, *Gwangyang*, Jeju, Seoguipo, Samcheonpo, Tongyoung, Gohyun, Okpo, Jangseungpo, Masan, Jinhae, *Busan*
		Eastern coast (7)	Ulsan, Pohang, Samcheok, Donghae, Mukho, Okgae, Sokcho
Coastal ports (23)	Construction: MOMAF	Western coast (7)	Yonggipo, Yeonpyungdo, Daecheon, etc.
	Operation: regional or city governments	Southern coast (11)	Shinma, Narodo, Hallim, etc.
		Eastern coast (5)	Wolop, Ulreung, Hupo, etc.
New ports (9)	MOMAF	Western coast (5)	Pyeongtaek, *Incheon North Harbour*, Mokpo New Port, Boryeong New Port, Saemangeum New Port
		Southern coast (2)	*Gwangyang, Busan New Port*
		Eastern coast (2)	Ulsan New Port, Pohang Youngilman New Port

Source: Compiled from Ministry of Maritime Affairs and Fisheries (2002).
[a]MOMAF, Ministry of Maritime Affairs and Fisheries.
[b]Ports in italics are major container ports in Korea.

development project to further boost its container throughput through physical development as well as governance restructuring. This will be discussed in the following section.

2.2. Legislation and Organisation

The development and management of Korean ports is controlled by the Harbour Act (since 1967). There are three acts in respect of Korean port development – the Harbour Act and New Port Construction Promotion Law (both under MOMAF), and the Law of Private Participation in Infrastructure (under the Ministry of Finance and Economy). In addition, two acts in association with the Busan and Incheon Port Authorities are to be

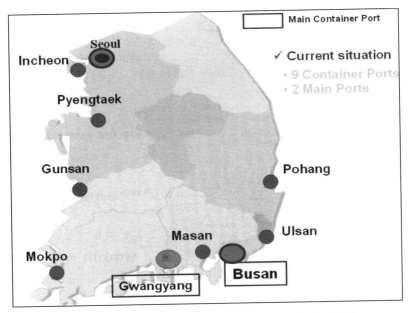

Fig. 16.2. Major Ports in Korea. *Source:* Drawn by the Authors.

established for the effective and efficient promotion and management of those ports. Details of these pieces of legislation are provided in Table 16.2.

2.3. Container Traffic in Korea

Korea has maintained its container business rather well, achieving greater success than most of the world. Busan dominates the market, handling about 80 percent of total container cargoes, while Gwangyang is expected to play a more important role as an assistant or complement to Busan. In particular, container throughput in Korea has swiftly grown in just a few ports, notably Busan, Gwangyang and Incheon, because of the rapid growth of the Chinese economy and the recovery of the Asian economies, alongside the growth of new Chinese ports. Under these circumstances, these Korean ports attract the lion's share of foreign port operators, such as Hutchison Port Holdings (HPH)[3] in Busan and Gwangyang, PSA Corporation in Incheon and Busan New Port, and Dubai International Ports in Busan New Port (Busan Port Authority, 2005).

Table 16.2. Port Legislation and Administrators.

Classification	Contents	Administrator
Harbour Act	• Designation, development and management/operation of ports • Port facility tariff • Non-managing agency port works • Management of port facility and equipment	Port Policy Division of MOMAF
New Port Construction Promotion Law	• Formulation of basic plan on new port construction • Designation and management of the area determined for new port construction • One-stop service for matters to be authorized, and permitted as prescribed in 25 laws on the promotion of new port construction	Port Development Division of MOMAF
Law of Private Participation in Infrastructure	• Formulation of basic plan on private investment facility project • Conclusion of concession agreement and designation of concessionaire • One-stop service for matters to be authorized and permitted as prescribed in 39 laws on the promotion of expanding infrastructure facilities	Private Investment Division of MOMAF
Act on Korea Container Terminal Authority	• Establishment and operation of Korea Container Terminal Authority except for Busan and Incheon • Development and operation of container terminals • Financing of development funds, etc.	Korea Container Terminal Authority (KCTA)
Act on Busan Port Authority	• Management and development of container port, facilities and areas (e.g. distripark, Inland depot) related to Busan port • Management and development of overseas container ports	Busan Port Authority (BPA)
Act on Incheon Port Authority	• Management and development of container port, facilities and areas (e.g. distripark, Inland depot) related to Incheon port	Incheon Port Authority (IPA)

Source: Compiled from Ministry of Maritime Affairs and Fisheries (2002).

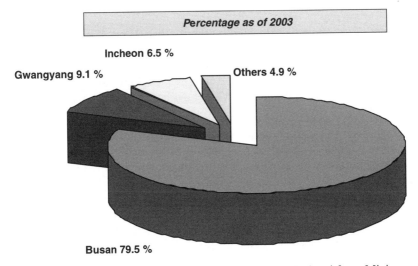

Fig. 16.3. Container Traffic in Korea (2003). *Source:* Calculated from Ministry of Maritime Affairs and Fisheries (2004).

More specifically, Korean container ports handled 13.2 million TEUs in 2003, with 79.5 percent of this figure handled in Busan, 9.1 percent in Gwangyang, 6.5 percent in Incheon and 4.9 percent in others (as shown in Fig. 16.3). Busan handled 10.4 million TEUs in the year, while its growth rate between 2002 and 2003 was 10.9 percent.

2.4. The Situation at the Three Major Ports

2.4.1. Busan
Located at the eastern tip of the Korean Peninsula, the Port of Busan is a gateway linking its state with the Pacific Ocean and the Asian continent. Busan, handling more than 10.1 million TEUs as of 2003, will be further boosted after 2011 by the construction of the Busan New Port, located next to the current port in Busan and developed as a cutting-edge container port with 30 berths for 50,000-ton ships and a handling capacity of 8 million TEUs per year. As the primary port in Korea, Busan handles a total of 40 percent of seaborne cargoes, 85 percent of containers and 40 percent of national marine products. It has four branches – North Harbour, South Harbour, Gamcheon Harbour and Dadepo Harbour.

2.4.2. Gwangyang

The port of Gwangyang is located on the south-west coast of the country. It has 16 container terminals and has operated since its opening in 1998. The Port of Gwangyang handled 1.2 million TEUs as of 2003 and is expected to handle 9.3 million TEUs annually if the planned 33 berths are in operation by 2011. It has three branches – West, East and Yulchon Harbours.

2.4.3. Incheon

The Port of Incheon, situated on the mid-west coast of the Korean Peninsula, has significantly contributed to the development of national economy and industries, as a gateway to Seoul, the capital of Korea. It is an artificially created port with lock gate facilities to cope with a tidal difference of 10 metres. It has internationally competitive equipments, with a variety of modern facilities for national, regional and international trade with Korea's main trading partners. Incheon is just behind Busan in terms of traffic volume handled, mainly due to its proximity to the capital city.

3. PORT GOVERNANCE IN KOREA

3.1. Brief History

Under the auspices of MOMAF, the public administrator of the country's ports, the 12 local sub-branches have the authority to control the nation's 50 international trading and coastal or local ports mentioned in the previous section (as illustrated in Fig. 16.4).

The Korea Maritime and Port Administration (KMPA), under the auspices of the Ministry of Construction and Transport (MOCT), controlled ports in the country until 1996. Thanks to the importance of the maritime industry, the Korean government enlarged the KMPA (a public hierarchy lower than ministry level) into the ministry-level MOMAF with seven bureaux in-house. At the same time, MOMAF became an independent ministry from the MOCT. In addition, the Korea Container Terminal Authority (KCTA) was established to promote the construction and management of container ports.

The independence of MOMAF from MOCT gave the former the right of development and management related to shipping, ports and associated facilities. Consequently, the port industry enjoyed a proportionally

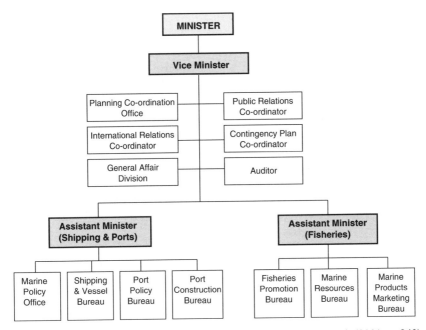

Fig. 16.4. Administrative Structure of MOMAF. *Source:* Song et al. (2001, p. 243).

increased budget allocated by the ministry and utilised it according to an established plan and policy. For the purposes of decentralisation and promoting competitiveness, the government has gradually handed over its right of port administration to local (regional or city) governments. Thus, Busan and Incheon became independent entities in 2003 and 2005 respectively. Eventually, the Korean port industry faced the concept of privatisation or its pseudo form of more private participation in the sector. The further independence of Gwangyang from the central authorities is currently under discussion between MOMAF and the local government concerned. Table 16.3 shows a brief history of the evolution of port administration that has taken place in Korea over the years.

3.2. Port Authorities

Korea currently has three public and semi-public (or corporatised) port authorities, holding authority and responsibility in the different parts of the country.

Table 16.3.   Korean Port Administration in History.

Year	Organisation	Main Content
1976	Korea Maritime and Port Administration (KMPA)	• Renamed from the Korea Port Authority • Under central government • Port management and development
1989	Korea Container Terminal Authority (KCTA)	• Operating and managing container ports • Under central government • Management and development of container ports for KMPA
1996	Ministry of Maritime Affair and Fisheries (MOMAF)	• Port management and development • Central government • Expanded to the Ministry Level (from KMPA to MOMAF)
2003	Busan Port Authority (BPA)	• Management and development of Busan's entire port facilities • Independent Corporation under the Port Authority Law • Current Busan and New container ports
2005	Incheon Port Authority (IPA)	• Management and development of Incheon's entire port facilities • Independent Corporation under the Port Authority Law • Current Incheon and Songdo New ports

Source: Compiled from Ministry of Maritime Affairs and Fisheries (2002).

3.2.1. Korea Container Terminal Authority

The KCTA, established in 1990, recognises changing trends that occur in and around a port and its related logistics industry and expedites the development of ports that are equipped to be internationally competitive. As part of its efforts, the KCTA has completed the development of 20 container wharves in the past 13 years, such as Gamman Wharf in Busan Port, New Gamman Wharf and the primary wharves of Phase 1 and 2 of Gwangyang Port. The Authority aims to enhance the national economy through efficient development, management and operation of container terminals in accordance with the Korea Container Terminal Authority Act. The Authority has managed 12 terminals in five ports, including Gwangyang Port. Since 2004, however, and in accordance with the Busan Port Authority Act, the KCTA has handed over all the container terminals in Busan to the newly established Busan Port Authority. Based on its achievements and the experience it has accumulated over the years, the KCTA continues to exert its best efforts in positioning Korea and its container ports as the central distribution centre for the North-East Asian region. It seeks to achieve this through its continued participation in the development of container terminals, such as the Busan New Port and the Level 3 terminals of Gwangyang Port, as well as through effective advancement of domestic terminal operation systems and overseas investments, especially in Northern China. This movement is thought of as a viable business diversification strategy in the long term.

3.2.2. Busan Port Authority

The Busan Port Authority (BPA) was established in December 2003 under the Port Authority Law and commenced operations in January 2004. BPA is a newly created organisation in the sector – that is, a public and private combined entity that is akin to the concept of a corporation as borrowed from the Singaporean experience of the creation of the PSA Corporation from its predecessor, the Port of Singapore Authority. BPA is fully responsible for port operations and development in Busan and its hinterland, as well as for the development of a vast logistics complex to be linked with the port. BPA was financially supported by 2 trillion Korean Won from the central government as an initial investment. BPA aims to apply 'private spirit' to port management and development in Busan's old and new terminal areas. The Authority controls the existing container terminals and facilities within and even outside the terminals and also the forthcoming facilities in Busan as described in the previous section. With the recent revision of the Act in early 2005, BPA is enabled to take part in the development of a distripark directly associated with Busan's terminals and to actively participate in a variety of overseas business activities to promote its ports and terminals, to enhance its competitiveness by attracting more cargoes and to maintain business and financial stability through a portfolio business strategy. Today, BPA is focused on developing Phase 2 terminals and distripark in the Busan New Port. It has already acquired half of the ownership stakes in the northern terminal distripark in the Busan New Port. For the purpose of its long-term prosperity, BPA will expand the scope of business in Busan and overseas alike.

3.2.3. Incheon Port Authority

Following the case of Busan, the Incheon Port Authority (IPA) was established in early 2005 under the Port Authority Law and is expected to commence its first operations in mid-2005. Mirroring exactly the case of BPA, IPA is a new corporation that is fully responsible for operations and the development of terminals and ports in Incheon and its hinterland, as well as for the development of vast logistics complexes to be linked with the port in the near future. IPA is financed with 1 trillion Korean Won from the central government as an initial investment, out of the earmarked 5 trillion for its total capital. The Authority controls the existing container terminals, related facilities and future ones in the Incheon Metropolitan Area and, with the same philosophy of BPA in mind, IPA aims to introduce a 'private spirit' into port management and development in Incheon's existing and future port/terminal areas.

3.3. Overall Port Administration and Organisation

What has been discussed in the previous sections will be summarised in a single picture in this section to show the overall administrative and organisational structure of Korea's ports. In respect of port administration systems, the KCTA has until recently played a central role in port management and operations. However, after the creation of the BPA in 2004 and the IPA in 2005, the KCTA has been forced to streamline its business and functional scope; only Gwangyang and other small and medium-sized ports are now managed, operated and supervised by KCTA. The Authority currently supervises six terminals in Gwangyang and two in other trading ports of the country. These terminals are all leased to KCTA by the government (viz. MOMAF) without payment. Then KCTA rents out these terminals to each terminal operator in return for payment.

Fig. 16.5 shows that there are three different entities controlling all the container terminals in Korea, except for Busan and Incheon; the terminal operators are subject to the supervision of the KCTA, and in turn KCTA is under the control of MOMAF (either central or local). This hierarchy gives much power to MOMAF acting on behalf of the central government, indicating that the government has deeply influenced the management, operation and development of container terminals in Korea.

One particularly interesting development in the Korean container port sector is that fully privatised terminal operating companies were born in 1997. It has led to greater competition among container terminal operators in Korea, mainly in the ports of Busan, Incheon and Gwangyang. It is envisaged that this scheme will be introduced into other ports as part of an endeavour to generate greater private spirit in the industry. This can materialise as port privatisation or as another form of promoting greater private sector participation in the port business (Cullinane & Song, 2002).

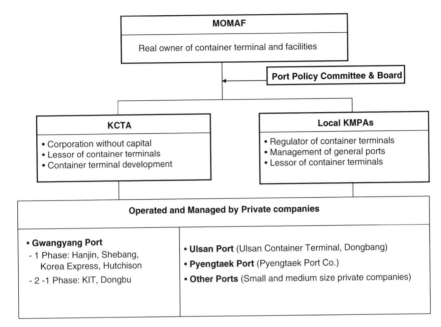

Fig. 16.5. Korean Port Administration (Excluding Busan and Incheon). *Source:* Drawn by the Authors.

Turning to the recent changes that have taken place in the country's port administration systems, BPA gained importance in terms of business and operational philosophy and its impact on other ports by taking over a central role from its predecessor, the KCTA, in the management and operation of Busan's container terminals. BPA holds the responsibility for management, operations and supervision of the container terminals in Busan, while regulatory and other public functions are still under central control through MOMAF. The BPA currently supervises six terminals in Busan and one in Busan New Port, which is scheduled to become operational by the end of 2005. As was the case with the KCTA, these terminals are all leased from the government (viz. MOMAF) without payment. BPA then rents out these terminals to individual terminal operators in return for payment.

Fig. 16.6 shows that there are three different entities controlling the container terminals in Busan; the terminal operators are subject to the supervision of the BPA, and in turn the BPA is indirectly under the control of MOMAF. Before 2003, Busan container terminals were controlled, operated and supervised by the KCTA as mentioned above. This hierarchy gives much

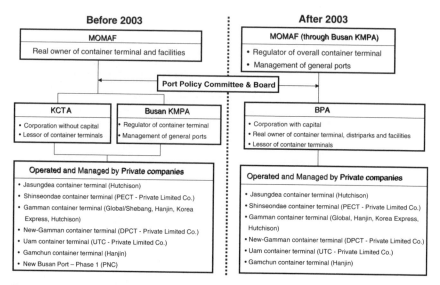

Fig. 16.6. Port Administration in Busan (Before and After 2003). *Source:* Drawn by
the Authors.

power to the BPA acting on behalf of the central government (viz. MOMAF).
However, this structure implies that the central government has still
influenced the management, operations and development of container terminals in Busan. In comparison with the previous system implemented by
KCTA, the level of governmental influence has been significantly reduced
over Busan's ports. The idea of privatisation seems eventually to be in place at
the BPA.

   Another newly established, more-autonomous authority, the IPA, follows
the line of the BPA from its way of business and operations. The IPA will
play a central role in the management and operation of Incheon container
terminals after taking up responsibility from the Incheon KMPA. The IPA
is going to manage, operate and supervise container terminals in Incheon,
but will not develop regulations or engage in other public functions. The
IPA is currently negotiating with MOMAF about how to take up the scope
of government property and debts. Incheon container terminals are all
leased from the central government (viz. MOMAF) without payment. Then
the IPA will rent out these terminals to each terminal operator in return for
payment, as in the case of the BPA.

   Fig. 16.7 shows that there are three different entities controlling the container terminals in Incheon; the terminal operators are subject to the

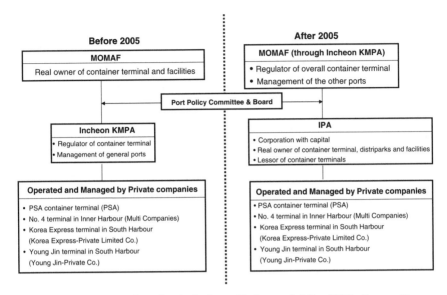

Fig. 16.7. Port Administration in Incheon (Before and After 2005). *Source:* Drawn by the Authors.

supervision of the IPA, and in turn the IPA is indirectly under the control of MOMAF. Before 2004, all the container terminals in Incheon were controlled, operated and supervised by Incheon KMPA. This hierarchy will give much power to the IPA acting on behalf of the government. In comparison to the previous system implemented by the KCTA, however, the influence of government has been decreased over Incheon's container ports by establishing this more decentralised organisation in Incheon.

4. KOREAN PORT GOVERNANCE MODEL

Baird (1995, 1997) put forward a conceptual framework for a port governance model, which is called the Port Function Matrix. Regardless of whether a port concerned is in the private or public sector, it is generally expected to carry out three fundamental functions, namely regulatory, landowner and operator functions. Subject to which side – private or public or joint – undertakes which of these three functions, the model can be classified into the following four types of model, which are summarised in Table 16.4.

Table 16.4.   Port Function Matrix (or Port Governance Model).

Port Governance Model	Port Functions		
	Regulator	Landowner	Operator
PUBLIC	Public	Public	Public
PUBLIC/private	Public	Public	Private
PRIVATE/public	Public	Private	Private
PRIVATE	Private	Private	Private

Source: Baird (1995, 1997).

- PUBLIC port – all three functions are under the public sector.
- PUBLIC/private port – the regulatory and landowner functions are under the public sector, while the operator function rests with the private side.
- PRIVATE/public port – only the regulatory function is under public responsibility, while the other two are in private hands.
- PRIVATE port – all three functions are under the private sector.

Given this classification and discussion in the previous sections, the Korean port governance model belongs to neither the PUBLIC nor PRIVATE model, but somewhere between the two extreme and rare cases – that is, either PUBLIC/private or PRIVATE/public port models. In other words, the port governance model in Korea is a mixture of private and public participation. More specifically, the three major ports – Busan, Incheon and Gwangyang – have the following types of governance model. In the case of Busan and Incheon, the central government MOMAF is still in charge of regulations, while the BPA and the IPA respectively, as half-private and half-public organisations, own the land. The operational functions are completely in the hands of private terminal operating companies. As shown in Table 16.5, in these two ports, the private sector has the dominant role and the public side supplements this.

As shown in Table 16.6, Gwangyang port has a slightly different picture in terms of its governance, which is a model where the public sector plays a dominant role and the private sector provides only a limited component of port businesses. In other words, the KCTA, as a proxy for central government, is involved in almost all port activities, while private terminal operators are carrying out their business with a limited capacity.

Table 16.5. Port Governance Model for Busan and Incheon.

Port Governance Model	Port Functions		
	Regulator	Landowner	Operator
PUBLIC	Public	Public	Public
PUBLIC/private	Public	*Public*	Private
PRIVATE/public	*Public*	*Private*	*Private*
PRIVATE	Private	Private	Private

Table 16.6. Port Governance Model for Gwangyang.

Port Governance Model	Port Functions		
	Regulator	Landowner	Operator
PUBLIC	Public	Public	Public
PUBLIC/private	*Public*	*Public*	*Private*
PRIVATE/public	Public	*Private*	Private
PRIVATE	Private	Private	Private

5. CONCLUDING REMARKS

Korean ports have passed through a variety of port governance stages. In the early years, the public sector was the only player in the industry, doing everything from planning and developing to operating. This type of governance was created by necessity – a government-focused economic development policy revolving around a strategically chosen trade-oriented economy, with seaborne dominated trading patterns. However, this approach to economic development caused a number of problems in the port sector, for example, inefficient management and operations, longer response times to an ever-changing business environment and an overly bureaucratic administration.

In dealing with these problems, the government gradually introduced the so-called private spirit into public dominated industry sectors. The efforts were made in a number of ways, for instance, allowing fully private operators (domestic as well as overseas) to take up their portion of the market, and moving closer to completely private ways of business by corporatising the public authorities. All of these movements are in large part based on the fact that privatisation can create greater efficiency in management, operations, and administration (Yarrow, 1986). However, an interim report of the impact of privatisation on port efficiency does not confirm this theoretical claim (Song et al., 2001).

Nevertheless, the alteration that has taken place in Korean ports has initiated a number of positive influences – for example, introducing the 'private business mind' into a sector that never had such an experience until recently, and attracting more and more investments from overseas partners that demand a more transparent governance model. At the same time, this movement creates a more complex picture of the port governance structure in Korea. Throughout its 50 ports, central government is still, to some extent, involved in daily business issues. However, its interference has become less and less as the overall governance structure evolves. All parties (private, public or joint) engaged in port businesses are required to cope with such a changed governance pattern in order to better equip themselves for stronger competitiveness in the long term.

NOTES

1. The 1996 Act is an updated and extended version of the previous 1967 Act. However, all the ports in Korea are classified under these two Acts.

2. The word *new* is used simply because these nine ports are all recently planned and developed.

3. The Korean subsidiary of HPH (the Hutchison Busan Container Terminal Co.) commenced its operations at the Jasungdae Container Terminals in 2003.

REFERENCES

Baird, A. (1995). UK port privatisation: In context. *Proceedings of UK port privatisation conference*, Scottish Transport Studies Group, 21 September, Edinburgh.

Baird, A. (1997). Port privatisation: An analytical framework. *Proceedings of international association of maritime economists conference*, 22–24 September, City University, London.

Busan Port Authority. (2005). About BPA. http://busanpa.com/service?id = eng_index

Chenery, H., & Syrquin, M. (1986). The semi-industrial countries. In: H. Chenery, S. Robinson & M. Syrquin (Eds), *Industrialisation and growth: A comparative study* (pp. 84–118). London: Oxford University Press.

Cullinane, K. P. B., & Song, D.-W. (2002). Port privatisation policy and practice. *Transport Reviews, 22*(1), 55–75.

Ministry of Maritime Affairs and Fisheries. (2002). *Port and port development.* Seoul: Government Printer.

Ministry of Maritime Affairs and Fisheries. (2004). *Statistics of maritime and fisheries.* Seoul: Government Printer.

Song, D.-W., Cullinane, K. P. B., & Roe, M. (2001). *The productive efficiency of container terminals: An application to Korea and the UK.* Aldershot, UK: Ashgate.

Yarrow, G. (1986). Privatisation in theory and practice. *Economic Policy,* April, 323–377.

PART III:
PORT GOVERNANCE AND DEVOLUTION

CHAPTER 17

PORT GOVERNANCE, DEVOLUTION AND THE MATCHING FRAMEWORK: A CONFIGURATION THEORY APPROACH

Ramon Baltazar and Mary R. Brooks

ABSTRACT

The purpose of this chapter is to apply contingency theory to the management and governance of ports. It describes port devolution as part of a larger attempt by governments to apply the "one best way" principles of new public management (NPM) to the transportation sector. As an alternative approach, the authors develop a Matching Framework that identifies and focuses attention on key contingency variables in port governance and management. An application of the Matching Framework to the Canadian port experience and its implications for port management and research are provided.

1. INTRODUCTION

Over the past two decades, governments have devolved responsibility from the public sector to the private sector for transport support industries and in

Devolution, Port Governance and Port Performance
Research in Transportation Economics, Volume 17, 379–403
ISSN: 0739-8859/doi:10.1016/S0739-8859(06)17017-1

many cases have privatized transport operators.[1] Often, the intention of such devolution is to secure the benefits of commercially driven business decision-making in organizations previously run by government, to acquire compensation for prior investments by taxpayers, to isolate marginal ports from access to scarce government capital, to separate the regulation of ports from the operation of ports, or some combination of these.

Port devolution programs are part of a larger attempt by governments to apply new public management (NPM) concepts to the transportation sector. NPM proponents attempt to apply commercial private sector principles to government operations. For example, Osborne and Gaebler (1992), who studied successful government agencies, note that although government can never be exactly like a business, successful agencies demonstrate many business-like qualities. These include being customer-driven, acting as catalysts for competition, being entrepreneurial, outcome-driven, decentralized and mission-driven, and having an earning (its keep) mentality. Similarly, Charih and Rouillard (1997) describe NPM as being client-focused, management-centred (as opposed to being administration-centred), entrepreneurial, innovative, decentralized, frugal and a user of quasi-market mechanisms. Other recommendations are directed to more specific types of NPM activities. For example, Dobell and Bernier (1997, p. 257) focus on the need for greater collaboration among agencies in serving clients, including true partnership that is more than just a "warm, fuzzy word for contracts," that is characterized by simultaneous "core policy 'steering' and operations 'rowing'," as well as clarity, transparency and openness among partners.

After at least two decades of devolution activities, port performance outcomes have been mixed. As Brooks notes in Chapter 25 of this book, there has been considerable dissatisfaction with the performance of post-devolved port entities and with devolution programs in general. Even fully privatized ports have not performed to expectations, as Baird (in Chapter 3 of this book) notes for the United Kingdom.

For the authors, at least a part of the explanation for the poor performance outcomes of devolution activities resides in an assumption underlying the NPM literature. A common theme in the NPM literature is that there is "one best way" to manage. In many countries, this thinking has been used to argue in favour of greater devolution in the form of port privatization, commercialization, or concessioning apparently without first questioning whether or not the devolution model proposed is appropriate to local conditions.

An alternative to the one best way approach for port devolution programs is to adopt contingency-based models of program delivery. Based on contingency theory (see Donaldson, 1996), policy makers would isolate variables on whose characteristics the "right" program depends, and make contingent program recommendations depending on the properties of the variables. In other words, contingency theory suggests that there is not one best way but an appropriate way to manage for a given context. Contingency theory does not refute the validity of making absolute recommendations with respect to some variables that affect a given organizational situation. However, the theory suggests that for a given situation, there will likely be sets of contingency variables whose characteristics should match if organizational performance is to be maximized. The challenge for contingency theorists is to identify and isolate these variables.

The purpose of this chapter is to apply contingency theory to the management and governance of ports. In the chapter, the authors identify the contingency variables for port authority management and develop a "Matching Framework" for use as a guide in the governance of ports. The next section provides background on contingency theory. Following that, the Matching Framework is developed along with a Canadian example of its application. Finally, implications of the Matching Framework for port governance and management are discussed.

Contingency theory is derived from literature in the fields of organization theory and strategic management. In organization theory, the starting point of organizational analysis is the firm's external environment. The environment consists of many sectors external to the organization, including the industry, raw materials, human resources, financial resources, market, technology, economic conditions, government, socio-cultural and international sectors (Daft, 2004). Of particular relevance is the organization's operating or "task" environment. A subset of the organization's external environment, the task environment includes the sectors that have a direct impact on the firm's ability to achieve its goals (Daft, 2004).

A commonly used variable for describing task environment is degree of environmental uncertainty, which refers to the extent to which information about the organization's task environment is perceived to be absent (Daft, 2004). The sources of this uncertainty include environmental complexity and environmental dynamism (Daft, 2004; Duncan, 1972; Jones, 1995). Complexity refers to the number of dissimilar elements in the environment that have to be dealt with by the organization. Dynamism characterizes the extent to which the relevant environmental elements are changing. The more complex and dynamic an environment is perceived to be, the greater the environmental uncertainty.

Organization theorists view environmental uncertainty as a contingency (Daft, 2004; Jones, 1995) that should be managed by way of the organization's structure. Organization structure consists of the organization's hierarchical reporting relationships, operating procedures and information and control systems (Bourgeois, Duhaime, & Stimpert, 1999, p. 260). According to the organization theory, successful organizations have structures appropriate to the level and source of environmental uncertainty facing them. For example, Lawrence and Lorsch (1967) found that successful organizations facing highly complex and dynamic environments were less formalized (i.e. had less documented routines), more decentralized and relied more on mutual adjustments between members of the firm than the average organization. Conversely, they found that firms whose environments were more certain performed better when they adopted more centralized, formalized and standardized structures. Similarly, Burns and Stalker's (1961) research showed the need for organic structures in dealing with uncertain environments and for mechanistic structures in dealing with stable environments.

While organization theorists assume the environment and its characteristics to be given, strategic management researchers do not. Instead, they assume that, within limits imposed by environmental characteristics, the organization may choose to operate within one of several alternative operating environmental domains.[2] The choice is made through organization strategy, which is defined as the pattern that integrates an organization's decisions and actions into a cohesive whole (Quinn, 1996, p. 3). For example, Porter (1980) contends that in any given industry, firms have the option of adopting either a cost leadership strategy or a strategy of differentiation. In cost leadership, the organization focuses on the efficient delivery of the basic product or service at lower costs and, often, at lower prices. In differentiation, the focus is on the effective delivery of augmented product and service characteristics that go beyond the basics and for which the market is willing to pay a premium.

Similarly, Miles and Snow (1978) contend that organizations may choose between prospector strategies that are innovation-focused and defender strategies that are efficiency-focused. The latter are best implemented using mechanistic structures, while innovation-oriented strategies require more organic forms. Were a government to impose a governance structure in its devolution legislation without regard for the port strategies that might be followed, the strategic objectives of the particular organization or its environment, it would likely find the outcome to be less than optimal.

Boschken (1988) applied this approach in his study of port managers. Based on Miles and Snow's (1978) typology of organizations, with four categories – prospectors, analysers, reactors and defenders – Boschken elaborated (with appropriate modifiers) on the categories and extended the typology to the evaluation of port managers on the US west coast; he classified Oakland, Long Beach and Seattle managers as "enthusiastic" prospectors. Ircha (1997, p. 134) used Boschken's approach in his assessment of Canadian port reform, concluding that

Although many Canadian port managers may act as "reluctant reactors" casting envious eyes at innovative approaches being introduced in other ports, the real need is for "enthusiastic prospectors" to cope with today's economy.

Ircha determined that the economy for Canadian ports created a highly uncertain task environment, one that would favour creative structures and innovative approaches instead of a more certain task environment that was well suited to mechanistic approaches and reactionary personnel.

These two streams of research (organization theory and strategic management) gave rise to a third stream that would integrate findings. Known as configuration theory, this stream sought matching environment–strategy–structure relationships that would influence organization performance. According to configuration theory, alternative strategies imply a choice of operating environment and concomitant characteristics (Miles & Snow, 1978; Miller, 1986; Porter, 1980). Environments where effective delivery to changing market needs is crucial will tend to be relatively uncertain. In contrast, environments where efficiency-oriented strategies thrive will tend to be more stable. Moreover, alternative strategies require matching structures to succeed. Efficiency-oriented strategies are best implemented using relatively mechanistic structures, while innovation-oriented strategies require more organic forms. Ketchen et al. (1997) performed a meta-analysis of the configuration research literature. In aggregating results across studies, meta-analysis is able to account for sampling and other comparison errors that may hinder attempts to synthesize the results (Hunter & Schmidt, 1990). The Ketchen et al. (1997, p. 233) study "removes any equivocality surrounding configurations' ability to predict performance."

It is important to note that the configuration literature makes no assumptions about the inherent superiority of one environment–strategy–structure configuration over the other. Rather, the fit exhibited by the configurations is seen as the primary determinant of organization performance. Put another way, organization performance is viewed as being contingent on the match between the characteristics of the organization's environment, strategy and structure. If this is valid, it would cast doubt on the universal application of the anti-centralization, anti-bureaucracy and pro-flexibility principles that, as discussed earlier, are normally espoused by proponents of NPM. Instead, contingency thinking would encourage stewards of public organizations who, in striving for good governance, may find the need for organizational designs that are centralized, bureaucratic and relatively inflexible, provided these characteristics are suitable to the organization's task environment.

2. THE MATCHING FRAMEWORK

The Matching Framework is an application of configuration theory to port organizations. It starts with two notions. First, obtaining positive economic returns requires the organization to be both efficient in delivering its products and services and effective in satisfying the needs of target customers. Second, following Thompson, Strickland, and Gamble (2005), Hitt, Ireland, Hoskisson, Rowe, and Sheppard (2006), Porter (1980, 1985) and others, obtaining superior returns in comparison to similar organizations requires excellence in being efficient or being effective and at least average competence in the other.

As illustrated in Fig. 17.1, the Matching Framework views port performance as a function (output) of the match (or fit) among the characteristics of the organization's external operating (or task) environment, strategies and structures. The greater the fit, the better the expected performance will be; the poorer the fit, the worse the expected performance will be.

A discussion of the four variables in the framework follows.

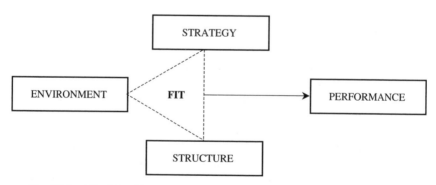

Fig. 17.1. The Matching Framework. *Source:* Baltazar and Brooks (2001).

2.1. External Environment

Like Pierce and Robinson (2000), the authors find it useful to divide the organization's external environment into three categories: the remote environment, the industry environment and the operating environment. The classification is based on the individual organization's ability to impact particular environmental factors.

The *remote* environment consists of external factors that affect all industries. These include the general political, technological, economic and socio-cultural environments. For example, rising interest rates impinge upon the ability of organizations in all industries to secure capital. Advances in computer technology alter the way organizations gather and assess important data and communicate across geographical boundaries. Like all other organizations in all industries, port organizations have little (if any) influence on these remote factors, but must respond to them.

The *industry* environment consists of those factors that affect an industry's participants. An industry is defined to consist of organizations whose products and/or services are close substitutes for one another (Thompson et al., 2005). As noted by Robinson (2005, p. 249), Porter's Five Forces analysis provides a useful framework for understanding the determinants of long-term industry profitability and how companies might influence these forces.[3] Compared to the remote environment, the characteristics of the industry environment are subject to greater influence by individual organizations. The degree of influence held by a particular organization depends on several factors including the number of industry participants, the relative size of the organization in question compared to other industry participants, and the extent to which industry participants normally collaborate through industry associations and other institutions. The industry environment is also constrained by the ground rules established by government in its ports policy, which may itself be influenced by the lobbying efforts of the firms in the industry.

The *operating* environment consists of the factors that a specific organization interacts with on a regular basis. Every organization also faces a particular mix of buyers, suppliers, direct competitors (if any), potential new competitors and the substitute products and/or services available to the organization's buyers. Compared to the industry and remote environments, organizations have the greatest amount of potential influence on the operating environment. To a large extent, the organization can influence the extent to which the operating environment exhibits uncertainty that, as noted earlier, constitutes complexity and dynamism. In the view of the authors, the primary mechanism through which the organization exercises influence on the degree of uncertainty in the operating environment is strategy, which is discussed next.

2.2. Strategy

Organization strategy is defined as the pattern in a stream of organizational decisions and actions (Mintzberg, 1987). Following Robinson (2005), the authors see strategy as the organization's response to external (operating) environmental conditions with its unique mix of resources and capabilities.[4] For the authors, two dimensions of strategy are particularly relevant to ports concerned with economic performance: product and market scope and competitive emphasis.

Product and market scope refers to the range of specific products and services an organization offers and the markets to whom the offerings are made. For ports, the choice of product service offerings include those mandated by law for safety reasons and those it chooses to make available for specific customers or as a whole in the form of common user services. A sample list of service options, while not comprehensive, appears in Table 17.1 and illustrates the scope concept. Market choices include those types of markets the particular port wishes to compete in, such as container (import/export, transhipment of feeder cargo), bulk (oil, dry bulk, specialty cargoes), automobile handling, cruise, recreation, land development and so on. The type of facilities the port or terminal builds, or seeks to develop via concession agreement, depends on the market choice decision.

In Table 17.1, port A's approach focuses on a strategy to serve one market (dry bulk) with a limited basic product that does not include stevedoring, line-handling or storage, elements of the product offering that many ports would consider part of the basic product. On the other hand, ports B and E seek to serve a wide range of markets with a wide range of product–service offerings. Table 17.2 confirms that most large ports choose to serve a wide variety of markets while only a few specialize in serving only one or a few.[5] The variations noted in Tables 17.1 and 17.2 confirm that port strategies may differ significantly in product-market scope.

Worth noting is that the wider the product-market scope, the greater the complexity of the operations and the higher the potential for dynamism. For example, the port of Taranto in Italy has opted to compete in the container market, and has entered into a 60-year concession agreement with Evergreen Marine Corporation of Taipei (Taiwan), to provide the product offering for

Table 17.1. Port Product and Market Scope Decisions.

Scope Decision	Port[a]				
	A	B	C	D	E
Product/Service					
Berths	Yes	Yes	Yes	Yes	Yes
Cruise terminal			NI	Yes	Yes
Line handling	No	Yes	Yes	Yes	Yes
Marketing support	Yes	Yes	No	Yes	Yes
Port access	Yes	Yes	No	Yes	Yes
Ship repair	Yes	Yes	Yes	Yes	Yes
Stevedoring	No	Yes	Yes	Yes	Yes
Storage	No	Yes	Yes	Yes	Yes
Market					
Breakbulk	No	Yes	Yes	Yes	Yes
Container	No	Yes	No	Yes	Yes
Dry bulk	Yes	Yes	Yes	Yes	Yes
Fishing	NI	No	Yes	Yes	NI
Liquid bulk	No	Yes	Yes	Yes	Yes
Port volume (in millions of metric tonnes)	4	100 +	23	24	100 +
Region	Middle East	Europe	North America	North America	Asia

Note: NI = No information or not provided by the port.
Source: Based on material provided by ports or from port web sites and/or Annual Reports.
[a]Ports are "renamed" with letters to meet requests for confidentiality.

the Taranto Port Authority. Through the concession agreement, the port authority at Taranto has effectively withdrawn from the port management business and entered into the port landlord business. By doing so, Taranto has devolved the impact of uncertainty in the port's operating environment onto a private sector operator and retained for the Authority a less complex environment where strategic response can be less flexible and structures mired in bureaucracy can be more successful. In contrast, Evergreen as the terminal operator has assumed an extremely complex and dynamic operating environment that includes being in competition with many other large and global terminal operators, and having to cope with rapidly changing technological and logistical trends in the business.

Competitive emphasis refers to aspects of the way the firm conducts business that are unique among competitors in delivering products and services to target markets. Of the many choices the organization has in selecting

Table 17.2.   Markets Served.

Market	Frequency (n = 39)	Number of Markets Served[a]	Frequency (n = 39)
Container[b]	35	One of five	1
Less than 750,000 TEU p.a.	24	Two of five	0
750,000–1.5 million TEU p.a.	0	Three of five	2
More than 1.5 million TEU p.a.	7	Four of five	14
Liquid bulk	29	All five	22
Dry bulk	35		
Breakbulk, neo-bulk, general cargo not otherwise categorized	34		
Other[c]	15		

Source: The Port Performance Research Network database, created from the survey of 42 ports in 2004–2005.

[a]Ports were given the five choices indicated in the first column, so this is of five. Of the 42 ports in the database, 3 declined to respond to this question.

[b]Four ports declined to indicate size of activity in TEU per annum although the volume measures in tonnes would likely classify them as larger than 750,000 TEU.

[c]Ro–Ro was a common addition with a few ports noting passenger/car ferry, cruise, military and fishing businesses.

competitive emphasis, none is more fundamental than the choice that must be made among cost leadership, differentiation and best-cost approaches to doing business. Distinguishing these choices starts with an understanding of what constitutes a product (or service) offering. Any offering can be broken down into a set of core characteristics, referred to as the basic product, and a set of supplemental characteristics, called augmented products or services. For a port, the basic product might constitute the availability of a berth and a dredged channel to get to it. The augmented product might include ship repair facilities, specialized cranes, 24-hour stevedoring labour, on-dock entertainment for passengers of cruise vessels and so on.

Because all comparable product offerings would include the basic product, superior performance with a basic product focus requires the firm to become more efficient than the average competitor. Porter (1980) called this cost leadership. For ports that have little product scope beyond the basic product, their interest in port efficiency is therefore obvious; they must be more efficient than the average if they expect to be in business long-term, and the average is, of course, a moving target.

Alternatively, the firm could go (way) beyond the basic product, and focus its efforts on some or all of the possible augmented offerings. The problem with this approach is that it adds costs that the average competitor focusing on basic product delivery would not have. For the approach to be fruitful, the firm's offering would have to be so valued by the customer as to command a premium price that yields a margin exceeding the average competitor's. To the extent the firm is able to renew the extra value in the eyes of the customer, it would have developed a profitable *differentiation* strategy. In contrast to the cost leadership approach, a differentiation strategy has an effectiveness focus. Here the port is seeking to create a sustainable, differentiated set of product offerings in a particular part of the market that sets it apart from the others against which it competes.

For example, as discussed by Cullinane, Yap and Lam in Chapter 13, the port of Singapore's competitive emphasis for some time has been to offer higher-priced premium services. With the opening of a lower-priced alternative across the Malaysian border at Tanjung Pelepas, Singapore lost two key clients in Maersk and Evergreen. Instead of lowering prices to match those of Tanjung Pelepas, the port authority changed tactics. Until 2003, the port would not lease land to a private operator. After losing Maersk and Evergreen, the port began to offer its existing and potential customers the option of dedicated or joint venture terminals in exchange for long-term leases. By augmenting its services in this manner, the port maintains its premium competitive emphasis while reducing the risk of defection and reducing the uncertainty in its environment.

The best-cost route seeks to balance the best of both strategies. Here, the quality of the firm's offerings would fall in between those of the cost leader and those of the differentiator. There is no consensus in the strategic management literature on the feasibility of the best-cost approach. At the heart of the discussion is whether or not the approach is sustainable, as it entails being stuck between two apparently opposed performance rationalities with the potential for sniping or poaching customers existing in the overlap.

The choices organizations make regarding competitive emphasis have implications for environmental uncertainty particularly in the organization's operating environment. The operating environment of an organization that chooses a strategy of differentiation tends to be more uncertain than the operating environment of a cost leader. First, the differentiator organization's environment is more complex because it provides extra services that the cost leader does not. Each extra product or service provided resides in a unique environment that comprises such factors as government regulations surrounding product or service standards, customer expectations and competitive standards; each of these requires organizational monitoring and periodic action if the organization is to succeed in delivering product or service. Second, because the potential for change exists within each environmental context the organization faces in delivering an extra product or service, the differentiator's operating environment as a whole tends to be more dynamic than the cost leader's operating environment.

It is important to note that because there is often more than one basis of cost leadership (e.g. economies of scale, automated back room systems) or differentiation (e.g. reliability, personalized service), several successful cost leaders and differentiators may co-exist in a sufficiently large market.

In summary, strategy represents the organization's choices of product and market scope and competitive emphasis. By making these choices, the organization carves out that portion of the external environment in which it will *choose* to operate. This environment will have uncertainty levels and characteristics that the organization will have to accommodate if the strategy is to succeed. The principal coping mechanism at the organization's disposal is structure, which is discussed next.

3. STRUCTURE

Organization structure consists of the organization's hierarchical reporting relationships, operating procedures and information and control systems (Bourgeois et al., 1999, p. 260). Structure affects the way people in the organization work. For example, people will tend to behave in ways that are reinforced by the organization's reward structure, and tend to avoid behaviour that is punished or ignored by the system. As such, structure is the principal means through which the organization implements strategy. As noted by Hitt et al. (2006), failure to match structure to strategy will cause port performance to suffer.

In the view of the authors, the principal structural concepts that port managers should consider in applying the Matching Framework are centralization and standardization. Centralization is the extent to which important operating decisions are made at higher levels in the organizational hierarchy; standardization is the extent to which there are behavioural rules, norms and operating procedures in the organization (Daft, 2004).

The organization's choice of product–market strategy has implications for the degree to which structure should be centralized and standardized. As noted in the previous section, complexity and the potential for dynamism increase as the organization adds more products and markets to its portfolio of operations. As product and market scope becomes wider, the need to customize operations to meet the needs of different products and market environments becomes greater. At some point bounded rationality (the limited ability of human beings to process and deal with ambiguous information) makes it necessary to decentralize responsibility. Lower levels of standardization across responsibility units come with decentralization, as unit managers gain the freedom to make decisions and vary operations according to their unit's specific needs.

The organization's choice of competitive emphasis also has implications for the degree of centralization and standardization. The goal of cost leadership is to sell large quantities of basic products (or services) to customers. Because the product (or service) of the cost leader is basic, the market's requirements are known by the organization. As such, the organization's efforts can focus on maximizing the efficiency through which the product is made and delivered. In reaching for efficiency, a centralized corporate staff and highly standardized rules emanating from this staff are appropriate (Hitt et al., 2006).

In contrast, the goal of a differentiation strategy is to customize products and services to customers with unique needs. Because needs vary customer to customer, the organization's focus is to maximize the effectiveness with which it can serve differing needs. As Hitt et al. (2006) note, this focus requires lower levels of standardization and centralization of structure, in comparison to the cost leader approach. Standardization would be lower to accommodate the greater variety of needs by the organization's target market. Centralization (of decision-making authority) would be lower because employees would be allowed to use some judgment in catering to the market's varied needs. As ports reach into the provision of ancillary services (such as logistics hubs, local transit and so on) as bases of differentiation, the stresses placed on port managers guide the port towards organizational structures and business processes that are more decentralized, flexible and mutually adjusting between structural units, to accommodate the myriad of customers and stakeholders and the diversity of regulatory agencies.

4. PERFORMANCE

From the discussion above, Fig. 17.2 summarizes generic configurations in the Matching Framework. For easy reference, the configurations are designated as *efficiency oriented* or *effectiveness oriented* configurations. Efficiency oriented configurations are characterized by low uncertainty in the organization's operating environment, a narrow product-market scope or a cost leadership approach to product and service delivery, and a high level of

	Efficiency-oriented configuration	Effectiveness-oriented configuration
Environment	Low uncertainty (Low complexity and dynamism)	High uncertainty (High complexity and dynamism)
Strategy	Narrow product market scope Cost leadership approach (Focus on delivery of the basic product or service)	Broad product market scope Differentiation approach (Focus on delivery of augmented products and services)
Structure	Mechanistic (Centralized decision making characterized by higher standardization and lower customization)	Organic (Decentralized decision-making characterized by higher customization and lower standardization)

Fig. 17.2. Alternative Configurations in the Matching Framework. *Source:* Adapted from Baltazar and Brooks (2001).

structural centralization and standardization. In contrast, effectiveness-oriented configurations face high levels of environmental uncertainty, a broad product-market scope or a differentiation approach in delivering products and services. In the Matching Framework, port performance is viewed as a function (output) of the match (or fit) among the characteristics of the organization's external operating (or task) environment, strategies and structures. The greater the fit, the better the expected performance will be; the poorer the fit, the worse the expected performance will be.

The Matching Framework assumes a desire to maximize the organization's economic performance, subject to the constraints imposed by the organization's other goals.[6] In the research being conducted by the Port Performance Research Network, 23 of 42 ports surveyed listed an economic performance measure as their primary strategic purpose (Table 17.3).[7] In fact, a significant share of the ports listed economic development as one of their strategic objectives and the largest group saw it as their only objective. This means that the strategic intent of these port authorities is not as much about maximizing port efficiency or its effectiveness in serving its customers, as it is about creating jobs and opportunities for businesses. In these cases, the Matching Framework will not directly help in achieving the primary goal of job or business creation, but will help in attaining the secondary goal of maximizing economic performance within the constraints imposed by the primary goal.

Table 17.3. The Port's Primary Strategic Purpose.

Goal	Frequency	Port's Only Goal[a]
Maximize profits for our shareholders	12	4
Maximize return on investment for government	11	7
Maximize traffic throughput	9	2
Maximize traffic throughput subject to a zero profit	0	0
Maximize traffic throughput subject to a maximum allowable operating deficit	6	1
Optimize economic development prospects (local or national)	18	11
Other[b]	4	4
Total (*n*)[c]	42	28

Source: The Port Performance Research Network database, created from the survey of 42 ports in 2004–2005.

[a] As ports were asked about their primary goal, and were expected to check only one of the options, 14 checked more than one. This column indicates the number for whom only one goal was checked.

[b] One port's only goal was to maximize throughput subject to financial self-sufficiency; one port's goal was to collect fees for the government and minimize environmental damage; the third had several goals including maximization of the use of the facilities; the fourth sought to maximize the value added that the port generates for the city and the region in the long term.

[c] Of these 42, 31 saw their primary role as continental or national gateway, 1 as primarily a transit port with the remainder serving local, regional or feeder roles. As only ports with more than 2 million metric tonnes of annual activity were approached, this is not surprising.

In applying the Matching Framework, it is important to note that the choice of configuration is a matter of focus or emphasis on efficiency or effectiveness, but not the abandonment of the other. Some minimum level of both efficiency and effectiveness is critical to organizational survival. Efficiency oriented configurations perform well under the assumption that the organization's products and services are targeted at a market whose needs are basic, and to that extent, the organization knows what the market desires. In this case, failure to effectively meet minimum customer expectations will result in poor economic performance. On the other side, effectiveness oriented configurations recognize that they must first and foremost cater to the needs of customers whose product and service expectations are more sophisticated and varied. However, the economically successful organization will meet these needs with efficiency to the possible extent. Put another way, a successful efficiency oriented configuration delivers products and services that effectively meet the expectations of customers with basic needs, while successful effectiveness oriented configurations deliver to the needs of more sophisticated customers as efficiently as possible.[8]

What is measured gets done. Thus, the selection of one configuration over the other requires the development and application of measures to monitor important aspects of the configuration being pursued. As Brooks notes (in Chapter 25 of this book), it is not enough to use only broad organizational performance measures such as volume throughput, sales volume and profitability. It is imperative that the port authority also develops indicators for, and applies measures of, port effectiveness and efficiency for the configuration being implemented.

5. APPLYING THE MATCHING FRAMEWORK: THE CANADIAN CASE[9]

The Canadian ports situation may be used as a brief example for using the Matching Framework to analyse Canadian port management and performance over time. Building on the materials presented by Brooks in Chapter 11, the change from NHB port to LPC port in the early 1980s did not really shift Canadian ports from their predominantly Efficiency Oriented Configuration position noted in Fig. 17.2. In the post-World War II era, traffic was predominantly commodity-based and therefore generally captive to the nearest port. The operating environment was relatively predictable. Because of this, centralized control worked and port managers were accepting of the efficiency-focused defender strategies proposed by Miles and Snow (1978); as a result, ports continued to function in the post-1982 reform period in the interests of all of Canada. They sent their operations-driven plans to the federal government crown corporation (Canada Ports Corporation) for approval and local capital expenditures were capped at C\$10 million. The systems and structures operated in a predictable manner, and decision-making processes were not very adaptable when the environment changed.

The remote environment did change. World trade grew, and with it the share carried in containers. Just-in-time production and delivery systems coupled with globalization of manufacturing operations meant that containerized trade grew far faster than either world GDP or world commodity output. This resulted in changes in the operating environment for container ports. In addition, Canada's overseas export mix shifted in proportion and introduced more complex port usage patterns. The rise of the cruise trade in North America, as cruise operators sought non-Mediterranean routings following the *Achille Lauro* incident, meant that single-use ports with industrial designs were contemplated as day stops for tourists. There was pressure on ports to reconsider their scope and market decisions, although any market decisions would of course be accompanied by the need for supporting infrastructure investment. Moreover, with the deregulation of surface transport, the signing of the *Canada US Free Trade Agreement* and, later, the *North American Free Trade Agreement*, Canada's inland transport systems shifted from a primarily east–west orientation to a north–south one. This trade realignment[10] pit three Canadian ports (Vancouver, Montreal and Halifax) against US ports (particularly Seattle, Tacoma, New York and Norfolk) in competition for North American container traffic as land bridging activities on the continent grew. The environment had shifted to the right, as shown in Fig. 17.3, becoming much more complex and more dynamic (and thereby more uncertain), demanding a reconsideration of strategy and structure in delivering port services as they were no longer aligned with the environment.

In addition, the fiscal difficulties facing the Canadian Federal Government had led to interest in alternative governance models. Its experiment with airport devolution had largely worked; here the government had transferred management and operating responsibility to newly created not-for-profit non-share capital corporations. These entities could embark on much-needed large-scale capital investments and could borrow the needed funds limited only by what the private capital markets were prepared to invest. New Airport Boards had proven themselves to be good prospectors, with a strong customer orientation, and adept at attracting the private capital necessary to rebuild the infrastructure, something the financially strapped federal government had been unable to do. Airports had become full effectiveness-oriented entities (Fig. 17.2).

The release of the *National Marine Policy* had reflected the government's recognition of this success with its plan to make Canada's ports more responsive to the local community – by instituting a new structure, one more suited to the environmental shift to the right. However, as the government was unhappy with the loss of control over what investment decisions airports made, and because it was requested not to make as great a change by port managers themselves, the corporatization of airports was not repeated and Canada commercialized ports; they remained federal agencies but without recourse.

However, as can be seen in Fig. 17.3, the government failed to go the distance necessary to align the three elements and gain the fit required for an optimized outcome. The operating environment had shifted to one of

1970s	Efficiency-oriented configuration	Effectiveness-oriented configuration
Environment	Low uncertainty (Low complexity and dynamism)	
Strategy	Efficiency-oriented (defender) Focus on delivery of the basic product or service	
Structure	Mechanistic Centralized decision-making characterized by high standardization and little local authority	

About 2003		
Environment		Higher uncertainty (Higher complexity and dynamism, both somewhat diminished by the landlord role)
Strategy	Efficiency-oriented (defender) Focus on delivery of the basic product or service	Some focus emerging on ancillary service development; some entry into additional markets like short sea feeder or cruise.
Structure	Mechanistic (large capital projects approved centrally) Governance of activities entrenched in legislation. High standardization in business processes (such as pricing of services)	Decentralized decision-making but still have to get approvals for large capital projects Moving towards some customization

Fig. 17.3. Canada's Devolution Experience (for CPAs) Using Matching Framework.

greater uncertainty and with north–south as opposed to east–west competition. The implementation of port devolution failed on two fronts. First, by only making the ports federal agencies, it decentralized decision-making but not in its entirety; ports had full responsibility without sufficient authority as would be found in wholly independent local bodies. Each port's responsibility for investment was hindered by the cap imposed by government on its ability to raise funds for capital projects (the limit being that which could be supported by the pledging of cash flow only). Ports did not have access to the myriad of capital funding opportunities that would be available if they were wholly independent, autonomous, private entities; they were unable to sell land (trading was possible). The government also failed to change the orientation of management from its established defender ways towards the necessary innovation-focused prospector (Ircha, 1997).[11] Without going the whole way to an independent Board of Directors keen to grow the facility, by placing token value on user input[12] and by retaining patronage appointments, the federal government set up the port Boards to fail not only to invoke the management change necessary, but also to deliver the effectiveness-oriented, customer-focused strategies necessary. The objective of the government was one of local responsiveness rather than economic performance. The performance expected was one of self-sufficiency (no drain on the public purse). If ports succeeded, it was because they fought for change and embraced the challenge despite the hurdles; success became a product of management skill. They also had help from the principal industry association, the Association of Canadian Port Authorities, in developing training on governance and policy positions on capital funding. At least before reform, the strategy and structure were internally consistent (although not matched to the turbulent environment). As can be seen in Fig. 17.3, there is progress towards customization and strategies examining ancillary service development, but they still have bureaucratic processes and political lobbying to do if they are to make the capital investments they need to respond.

Currently, ports in Canada sit in the unenviable position of being neither efficiency- nor effectiveness-focused. The level of uncertainty in the competitive environment suggests that an appropriate fit could be developed if the government relaxed restrictions on port processes and structures; one of these might be a reduction or elimination of the restrictions on port activities and port funding in the letters patent. Such a move could encourage ports to develop more nimble and agile processes with less standardization. Port strategies could focus on meeting customer needs, delivering the activities that make the most sense for the circumstances rather than what is *proscribed* to be "port activities." If the lead comes from government, the current nature of competition between Canadian and US ports will drive the strategy rethink once the product and market possibilities are fully available.

Overall, the Canadian case illustrates a configuration in transition from an efficiency orientation to an effectiveness orientation. The configuration in 1970 was internally consistent in its efficiency orientation, and performance goals were being achieved. Since then, changes have occurred within the operating environment that has forced some response in favour of an effectiveness orientation. However, by 2003 the response had been only partial and so the organization was stuck in the middle between two competing configuration approaches. This being the case, organizational performance was being undermined, and will continue to be undermined until organizational characteristics are brought more in line with environmental requirements.

6. IMPLICATIONS FOR PORT GOVERNANCE AND MANAGEMENT

Interest in the economic performance of ports continues to grow.

The decision to devolve ports partially or fully should rest on the assessment by government of its own ability to manage the port's performance-critical factors. Some of these factors will encourage management (a case of appropriate incentives) in a common direction for all ports, such as upgrading organizational systems and capabilities to keep up with rising global standards in information processing capability, equity and participation of the workforce and ethics. However, there are also factors for which there are few, if any, absolute standards, and for which appropriate recommendations depend on other factors. In this chapter, the authors developed a Matching Framework that recognizes the contingency relationships that exist in the organization's operating environment, strategy and structure, and that predicts the impact of the fit (or misfit) among the variables' characteristics on organization performance.

The Matching Framework is an analytical tool that is useful for practitioners and researchers alike. Whether or not the port is partially or fully privatized, those who govern ports and those who manage port operations will benefit the port economically from applying the Framework. For practitioners, the Framework is particularly useful in situations where current economic performance is unsatisfactory or organization re-engineering is required by situations such as privatization or merger, and the practitioner has at least some influence over the strategy, structure and operating environment variables in the Framework.

The Framework's utility is in focussing the manager's attention on *relationships between variables* that might otherwise escape his or her attention as a possible source of organizational problems or opportunity creation while facing day-to-day operations. Although the intent of the framework is to maximize fit in the characteristics of the port's operating environment, strategy and structure, as a practical matter the practitioner should start applying the Framework by first identifying the *misfits* between variables.

Three categories of misfit are possible: environment–strategy, strategy–structure and environment–structure. Environment–strategy misfits occur when the selected strategy does not have a large enough market base to be feasible. Strategy–structure misfits occur when the intent of the strategy and of the chosen structure are at odds. For example, a strategy that focuses on providing high customization to meet varying customer needs is at odds with a structure in which extensive rules guide the behaviour that the service provider is unable to bend in order for the customization to happen. Environment–structure misfits occur when there is a mismatch between environmental uncertainty and the structure's ability to deal with it. For example, an environment that requires dealing with multiple stakeholders will tend to be ill-served by a structure with only a few units, each of which is assigned to monitor and deal with the needs of more stakeholder groups than it is capable of handling.

It is the process of examining the relationships between variables that assists the decision-maker in determining the particular alignment that will be adopted. Once a focus is determined, then the details will need to be elaborated to improve fit, and an implementation plan developed. How many ancillary services will be offered in an efficiency-oriented port? Will it be a broader product scope for an effectiveness-oriented port, and will the broadened scope provide a competitive advantage over other ports the customer may consider? Once a port has determined its focus and approach, it can then benchmark itself against competitors and/or those with similar configuration focuses.

After the choice of configuration is made, the practitioner's task should turn to ensuring that the principles of the configuration selected are understood and acted upon throughout the port organization. Without this effort, the chosen configuration will remain trapped within the cage of a Matching Framework located in the minds of the configuration planners. Unleashing the configuration's potential requires a rigorous goal-setting process that will translate the goals and principles of the chosen configuration into meaningful behavioural guidelines. As Mealiea and Baltazar (2004) note, the process should encompass the team (or unit) level as well as the level of individual port employees.

For port researchers, the Matching Framework is useful for exploring the performance implications of management decisions in areas that affect the Framework variables. The main issue for researchers is the choice of methodology. The characteristics of the variables in the Matching Framework (environment, strategy, structure, performance) and the concept of fit among them are macro organization variables for which information content is voluminous, rich and often ambiguous. As such, large-sample, survey-based, quantitative research methods are difficult and may be resource prohibitive. More feasible (and suitable for researching rich, ambiguous situations) are small-sample, case-based, longitudinal and/or qualitative methods of investigation. The Framework is most useful in studies that compare and contrast the performance and configuration characteristics of different organizations for a common time frame, or in studies that examine the configuration and performance of single organizations over the course of different time frames.

NOTES

1. Many governments devolve ports by transferring the authority functions to another level of government (decentralization) or to a new entity (through commercialization, corporatization or privatization programs).

2. Institutional theory (see Scott, 1995) and co-evolution theory (e.g. Dijksterhuis, Van den Bosch, & Volberda, 1999) go even further than the notion that organizations select their operating environments (through strategy). Institutional theory posits that the operating environment can be influenced (and thus changed) by the strategy-making activities of an organization through mimetic processes that tend to institutionalize at least some of the said activities. In co-evolution theory, managerial intention and environmental selection are seen as primary determinants in the co-evolution of the organization and its operating environment.

3. Five Forces analysis may be applied at the industry level, the industry segment level and at the level of the individual organization.

4. This view supports the primacy of external considerations over internal organization considerations in the determination of strategy. The resource-based view (see Barney, 1995; Collis & Montgomery, 1997) argues that strategy rests on the resources of the firm in the form of assets, skills and capabilities. The focus of the resource-based view is "not what a firm *wants to do*, but what *it can do*" (Collis & Montgomery, 1997, p. 9). The present authors view strategy principally as a response to the external environment, and resources and capabilities as time-bound constraints that should be considered in formulating that response. A selected strategy implies the existence or development of required levels of resources and capabilities without which the strategy is infeasible. The acquisition or development of the required resources and capabilities over time is seen (by the authors) more as a concern of strategy implementation than strategy formulation.

5. This is not surprising as ports surveyed were limited to those with more than 2 million metric tonnes in annual throughput and a broader market scope would be anticipated in larger ports.

6. Not all ports have similar goals and objectives. Talley (1996), for example, suggested that some ports, rather than having a profit motive, may seek to maximize throughput subject to a zero profit constraint (in the case of a non-profit public port), or subject to a maximum allowable operating deficit (as in the case of a subsidized port). The research presented in Table 17.3 indicates that few ports

maximize traffic subject to zero profit or an allowable operating deficit but confirms Talley's suggestion that factors other than profit motives are at play.

7. Members of the network who assisted Mary R. Brooks (Dalhousie University) with data collection were Khalid Bichou (Imperial College London), Dionisia Cazzaniga Francesetti (Universita Di Pisa), Guldem Cerit (Dokuz Eylul University), Peter W. de Langen (Erasmus University), Sophia Everett (Melbourne University), James A. Fawcett (University of Southern California), Ricardo Sánchez (Austral University/ECLAC), Marisa Valleri (University of Bari), and Thierry Vanelslander (Universiteit Antwerpen).

8. Katz and Kahn (1978) suggest that best performers are concerned about both efficiency and effectiveness. The authors of this paper concur, but add that success requires that either effectiveness or efficiency considerations should have primacy over the other, but without the other failing to meet a minimum standard of performance.

9. The Canadian case was selected on the basis of convenience and analysed for illustrative purposes only. The analysis is based entirely on the facts presented in Chapter 11 of this book. The reader is encouraged to read the chapter prior to the present analysis. The reader should also note that Sánchez and Wilmsmeier (Chapter 9) use the Matching Framework in their analysis of the case of the River Plate Basin in Latin America.

10. Canada's trade dependence on the US market grew from 63% of Canada's exports in 1980 to 87% 20 years later.

11. In his study of Canadian port reform, Ircha (1997, p. 134) concluded, "the port managers' perspectives may reflect imprinted institutional values developed from decades of strong federal presence ... in Canada's ports." Dick and Robinson (1992) too talked about the culture within port authorities as discouraging innovation and initiative on the part of port managers.

12. The governance structure imposed requires that four of seven Board members come from "user groups" but the nominations are ranked and local party interference in the process can mean that the Minister's favourites can appear at the top of the list. Politics may not intervene as well. Of course it could be argued that Board Member selection could be easily derailed on the independent side of old "business establishment" buddies appointing their friends.

REFERENCES

Baltazar, R., & Brooks, M. R. (2001). *The governance of port devolution: A tale of two countries.* Seoul, Korea: World Conference on Transport Research, July.

Barney, J. (1995). Looking inside for competitive advantage. *Academy of Management Executive, 9*(4), 49–61.

Boschken, H. L. (1988). *Strategic design and organizational change: Pacific Rim seaports in transition.* Tuscaloosa: University of Alabama Press.

Bourgeois, L. J., III, Duhaime, I. M., & Stimpert, J. L. (1999). *Strategic management: A managerial perspective.* Fort Worth: Dryden Press.

Burns, T., & Stalker, G. M. (1961). *The management of innovation.* London: Tavistock.

Charih, M., & Rouillard, L. (1997). The new public management. In: M. Charih & A. Daniels (Eds), *New public management and public administration in Canada*. Toronto: Institute of Public Administration of Canada.

Collis, D. J., & Montgomery, C. A. (1997). *Corporate strategy: Resources and the scope of the firm*. Chicago: Irwin.

Daft, R. L. (2004). *Organization theory and design* (8th ed.). Mason, OH: Thomson South-Western.

Dick, H., & Robinson, R. (1992). Waterfront reform: The next phase. Presented to the National Agriculture and Resources Outlook Conference.

Dijksterhuis, M. S., Van den Bosch, F. A. J., & Volberda, H. W. (1999). Where do new organizational forms come from? Management logics as a source of coevolution. *Organization Science, 10*(5), 569–582.

Dobell, R., & Bernier, L. (1997). Citizen-centred governance: Implications for inter-governmental Canada. *Alternative service delivery: Sharing governance in Canada*. KPMG Centre for Government Foundation and Institute of Public Administration of Canada.

Donaldson, L. (1996). The normal science of structural contingency theory. In: S. R. Clegg, C. Hardy & W. R. Nord (Eds), *Handbook of organization studies*. London: Sage.

Duncan, R. B. (1972). Characteristics of organizational environment and perceived environmental uncertainty. *Administrative Science Quarterly, 17*, 313–327.

Hitt, M., Ireland, R., Hoskisson, R., Rowe, W. G., & Sheppard, J. (2006). *Strategic management: Competitiveness and globalization* (2nd Canadian ed.). Toronto: Nelson.

Hunter, J. E., & Schmidt, F. L. (1990). *Methods of meta-analysis: Correcting error and bias in research findings*. Newbury Park, CA: Sage.

Ircha, M. C. (1997). Reforming Canadian ports. *Maritime Policy and Management, 24*(2), 123–144.

Jones, G. R. (1995). *Organization theory: Text and case*. Reading, MA: Addison-Wesley.

Katz, D., & Kahn, R. L. (1978). *The social psychology of organizations*. New York: Wiley.

Ketchen, D. J., Combs, J. G., Russell, C. J., Shook, C., Dean, M. A., Runge, J., Lohrke, F. T., Naumann, S. E., Haptonstahl, D. E., Baker, R., Beckstein, B. A., Handler, C., Honig, H., & Lamoureux, S. (1997). Organizational configurations and performance: A meta-analysis. *Academy of Management Journal, 40*(1), 223–240.

Lawrence, P. R., & Lorsch, J. W. (1967). *Organization and the environment*. Boston: Graduate School of Business Administration, Harvard University.

Mealiea, L., & Baltazar, R. (2004). A strategic guide for building effective teams. *Public Personnel Management, 34*(2), 141–159.

Miles, R., & Snow, C. (1978). *Organizational strategy, structure and process*. New York: Mc Graw-Hill.

Miller, D. (1986). Configurations of strategy and structure: Towards a synthesis. *Strategic Management Journal, 7*, 233–249.

Mintzberg, H. (1987). Strategy concept I: Five Ps of strategy. *California Management Review, 30*(1), 11–24.

Osborne, D., & Gaebler, T. (1992). Introduction: An American perestroika. *Reinventing government: How the entrepreneurial spirit is transforming the public sector*. Reading, MA: Addison-Wesley.

Pierce, J., & Robinson, R. (2000). *Strategic management: Formulation, implementation, and control* (7th ed.). Boston: Irwin McGraw-Hill.

Porter, M. (1980). *Competitive strategy*. New York: Free Press.

Porter, M. (1985). *Competitive advantage: Creating and sustaining superior performance.* New York: Free Press.

Quinn, J. B. (1996). Strategies for change. In: H. Mintzberg, J.B. Quinn (Eds), *The strategy process* (3rd ed.), Englewood Cliffs, NJ: Prentice-Hall.

Robinson, R. (2005). Liner shipping strategy, network structuring and competitive advantage: A chain systems perspective. In: K. Cullinane (Ed.), *Research in transport economics* (Vol. 12, pp. 247–289). Amsterdam: Elsevier.

Scott, W. R. (1995). *Institutions and organizations.* Thousand Oaks, CA: Sage.

Talley, W. K. (1996). Performance evaluation of mixed cargo ports. Presented to International Association of Maritime Economists Annual Conference, Vancouver.

Thompson, A., Strickland, A. J., III, & Gamble, J. (2005). *Crafting and executing strategy: The quest for competitive advantage.* Boston: Irwin McGraw-Hill.

CHAPTER 18

GOVERNANCE MODELS DEFINED

Mary R. Brooks and Kevin Cullinane

ABSTRACT

The recent worldwide trend towards devolution in the port industry has spawned considerable variety in the types of governance structures now in place around the world. This chapter discusses the range of devolution models in the global ports sector, as identified by the World Bank and academic researchers, and then attempts to validate the existence of these models. It identifies key strategic objectives for port governance and proxies for the range of approaches to the delivery of port activities, identifying nine for use in further research. The chapter concludes that the data do not validate the World Bank or other models proposed as in use, and that these models are oversimplified. The chapter lays the foundation for further research into port governance.

1. INTRODUCTION

As noted in Chapter 1, the definition of governance is a topic fraught with debate, perhaps because it can be approached from two perspectives. At the firm level, corporate governance is a system of rules and structures applicable to the firm's managerial decisions. At the government level, it is the rules and structures imposed on firms, influencing the scope and the manner of those managerial decisions.

Devolution, Port Governance and Port Performance
Research in Transportation Economics, Volume 17, 405–435
Copyright © 2007 by Elsevier Ltd.
ISSN: 0739-8859/doi:10.1016/S0739-8859(06)17018-3

The appropriate role for port authorities, and the means by which government might establish the governance environment in which ports operate, were discussed heatedly over the 1990s and continue today (e.g., Goss, 1990; Baird, 1995, 1999; Everett, 2003). As a result, governments have developed state and local models in some cases while, in others, strong national approaches have been implemented. Chapters 3–16 provide a rich history of what has happened and, in most cases, have explained why. For several, the impact of political interference on port performance has resulted in a sub-optimal outcome; Greece, for example, continues to have this problem as noted by Pallis in Chapter 7.

While these chapters provide the history we hope is not to be repeated, they also raise questions about the choices made by governments. As one illustration (Chapter 12), Everett and Robinson have provided a history of port reform in Australia and the intention of the state governments to find port governance models that would deliver improved efficiency, many but not all opting for corporatization. This arose because, under the Government Business Enterprise model, Australian ports did not deliver aggressive port management. Dick and Robinson (1992) concluded that they failed to achieve the desired outcome because the government had not altered the fundamental structure of port authorities, and the existing culture discouraged innovation and initiative. The purpose of this chapter is to examine governance models in ports today, based on the research data collected by the Port Performance Research Network[1] on 42 ports in nine countries (Argentina, Australia, Belgium, Canada, Chile, Italy, Turkey, United Kingdom and United States) and to establish a foundation on which future port governance research can be based. However, it starts with a review of the existing literature on port governance models, beginning with the World Bank typology.

2. ADMINISTRATIVE MODELS UNDER THE WORLD BANK PORT REFORM TOOLKIT

The World Bank provides a Port Reform Toolkit for governments to use in examining the options for devolution, and Module 3 provides a typology of models in use and general principles to be observed in planning devolution to those models. Which of these models provides the best fit will depend, according to the theory of the Matching Framework (see Chapter 17), upon the environment (including cultural and political components) and the strategies and structures (governance models) employed in implementation. According to the Matching Framework, and confirmed by Sánchez and

Wilmsmeier (Chapter 9), the willingness of the private sector to invest in port activities in less than stable environments may have as much to do with the perception of the capital markets in the country, the political and economic stability of the country and the general investment climate as it has to do with the governance model per se.

As much of the discussion of port governance starts with Module 3 of the World Bank Port Reform Toolkit (WBPRTK), so too will this chapter. However, the reader should not assume that the WBPRTK models are governance models. They reflect the broad allocation of responsibilities for port activities but they do not indicate the complete set of indicators found in a full governance model, one that indicates who assumes the risk and what the lines of accountability are.

Module 3 of the WBPRTK (World Bank, undated) outlines four port administration models – the Service Port, the Tool Port, the Landlord Port and the Private Service Port – and assesses the strengths and weaknesses of each model. These models differ in several ways: by whether port services are provided by the public sector (the government), private sector or mixed ownership providers; their orientation (local, regional or global); who owns the superstructure and capital equipment; and who provides dock labour and management. The choice of model by government is influenced by the way the ports are organized, structured and managed. These factors include the socio-economic structure of a country, the historical development of the port, the location of the port (urban area or isolated region) and the types of cargo typically handled (liquid or dry bulk, containers). Decisions about the balance of public sector and private sector participation are also a product of the strength of the capital markets in the country and the philosophy of the country with respect to alternative service delivery. Each of these "models" will be presented in this section before we discuss why the approach requires validation.

2.1. Service Port Model

Used in many developing countries prior to the devolution pressures of the last two decades, this is a predominately public administrative model in which the Port Authority owns the land and all available assets (fixed and mobile) and performs all regulatory and port functions including vessel handling and stevedoring operations. Usually controlled by the Ministry of Transportation (and/or Communications), the Port Authority directly employs the labour that performs cargo-handling operations in the port, although in some cases, cargo-handling services are performed by separate public entities; this division of operations between separate public entities can present unique management challenges. The Chairman of the Port Authority is usually a civil servant responsible for port administration, who reports directly to the appropriate minister. Under this model, the same organization has the responsibility for performing regulatory functions, developing infrastructure and superstructure, and executing operational activities. Generally, there is an absence of private sector involvement in port activities.

Because facilities development and operations are the responsibility of only one entity, there is the potential for a streamlined and cohesive approach to growth. However, there is also potential for inefficient port administration as a result of limited internal competition; there may also be little incentive for innovative management, and this may lead to services that are neither user-friendly nor market-oriented. Furthermore, the dependence of the authority on the government for funding may lead to a wasteful use of resources or under-investment. The incentives for efficient operations may not be fully functional and non-economic motivations may dominate.

2.2. Tool Port Model

In this model, as with the next, there is a division of operating responsibilities between the public and private sectors. The public sector Port Authority owns, develops and maintains the port infrastructure and superstructure, including cargo-handling equipment such as quay cranes, forklift trucks and so on. The operation of Port Authority equipment is usually performed by Port Authority labour, but other operations are performed by private cargo-handling firms, on board vessels as well as on the quay and apron. The private operators are usually small companies, and the fragmentation in responsibility for cargo handling can lead to conflict between small operators and between the stevedoring companies and port administrators. Strong stevedoring companies are seldom found in these ports.

The approach avoids duplication in facilities provided because investment in infrastructure and equipment is provided by the public sector. However, this approach also runs the risk of under-investment.

2.3. Landlord Port Model

The landlord model is stated by the World Bank to be the most common model of allocating public versus private sector responsibilities in the provision of port services. In this model, the Port Authority retains ownership of the port land while the infrastructure is leased to private operators. The private companies provide and maintain their own superstructure, purchase and install their own equipment and employ the stevedoring labour. The responsibilities of the Port Authority as landlord include economic exploitation, the long-term development of the land and the maintenance of basic port infrastructure such as access roads, berths and wharves.

In this approach, the entity that executes operations also owns the cargo-handling equipment; therefore, operational planning is likely to result in appropriate superstructure investment, efficiency and be more responsive to changing market conditions. This model also has risks, the greatest of which is the risk of excess capacity in infrastructure as more than one private operator may press for expansion. In addition, a duplication of marketing effort can occur as both terminal operators and the port authority visit potential customers. Therefore, greater co-ordination of marketing and planning is required with this model.

2.4. Private Service Port

In this allocation of responsibilities, the government no longer has an interest in port activities. All regulatory, capital and operating activities are provided by the private sector. Port land is also privately owned by the private sector. This is the model used in many ports in the United Kingdom.

This model often results in investment in port operations that are flexible and market-oriented. However, this approach may also result in monopolistic behaviour and the potential for abuse of the natural monopoly position that some ports may enjoy increases dramatically; the ability of the public sector to influence economic development is diminished as the public sector role is minimal. In cases where companies are not community-minded, public involvement in developing long-term economic policy and strategies is also lost.

2.5. Discussion of the World Bank Typology

The allocation of responsibilities implicit in the characteristics of the four WBPRTK models is summarized in Table 18.1.

These four models present a simple approach to classifying port responsibilities, but they fail to provide adequate guidance to a government faced

Table 18.1.   Allocation of Responsibilities (World Bank).

Responsibilities	Service	Tool	Landlord	Private
Infrastructure	Public	Public	Public	Private
Superstructure	Public	Public	Private	Private
Port labour	Public	Private	Private	Private
Other functions	Majority public	Mixed	Mixed	Majority private

Note: Mixed = public/private.
Source: Adapted from World Bank Port Reform Toolkit, Module 3, p. 21. Available at http://www.worldbank.org/transport/ports/toolkit/mod3.pdf/

with pressure to devolve port administration as to which approach(es) to take for a given local situation. For that, there needs to be greater understanding of the antecedents of improved port performance (the topic of the last section of this book). These classifications are highly regarded as a first step in understanding the allocation of responsibility for capital investment at a port (infrastructure and superstructure), and for the management and operations at a port (from a labour perspective), but fail to fully provide an understanding of the strategic intent of a port, its role in the economy as seen by government and the allocation of responsibility for regulatory monitoring (such as environmental and safety monitoring). Furthermore, if we return to Table 1.2 in Chapter 1 (repeated here as Table 18.2), can we add some validation to the rough allocation of public, private and mixed in terms of both a greater detail in the type of responsibilities, and in the levels of government responsible (national, regional and local) as well as a better understanding of the meaning of "mixed" responsibility?

Before we explain our thinking on extending the model used by the World Bank, the next section will examine other approaches to port governance and port devolution.

3. OTHER APPROACHES

In their assessment of EU port policy, Chlomoudis and Pallis (2002) noted the variety and diversity of arrangements for port management and operations. They built on the earlier work by Pallis (1997). He noted the diversity of management practices from the local municipal management typical of northwest Europe (with examples like Hamburg, Rotterdam and Antwerp) to the Latin approach where an influential central government is a feature (common in Mediterranean ports; we might think of Italian ports as

Table 18.2. Characteristics of Types of Devolution.

Characteristics	Decentralization	Commercialization	Corporatization	Privatization
Changed government objectives	From executing national policy to local responsiveness	To improve efficiency and responsiveness	To improve efficiency and responsiveness	Private sector-led efficiency; market competition
Changed organizational structure (new lines of accountability)	Yes	Yes	Yes	Yes
Establishment of a legal entity	No	Yes; there is no share capital	Yes; share capital may be owned in part or in full by government	Yes (if not sold to an existing private entity)
Control of operations and management	Local government	Transferred from government	Transferred from government	Transferred from government
Ownership of existing capital assets[a]	May be transferred to another level of government	Not transferred	Usually not transferred	Transferred
Ownership of new capital assets[a]	Dependent on negotiated arrangement	Dependent on negotiated arrangement	Resides with new entity	Privately owned by new entity
Responsibility for risk[a]	Remains with public sector	Transferred to new entity; there may still be recourse to government, depending on contract terms	Transferred to private sector	Transferred to private sector
Right to borrow money[a]	May require national approval	Yes, there may be caps imposed	Yes	Yes
Ability to sell the assets[a]	Usually not without national government approval	Usually limited, depending on contract terms	Usually limited; depending on contract terms	Yes

[a]Critical control elements.

examples) to the trust ports found in the United Kingdom (Pallis, 1997, p. 375). He also reported that, in addition to the operational characteristics inherent in the WBPRTK type of classification, ports often had considerable variety in the way they organized port labour (from piecework approaches to labour pools or permanent employees). In many cases, the authority and responsibility for the allocation of an activity reflects the

Table 18.3.   Allocation of Responsibilities as Described by Baird's Port
Function Privatization Matrix.

Port Models	Port Functions		
	Regulator	Landlord	Utility
PUBLIC	Public	Public	Public
PUBLIC/private	Public	Public	Private
PRIVATE/public	Public	Private	Private
PRIVATE	Private	Private	Private

Source: Baird (2000, Table 1, p. 180).

flexibility inherent in non-prescriptive imposed governance approaches. For
example, port operators may opt to use piecework or labour pools as best
suits their managerial structure and corporate philosophy. In other cases,
the approaches used are heavily prescribed in national, regional or local law
or regulation, and the provision of activities follows strict guidelines. What
is missing is the determination of whether highly prescriptive or loosely
guided approaches are more effective in generating strong performance
outcomes.

Through the 1990s, Baird undertook a significant amount of work (Baird,
1995, 1999, 2000) in which he specified the governance models used in port
privatization by developing a port function privatization matrix. Recognizing that there are varying degrees of emphasis in the public–private provision of port functions (as can be seen below in Table 18.3), his four models
allocated specific port activities to the three port function roles (regulator,
landlord and utility). However, when Baltazar and Brooks (2001) attempted
to use Baird's allocation of activities to assess governance in two countries,
Canada and the Philippines, they found that these activities did not always
fit the columns to which they had been assigned. To clarify the distinctions
between differing types of devolution for the port industry, Baltazar and
Brooks (2001) classified port-related activities into regulator, landlord and
operator activities based on Baird's work (see Table 18.4). This problem
with allocation has been further reinforced by Brooks' work on Canadian
governance models, as evident from the Appendices to Chapter 11 of this
book. As Brooks notes, in Canada there are three governance models and,
for the most part, the allocation problem lies in the balance between private
and public sector provision of services. Most important, however, is the idea
that the functions may not always be provided as neatly as Baird or his
predecessors (de Monie, 1994, 1995; Goss, 1990) suggest.

Table 18.4. Baltazar and Brooks' Port Devolution Matrix.

Governance	Regulator Functions	Port Functions	
		Landlord	Operator
Public	• Licensing, permitting • Vessel traffic safety • Customs and immigration	• Waterside maintenance (e.g., dredging) • Marketing of location, development strategies, planning • Maintenance of port access • Port security • Land acquisition, disposal	• Cargo and passenger handling • Pilotage and towage • Line handling • Facilities security, maintenance, and repair • Marketing of operations • Waste disposal Landside and berth capital investment
Mixed public/ private	• Port monitoring • Emergency services • Protection of public interest on behalf of the community		
Private	• Determining port policy and environmental policies applicable		

Note: Depending on the particular regulatory regime of each country, not all functions assigned to the Regulator column may be relevant. All functions assigned to the Operator column may take place, but be provided by the Landlord or the Regulator as determined by local practice.
Source: Baltazar and Brooks (2001) reclassified the functions, building on earlier work by Baird (2000) and Goss (1990).

Perhaps the contribution made by Baltazar and Brooks (2001) is that they clearly separated regulatory functions from port functions (both landlord and operating), and sought to distinguish landlord activities (like land acquisition and disposal) as specifically different from the day-to-day operating activities (like cargo and passenger handling and facility security), which might be organized port-wide or on a terminal-specific basis. The Baltazar and Brooks' approach also anticipated that various models may incorporate differing levels of responsibility and that responsibilities may be shared among providers. They assumed that the allocation between public and private and the degree of "public" in mixed arrangements would vary widely across countries, as did Baird,[2] but they did not draw lines horizontally across the functional columns. It would be the pattern of allocation that would identify a particular model as distinct. This conceptual approach has not been validated.

Are there only four models of port governance? Why not five or three? Most ports in the world appear to belong to one of three groups: fully public, fully private or mixed. We would argue that most fall into the last category. While the private model privatizes everything theoretically, the reality is that this is only true for some UK ports. Full privatization of public ports seems unpalatable to most governments, no matter what the tenets of new public management. On the other hand, the number of fully public ports has steadily diminished as governments globally have sought to garner greater efficiency or effectiveness or local responsiveness from ports through new governance arrangements.

A simple look at Baird (2000) suggests more than four models. Under the philosophy of alternative service delivery and the principles of new public management, there is the possibility that the government may act solely as regulator, seeking to secure that which it deems to be in the public interest, such as public safety and security and the prevention of maritime pollution. The WBPRTK approach has no provision for this, yet many governments would argue that monitoring and enforcement of regulatory and treaty obligations are their primary obligations to the citizens of the country. The WBPRTK approach does not recognize that differing governance models may reflect differing strategic purposes, be they defined by government or the port entity. As already indicated in Table 18.2, local responsiveness, non-economic objectives like employment growth and so on may be viewed as the objectives of a government's devolution program. While the WBPRTK models include the option of fully private, economic performance-driven approaches, the other three governance models and their purposes (and accompanying allocation of responsibilities and risks) seem less well defined.

Bichou and Gray (2005) note the problem faced when researchers attempt to identify governance models applicable to seaports. They report on the five core functions identified by UNCTAD (1992) in its improving port performance (IPP) model – cargo planning, ship operations, quay operations, storage operations and receipt/delivery operations. While these operations do indeed take place in ports, and their optimization will lead to production efficiency, there is more to port activities than just production of services. The research here looks beyond functional production to a broader view that incorporates all port activities reflecting the current supply chain developments and the fact that many ports have moved to provide more services than just the traditional port activities.

According to the Matching Framework (see Chapter 17 of this book), it is the fit of the environment–strategy–structure configuration that leads to positive economic performance. It is quite possible that organizational designs that are centralized, bureaucratic and inflexible are the ones best suited to a particular strategy–environment combination. This means that, depending on the existing environment and compatible strategic intent, government can develop governance models that have the potential to maximize the desired performance outcomes. To validate the models possible, we must determine if there are allocation combinations that match existing theory. Before we do this, we (1) expand the range of governance arrangements to capture greater detail, and (2) revisit the issue of strategic intent.

The first step is reconfiguration of the mix of public arrangements (from one to two reflecting local or decentralization versus national or centralization approaches) and of public–private arrangements from two (described as *public*–private and *private*–public) to two that more accurately reflect the governance models imposed (according to the government's strategic purpose as proposed in Table 18.3), thereby yielding five governance combinations. These five ownership and management combinations have been explored to expand the basic WBPRTK combinations to more accurately reflect the trend to decentralization of control with retention of ownership being reflected in many countries' devolution efforts. These five are

- Central government owned with central government management and control.
- Government owned but management and control are decentralized to a local government body.
- Government owned (federal, regional or municipal) but managed and controlled by a corporatized entity.
- Government owned but managed by a private sector entity via a concession or lease arrangement, or owned and managed via a public-private partnership agreement.
- Fully privately owned, managed and controlled.

The activities are not just core management activities but include operations, planning, financing and so on. The intention in this chapter is to find groupings of regulatory, managerial and operating activity, including capital investment and cost recovery requirements, to identify more thoroughly a typology of port governance models that represents the full spectrum of activities, not just those posed by either Baird (2000) or the World Bank in its Port Reform Toolkit. On the one hand, this research may validate the work of one of these two or, on the other, provide a richer tapestry of possibilities leading to a better understanding of devolution models and their ultimate effectiveness within the context of the other configuration variables.

4. ALLOCATION OF RESPONSIBILITIES

4.1. Description of Survey and Respondents

The determination of how, and what combination of, port activities are allocated between the public and private sectors is based on a questionnaire survey of 202 ports in 21 countries that was conducted in 2005. Valid responses were received from 42 ports in 10 countries, amounting to a response rate of 20.8 per cent. The individuals who responded on behalf of the port organizations surveyed were primarily from the senior management of the organization, with 32 of the 42 respondents (76 per cent) being at CEO level or just below. Partially as a consequence of this, a relatively high proportion (67 per cent) of these individuals had been in the position for five years or more.

Table 18.5 shows the number of responses received from port organizations in each of the countries included in the sampling frame.

4.2. Strategic Intent

Models imposed by government need to clarify the strategic intent of government as well as match the imposed model to that anticipated outcome.

Table 18.5. Survey Returns by Country.

Country or Region	Approached	Responses
Argentina and Uruguay	7	1
Chile	3	1
Caribbean Basin	4	0
Canada	11	4
United States	55	5
Other Americas	3	0
Australia	23	3
China and Hong Kong	19	0
Belgium	3	2
Greece	5	0
Italy	22	4
Korea	10	0
Netherlands	3	2
Turkey	8	8
UK	26	12
Total	202	42

Goss (1990) notes that these "strategies" may be manifest in the way port authorities structure tender requirements: if the intent is to encourage competition (to maximize return or profit), they may seek terminal bidders able to engage in a variety of rent and pricing schemes to deliver that competitive climate. The survey asked respondent port managers to allocate the strategic objectives of their port to any number of six categories (already reported in Table 17.3). In Chapter 17, Baltazar and Brooks argue that the first two of these categories represent forms of economic performance objectives, while the rest are non-economic. Applying this broad dichotomous classification, the results suggest that the sample comprises three groups of respondents: 19 that have solely non-economic objectives, 11 that have strictly economic objectives and 11 that have a mixture of both economic and non-economic objectives.

No relationship is found between these groupings and whether the cargo-handling activity within a port is provided either by a government owned and managed entity or, irrespective of ownership, by a privately managed entity ($\chi^2 = 2.476$, DF $= 2$). Although the data are not suited to a formal test, there does, however, appear to be some relationship to the country in which a port is located. The sample seems to divide into two. Ports in Australia, Turkey and the UK exhibit the complete range of possible strategic objectives, with some ports pursuing solely economic objectives, others pursuing solely non-economic objectives and yet others pursuing a mix of economic and non-economic objectives. In the case of Australia and the UK, this is likely to reflect the different possible governance models following the implementation of a policy of port corporatization at the state level (Australia, Chapter 12) or port privatization (UK, Chapter 3), while in Turkey this is probably due to the transitional phase that the sector finds itself in as the result of an ongoing privatization process that is not yet complete (see Chapter 8).

By subtly altering the classification of strategic objectives so that they are identified as (a) economically oriented, (b) mainly concerned with the maximization of throughput and (c) those where the wider macroeconomic benefits are the main concern, it is then possible to test again for the degree of association with private sector participation in the main economic activity of the respondent ports – the provision of cargo-handling services. Using a dichotomous classification whereby the cargo-handling activity within a port is either provided by a government owned and managed entity or by a privately managed entity (irrespective of ownership), a χ^2-test of association indicates a relationship between these two variables that is significant at the 5 per cent level ($\chi^2 = 6.414$, DF $= 2$). Indeed, by decomposing the test statistic, the responses suggest that more than the number of government owned and operated enterprises than might be statistically expected actually have financial objectives as their prime strategic purpose, and significantly less than might be expected have either maximum throughput or wider economic benefits as their main strategic purpose.

4.3. Services to Vessels or Terminals

Fig. 18.1 shows the average percentage of the activities within a port that can be characterized as "services to vessels or terminals" supplied by different governance arrangements.

It can clearly be seen that there is great diversity in the nature of service provision across all of these activities. Many of the results are exactly what would be anticipated in a new public management environment, with activities such as container handling, chandlery and vessel re-supply, container terminal operation, on-dock storage, towage and terminal equipment maintenance all being services that are easily devolved through contracting out to service providers who are privately owned, managed and controlled. With the worldwide tendency towards the privatization of port services, this is likely to become increasingly the case in future. In counterpoint, for non-commercial activities or what might be perceived as public goods, supply tends to be retained by the public sector. Examples of these activities include vessel traffic management, and the provision of anchorages, emergency services, fire protection and so on.

In seeking to validate established models of port governance, the sheer variation in the provision of these services is itself sufficient evidence to imply that any port governance model that is based on a discrete and absolute apportionment of activities categorized under function headings and with minimal assumptions as to environmental configuration will be too simplistic to capture the obvious complexity that exists. As such, these aggregate results provide support for the finding of Baltazar and Brooks (2001) (and of Baltazar and Brooks' assertions in Chapter 17) that even the allocation of activities to the standard functional headings of existing port governance models is not necessarily a simple and straightforward task.

At the aggregate level, there is clearly great variation in the provision of services to vessels and terminals. When attempting to gain a deeper insight into the development of appropriate port governance models, however, it is important to focus on the degree of correlation that exists between the governance arrangements that have been put in place for each of these services by all the ports in the sample. While it is extremely interesting to find that 68 per cent of the container handling services in the sample ports

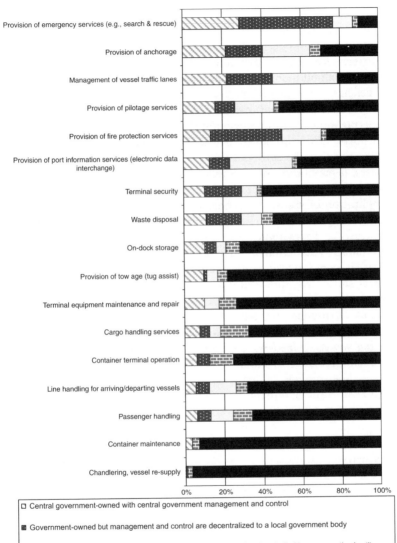

Fig. 18.1. Provision of Services to Vessels or Terminals.

are provided by organizations that are fully privately owned, managed and controlled and that the same figure for container terminal operation is 76 per cent, in order to develop a more realistic and generally applicable set of port governance models, it is important to determine to what extent the same ports exhibit similar patterns of operation. For example, if a port's container handling services are provided by the private sector, is it the case that its container terminal operation will also be a private sector concern? Considering simply the number and diversity of activities that take place within ports, this is obviously a difficult and complex task. However, by grouping together the activities that exhibit a high degree of correlation in responses, it may be possible to render the complex simple.

For the data in the "services to vessels or terminals" category of activities, the application of statistical techniques led to the groupings of highly correlated activities shown in Table 18.6. While the number of groups is relatively few at four, there does not appear to be any really obvious underpinning explanation for why particular activities should be grouped together. In consequence, while the grouping of all the activities into four groups is statistically valid, the delineation of activities within each group has not emerged as entirely satisfactory. For this reason, no attempt has been made to provide each group with a summary label. It just would not be logical to do so. What is clear, however, is that "line handling for arriving/ departing vessels", "provision of fire protection services", "terminal security" and "provision of pilotage services" provide proxies for how the

Table 18.6. Activity Groupings for "Services to Vessels or Terminals".

Group 1	Group 2
Cargo-handling services	Management of vessel traffic
Chandlery, vessel re-supply	Container maintenance
On-dock storage	*Provision of fire protection services*
Terminal equipment/repair	
Container terminal operations	
Line handling for arriving/departing vessels	
Passenger handling	
Anchorage	
Port information (EDI)	
Group 3	Group 4
Towage	Emergency services
Terminal security	*Provision of pilotage services*
Waste disposal	

Note: Those activities in italic are selected as representative of the group (proxies) as they exhibited the least statistical deviation from the mean response of the group.

activities will group in the model, enabling future researchers to substitute these four activities for the full set of 17 activities provided in the survey and summarized in Table 18.6.

4.4. Other Activities

Fig. 18.2 shows the activities within a port that have been categorized as "other activities" for the purpose of this analysis. The average percentage supplied by different forms of provider is shown for each of the individual activities.

For this range of activities, it is clear that "contracting with stevedoring labour" is very much an activity that is controlled by the private sector. Quite interesting is the high level of private ownership in port security provision, something not anticipated! At the other extreme and exactly as one might expect, central government has the major involvement in "customs and immigration services". Between these two, the other three activities exhibit a relatively shared responsibility across the range of potential providers. However, this does not provide any real insight as to how correlated is the provision within individual ports. For example, if the "general marketing of the port" is undertaken by central government, does it necessarily follow that there is a high likelihood that all four of the other activities are also provided by central government? A statistical analysis of the responses reveals two significantly correlated groupings with respect to this category of activities (see Table 18.7).

Assuming the sample is representative of the entire port sector, what this suggests is that, within any individual port, if the private sector (broadly) is responsible for "contracting with stevedoring labour", then it will also tend to be responsible for "general marketing of the port", "port security" and the "provision of port information services". Since these activities comprise one of the two derived significantly correlated groups, it is also equally the case that the public sector (broadly) may also be responsible for all four of the activities. In addition, since the activity identified in "Group 2" (i.e., "customs and immigration services") is not significantly correlated with the other four activities, then it can be deduced that this may be provided by either the public or the private sector, irrespective of which sector provides the other four grouped activities within a port. Thus, even for just these five "other activities" and utilizing merely a dichotomous public–private delineation of responsibilities, there are potentially four permutations of governance model across the two groups (i.e., both public, both private, "Group 1" public with "Group 2" private, and "Group 1" private with "Group 2" public).

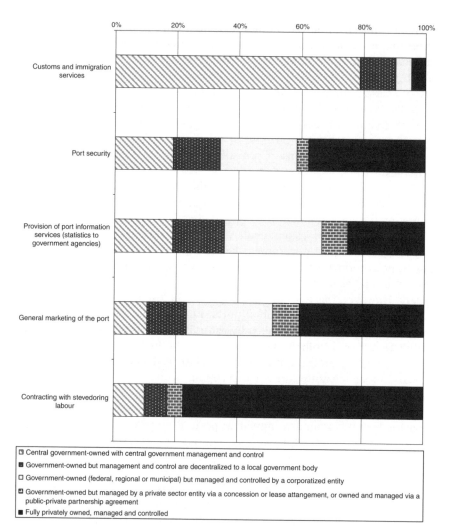

Fig. 18.2. Provision of "Other Activities".

4.5. Policies, Regulation and Planning for the Port

Fig. 18.3 shows the activities related to the category of "policies, regulation and planning" for a port. The average percentage by which different generic forms of entity are responsible for monitoring or managing each of the

Table 18.7. Identified Groupings for "Other Activities".

Group 1	Group 2
Contracting with stevedoring labour General marketing of the port	*Customs and immigration services*
Port security	
Provision of port information services (statistics to government agencies)	

Note: Those activities in italic are selected as representative of the group (proxies) as they exhibited the least statistical deviation from the mean response of the group.

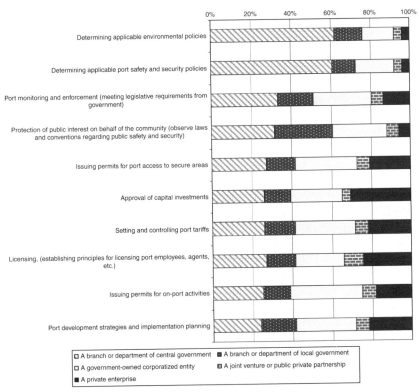

Fig. 18.3. Provision of "Policies, Regulation and Planning for the Port".

different activities is also shown. Again, from a simple visual inspection of the figure, it would seem that the determination of "environmental policies" and "port safety and security policies" are activities that are distinct from the others. The major responsibility for these two activities lies solidly with a

Table 18.8.   Identified Groupings for Policies, Regulation and Planning
for the Port.

Group 1	Group 2
Approval of capital investments	Determining applicable
Licensing, (establishing principles for licensing port	environmental policies
employees, agents, etc.)	*Determining applicable port*
Issuing permits for on-port activities	*safety and security policies*
Issuing permits for port access to secure areas	Protection of public interest on
Port development strategies and implementation planning	behalf of the community
Port monitoring and enforcement (meeting legislative	(observe laws and
requirements from government)	conventions regarding public
Setting and controlling port tariffs	safety and security)

Note: Those activities in italic are selected as representative of the group (proxies) as they exhibited the least statistical deviation from the mean response of the group.

"branch or department of central government". For the other activities under this category, however, there appears to be a reasonably consistent distribution of responsibility across the full range of possible entities, with a "branch or department of central government", a "government owned corporatized entity" and a "private enterprise" having fairly substantial and even roles and with a "branch or department of local government" and a "joint venture or public–private enterprise" having smaller parts to play.

Applying statistical techniques to the data to pull together activities where the responsibilities are significantly correlated, the groupings shown in Table 18.8 are derived. The result is interesting in that what is not actually evident from Fig. 18.3 is that the allocation of responsibility for the "protection of the public interest on behalf of the community" is actually found to be significantly correlated with the determination of applicable "environmental policies" and "port safety and security policies". Thus, although the distribution of responsibility for protecting the public interest is similar to the bulk of other activities within this category, it is not generally the case that within a given port, the entity responsible for these other activities will also be responsible for the protection of public interest.

4.6. Investment Activities

The cross-sample distribution of specific sources of finance for six individual port "investment activities" are shown in Fig. 18.4. A fairly even distribution for the provision of investment funds is exhibited, with central

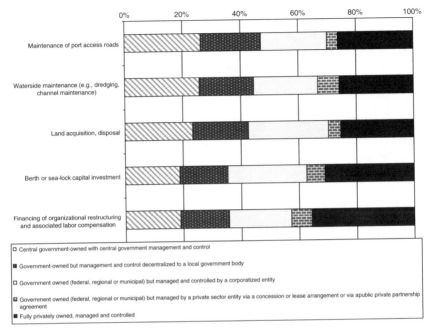

Fig. 18.4. Provision of "Investment Activities".

government providing about 20 per cent and mainstream private enterprise about 30 per cent of the sector's investment funding.

In this instance, the application of statistical techniques reveals that all activities under this heading actually fall into a single grouping of significantly correlated responsibility for the provision of finance (Table 18.9). Therefore, within any given single port, the finance for all six of these individual "investment" activities is likely to come from either the same source or exhibit the same mix of responsibility in the provision of finance across all activities.

As an adjunct to this part of the analysis, the survey also collected data on cost recovery for each of the six "investment activities". While the single-derived grouping for this range of activities suggests a common approach to financing for each activity, there was a little more diversity with respect to the issue of cost recovery. The results (Fig. 18.5) are generally consistent across five of the six activities with no cost recovery in approximately 50 per cent of cases, full cost recovery in 20–25 per cent of cases and partial cost recovery in 25–30 per cent of cases. Only for the "leasing or concessioning of terminal operations" is there any significant deviation from this norm, with

Table 18.9. Identified Groupings for "Investment Activities".

Group 1
Berth or sea-lock capital investment
Financing of organizational restructuring and associated labour compensation
Land acquisition, disposal
Leasing, concessioning of terminal operations
Maintenance of port access roads
Waterside maintenance (e.g., dredging, channel maintenance)

Note: The activity in italic is selected as representative of the group (proxy) as it exhibited the least statistical deviation from the mean response of the group.

a significantly greater number of ports in the sample reporting both full and partial cost recovery with respect to this activity. A χ^2-test of the level of cost recovery against a dichotomous classification of the source of finance as being either public or private revealed no statistical association.

4.7. Summary of Survey Analysis

The analysis reported above identifies groupings where all the activities within it are significantly correlated in terms of the nature of the provider. Thus, for a particular group, responsibility for the provision of all the activities within it will be located in and around (i.e., grouped around) the same point on a public–private continuum. At this juncture, it is worth reiterating that the nature of this correlation is in all cases positive. It has been determined that separate groups under the same generic function category are not well correlated. This does not necessarily mean, however, that the structure of provision of the activities within one group will inevitably differ from that of another group under the same functional heading. Since there is no correlation between such groups, all that can be said is that provision could be similar, but might equally well differ.

What we are left with, therefore, is a modular port governance model comprising a number of separate groups of activities that can be fitted together like a jigsaw puzzle (see Fig. 18.6), with the appropriate decision-makers selecting and imposing a particular form of governance for each of the nine groups of activities.

Within the survey and subsequent analysis reported herein, five forms of governance along the private–public continuum have been assumed (see Figs. 18.1–18.4). Across nine groups of port activities, this implies that there

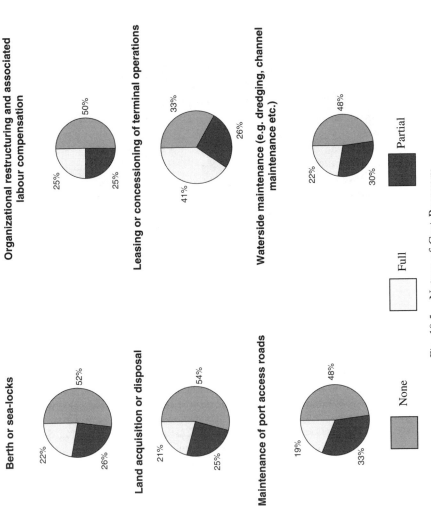

Fig. 18.5. Nature of Cost Recovery.

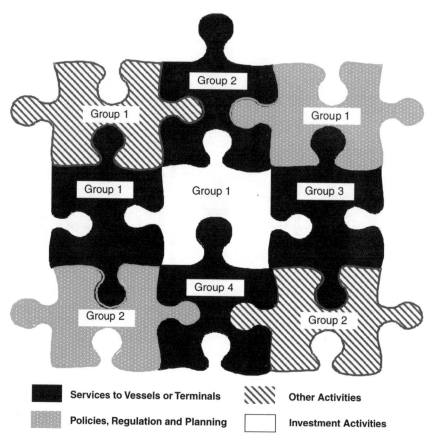

Fig. 18.6. Conceptualization of a Modular Port Governance Model.
Note: Proxies for these Groups are Nine Activities: Line Handling for Arriving/
Departing Vessels, Provision of Fire Protection Services, Terminal Security,
Provision of Pilotage Services, Port Security, Customs and Immigration Services,
Setting and Controlling Port Tariffs, Determining Applicable Port Safety and
Security Policies, and Land Acquisition and Disposal.

are 1,953,125 possible combinations of governance – activity structures for
the management and control of port activities. While this undermines the
validity of existing oversimplified port governance models, it does not necessarily leave the analyst without a model that is tractable to apply in practice. It may be that only a limited number of the full range of possible
combinations are actually selected in practice and that, furthermore, there

may even be some convergence among ports in the combinations that are actually implemented.

To investigate this possibility, data from our sample of 42 ports were analyzed with respect to the actual governance decisions that each of the ports (or its government through legislation) had taken across each of the nine groups of activities. In order to facilitate such an analysis, a necessary first step was to constrain the number of possible governance choices by reducing this from five to just three. This was done by (1) combining the two choices where management and control of the port remains within the purview of either central or local government and referring to this as "public"; (2) treating a port that was fully privately owned, managed and controlled as a single category of "private" and; (3) combining the two choices where management and control (and possibly ownership under a private–public partnership agreement) lies with either a private concern operating under a lease (or concession agreement) or with a quasi-private (e.g., corporatized or commercialized) concern and referring to this as "mixed". This latter category was also used for those situations where the respondent sample ports had indicated shared responsibilities for particular activities across the available governance forms. Thus, for example, if a respondent port had stated that the responsibility for port security was 60 per cent undertaken by a central government agency and 40 per cent by a private company, then this would be categorized as "mixed" governance.

In order to further facilitate this analysis, it was also necessary to identify the most representative individual port activity from each of the nine groupings of activities in order to act as a proxy for the group as a whole. The results of this analysis can be seen in Table 18.10, where the range of possible governance configurations that are actually applied in practice is graphically illustrated, with identified proxy activities emboldened within the column headers of Tables 18.6–18.9.

Out of a possible 512 configurations available to each of the ports in the sample, a total of 34 combinations have actually been adopted in practice by the sample of 42 ports, with only some of the ports in the UK, Italy and Turkey exhibiting any convergence. Table 18.10 reveals that there is very little convergence in governance structures among the ports in the sample, but that it does occur most at the "public" end of the governance continuum. Bearing in mind that this finding assumes merely a rather simplistic and aggregate classification of governance as either "public", "private" or "mixed", the further refinement of governance choices as argued for earlier would inevitably result in even less convergence. It would seem, therefore, that when port activities are considered at a much greater level of detail than

Table 18.10. Port Sample Governance Configurations.

Services to Vessels or Terminals				Other Activities		Policies, Regulation and Planning		Investment Activities	Countries (Number of ports)
Group 1 Line handling	Group 2 Fire protection	Group 3 Terminal security	Group 4 Pilotage services	Group 1 Port security	Group 2 Customs	Group 1 Setting tariffs	Group 2 Safety policy	Group 1 Land acquisition	
Public	Public	Public	Public	Public	Public	Public	Public	Public	Turkey (4)
Public	Public	Public	Private	Public	Public	Public	Public	Public	Italy (2)
Mixed	Public	Mixed	Mixed	Public	Public	Private	Public	Private	UK
Mixed	Public	Mixed	Private	Mixed	Public	Mixed	Public	Mixed	USA
Mixed	Public	Private	Mixed	Mixed	Public	Mixed	Public	Mixed	Belgium
Mixed	Mixed	Mixed	Public	Public	Public	Public	Public	Public	UK
Mixed	Mixed	Mixed	Public	Mixed	Public	Private	Public	Mixed	UK
Mixed	Mixed	Mixed	Mixed	Mixed	Public	Mixed	Mixed	Public	Turkey
Mixed	Mixed	Mixed	Private	Mixed	Public	Mixed	Public	Mixed	UK
Mixed	Mixed	Private	Private	Mixed	Public	Mixed	Mixed	Mixed	Argentina
Mixed	Mixed	Private	Private	Private	Public	Mixed	Public	Public	Turkey
Mixed	Private	Private	Public	Private	Public	Private	Public	Mixed	UK
Private	Public	Mixed	Public	Mixed	Public	Public	Public	Mixed	USA

Private	Public	Mixed	Public	Mixed	Public	Mixed	Mixed	Canada
Private	Public	Private	Public	Mixed	Public	Mixed	Mixed	Canada
Private	Public	Private	Mixed	Private	Public	Public	Public	Canada
Private	Public	Private	Private	Public	Public	Mixed	Public	USA
Private	Public	Private	Private	Mixed	Public	Mixed	Mixed	Netherlands
Private	Public	Private	Private	Private	Public	Public	Public	Netherlands
Mixed	Mixed	Public	Mixed	Public	Mixed	Private	Private	Italy
Mixed	Mixed	Mixed	Mixed	Private	Public	Private	Public	UK
Mixed	Mixed	Mixed	Private	Mixed	Public	Private	Public	UK
Mixed	Mixed	Mixed	Private	Private	Public	Private	Public	UK
Mixed	Mixed	Mixed	Private	Private	Public	Mixed	Mixed	UK (2)
Mixed	Mixed	Private	Mixed	Mixed	Public	Mixed	Mixed	Belgium
Private	Mixed	Private	Mixed	Mixed	Private	Mixed	Mixed	Australia
Private	Private	Mixed	Private	Private	Public	Public	Public	Australia
Private	Private	Private	Public	Mixed	Public	Public	Mixed	USA
Private	Private	Private	Mixed	Mixed	Public	Mixed	Mixed	Chile
Private	Private	Private	Private	Mixed	Mixed	Mixed	Mixed	Australia
Private	Private	Private	Private	Mixed	Public	Mixed	Mixed	Canada
Private	Private	Private	Private	Private	Public	Public	Public	Turkey
Private	Private	Private	Private	Private	Public	Public	Public	UK
Private	Private	Private	Private	Private	Private	Private	Private	Turkey

is accounted for in existing port governance models, a picture of "infinite variety" in the governance choices actually made in practice emerges.

5. CONCLUSIONS

As noted in Chapter 1, discussed in Chapters 3–16 and confirmed in this chapter, ports in many countries generally have been moving away from the public model, even the decentralized one. Perhaps the only country of those studied with a stable port policy in the last decade is the US, a country without a national ports policy and a system that according to Fawcett (Chapter 10) "seems to function well". However, even the US is under pressure and appears to be facing a capacity crisis, as its burgeoning trade with China floods both east and west coast ports; convincing public authorities to invest in new capacity is the topic of the day. That said, access to public funding is not necessarily any easier for public ports to acquire. The requirement for full or partial cost recovery may be a feature of a government's ports policy and, as illustrated by Fig. 18.5, funding for some ports is not necessarily fully repayable (as it would be in a fully private system), although it can be in those countries requiring it.

> The new distribution of roles between public and private actors, in particular, calls for an appropriate allocation of duties and responsibilities, of risks and rewards, to make the global transportation system work to its best efficiency. (Juhel, 2001, p. 174)

The balance between public and private sectors in those industries that may serve the citizens of the country, contribute to economic development or facilitate trade is the critical issue in devolution outcomes irrespective of the industry. What is that balance and where is the tipping point? Based on the research here, even the regulator functions Baltazar and Brooks (2001) posited as likely to be provided by the public sector (e.g., safety policy) are sometimes provided by the private sector. There does not appear to be agreement among governments as to what governance models should be implemented based on what has been imposed through legislation or, in the case of *laissez-faire* nations, adopted by the port authorities in the absence of legislation. Within the 42 study ports, there exists a full spectrum from fully public to fully private management of port activities, with four ports being the only ones operating under a fully public model, and only one port being fully private. Perhaps even more interesting is that these five were all in Turkey, reflecting its period of transition. Based on previous academic research, it was expected that the UK ports would be the most privately managed. In fact, we found that most ports, including those in the UK, had significant "mixed" governance of port activities.

The "mixed" nature of our findings was not just limited to the management and provision of port activities but also to the strategic purpose of the port. Baltazar and Brooks in Chapter 17 proposed that the strategic intent of ports could be divided into two groups – those with economic objectives and those seeking non-economic ones – as a means to evaluate performance. In this chapter, we suggest that there are three logical groups. Restating the findings in Section 4, three groups are the 19 that have solely non-economic objectives (including wider economic benefits), 11 that have strictly economic objectives (of profit maximization and/or maximization of return on investment) and 11 that have a mixture of both economic and non-economic objectives. What is also clear from this analysis is that there is no relationship between the way authorities structure their cargo-handling activities and the strategic purposes they are trying to achieve. In other words, ports do not appear to match methods of governing this activity with primary purpose, contrary to the principles proposed by the strategic management literature presented by Baltazar and Brooks in Chapter 17. Such a conclusion needs to be validated with a much larger sample and a streamlined research instrument.

We can conclude from Section 4 that some ports are clearly trying to achieve what may be incompatible goals. For example, efforts focused on the maximization of throughput or of profit may result in a lagged investment in capacity that, in the long term, could lead to a loss of customers or less economic development in logistics activities landside. We have found that all ports do not focus on achieving similar strategic goals, and therefore future research into governance models and performance will need to recognize this in the research design if the Baltazar and Brooks model is to be validated.

In examining the environment facing port authorities, and the strategy or strategies they employ to meet their objectives, it is not sufficient merely to identify the activities that may be provided within a port, and whether they are to be provided by the public sector, the private sector or some mixed arrangement. It is necessary to determine whether they are to be locally or nationally responsive, and who will hold the managers accountable for results (shareholders, governments, local stakeholder groups and so on). Who will define the desired outcomes? That is, governance structure is not just about ownership, but also about management responsibility, accountability and control. Table 18.3 is a useful means for governments to identify their intentions and then to develop models that are consistent with those objectives. The question of balance will then be one of determining if the government-imposed framework will be highly prescriptive or loosely guided. In cases where the strategic intent adopted by port authority management varies from that desired by government, political interference from disgruntled politicians can be expected to be the by-product of a devolution program.

Brooks (2004) noted that most port devolution programmes have been fraught with difficulty, and that none of the approaches is without its detractors. Baltazar and Brooks (2001) argued that, for all sizes and types of ports, governance-performance links should be examined and governance models tested against performance outcomes, for varying port strategies. This chapter has gone some way in attempting to define governance models that may be used in further research on the relationship between governance and performance. It has identified the existence of multiple strategic goals, and wide variance in the way activities are managed and controlled. However, it has also identified nine proxy activities that can substitute for the great variety of port activities in any future research on governance approaches. It is disappointing that the research did not result in a simple set of models for such governance performance research; it did confirm that the existing models by Baird (2000) and the World Bank (undated) are oversimplified, cannot be validated, and do not reflect the hodgepodge of "infinite variety" implemented in today's highly competitive port environment.

This research has confirmed that devolution has been the practice of government in the past 20 years. Both governments and port authorities today supply less of the operations than they used to and are predisposed to contract out, licence or grant concessions to private sector operators for the delivery of port activities, including those once thought to be the purview of governments; this situation underscores the need for sharing experience on a global basis between port authorities. Without such a concerted effort to stay abreast of best practices in such contracting, and with the global consolidation of many of the providers of these activities, the future strategic purposes of ports may not be realized, as global service providers are not necessarily interested in delivering that which a government seeks.

NOTES

1. Members of the network who assisted with data collection were Khalid Bichou (Imperial College London), Mary R. Brooks (Dalhousie University), Dionisia Cazzaniga Francesetti (Universita Di Pisa), A. Güldem Cerit (Dokuz Eylül University), Peter W. de Langen (Erasmus University Rotterdam), Sophia Everett (Melbourne University), James A. Fawcett (University of Southern California), Ricardo Sánchez (Austral University/ECLAC), Marisa Valleri (University of Bari) and Thierry Vanelslander (Universiteit Antwerpen).

2. Baird (2000, p. 181) noted that the actual proportion of public and/or private sector participation with regard to those individual activities within a port associated with each of the regulator, landowner and utility function may also vary.

REFERENCES

Baird, A. J. (1995). Privatisation of trust ports in the United Kingdom: Review and analysis of the first sales. *Transport Policy*, *2*(2), 135–143.

Baird, A. J. (1999). Analysis of private seaport development: The case of Felixstowe. *Transport Policy*, *6*(2), 109–122.

Baird, A. J. (2000). Port privatisation: Objectives, extent, process and the U.K. experience. *International Journal of Maritime Economics*, *2*(3), 177–194.

Baltazar, R., & Brooks, M. R. (2001). The governance of port devolution: A tale of two countries. *World conference on transport research*, Seoul, Korea, July.

Bichou, K., & Gray, R. (2005). A critical review of conventional terminology for classifying seaports. *Transportation Research Part A: Policy and Practice*, *39*(1), 75–92.

Brooks, M. R. (2004). The governance structure of ports. *Review of Network Economics: Special Issue on the Industrial Organization of Shipping and Ports*, *2*(2), 169–184. Available at http://www.rnejournal.com

Chlomoudis, C. I., & Pallis, A. A. (2002). *European union port policy: The movement towards a long-term strategy*. Cheltenham, UK: Edward Elgar.

de Monie, G. (1994). Mission and role of port authorities. *Proceedings of the world port privatisation conference,* London, 27–28 September.

de Monie, G. (1995). Restructuring the Indian ports system. *Maritime Policy and Management*, *22*(3), 255–260.

Dick, H., & Robinson, R. (1992). Waterfront reform: The next phase. Presented to the national agriculture and resources outlook conference, February.

Everett, S. (2003). Corporatization: A legislative framework for port inefficiencies. *Maritime Policy and Management*, *30*(3), 211–219.

Goss, R. (1990). Economic policies and seaports – Part 3: Are port authorities necessary? *Maritime Policy and Management*, *17*(4), 257–271.

Juhel, M. H. (2001). Globalization, privatisation and restructuring of ports. *International Journal of Maritime Economics*, *3*, 139–174.

Pallis, A. A. (1997). Towards a common ports policy? EU-Proposals and the ports industry's perceptions. *Maritime Policy and Management*, *24*(4), 365–380.

UNCTAD. (1992). *Development and improvement of ports: The principles of modern port management and organisation*. Geneva: United Nations Conference on Trade and Development.

World Bank. (undated). *World Bank Port Reform Toolkit*. Available at http://www.worldbank.org/transport/ports/toolkit/

CHAPTER 19

CONCESSION AGREEMENTS AS PORT GOVERNANCE TOOLS

Theo Notteboom

ABSTRACT

This chapter discusses the role of concessions as a tool in port governance under the landlord port authority model. The specific design of the concession agreement, its regulatory regime, the tariff regime and the way the concession is awarded reveal the priorities of port authorities and government agencies. Through concession policy, port authorities can retain some control on the organization and structure of the supply side of the port market and can encourage port service providers to optimize the use of scarce resources such as land.

1. INTRODUCTION

In many countries around the world, governments and public port authorities have retreated from port operations in the belief that enterprise-based port services and operations would allow for greater flexibility and efficiency in the market (through more competition) and a better response to consumers' demands. Ports, which have traditionally been run like a government department, are witnessing an infusion of private money that promises greater competition, higher productivity and eventually lower costs, which will be

Devolution, Port Governance and Port Performance
Research in Transportation Economics, Volume 17, 437–455
Copyright © 2007 by Elsevier Ltd.
ISSN: 0739-8859/doi:10.1016/S0739-8859(06)17019-5

passed on to importers and exporters. Often this involves the transfer of provision of services from public bodies to private enterprise (Braddon & Foster, 1996; Everett, 1996). Ports have become an interesting business, attracting the attention of large investment groups and equity fund managers. Ports are inclined to develop new governance structures, which should be tailored to the specific local conditions in terms of culture and port objectives.

In this new environment, the concessioning of port services to private operators has become common practice. A concession is a grant by a government or port authority to a (private) operator for providing specific port services, such as terminal operations or nautical services (e.g. pilotage and towage). Concession policy has become a powerful governance tool to port managers in particular in the terminal operating business. First, market developments such as vertical and horizontal integration are challenging (landlord) port authorities to redefine their role in the competitive process (for further readings see e.g. Comtois & Slack, 2003; Heaver, Meersman, Moglia, & Van de Voorde, 2000; Martin & Thomas, 2001; Notteboom & Winkelmans, 2001). Through concession policy, port authorities can retain some control on the organization and structure of the supply side of the port market. Second, through concession policy port managers can encourage port service providers to optimize the use of scarce resources such as land.

This chapter discusses the role of concessions as a tool in port governance. The scope of the paper is narrowed to the concessioning of land to private terminal operators under the landlord port authority model. Under this model, the landlord port authority is normally a separate entity under public law established by specific legislation (World Bank, n.d.). It has the capacity to conclude contracts (including concession agreements), enforce standards and make rules and regulations applicable within the port area. Port operations (especially cargo handling) are carried out by private companies. Today, the landlord port is the dominant port model in large and medium-sized ports.

Van Niekerk (2005) argued that the privatization of terminals through concession contracts is a valuable option where port competition is effective, but not necessarily in cases where competition needs to be created by regulation. This chapter does not intend to discuss the suitability and effect of concessioning under circumstances of limited competition (as is often the case in developing countries). Instead, the discussion in this chapter focuses in particular on ports in developed countries operating in a (highly) competitive environment. Valuable lessons from cases in developing countries are not discussed.

Section 2 discusses ways for private companies to get involved in terminal operations in public ports. Section 3 offers insight into bidding procedures related to terminal concessions under the landlord port model. Section 4 deals with concession agreements and how these contracts can be deployed to achieve the objectives of the port authority. Section 5 offers concluding remarks.

2. PRIVATE TERMINAL OPERATIONS IN PUBLIC PORTS

Four major types of combinations of port/terminal ownership and port/terminal operations can be distinguished: (a) public/government ownership and public participation in operations; (b) public/government ownership and private participation in port/terminal construction, operations and management; (c) public/government ownership and private participation in superstructure installation (e.g. cranes) and operations and (d) private ownership and operations. The use of the model of Public Ownership and Private Operations (POPO) is widespread (Murthy & Notteboom, 2002). Under the POPO model, a concession agreement is signed between a private terminal operator and a landlord port authority or empowered government agency. Concessions can take the form of a long-term lease or an operating license. Under the long-term lease system, a private company is allowed to operate a specified terminal for a defined period of time. The government or a public authority holds the property rights of the facilities throughout the concession period and receives lease payments on the assets. The concession/lease fees paid by the private terminal operator are used to upgrade and expand the facility.

In case of greenfield developments, more alternatives are available to structure the responsibilities of terminal operator and port authority/government with respect to the construction, financing and operations of the terminal facility. These include:

Build-Lease-Operate (BLO). The port authority leases the construction and operation of the whole port or part of it to a private company through a long-term concession. The private company constructs the terminals/berths/other facilities. In turn, the port authority controls the rights throughout the concession period and receives a lease payment annually. An example of BLO is Fuzhou Qingzhou Container Terminals of Fuzhou Port in China. PSA Corporation has contracted in 1998 a BLO for a period of 20 years. BLO is quite popular in China as port authorities are directly leasing the ports or terminals.

Build-Operate-Transfer (BOT). In case of the BOT technique, a government or public authority grants a concession or a franchise to a private company to finance and build or modernize a specific port facility. The private company is entitled to operate the facilities and to obtain revenue from specified operations or the full port for a designated period of time. The private sector takes all commercial risks during the concession. At the end of the concession period, the government retakes ownership of the improved assets. Arrangements between the government and the private operator are set out in a concession contract that may or may not include regulatory provisions. A major BOT project is Tanjung Pelapas Port (PTP) in Malaysia. Pelabuhan Tanjung Pelapas Sdn Bhd is a joint venture between Seaport Terminals and Indra Cita Sdn Bhd and Koperasi Permodalan Nageri Johor. The port started its operations in July 1999 and has become a major competitor of Singapore.

Rehabilitate-Operate-Transfer (ROT). The government or public authority grants a concession to a private company to finance and rehabilitate or modernize a specific terminal or an entire port. This company is entitled to operate and obtain revenue from the rehabilitated port for a specific period. The private company takes all commercial risks, and at the end of the concession period, the government retakes ownership of the improved asset.

Build-Rehabilitate-Operate-Transfer (BROT). The government or public authority grants a concession to a private company to finance and rehabilitate or modernize a specific terminal or an entire port. This company is entitled to operate and obtain revenue from the rehabilitated port for a specific period. The private company takes all commercial risks, and at the end of the concession period, the government retakes ownership of the improved asset.

Build-Operate-Share-Transfer (BOST). BOST is similar to BOT, when a government grants a concession or a franchise to a private company to finance and build or modernize a specific port/terminal for a designated period of time. The revenue obtained from terminal operations is shared with a designated public authority throughout the concession period. The government/public authority should assure a specific quantity of throughput for revenue. The commercial risks are shared among the government and the concessionaire. At the end of the concession period, the government retakes ownership of the improved asset. An example of BOST is the proposal of BCC Shipping & Shipbuilding Ltd and its UK partners for developing Tadri Mini Seaport at Karnataka State in India.

Long-term leases, operating licenses and BOT schemes are the most commonly used forms of concession. The following sections focus on long-term lease contracts for terminals in the landlord port model.

3. THE BIDDING PROCEDURE

Terminal concession agreements may be awarded by several methods, including without limitation, by direct appointment, private negotiation, from a qualified pool or using a competitive process. The concession policy in many ports around the world has undergone some changes in recent years. Where before there were often no formal conditions required, the granting nowadays consists of a thorough inquiry of the different candidates for obtaining a concession. National and supranational legislation (e.g. EU legislation), port privatization schemes and legal disputes with regard to irregularities in concession policy have made competitive bidding the most common procedure used in concession granting today. Any competitive bidding should comply with the principle of equality, which states that every candidate should be equally treated and compared and that there will be no favoritism in the awarding of the concession or no substantial reduction of competition. A competitive bidding procedure for berths and terminals typically consists of two stages: a qualification stage and a selection stage.

3.1. Stage 1: Qualification

The first stage encompasses a qualification of the candidates based on experience and financial strength. The experience of the candidate can, for instance, be demonstrated by the management of facilities for similar cargo in the same or other ports. The candidate thus has to credit his experience in the activities related to the project by giving proof of specific antecedents in the exploitation of terminals. The bidding procedure typically contains thresholds on the financial strength of the bidders. For example, it can be stipulated that 40% of net worth of the bidder (combined net worth in the case of consortium) should at least equal the estimated project cost and that the aggregated net cash accruals of the bidder during the proceeding three-year period should be at least equal to half of the estimated project cost. In view of evaluating the financial capacity, the candidate will have to present balance sheets and auditor's notes for a number of fiscal years, the net worth at the closure of the last fiscal year and or invoicing details for a number of fiscal years.

The first stage in the bidding procedure reduces the number of potential bidders, which is bad for competition, but it also reduces the risks of non-compliance by unreliable bidders.

3.2. Stage 2: Selection

The second stage typically comprises a *technical and financial proposal* and a *price bid*. Every qualified bidder will present only one offer, without variants or alternatives. Based on the technical and financial proposal and or the price bid, a bidder among the bidders that have been prequalified, is selected.

3.2.1. The Technical and Financial Proposal

One way of deciding who will get the concession is to base the decision primarily on the *technical and financial proposals* submitted by the bidders. Although the required contents of such a technical proposal tend to differ significantly from case to case, it usually consists of: implementation details, financing details, a marketing plan, operational and management details, employment impact, an environment plan and an organizational plan.

The development of the terminal will follow an implementation plan, if possible ordered by stages according to the growth of the traffic. The implementation plan is backed by the results of a market study. The calculations of handling capacity will have to demonstrate that the installations offered will have the necessary capacity to serve adequately the needs of the projected throughput. The timetable of investments will have to include the foreseen cost of works and equipment. Each bidder has to present a complete list of fixed or mobile equipment he foresees to operate on the terminal, with definition of type, capacity, specifications, life span and date it will enter the operation.

Where the bidding procedure concerns the development of a greenfield or brownfield terminal by the bidder (e.g. following a BOT arrangement), each bidder has to present studies and preliminary sketches of the works, estimations of the costs of building and a plan of investments of all the works that his proposal contemplates. These works typically include the marine infrastructure necessary for the creation of the terminal (e.g. dredging of access channel, breakwaters) and the terrestrial infrastructure of the terminal and related installations.

The marketing plan typically includes a market study that defines terminal demand and justifies the provisions about the magnitude and requirements of the installations, including projections of yearly throughput for a number of years.

Each bidder has to explain his intended strategy to attract the highest market share and to reach the highest efficiency in his operations. This includes calculations of cash flow based on projected throughput. The candidate will have to indicate the prices and maximum charges for any services offered to users of the terminal, and the costs of operation (including labor, equipment, fuels and other inputs and supplies), maintenance, supervision and administration.

Each bidder typically has to present studies of environmental and territorial impact covering aspects such as the impact of the terminal operations on the environment and the alternatives available to eliminate, reduce or mitigate certain effects.

Each bidder typically has to quantify the requirements of staff for the concession and has to project its evolution according to the cargo projections.

In a well-prepared proposal, there should be a continuous chain of logic, which flows from the market demand and cargo projections through to the physical layout, equipment purchases, staffing levels and operating assumptions. In view of the final selection, it is quite common to rate various aspects of the technical and financial proposal and add up the results in a weighted or unweighted score, based on a weighted score on each of the evaluation criteria related to the elements in the proposal. The weights and passing criteria are included in the bidding documents if applicable. This process may lack transparency for potential bidders.

3.2.2. Price Bid

An alternative way of awarding a concession consists of putting a strong focus on the price. The stipulations on the *price bid* depend on the price bidding systems used. Goss (1990) described quite extensively the price bidding systems that can be used in concession and lease agreements. The alternatives available range from a given rent but minimal charges to a maximum rent and freedom to the private operator to set his own charges. Given a set level of investment and quality requirements, the winner is either (a) the highest payer for the right to provide terminal services, or (b) the bidder offering the lowest price to be paid by the terminal users. In the first option the port authority or government agency aims at maximizing the revenue. The payments are typically made annually. The second option focuses more on the interest of the port users and ensures price minimization rather than revenue maximization. The concession fee under the first option is seldom renegotiated while the fee under the second option is often renegotiated in such a way that the concessionaire ends up with a larger share of the rent created by the efficiency gains achieved by the concessionaire.

In some cases, particularly with new terminal developments in developing countries, bidders have to quote the percentage of their revenue to be passed on to the government. The competition is then based on bidding and priority will be given to the party offering the highest percentage to the government.

3.3. Additional Elements in the Bidding Procedure

Bidding procedures typically include stipulations on all the conditions of the actual bidding, such as the ability to cancel the bidding at any stage of the process if an offer is considered unacceptable.

Candidates for a concession typically are given the opportunity to visit the land until the moment of presentation of the offers. During such visits they will be able to verify the state of the premises and, if needed, perform surveys and/or field studies considered necessary to the preparation of their offer. On top of this, the candidates will be able to access available information related to the physical, operative, commercial, juridical, labor, etc., aspects of the proposed concession.

4. CONCESSION AGREEMENTS AS PORT GOVERNANCE TOOLS

4.1. Contractual Arrangements and Information Asymmetries

In a terminal leasing/concession system, a private company is allowed to operate a specified terminal for a given duration. The design of the concession agreement or contract, starting with the rights and obligations of the concessionaire, is a key element. Table 19.1 depicts the typical structure of a concession agreement under the landlord model.

One of the main reasons behind particular contractual arrangements in a concession agreement relates to potential information asymmetries as described in the principal/agent theory. Akerlof (1970) and Spence (1973) have done ground-breaking research with respect to asymmetric information distributions in markets. Rasmussen (1994) distinguishes three models of principal-agent problems (see also Bergantino & Veenstra, 2002), moral hazard problems with hidden information, moral hazard problems with hidden action and adverse selection problems. Where the principal (in this case the port authority) is not informed about a certain characteristic of the

Table 19.1. Typical Content Structure of a Concession Agreement in the Landlord Port Authority Model.

1. General background, definitions and notes
2. Description of business
   General description
   Promotion of the business of the container terminal
   Nature of terminal (common user, dedicated)
3. Overall scope
   Technical description of the facility (berths, container yard, container handling equipment, ...)
4. General conditions
   Services to be provided by licensee
   Ownership and maintenance stipulations (land, non-land assets, assets of licensor, maintenance of assets)
   Rights of licensor and licensee with respect to licensed premises
5. Prices
   Setting of prices by the licensee
   Regulation and review of prices
   Currency adjustments
6. Major commercial terms
   Initial payment for bidders
   Royalty payment (e.g. per TEU)
   Performance indicators (cf. minimum average crane productivity)
7. Termination provisions
8. General provisions
   Confidentiality, permits, amendments, etc.

Source: Based on Drewry Shipping Consultants (2002).

agent (the terminal operator), an adverse selection problem may arise. The term 'adverse selection' was originally used in insurance. It describes a situation where the people who take out insurance are more likely to make a claim than the population of people used by the insurer to set their rates. In moral hazard models, information asymmetries result from the principal's inability to observe the agent's action (moral hazard problems with hidden action) or to uncover hidden information about the agent (moral hazard problems with hidden information).

The adverse selection problem only partly applies to terminal concession agreements. As mentioned earlier, candidates for a concession typically have the right to verify the state of the premises and have access to all available information. As such, both port authority and (candidate) terminal operator generally start with the same information as regards the object of the concession.

Moral hazards problems are more prominent in the concession procedure. In principle, a port authority (the principal) bases its decisions about the granting of a concession on information available at the time of the granting. Although the information about the candidate might be very elaborate (e.g. financial status, performance in other ports, client base), the port authority has no guarantee that the agent will meet its objectives in terms of cargo generation. As such, concession agreements often take the form of performance-based contracts to create incentives for the agent (terminal operator) to act in the principal's interest.

The specific design of the concession agreement, its regulatory regime, the tariff regime and the way the concession is awarded reveal the priorities of port authorities and government agencies and, as such, play an important role in port governance. In the next sections, I identify key issues in concession agreements and reveal how port authorities deal with potential information asymmetries. Moreover, it will be demonstrated how concession agreements can be employed as port governance tools.

4.2. Duration of the Concession Agreement

In most cases the term of the concession is preset by the port authority or government agency. In some cases, the port authority leaves it up to the bidder to indicate the term in years that it requires for the concession respecting the valid regulations. The term will be counted from the date of taking over the concession.

In many parts of the world, legislators have developed rough guidelines on concession durations with a view to safeguarding free and fair competition in the port sector. For instance, in its proposal for a Directive on Market Access to Port Services, the European Commission proposed a series of limited and renewable periods of time for authorizations for providers of port services (European Commission, 2004). An 'authorization' is defined as any permission, including a contract, allowing a natural or legal person to provide one or more categories of port services. A concession agreement as such is a form of authorization. Article 12 of the Commission's proposal sets maximum durations for authorizations depending on the type of investments made. In case the authorization relates to investments by the concessionaire in immovable assets and comparable movable capital assets (such as gantry cranes) the maximum period is 30 years, irrespective of whether or not their ownership will revert to the port authority after the concession period.

A recent internal survey by the European Sea Ports Organization revealed a large variance in concession duration in European ports: domain concessions for cargo handling can range from 3 to 65 years. It is not realistic in the port sector to try to fit everything into one set of durations and, indeed, there are hardly any rules of thumb on the matter. In general, the duration of the concession will vary with the amount of the initial investment required, the compliance with the development policy of the port and land lease and other easement rights. A number of ports apply phased concession terms, with a base duration of, for example, 15 years and consecutive renewals (based on specific criteria) every five years. Many concession agreements contain stipulations on a possible prolongation of the concession beyond the official term.

The duration of the agreement is of crucial importance both to terminal operators and port authorities. In general, long-term agreements allow private port operators to benefit from learning-by-doing processes and to achieve a reasonable ROI. Port authorities try to find a balance between a reasonable payback period for the investments made by terminal operators, on the one hand, and a maximum entry to potential newcomers, on the other. As long-term agreements limit market entry, intra-port competition will only take place among the existing local port operators.

However, even when concession periods are long, new players can still enter the market either through a merger or acquisition of a local operator or when a long-term concession or lease of a new terminal expansion is allocated to them. As discussed in the next section, port authorities can even build safety valves in concession agreements, so as to make the concession available to other candidates in case the existing concessionaire does not meet specific performance thresholds.

4.3. Throughput Guarantees

A concession agreement typically contains provisions in view of protecting terminal operator and port authority against arbitrary and early cancellation. However, it can also include paragraphs that empower the port authority to (unilaterally) end the concession in case the terminal operator does not meet certain preset performance indicators.

The most common indicator relates to cargo throughput. The port authority or government agency can indicate upfront a minimum throughput to be guaranteed by the concessionaire (especially in case of existing berths/terminals). This should encourage the operator to market the port services to attract maritime trade and to optimize terminal and land usage. Should the terminal operator not meet the objectives set out in the concession agreement, a penalty will be payable to the port authority (e.g. a fixed amount per ton or TEU short) or, in the most extreme case, the concession will be cancelled.

For instance, in the concession agreements for the Deurganck Dock in the port of Antwerp, stringent demands were made regarding the use of space by the concessionaires. Taking into account the real use, certain parts of the terminal could be retracted and reallocated due to underutilization. Furthermore, should the Maersk Sealand shipping company meet with the proposed conditions for further volume growth in Antwerp, then (as from 2007) Maersk would be able to dispose of a to-be-defined dedicated terminal site at the Deurganck Dock, if needed and at the expense of existing concessionaires, i.e. PSA at the western side of the dock and the Antwerp Gateway consortium (headed by P&O Ports) at the eastern side.

Throughput guarantees are a powerful governance tool to port authorities for three reasons. First, from a land management perspective, throughput guarantees secure a reasonable level of land productivity. Land for port development is scarce in most parts of the world and efficient use is necessary to safeguard it for the future. Moreover, it sends out a positive signal to environmentalists and community groups about the port community's intentions vis-à-vis rational land use. Environmental considerations are very prominent in the relationship of community stakeholders with port authorities.

Second, imposing specific performance targets is particularly useful in cases where the level of the concession fees is not giving enough incentive to terminal operators to reach high terminal utilization rates. By integrating throughput guarantees, efficient land use becomes a shared responsibility between port authority and terminal operator. The question remains whether or not specific terminal throughput targets should be used to modify the behavior of terminal operators. It may be more appropriate in some cases to achieve the objective through the basic allocation of responsibilities or a modification of the concession fees. Moreover, specific throughput guarantees will deliver the expected results only if the performance of the terminal operator can be adequately monitored and if the targets can be enforced (see Section 4.6).

Third, stringent demands regarding the use of space by the concessionaires can lower the entry barriers to newcomers. The port authority could retract or reallocate certain parts of the terminal due to underutilization. These kinds of stipulations in concession agreements contribute to an improved contestability in the container handling industry (Notteboom, 2002).

Moreover, port authorities can move an intra-port monopoly toward a local oligopoly by allocating a long-term concession or lease of a new terminal expansion to a new operator.

4.4. Concession Fees

When leasing land for terminal operations, a port authority has to address three financial questions. First, it has to decide how much the terminal operator will have to pay for the use rights received through the concession (i.e. the concession fee). Second, it has to determine the amount the terminal operator should be paid by the users to cover the cost of providing cargo handling services on the terminal. Third, it has to decide on the way to collect the concession fees.

In many cases, the port authority imposes a fixed concession fee and the private operator has the freedom to set its own charges. For instance, the lease to be paid to the port authority takes the form of a fixed sum per square meter per year, typically indexed to some measure of inflation. The level of the lease amount is related to the initial preparation and construction costs (e.g. land reclamation and quay wall construction), the location in the port and the type of activity. This system is clear and straightforward, easy to manage and leaves little room for misinterpretation.

The landlord port authority possesses a detailed overview of all land concessions in the port with, for each concession, a description of the type of port activity (i.e. type of cargo handling, port industry or other port service), the current concessionaire and the fee per square meter per year. The highest fees will apply to the most valuable land in the port area, characterized by a large terminal space and excellent access to waterways and the hinterland. Often, the fees also differ according to whether the land is used as terminal area or for buildings and similar constructions. In the port of Antwerp, for example, the concession fees amount to ca. €5/m^2/year for the container terminal areas in the northern part along the river Scheldt (i.e. Europaterminal and Noordzee Terminal, €10/m^2/year for buildings and similar constructions) to €2.5/m^2/year for terminal areas in the older dock system near the city.

Occasionally, concession agreements grant the terminal operator an extension of payment in stages: for instance, 25% of the total area in year one, 50% in year two, 75% in year three and the total in year four. This kind of arrangement is often applied in the case of a phased terminal development.

In some cases, the port authority uses a two-part payment system, combining a fixed concession fee and a variable royalty fee per ton or TEU.

A system of royalty payments related to a minimum guaranteed throughput tempts port authorities to redefine the terms of the concession after a few years in order to gain more revenue. The system is sensitive to renegotiation, and as such decreases the legal certainty of the concession agreement in the eyes of the terminal operator.

Another type of concession payment relates to revenue sharing between terminal operator and port authority. Such arrangements are used particularly in the case of new terminal developments under BOST in developing countries (as mentioned earlier in this chapter).

No matter the payment base, the financial side of a concession agreement remains a balancing act. High concession fees, royalty payments and/or revenue-sharing stipulations are detrimental to the terminal operator's return on investment and could as such decrease the investment potential of the incumbent terminal operator and scare away future investors. Low payments could negatively affect the revenue base of the port authority in such a way that it can no longer properly perform its landlord functions.

The question remains whether concession fees actually cover the costs incurred. The general policy prescription should be to charge cost-covering concession fees. In a port context, the application of this basic principle is not always straightforward.

First, in European ports, but also elsewhere, there still exist a variety of subsidies. This raises the issue of how to deal with the level of concession fees in the case of partly subsidized port infrastructure. For example, a government body might fund a share of the total construction costs of commercial quay walls. In that case, the port authority reaches full-cost recovery (based on its contribution to the total investments costs) with concession fees, which in the longer term do not cover the total investment costs. When subsidies are deemed absolutely necessary from a (national) port policy perspective, they should minimize distortions in resource use, maintain incentives for the efficient use of the port infrastructure and be implemented in a transparent manner so that the direction and magnitude of subsidies can be kept under close scrutiny when, for example, fixing concession fees.

Second, there is also a strategic factor at play. Port authorities might lease out land at a price below the intrinsic land value in the hope the container terminal might attract high value-adding logistics and industrial activities to the port. The local community might wonder whether it is getting a fair input payback for the scarce local resources used by ports. Port authorities should make the relationship between the price for scarce resources and the socioeconomic payback more transparent both to port users and community groups.

4.5. Structuring Intra-Port Market Organization through Concession Policy

Through specifications in concession agreements port authorities can actively shape the structure and market organization of the terminal handling business in the port area, both in spatial and functional terms. To widen the private sector's participation and provide competition, the port authority or government agency can require that the operator not be allowed to participate in more than one contract at the same port. In smaller ports, a concession agreement could state that no other stevedore can develop container handling activities in the same port. As such, port authorities can partially design an intra-port market configuration they prefer through the bidding procedures and concession agreements used. Port authorities can have good reasons to opt for a market configuration of only one or two container terminal operators within a specific port area, for example, to provide a better answer to carrier power and carriers' demand and to guarantee a larger financial base for investments in expensive terminal infrastructure.

Concession agreements can have a great influence on overall port performance. A poorly designed lease contract can lead to situations where the port authority has no legal means to penalize inefficient companies or to end the contract unilaterally. The resulting underutilization of valuable land and infrastructure undermines the growth potential of the port. A good example is the legal dispute that took place between MBZ, the port authority of the Belgian coastal port Zeebrugge, and the private company Katoen Natie concerning the concession of the Flanders Container Terminal (FCT). Katoen Natie obtained a concession at FCT in the 1990s but never succeeded in reaching a reasonable utilization rate for the terminal (at times, occupation in the stacking area was less than 10% of the design capacity). This situation conflicted with the aim of the port authority to attract more intercontinental container traffic by manifesting itself as an alternative for the neighboring load centers, Antwerp and Rotterdam. However, Katoen Natie had a watertight concession contract that enabled the company to keep the terminal underutilized for several years. After a long legal battle, Katoen Natie finally pulled out of container terminal activities in Zeebrugge. In October 2004, MBZ announced that APM Terminals had been named the preferred bidder for the concession to manage and operate a container terminal on the Albert II dock south in Zeebrugge, formerly known as FCT. The decision was partly based on investment commitments and the related opportunities for increasing the volume in the port. The terminal will resume operations in 2006.

Turner (2000) demonstrated that the position of terminal leasing policy as regards single user terminals has a clear impact on system performance. The question of whether or not to allow dedicated terminals forms an integral and crucial element in terminal leasing policy. Musso, Ferrari, and Benacchio (2001) and Cariou (2003) provide a more in-depth analysis on the concept and pros and cons of dedicated terminals. Today, dedicated terminals are widespread in Asia and North America. In Europe, shipping lines more recently entered the market via the development of dedicated terminals at major load centers. Market reality has made some world ports that were previously reluctant to allow dedicated terminals embrace the concept in view of binding cargo to the port. Typical examples are Antwerp with the MSC Home Terminal (a joint venture between shipping line MSC and global terminal operator PSA), and Singapore where shipping lines are slowly given more opportunities to enter the terminal business (e.g. via minority shareholdings in newly developed PSA terminals). The changing attitude of ports with regard to dedicated terminals underlines that terminal concession policies as part of governance structures are not static but evolve constantly in line with the requirements imposed by the market players and the legislators.

4.6. The Effectiveness of Sanctions in Concession Policy

Whatever the explicit objectives of the concession, each lease contract should contain an explicit list of penalties that will be incurred if the rules are not respected. With the emergence of international terminal operator groups and shipping lines, port authorities are confronted with powerful and footloose players. Fierce competition and the fear of traffic losses might make port authorities less observant and strict with regard to the editing and the enforcement of the rules in the concession agreement.

In a well-tailored concession contract, sanctions are such that the parties do not respect the rules when, and only when, it is optimal not to respect them. Too high penalties give the contract a rigidity that is inefficient. Only credible threats should be included in a concession agreement, i.e. threats that the port authority will have an interest in implementing if a violation occurs. An example is a penalty imposed on the terminal operator for not meeting a minimum throughput. Non-credible threats and sanctions are not effective because their presence in the concession agreement weakens the reliability and legal certainty of the whole text. As terminal operators do not take them into account and as the port authority does not punish violation of them, the port authority acquires a reputation of non-enforcement that induces the terminal operator to violate the rules or try to renegotiate any decision it does not like. In a port environment with powerful and footloose market players, this is not the position a port authority would like to be in.

Threat of severe penalties, such as termination, could be detrimental to the relationship between the port authority and terminal operator and may lack credibility. A way out might be to include a range of penalties in the concession agreement, for example, different levels of financial penalties in addition to the ultimate sanction of contract termination. A port authority might also consider rewarding terminal operators who are doing much more than expected. Penalties and bonuses should ideally reflect the economic costs and benefits of the behaviors they are trying to prevent or promote. For example, penalties may be related to the economic loss caused by not reaching the throughput levels as set out in the concession agreement. The economic loss includes terminal-related elements (terminal underutilization) and broader economic elements (missed added-value and employment).

5. CONCLUDING REMARKS

In today's port environment, the concessioning of port services to private operators has become one of the most important tools for port authorities to influence the prosperity of the port community. Through concession policy, port authorities can retain some control of the organization and structure of the supply side of the port market, while optimizing the use of scarce resources such as land.

Market considerations and new legal guidelines imposed by national and supranational legislation have dramatically changed concession policies in many ports around the world, both for bidding procedures and contract stipulations. Landlord port authorities have embraced concession policy as a means to promote free and fair competition in the cargo handling industry, to increase efficiency in port operations and land use, to further develop their role as (local) regulator and to deal with potential information asymmetries in the relation between port authority and terminal operator.

The dynamics in the port environment induce port authorities to continuously evaluate the effectiveness of their concession policies in light of market trends and advances in the legal framework. At present, port authorities around the world often rely on ad hoc arrangements, compliant with national or supranational regulations, while dealing with terminal concession issues in the highly competitive and dynamic market.

The economic analysis and implementation of dynamic terminal concession contracts in ports remains an unexplored study field for maritime economists, although the renegotiations of contracts is a well-studied topic in economic literature (e.g. Beaudry & Poitevin, 1993; Hart & Moore, 1988; Wang, 2000; Wernerfelt, 2004; Aumann, 2000, on cooperative game theory) and in empirical research on infrastructure services (e.g. Guasch, 2004).

REFERENCES

Akerlof, G. (1970). The market for lemons: Quality uncertainty and the market mechanism. *Quarterly Journal of Economics, 84*, 488–500.

Aumann, R. J. (2000). *Collected papers*. Cambridge, MA: MIT.

Beaudry, P., & Poitevin, M. (1993). Signalling and renegotiation in contractual relationships. *Econometrica, 61*, 745–782.

Bergantino, A. S., & Veenstra, A. (2002). *Principal agent problems in shipping: An investigation into the formation of charter contracts.* IAME Annual Conference. International Association of Maritime Economists, Panama City (proceedings on CD-ROM).

Braddon, D., & Foster, D. (1996). *Privatisation: Social science themes and perspectives.* Alderschot, UK: Dartmouth Publishing Company.

Cariou, P. (2003). *Dedicated terminals in container ports: A cost-benefit analysis.* Research seminar: Maritime transport, globalisation, regional integration and territorial development, Le Havre.

Comtois, C., & Slack, B. (2003). Innover l'autorité portuaire au 21e siècle: Un nouvel agenda de gouvernance. *Les Cahiers Scientifiques de Transport, 44*, 11–24.

Drewry Shipping Consultants. (2002). *Global container terminals: Profit, performance and prospects.* London: Drewry.

European Commission. (2004). *Proposal for a directive of the European Parliament and of the Council on market access to port services.* Brussels, COM (2004) 654.

Everett, S. (1996). Corporatisation strategies in Australian ports: Emerging issues. IAME 1996 Conference – Shipping, ports and logistics services: Solutions for global issue. IAME, Vancouver.

Goss, R. (1990). Economic policies and seaports – part 3: Are port authorities necessary? *Maritime Policy and Management, 17*, 257–271.

Guasch, J. L. (2004). *Granting and renegotiating infrastructure concessions: Doing it right.* Washington, DC: World Bank Institute Development Studies.

Hart, O., & Moore, J. (1988). Incomplete contracts and renegotiation. *Econometrica, 56*, 755–785.

Heaver, T., Meersman, H., Moglia, F., & Van de Voorde, E. (2000). Do mergers and alliances influence European shipping and port competition? *Maritime Policy and Management, 27*, 363–373.

Martin, J., & Thomas, B. J. (2001). The container terminal community. *Maritime Policy and Management, 28*, 279–292.

Murthy, N. T. R., & Notteboom, T. (2002). Mechanisms and degree of participation in private financed port infrastructure projects in Asia – Case in India. In: K. Misztal & J. Zurek

(Eds), *Maritime transport in a global economy* (pp. 31–49). Gdansk: Institute of Maritime Transport and Seaborne Trade.

Musso, E., Ferrari, C., & Benacchio, M. (2001). Co-operation in maritime and port industry and its effects on market structure. Paper presented at WCTR Conference, Seoul.

Notteboom, T. (2002). Consolidation and contestability in the European container handling industry. *Maritime Policy and Management, 29,* 257–269.

Notteboom, T., & Winkelmans, W. (2001). Structural changes in logistics: How do port authorities face the challenge? *Maritime Policy and Management, 28,* 71–89.

Rasmussen, E. (1994). *Games and information.* Cambridge: Blackwell.

Spence, A. M. (1973). Job market signaling. *Quarterly Journal of Economics, 90,* 225–243.

Turner, H. S. (2000). Evaluating seaport policy alternatives: A simulation study of terminal leasing policy and system performance. *Maritime Policy and Management, 27,* 283–301.

Van Niekerk, H. C. (2005). Port reform and concessioning in developing countries. *Maritime Economics and Logistics, 7,* 141–155.

Wang, Ch. (2000). Renegotiation-proof dynamic contracts with private information. *Review of Economic Dynamics, 3,* 396–422.

Wernerfelt, B. (2004). *Incomplete contracts and renegotiation.* MIT Sloan Working Paper no. 4506-04.

World Bank. (n.d.). Port reform toolkit. Retrieved from www.worldbank.org

CHAPTER 20

STAKEHOLDERS, CONFLICTING INTERESTS AND GOVERNANCE IN PORT CLUSTERS

Peter W. de Langen

ABSTRACT

In this chapter, an analysis is made of conflicting interests in seaports. Such conflicting interests are relevant since various stakeholders influence port development and have different goals. Based on an overview of the interests of different stakeholders, five conflicts of interests that are relevant in most seaports are identified. Each of these is briefly discussed and the concept of accommodations is presented to analyze these conflicts. As an illustration, the five accommodations in Rotterdam are discussed in some detail. In a concluding section, a research agenda for research on conflicting interests in seaports is discussed.

1. INTRODUCTION

The application of cluster theories to seaports yields a number of new insights (see de Langen, 2004a; Haezendonck, 2001). First, agglomeration economies frequently lead to the presence of clusters of economic activities related to the arrival of ships and cargo (de Langen, 2004a). These port

Devolution, Port Governance and Port Performance
Research in Transportation Economics, Volume 17, 457–477
Copyright © 2007 by Elsevier Ltd.
All rights of reproduction in any form reserved
ISSN: 0739-8859/doi:10.1016/S0739-8859(06)17020-1

clusters are often important specializations of the regional economy. Thus, the performance of port clusters is relevant for regional economic development. Second, traditional port activities, such as terminal operations and transport services, are mature activities with limited growth prospects in terms of employment and value added, even though in the container market, cargo volumes are growing substantially. Thus, a prospective development of port clusters requires growth in related economic activities, such as manufacturing and logistics (de Langen, 2005). This requires a transformation of ports to become attractive locations for logistics services. Third, the performance of port clusters depends on the quality of the governance of the cluster (de Langen, 2004b). Cluster governance issues (e.g. stakeholder management, investments in marketing, education and innovation and role of leader firms) are increasingly relevant.

This chapter develops a framework for analyzing conflicts of interests in seaports. First, cluster governance is defined and the relevant literature is discussed. Second, an overview will be given of the various stakeholders of a port cluster. Third, based on this overview and a review of the relevant literature, important conflicts of interests in port development are identified and briefly discussed. Fourth, a framework for analyzing such conflicts of interests is presented. Fifth, this framework is linked to discussions on governance models of port authorities (PAs), then some tentative illustrations of this approach are presented. A small survey in Rotterdam is used as illustration. A concluding section with a research agenda finalizes the chapter.

2. CLUSTER GOVERNANCE IN SEAPORTS

The concept of cluster governance differs substantially from the frequently used concept of corporate governance because, unlike a corporate hierarchy, a cluster consists of independent organizations with few formal control relations to govern their interactions.[1] In seaports, a distinction needs to be made between PA governance and port governance. The governance of the PA is closely linked with corporate governance issues, such as shareholder influence, structure of the board of governors and corporate social responsibility. Port governance, on the other hand, is more related to cluster governance since a port consists of a variety of actors. In some ports, the PA plays a large role in cluster governance, while in other ports, private (leader) firms are more important or port associations play a large role. For instance, firms may take the lead in establishing standards for exchanging information in some ports, while in others the PA takes the lead.

The quality of cluster governance depends primarily on two factors: the level of transaction costs and the scope of coordination in a cluster (see de Langen, 2004b). Lower transaction costs increase the competitiveness of firms in the cluster. The level of transaction costs is influenced by, among others, the presence of trust and intermediaries (such as a cluster association or a forwarder). An increased scope of coordination also increases the competitiveness of firms in the cluster. Owing to free-rider and monitoring problems, cooperation frequently does not develop, even though benefits are higher than costs. Collective action regimes and leader firms contribute to the scope of coordination in clusters (de Langen, 2004b). The analysis of how actors in ports aim to reduce transaction costs and cooperate (see de Langen & Chouly, 2004) adds to our understanding of governance in ports.

However, such an analysis only deals with the shared interests of the firms in the port cluster; all firms benefit from lower transaction costs and more cooperation. However, actors in ports are also faced with conflicting interests. Thus, an approach to understanding the interaction between ports and their environments requires attention for conflicts of interest that arise because the interests of the port cluster differ from, and may even be opposed to, the interests of other relevant stakeholders.

3. STAKEHOLDERS OF PORT CLUSTERS

Port clusters have a variety of stakeholders, defined as all actors that can affect or are affected by the achievement of the firm's objectives (Freeman, 1984). Each of these has different interests in the port and different sources of influence. Important stakeholders of ports include transport firms that are primarily interested in limited costs due to regulations (vehicle tax, emission standards and security regulations), which have a direct effect on their competitive position compared to transport services providers in other regions and countries. Furthermore, transport firms lobby for transport-friendly policies (infrastructure construction/limited congestion charges). The interests of port labour are job security, career opportunities and a high wage level. Port-related manufacturing industries are interested in the attractiveness of the seaport as location for manufacturing activities. Especially relevant are strong agglomeration economies, such as the presence of a large supplier and customer base, a good labour market and knowledge and information. Furthermore, they have an interest in securing a level playing field for environmental regulations, in order not to be disadvantaged compared with competitors in other locations.[2]

End users of ports (importers and exporters) are primarily interested in low generalized transport costs. Their interests are somewhat different from those of transport service providers, since end users are more sensitive to indirect cost factors, such as reliability, flexibility and damage prevention (Murphy, Daley, & Dalenberg, 1992). Local environmentalist groups are interested in minimal negative externalities, such as noise, pollution and safety. Environmentalists have more specific interests than local residents, who seek minimal negative externalities, job creation and waterfront housing and leisure development.

Regional government is especially interested in the contribution of the port complex to the regional economy and in the tax revenues of this complex. Furthermore, regional governments are interested in waterfront (re)development. National governments are interested in maximum welfare effects through low generalized transport costs and limited negative externalities. Furthermore, national governments (may) seek recovery of infrastructure investment costs.

Not all of these stakeholders have the same power. For instance, national governments may hardly be involved in some ports. Furthermore, the power of these different stakeholders differs substantially between seaports. Finally, the power of stakeholders also changes over time. Table 20.1 summarizes the main interests of stakeholders and their sources of influence.

4. CONFLICTING INTERESTS IN SEAPORTS

Based on the overview of interests of different stakeholders, a review of relevant literature and an analysis of annual reports of various PAs,[3] five important conflicts of interests can be identified. All these conflicts arise because interests of some stakeholders are not aligned with the main interest of actors in the port: economic development of the port cluster.

- Environmental protection versus port development
- Urban development versus port development
- Labour conditions (including wage, job security, union monopolies) versus port development
- Resident interests (safety, quality of life) versus port development
- Overall economic development versus port development

These conflicts, relevant in most seaports, are described in some detail below.

Table 20.1. Port Stakeholders, Interests and Different Sources of Influence.

Stakeholder	Interests	Sources of Influence	Indicators of Stakeholder Influence
Transport firms (including terminal operators)	Low generalized (trans)port costs, high quality of infrastructure, no (or limited) interference with logistics chain due to safety, security, product quality regulation and customs procedures	Lobbying through branch associations, diverting cargo to other ports	Existence of port-specific industry association Level of subsidies to sector (infrastructure pricing, projects to reduce congestion, etc.) and financing and functioning of customs and safety regulation
Port labour	High wages, job security	Port strikes, impact on image of working in seaports	Wage level Collective port wide labour agreements
Local port related manufacturing industries	Strong 'agglomeration economies' in cluster, space for manufacturing activities, level playing field regulation with regard to noise and environmental standards	Lobbying through branch associations, investing outside port cluster	Existence of port specific industry association
End users of ports	Low generalized transport costs, including factors such as reliability and damage control	Lobbying through branch associations, diverting cargo to other ports (limited)	Existence and role of port users association
Local environmental groups	Regulations preventing excessive negative local externalities, such as noise and pollution and spatial quality	Use of procedures to postpone/prevent investments such as port expansion of capital investments. Political pressure	Existence of local environmental groups Power derived from threatening to go to court/court actions
Local residents	Job creation in line with local labour market, limited traffic congestion, no reduction of 'quality of life' due to port	Political pressure	Existence of resident groups
Local and regional government	Contribution to regional economy, contribution to regional tax income, effective transformation of port/city interface	Regional planning, public investments in ports	Public land ownership Ownership and governance structure port authority
National government	Low generalized (trans)port costs for residents and firms, cost recovery of infrastructure	National investments in ports, creation of port laws	National role in infrastructure planning

Source: Author.

4.1. Environmental Protection versus Port Development

The tension between port development and environmental protection manifests itself in virtually every port. The issue is frequently mentioned in annual reports and in websites of PAs. The organization Ecoports, a network of European ports, is specifically established to share experiences in managing the environmental effects of seaports (Ecoports, 2005). Greenhouse gases, noise, water pollution and stench are produced in ports. Furthermore, ports generate a substantial volume of environmentally unfriendly (road) transport. Therefore, in most ports, environmental groups are active in promoting sustainability and a green living environment. Their interests are frequently opposed to those of the PA and port community. Already in 1979, environmentalism was found to be the most important problem faced by organizations involved in the development of seaports (Bird, 1984). In a more recent article, Gilman (2003) discusses the effects of environmental regulation on port development projects. He discusses how port development in the UK is influenced by national sustainability policies and emphasizes the conflict of interests:

> The appraisal system ... can accommodate measures of environmental mitigation and compensation and generally seeks such agreement between opposing interests as is likely to be available. (Gilman, p. 289)

His overall conclusion is that the public influence on port development (through various forms of regulation) in the UK is too large, given the fact that the port industry is a fully private one. Conflicts between environmentalism and economic interests can lead to delays in port expansion projects (a potential problem also raised by Gilman, 2003).

Environmental groups, PAs and port users can, and in some ports do, cooperate to develop a shared strategy. This does not imply that they do not have conflicting interests; it shows that they choose a cooperative instead of a confrontational strategy. Such an approach can have benefits, such as more support for development projects, smoother decision-making processes and more balanced projects. The port expansion project *Second Maasvlakte* in Rotterdam was characterized by such a dialogue and a joint vision by environmental groups and the PA (van Steekelenburg, Bilius, & van Muijen, 1998).

4.2. Urban Development versus Port Development

The tension between port and city has been discussed from different perspectives, such as spatial planning (Desfor & Jørgensen, 2004; Meyer, 1999), geography (Hoyle, 1996) and economics (Koide, 1990). Most seaport cities have witnessed a declining use of old docks and quays located in or close to the city centre. This is mainly explained by scale increases in terminals, storage facilities and ship sizes. Inner city port areas, sometimes dilapidated, have been rediscovered in the 1980s and 1990s as areas for urban redevelopment. In some ports (e.g. London, Vancouver, Antwerp), however, port facilities remain in the city centre; in this case, there is a conflict of land use, because land can be used for port functions as well as urban functions. Charlier (1992) describes a life cycle of port facilities and explicitly states that redevelopment for new port functions, also of inner city ports, may be viable. Pellegram (2001) describes the problems the Port of London Authority faces with preserving at least some facilities for port activities. He describes a process of crowding out of port facilities by real estate projects, arguing that this process has severe disadvantages, since it reduces use of environmentally friendly transport modes. This explains the involvement of the Secretary of State to secure land for port facilities.

Some PAs have moved into real estate development (Fleming, 1987; Port of Amsterdam, 2005; Port of Copenhagen, 2005; Rodrigue, 2004). This may be a sensible strategy for PAs to capitalize on high land values for urban development and a diversification of investments. The benefits may be re-invested in the port, making such a solution perhaps ideal from a port development perspective. However, it cannot be assumed revenues are always reinvested in port areas.

4.3. Labour Conditions versus Port Development

Many seaports have witnessed labour unrest. The British National Dock Labour Scheme, abolished in 1988 after a lengthy period of labour unrest, is perhaps the most famous example (Turnbull, 1991). Talley (2002) describes the relatively strong bargaining position of labour unions, especially in US West Coast ports. Both authors present arguments as to why this bargaining power is likely to diminish. The power of the union clearly influences the competitiveness of the involved ports: for some terminal operations, labour costs constitute as much as 50 percent of total operating costs (de Langen, Nijdam, & Van Enthoven, 2002). The issue of port labour is not relevant for dock labour alone. The conflict is also relevant for activities such as ware-housing and port-related manufacturing.

The presence of economic rents may partly explain the strong bargaining position of port labour and resulting wage level and labour conditions (Goss, 1999). Since ports frequently face limited competition, an economic rent may exist. PAs (excessive port dues), port service providers (excessive profits) as well as port labour (excessive wages), may compete to collect these economic rents.

4.4. Resident Interests versus Port Development

This conflict is in most cases related to conflicts of environment versus port development, and urban versus port development, but is treated separately because (a) environmental stakeholders are mostly interested in the environmental quality of port regions, not (only) in negative impacts for residents in the vicinity of the port and (b) the conflict of urban development versus port development generally focuses on restructuring old port areas for new functions. This issue also differs from interests of residents living in the vicinity of port areas.

Most ports have developed over a long period of time and were partly built when car ownership was not widespread and port activities were labour-intensive. Thus, housing was developed in the vicinity of port areas (Meyer, 1999). This has resulted in a mixture between housing and port functions,[4] and housing close to port activities. Such proximity leads to conflicting interests, for instance, with regard to noise levels, stench and safety. This can lead to a relocation of housing functions (e.g. Antwerp, where a small village was demolished to enable further port development) and/or to a relocation of port functions (e.g. Bombay, where new port facilities were developed outside the city).

4.5. Overall Economic Development versus Port Development

A final conflict is between regional economic development versus port development. Regions *have to* make choices that impact the development of various industries in the economy. For instance, an investment in freight infrastructure will benefit the port industry and manufacturing industries, while investments in passenger transport are more favourable for service industries. Interest groups try to influence these choices, to enhance the development of their industry. In many North European ports with a Hanseatic tradition (Kreukels & Wever, 1998), port interests traditionally are taken very seriously. They argue that the economic impact of the port is substantial and the port is an enabler of economic development in other sectors. These arguments may or may not justify continued investments in port development; the issue is that there is such a conflict of interests.[5]

5. ANALYZING ACCOMMODATIONS IN SEAPORTS

These conflicting interests are relevant in most seaports. The outcomes of such conflicts differ between ports, with substantial effects on the competitive position of the port. In some ports, governments may invest in a labour pool to solve labour conflicts, while in other ports such public investments are not made. As another example, in some ports, environmentalists may manage to prevent or delay port expansion (examples include Zeeland Seaport in the Netherlands and Dibden Bay in the UK), while port expansion goes ahead elsewhere. The different resolutions to these conflicts frequently alter the nature of the playing field. For instance, a government that subsidizes a labour pool clearly distorts the playing field. Such distortions can be prevented to some extent through international regulations (e.g. the European Union port package). However, it is virtually impossible to create a completely level playing field; in one port a government may be willing to compensate for the negative effects of a port expansion project, while a government in another port may not be willing to do so. More attention to conflicts of interests and consequences for port development is thus relevant. One could even go one step further to argue that the extent to which such conflicts can be resolved is a central explanatory factor in port competitiveness (in general, such an argument for proactive stakeholder management is taken by authors such as Burke (1999) and Jensen (1988)).

The concept of *accommodations* is proposed for analyzing conflicting interests.[6] These interests are accommodated in the sense of a settlement of differences (Webster's Online Dictionary, 2005). An accommodation is not a permanent resolution of a conflict. Because goals, strengths and strategies of the involved stakeholders change over time, an accommodation changes over time.

It is assumed accommodations can be analyzed fruitfully for a port as a whole. In some cases, micro-level accommodations (for instance, a labour agreement for one specific firm in the port) may be different from the port-wide accommodation. Port employers and unions may have a conflictive accommodation for the port as a whole, but some service providers may have no labour disputes. Thus, it may be necessary to analyze micro-level accommodations rather than analyzing accommodations for the port as a whole. Examples can include the redevelopment of a particular site, a particular port expansion project, and a specific agreement between residents and firms in the port. However, port-wide institutions are regularly created to accommodate these five conflicts of interests. These institutions influence various micro-accommodations. Examples of such institutions include specific laws governing port development, a labour pool, collective labour agreements, discussion platforms, general policy principles and port development plans. These institutions are relatively inert; they change, but slowly. For instance, once a shared labour pool is created, it is difficult to dismantle.

Fig. 20.1 shows four stylized accommodations, based on two variables. First, interaction can be either limited or frequent and second, behaviour of the relevant stakeholders can be either cooperative or non-cooperative. This leads to four different stylized accommodations.

All accommodations (temporary settlements of conflicts) can be classified in this framework. This applies both for the port-wide accommodations and the micro-level accommodations. It is important to note, however, that these accommodations change over time.

A *negotiated accommodation* arises if stakeholders are non-cooperative and interact regularly. Institutions for collective bargaining and interest representation are likely to develop, since stakeholders seek to maximize their influence and thus need to organize themselves. Thus, in such cases, unions, employers' associations, other industry associations, local environmentalist groups and resident platforms can be expected to be active. An *incidents accommodation* arises when interaction between different stakeholders is limited, and a non-cooperative attitude prevails. In this case, stakeholders have fewer incentives to organize themselves. An example could be the relationship between residents and adjacent firms that incidentally clash over negative externalities. A *silent accommodation* denotes

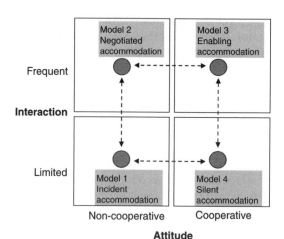

Fig. 20.1. Four Stylized Accommodations. *Source:* Author.

Table 20.2. Port Interests in Accommodations.

Conflict	Port Interests (Goals for an Accommodation)
Environmental protection versus port development	Securing port land
	Revenues of redevelopment of port areas accrue to port community
Urban development versus port development	Enabling port expansion
	Level playing field vis-à-vis competing ports
	Long-term license to operate of port cluster
Labour conditions versus port development	Reasonable labour costs
	Quality of labour, in terms of skills, flexibility and motivation
Resident interests versus port development	Space for port activities
	24/7 operations for port activities
Overall economic development versus port development	Public investments in port
	Investments in innovation, education, etc

that interaction is limited, and a cooperative attitude prevails. In such a case, institutionalized meetings/confrontations are limited, and actors only get together when there is an immediate cause. A *prospective accommodation* finally exists when actors cooperate and meet frequently. In such an accommodation, interaction is planned and supported by institutions.

The effects of accommodations on port performance and development have to be assessed case by case. One could assume cooperative accommodations to be more effective than conflicting ones, especially in ports where different stakeholders are increasingly capable of organizing themselves. Examples of conflicting accommodations (frequent strikes of port labour, environmentalists using courts to prevent port development, municipal governments constructing houses in the vicinity of the port) justify such an assumption, but clear scientific evidence is lacking. Table 20.2 shows the most important requirements for accommodations from the perspective of port performance.

5.1. Stakeholder Influence

Stakeholders influence the accommodation that is created and thus influence the development of the port. The analysis of stakeholder influence can lead to an overview of the stakeholder configuration in a specific port. Such a stakeholder configuration shows the power balance between various different stakeholders. Fig. 20.2 shows, as an illustration, four stylized port

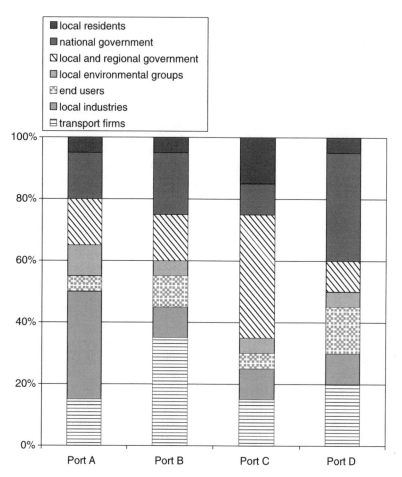

Fig. 20.2. Four Stylized Port Stakeholder Configurations.

stakeholder configurations. In the first configuration, Port A, local port industries are especially powerful. This may be the case in ports with many user-owned and managed terminals. As a consequence, such ports may develop an attractive utility infrastructure (water, energy, heat, etc.) and aim at attracting manufacturing investments. In the second configuration (Port B), transport firms are the most powerful stakeholders. Thus, the focus may be on infrastructure expansion and other means to reduce congestion. In the third configuration (Port C), local and regional governments are especially powerful. This is the case when the PA is a municipal department, such as in

Hamburg[7] and Amsterdam. This may lead to rent extraction (Goss, 1999), a focus on investments to create local economic benefits, or a reluctant attitude towards investments outside the region. The fourth configuration (Port D) shows a dominant national government. This may lead to introducing cost recovery in all ports across the country and port investments as an instrument to develop disadvantaged regions.

6. THE PA'S ROLE IN SHAPING ACCOMMODATIONS IN SEAPORTS

The PA is in most ports deeply involved in accommodating conflicting interests. Most PAs engage in port planning and deal with such aspects as land use policy and environmental standards. Furthermore, PAs in many cases regulate conditions under which firms can establish in the port. Finally, PAs may subsidize labour pools or develop special rules concerning port labour.

The governance model of the PA has direct consequences for the influence of different stakeholders, and thus on the accommodations in seaports. If, for instance, a port is municipally owned, it is likely to emphasize regional economic effects in the land use conflict. On the other hand, privately owned ports (PAs) would generally dislike environmental constraints and would be less inclined to invest in a prospective accommodation unless the influence of other stakeholders is large.

The development trajectories of different ports may be explained to a large extent by the quality of accommodations.[8] Especially interesting in relation to other contributions in this volume is the question of how the governance model of the PA influences the power of the different stakeholders, and consequently the development path of a port change over time. How do changes in the governance model (for instance, a transfer of responsibilities from public organizations to private firms or the appointment of a non-political board of directors instead of a political one) affect accommodations for important conflicts like land use? Does privatization change the labour accommodation in seaports? Such questions are addressed in the next section.

7. ILLUSTRATION: ACCOMMODATIONS IN THE PORT OF ROTTERDAM

In this section, the concepts discussed above are illustrated with an example from the port of Rotterdam. This explorative illustration is based mainly on desk research using relevant documents and previous research dealing with accommodations in the port of Rotterdam. Furthermore, relevant insights were derived from senior port experts that cooperated with this study.

The land-use accommodation in Rotterdam has changed radically over the last couple of years. Previously, the Port Authority of Rotterdam (PoR) was in charge of outdated and virtually abandoned port areas in the city centre. It had no incentives to transfer land in its jurisdiction to the city development corporation. As a consequence, the redevelopment of the waterfront started relatively late in Rotterdam (see Meyer, 1999). In the time between the disappearance of port activities and first redevelopments, port areas became wastelands in the centre of the city. This had a severe negative effect on the quality of life in nearby residential areas. Starting with the Kop van Zuid project, the waterfront was redeveloped and is slowly being transformed into an attractive location for leisure, housing and offices.

To better manage the transformation of other upriver port areas, Stads-havens was formed. This organization is a joint venture between PoR and the city development corporation and aims at a smooth transformation of the area from fully occupied with port activities to mixed land use. This new organization enables close cooperation between relevant port and urban stakeholders. This case shows the relevance of institutions; the joint venture is specifically set up to manage the transformation of the area and ensure that cooperation continues over time.

The labour accommodation in Rotterdam was analyzed by van Driel (1988). In the 1980s, the continuing decrease in demand for jobs led to labour redundancy, and threatened the jobs of many port workers who were employed by a labour pool, jointly owned by the large port companies. This led to social unrest, strikes and ultimately government subsidies to restructure the labour pool. Owing to the social unrest, Rotterdam lost most of its labour-intensive breakbulk activities to Antwerp. Collective agreements were made between port employers and port unions until the early 1990s. The collective bargaining gave labour unions a strong bargaining position and resulted in various costly arrangements, such as the agreement not to lay off employees involuntarily and the agreement to use only the existing labour pool. These arrangements drove up labour costs and reduced productivity and flexibility. By the end of the 1990s, employers refused to create new port-wide arrangements. Currently, labour costs of many terminal operating companies are still relatively high and productivity lags behind that of other ports (de Langen et al., 2002). Since 2000, interaction is limited to incidental conflicts, and there is no platform for addressing joint issues.

The port labour pool still exists and discussions on its restructuring are ongoing. This shows the resilience of institutions. The port-level accommodation has deeply influenced the labour relations of individual firms, and most firms still struggle to create better labour relations.

The sustainability accommodation in Rotterdam is deeply rooted in the Dutch tradition of creating consensus among all relevant stakeholders. A cornerstone of this accommodation is the twin goal of economic development of the port and improved sustainability. This twin goal was the basis for the initiative ROM-Rijnmond (http://www.rom-rijnmond.nl), a project that aimed to increase both the economic performance and the ecological quality in the port of Rotterdam. This project is cofinanced substantially by the province of Zuid-Holland and the municipality of Rotterdam. The success of this initiative led relevant stakeholders to the development of a joint vision and action statement that describes how port expansion can be realized, while improving sustainability and environmental quality of the port area. This joint vision was the basis for the port expansion project, *Second Maasvlakte*, and led to support for this project from environmental groups – a rather unique state of affairs. This support is partly attributable to substantial compensation projects to improve the environment in the region, funded by the national, regional and municipal governments. This support continued even after a court ruling halted the expansion plans after complaints by farmers and fishers. This case shows both the scope for creating effective accommodations (even though interests differ substantially), and that effective accommodation can be supported by funding from outside the port community.

The residents versus port development accommodation is illustrated by the village of Heyplaat, seen as a hindrance to port development by port planners in the 1970s, who argued the whole village had to be destroyed. This never happened and, because of the transformation of the inner city harbours, Heyplaat is now regarded as a basis for new urban development. A second small city in the port area, Pernis, is surrounded by chemical plants, which leads to very tight safety regulations for these plants. Conflicts hardly occur; residents are somewhat used to living in the vicinity of the port, and firms in the port (see e.g. Shell, 2005) invest in a license to operate.

The economic development versus port development accommodation is relatively cooperative in Rotterdam. The port cluster is one of the largest clusters in the region (de Langen, 2004a) and its economic impact is substantial. The municipality of Rotterdam has, over the last decades, invested substantially in the development of the port; it owns the land in the port and the PA. The PA financially contributes to the municipal government (on average about EUR 30 million). Furthermore, the municipality invests sub-stantially in transport infrastructure, marketing and education for the port cluster. Finally, the municipality has agreed to sell a minority share of the PA to the national government to enable the expansion project, *Second Maasvlakte*. Overall, the political and societal support for the port cluster is substantial.

Table 20.3 and Fig. 20.3 summarize this description of the five accommo-dations in Rotterdam. The position in the quadrant in Fig. 20.3 is not cal-culated on the basis of detailed data or comparison, but estimated based on qualitative analysis of the accommodations. The size of the circle indicates relative importance of the accommodations, larger being more important.

A thorough analysis of changes in the role and influence of stakeholders over time is beyond the scope of this paper. Results from a very small survey on changes in the stakeholder configuration are presented. Four managers of PoR were asked to indicate the importance of various stakeholders over time by dividing 100 points over these stakeholders. The results are given in Table 20.4.

A clear, common trend emerges from the four responses: the importance of local and regional government declines, as does the importance of trans-port firms, while local industries, end users and the national government gain importance. These changes can partially be explained by changes in the governance model of the PA; PoR was commercialized, albeit still publicly owned in 2004. These results provide an indication of likely changes in investment policies of PoR; less investment in projects aimed at generating regional economic growth and more investment intended to reduce gener-alized transport costs for port users (and transport firms), and to improve the location profile for manufacturing activities. Furthermore, the economy versus environment accommodation can be expected to change as a result of the increased influence of local residents.

8. PORT CLUSTERS, STAKEHOLDERS AND GOVERNANCE: A RESEARCH AGENDA

The issues addressed in this chapter have not been explored empirically. Loose descriptions of how conflicting interests are played out in ports have been presented (Barton & Turnbull, 2002; Beresford, 1995; Finney & Young, 1995). Such studies yield new insights for port policy and management and improve our understanding of determinants of port competitiveness, but be-come outdated and do not provide much theoretical advancement. Thus,

Table 20.3. Accommodations in Rotterdam.

Conflict	Relevant Organizations	Attitude	Interaction	Importance for Port Development	Role PoR
Environmental protection versus port development	Regional environmental organization (ZMF) DCMR, environmental agency	Cooperative, joint vision developed in an early stage of planning process, DCMR aims to be a firm-friendly agency	Infrequent, after joint vision was developed	Very important, port expansion is very important	Signed agreement with environmental stakeholders
Urban development versus port development	City government, Development agency 'Stadshavens'	Cooperative, port development is one of the goals for the regional economy	Frequent, institutionalized in 'Stadshavens' a partnership of the port authority and urban-development corporation	Important, city is a key partner in port development	50% share in 'Stadshavens', effort to relocate firms to downstream locations
Labour conditions versus port development	Labour unions (FNV and CNV)	Conflictive, disagreement on role of labour pool and labour regulation	Infrequent, only reactive	Important, labour costs are substantial, especially for terminal operators and transport firms	Funding for education and training, no active involvement in accommodation
Resident interests versus port development	Municipalities in port area	Cooperative, municipalities and residents are aware of importance of port development Actors in port arrange for visits of schools in the area	Infrequent, some firms arrange meetings	Important, a license to operate is necessary for port development	Investments in port events and visual quality of the port area
Overall economic development versus port development	City and province government	Cooperative, strong government support for port cluster	Frequent, city is actively involved in securing port interests	Important, city creates important conditions for port development.	Safeguarding support from city government, a.o. by focus on local economic development

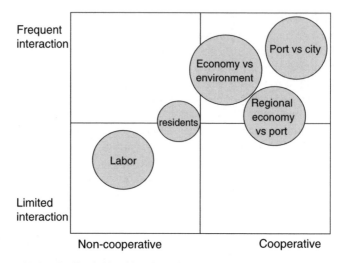

Fig. 20.3. Stylized Classification of Accommodations in Rotterdam.

Table 20.4. Stakeholder Importance in the Port of Rotterdam; An Explorative Survey ($n = 4$).

Stakeholder	Importance 5 Years Ago	Current Importance[a]	Importance in 5 Years
Transport firms	32.5	26.8	22.5
Local industries	15.0	18.3	20.0
End users	2.5	2.5	8.8
Local environmental groups	7.0	8.8	7.8
Local and regional government	30.0	21.3	14.3
National government	7.5	16.3	19.3
Local residents	5.5	6.3	7.5
Total	100	100	100

[a]Survey conducted in January 2005.

international comparative empirical studies using a well-developed research design would be interesting. Accommodations, stakeholder configurations, governance models and investment policies of the PA obviously differ but can be fruitfully compared. Such an analysis could reveal best practices in port management. Relevant hypotheses could be:

- National PAs are more likely to engage in conflictive land use accommodations than municipally owned PAs.
- Private PAs invest less in a prospective economy/environment accommodation.
- Municipally owned PAs in relatively port-dependent regions can successfully attract municipal resources to port development, even though this allocation of public resources is not effective for regional economic development.

Second, the relationship between accommodations and port performance could be an interesting avenue for further research. How do accommodations influence port performance? Under what conditions should port managers aim to create a prospective accommodation? Changes of the importance over time of stakeholders and their effects on accommodations and port performance also deserve attention.

Finally, such comparative studies could address the issue of how accommodations change over time and how such changes affect port performance. One particularly relevant question seems to be, "What capabilities do PAs need to develop to ensure accommodations are adapted to the changing environment?"

NOTES

1. Internationalization of the port industry does *not* reduce the importance of 'local' governance (see Wolfe & Gertler, 2004). Port clusters consist of increasingly international firms (terminal operators, shipping lines, forwarders, manufacturing firms). These global players are in most cases hardly 'embedded' in one port but serve customers in different ports and have access to international (standards in) technology (e.g. ICT systems, port equipment) and management skills. Thus, firms in different port clusters are increasingly similar and consequently *differences in firm specific capabilities diminish*. This may imply the relevance of cluster governance as a potential source of competitive advantage increases.

2. Manufacturing industries and other business interests generally prefer no regulation whatsoever, but such a position is seldom officially taken. Instead, they urge for a level playing field.

3. Thirty-two annual reports were studied, mostly from ports in Europe (12), North America (14) and Australia and New Zealand (6). Most Asian ports do not publish an (English) annual report.

4. In Rotterdam, for instance, a small village (Heyplaat) was built in the middle of the port area, to provide housing near the shipyard Rotterdamse Droogdok Maatschappij. In other ports, similar mixtures of port industries and housing were developed.

5. One could argue that governments decide rationally on economic development policies, but most scholars of public management acknowledge the importance of 'interest groups' and coalitions (Posner, 1974).

6. See Bowles and Gintis (1986) for a similar use of the term accommodations.

7. The institutional structure of the Port of Hamburg changed at the end of 2005, when the PA became an autonomous publicly owned organization.

8. This argument is used in a much wider context by writers such as North (1990).

REFERENCES

Barton, H., & Turnbull, P. J. (2002). Labour regulation and competitive performance in the port transport industry: The changing fortunes of three major European seaports. *European Journal of Industrial Relations, 8,* 133–156.

Beresford, A. K. C. (1995). Redevelopment of the port of Cardiff. *Ocean & Coastal Management, 27,* 93–107.

Bird, J. H. (1984). Seaport development: Some questions of scale. In: B. S. Hoyle & D. Hilling (Eds), *Seaport systems and spatial change; Technology, industry and development strategies* (pp. 21–42). Chichester, UK: Wiley.

Bowles, S., & Gintis, H. (1986). *Democracy and capitalism: Property, community, and the contradictions of modern social thought.* London: Routledge & Kegan Paul.

Burke, E. M. (1999). *Corporate community relations: The principle of the neighbor of choice.* Westport, CT: Quorum Books.

Charlier, J. (1992). The regeneration of old port areas for new port users. In: B. S. Hoyle & D. A. Pinder (Eds), *European port cities in transition.* London: Belhaven Press.

de Langen, P. W. (2004a). *The performance of seaport clusters: A framework to analyze cluster performance and an application to the seaport clusters in Durban, Rotterdam and the lower Mississippi.* Ph.D. thesis, ERIM, Rotterdam.

de Langen, P. W. (2004b). Governance in seaport clusters. *Journal of Maritime Economics and Logistics, 6,* 141–156.

de Langen, P. W. (2005). Relevant trends and resulting opportunities for the long term development of Rotterdam's port complex. *Coastal Management, 33,* 215–224.

de Langen, P. W., & Chouly, A. (2004). Hinterland access regimes in seaports. *European Journal of Transport and Infrastructure Research, 4*(4), 361–380.

de Langen, P. W., Nijdam, M. H., & Enthoven, D. Van. (2002). *Arbeid in de Rotterdamse haven.* Unpublished research report. Erasmus University, Rotterdam.

Desfor, G., & Jørgensen, J. (2004). Flexible urban governance: The case of Copenhagen's recent waterfront development. *European Planning Studies, 12,* 479–496.

Driel, H. van (1988). *Strategie en sociaal beleid in het stukgoed.* Unpublished research report. Erasmus University, Rotterdam.

Ecoports. (2005). Information retrieved from www.ecoports.com

Finney, N., & Young, F. (1995). Environmental zoning restrictions on port activities and development. *Maritime Policy and Management, 22,* 319–329.

Fleming, D. K. (1987). The port community: An American view. *Maritime Policy and Management, 14,* 321–336.

Freeman, R. E. (1984). *Strategic management, a stakeholder approach.* London: Pitman.

Gilman, S. (2003). Sustainability and national policy in UK port development. *Maritime Policy and Management, 30*, 275–285.

Goss, R. O. (1999). On the distribution of economic rent in Seaports. *International Journal of Maritime Economics, 1*, 1–18.

Haezendonck, E. (2001). *Essays on strategy analysis for seaports*. Leuven: Garant.

Hoyle, B. (Ed.) (1996). *Cityports, coastal zones and regional change: International perspectives on planning and management*. Chichester: Wiley. J

Jensen, M. (1988). Takeovers, their causes and consequences. *Journal of Economic Perspectives, 2*, 21–44.

Koide, H. (1990). General equilibrium analysis of urban spatial structure: The port-city model reconsidered. *Journal of Regional Science, 30*, 325–347.

Kreukels, A. M. J., & Wever, E. (Eds) (1998). *North Sea ports in transition: Changing tides*. Assen, Netherlands: Van Gorcum.

Meyer, H. (1999). *City and port: Urban planning as a cultural venture in London, Barcelona, New York, and Rotterdam: Changing relations between public urban space and large-scale infrastructure*. Utrecht: International Books.

Murphy, P., Daley, J., & Dalenberg, D. (1992). Port selection criteria: An application of a transportation research framework. *Logistics and Transportation Review, 28*, 237–255.

North, D. (1990). *Institutions, institutional change and economic performance*. Cambridge: Cambridge University Press.

Pellegram, A. (2001). Strategic land use planning for freight: The experience of the Port of London Authority 1994–1999. *Transport Policy, 8*, 11–18.

Port of Amsterdam. (2005). Annual report. Available at www.portofamsterdam.com

Port of Copenhagen. (2005). Annual report. Available (in Danish) at www.cphport.dk

Posner, R. A. (1974). Theories of economic regulation. *Bell Journal of Economics and Management Science, 5*, 335–358.

Rodrigue, J. P. (2004). Appropriate models of port governance: Lessons from the Port Authority of New York and New Jersey. In: D. Pinder & B. Slack (Eds), *Shipping and ports in the 21st century*. London: Routledge.

Shell. (2005). Milieuverslag Shell Pernis 2004. Available (in Dutch) at www.shell.com

Steekelenburg, A. van, Bilius, M., & Muijen, M. van (1998). Spiegelnota: t.b.v startnotitie PKB/MER mainportontwikkeling Rotterdam/Projectteam Duurzaam Rijnmond.

Talley, W. K. (2002). Dockworker earnings, containerisation, and shipping deregulation. *Journal of Transport Economics and Policy, 36*, 447–467.

Turnbull, P. (1991). Labour market deregulation and economic performance: The case of Britain's docks work. *Employment & Society, 5*, 17–35.

Webster's Online Directory. (2005). Available at http://www.m-w.com

Wolfe, D., & Gertler, M. (2004). Clusters from the inside and out: Local dynamics and global linkages. *Urban Studies, 41*, 1071–1093.

CHAPTER 21

EU PORT POLICY: IMPLICATIONS FOR PORT GOVERNANCE IN EUROPE

Athanasios A. Pallis

ABSTRACT

This chapter analyses the consequences of the European Port Policy (EPP) for the ports in Europe. A series of European Union (EU) policy proposals aims to redefine both intra- and inter-port competition, in order to assist the adjustment of EU ports to the challenges of a complex environment. While the search for EU rules that would satisfy a sector of remarkably diverse structures continues, the analysis of the initiatives under discussion hints at contradictory potential impacts on port governance. Apart from the generated tensions between different levels of authorities, or the diverse demands of stakeholders, the risk is that each EU initiative may still pull the configuration of the environment–strategy–structure triangle into an alignment that does not enhance the performance of ports in Europe.

1. INTRODUCTION

Unlike other parts of the transport sector, European Union (EU) ports have developed in their own diverse ways, reflecting their markets and national

Devolution, Port Governance and Port Performance
Research in Transportation Economics, Volume 17, 479–495
Copyright © 2007 by Elsevier Ltd.
ISSN: 0739-8859/doi:10.1016/S0739-8859(06)17021-3

characteristics. The setting up of a single European market in the 1990s was a catalyst for the development of a European Port Policy (EPP). EU ports handle on average 3.5 billion tonnes of cargo annually. Growth statistics aside, their potential as reliable handling, service, distribution and logistics centres coordinating maritime and inland transport is of great importance. Several ports are also key local economic generators connecting peripheral regions and increasing the use of maritime transport at the expense of less environmentally friendly modes.

At the same time, competition between ports intensifies due to various factors, such as economic globalisation and the increasing use of shipping, transhipment and logistics. The disappearing national character of hinterlands means that powerful users condition their loyalty on the presence of flexible forms of specialised services (Notteboom & Winkelmans, 2004) and global terminal operators seek to increase their stake in EU ports (Slack & Fremont, 2005). Decisions of a port regarding internal structures, development, pricing and financing have market effects on an international scale.

The scale of these developments contributes to blurring of the national boundaries of port policy making. Initially aiming "to give special attention to the maritime economy as a whole" (Commission of the European Union – CEU, 1991, p. 3), later in the context of promoting shortsea shipping (CEU, 1995), policy actors embarked on a discussion to develop an EPP. In 1997, the Commission launched a wide-ranging debate that remains the core of the EU port policy agenda (CEU, 1997a). Its themes include intra- and inter-port competition, such as (free) market access, (transparency of) financing, and (harmonisation of) charging systems, the modernisation of infrastructure and the integration of ports in the multi-modal trans-European transport networks (TEN-T), and the enhancement of maritime-traffic-related safety and security. While a comprehensive, all-embracing EU policy does not exist, this debate signifies the search for a long-term strategy (Chlomoudis & Pallis, 2002).

Tensions are high and the answers are as diverse as the sector itself. Nonetheless, the resulting legislative proposals will modify the operating environment of European ports with international traffic. The solutions that have been offered have certain implications for port governance. This chapter aims to shed light on these implications, concentrating on the consequences that the main EU policy initiatives have for the configuration of the environment, structures and strategies of European ports.

2. ADVANCING INTRA-PORT COMPETITION

2.1. Structural Changes and Strategy Implications

The cornerstone of the EU strategy is a proposal for a port services directive that has been in discussion since 2001 (CEU, 2001a) and aims to establish the widespread application of intra-port competition (currently limited in some EU member states) and the creation of a level playing field between EU ports.

In practical terms, the directive will open access to the provision of techno-navigational, cargo-handling and passenger services, on the basis of transparency and non-discrimination. The right to grant authorisations is maintained by national governments while the level of investment contemplated by the service provider will determine the maximum length of each authorisation. The number of service providers might be limited only in certain circumstances (e.g. public service on safety grounds).

The aim is to ensure free competition among providers to offer services ranging from pilotage to cargo handling within each port and, in particular, allow for self-handling by users. The right of port users to cargo self-handling remains a controversial issue. Under current national rules, port authorities in Europe are required to employ only local union labourers in cargo handling, the exemption being dedicated terminals. Shipping companies and some ports would like casual labourers or ship crews to perform these services. A second main area of dispute relates to pilotage, with stakeholders expressing concerns regarding degraded safety implications of a potential liberalisation of this service.

Following a lengthy consultation, the European Parliament rejected a compromise agreement (2003) and the Commission published a new proposal (CEU, 2004). New elements consist of a stricter and mandatory translational regime regarding authorisations, shorter maximum durations for each authorisation and a new and broader definition of self-handling. If adopted, it will be applied to all EU ports with average annual maritime traffic of not less than 1.5 million tonnes of freight and/or 200,000 passengers.

Although this new proposal increases the complexity of the operating environment, the easier access to port services market and the removal of discrimination during the process of granting and implementing concessions may bring in new players and oblige the introduction of competition between providers of the same service within a port.

Intra-port competition might redefine European ports' strategies in several positive ways (de Langen & Pallis, 2006). On one hand, it minimises any (private or public) strategy of monopolistic rent seeking, which is not rare as inter-port competition is imperfect (Goss, 1999). On the other, it eliminates the potential of price discrimination (i.e. according to demand elasticity). Switching to a second-best port leads users to higher transport costs (Bichou & Gray, 2004), so in the absence of intra-port competition, port service providers retain excessive bargaining power versus those shippers located in the captive hinterland.

Intra-port competition also favours strategies based on innovation and specialisation by enabling different service providers to operate within multiple post-Fordist worlds of production and offer both generic and dedicated services (Chlomoudis, Karalis, & Pallis, 2003). This is important because integrating ports with the land-based segments of the transportation chain requires specialised service providers that are effectively networked rather than being focused merely on operations and the ownership and control of the assets (Robinson, 2003). This presence helps the essential proactive approach to user demands (Paixao & Marlow, 2003), while the networking of these actors creates a competitive spirit that minimises difficulties in integrating ports with supply chains (Fleming & Baird, 1999).

The proposed EU framework also redefines the role of most port-managing bodies in their dual function of port authority and service provider. First, there is the legislative request for ports to allow competing service providers to enter the market should they wish to do so. Then, whenever the managing body provides services, it should be treated like a competitor: it would neither be involved in the selection of providers, nor discriminate between service providers in which it holds an interest.

The request for a separate competent authority to administer the whole process alters the conventional port authority concept. In the short term, the transfer of critical decisions away from the governing body of the port, which is the party with practical experience with the port it supervises, might result in duplication of functions and conflicts of interest. In the medium term it might also result in delays due to the lack of know-how.

In the long term, however, port authorities have to act as the smart institution that governs the implementation of new organisational forms (Chlomoudis & Pallis, 2004). EU ports will be in a situation in which a variety of actors need autonomy to develop various frameworks of action. At the same time, those involved in the provision of port services will be in a locked-in situation wherein they need an actor to ensure that these frameworks of action are well developed in a coordinated manner. This governor will guarantee that the autonomy of both individual and collective actors is respected through the presumption that their resources and, most important, their frameworks of action deserve equal consideration.

From an organisational point of view, the port authority has to undertake the responsibility to administer coherent (collective) attempts to overcome both inefficient operations and the failures of the market mechanism. It has to set the targets in cooperation with several partners, direct the process by defining the operational framework and advance networking. The latter role emerges as one of the most critical functions when policy-makers favour the idea of allowing several independent enterprises to operate within a single port, as networking of related firms may gain or sustain a competitive advantage vis-à-vis their competitors.

2.2. Difficulties of Implementing a 'One Size Fits All' Approach

Intra-port competition is already taking place in some EU countries (e.g. UK). In some cases, the right of autonomous port authorities to invite additional providers, and the consequential presence of intra-port clusters (de Langen, 2002) are part of a long-term port tradition. Still, intra-port competition in other member states is still just a possibility (e.g. Greece). In a context of different models of port governance throughout Europe, the directive provides well-defined market rules, diminishes uncertainty and ensures transparency.

The proposed reform creates several concerns. For instance, the pro-free-market private port operators and port authorities have advocated flexibility in the transposition of the rule into national legislation (ESPO & FEPORT, 2002). The problem is not only that each port is unique, but also that the different port traditions (Fig. 21.1) generate diverse contemporary port management and organisation strategies.

Different national frameworks denote heterogeneous consequences by the potential implementation of the same directive in 368 ports in 18 different countries (Table 21.1). Controversies began as a Continental versus Anglo-Saxon conflict (McConville, 2002), a conflict that is also apparent in other port policy discussions (i.e. port pricing). The existence of (quasi)open-market access regimes in some North European ports adds to the Anglo-Saxon thoughts that the directive might induce a number of disturbing structural changes.

Member states that have progressed to an open market (e.g. UK) would like to avoid the potential crowding out of private investments, due to a

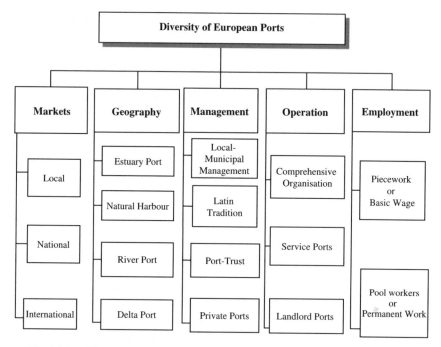

Fig. 21.1. Diversity of European Ports. *Source:* Adapted from Pallis (1997).

restart of the authorisation process (Farrell, 2001). Other member states (e.g. Belgium, the Netherlands, Scandinavian EU member states) have opted for devolution regimes, wherein the regulation of port services falls under the autonomous port authorities. These national administrations are reluctant to have structural changes imposed by a third party. Existing long-term contracts will have to be prematurely terminated depending on external circumstances, and this could reduce capital mobilisation towards port investments and may lead to legal uncertainty (van Hooydonk, 2005).

There are fewer problems in those cases where national reforms have favoured regimes similar to the one proposed by the port services directive. Conforming to the initial proposal, Spain introduced a law in 2003 that distinguishes between basic services (not provided by port authorities) and services of general interest (Orru, 2005). Italy's policy reform introduced a framework where port authorities manage but, in general, do not provide port services. National administrations in the early stages of restructuring state-controlled port systems (i.e. Greece and new EU members) favour the proposed (de)regulatory framework; it purposefully allows them to use the

Table 21.1. Total Freight and Passenger Movement by Port Services Directive (PSD) and All Ports, by EU Member States (in '000).

Member State	No of PSD Ports	Total Freight Tonnage By			No of PSD Ports[a]	Total Passenger Movements By		
		PSD Ports	All Ports	PSD as Percent of Total		PSD Ports	All Ports	PSD as Percent of Total
Belgium	4	178,689	181,110	99	0	591	739	80
Cyprus	2	4,823	7,220	67	0	287	287	100
Denmark	14	77,061	103,954	74	36	47,352	48,653	97
Estonia	4	43,467	44,682	97	0	5,172	5,172	100
Finland	16	90,633	104,439	87	2	15,865	16,341	97
France	19	308,767	319,032	97	8	27,148	27,405	99
Germany	16	246,349	254,834	97	22	30,009	32,146	93
Greece	20	132,870	–	–	0	–	–	–
Ireland	4	36,927	44,919	82	2	3,567	3,747	95
Italy	38	457,789	477,028	96	27	79,831	82,576	97
Netherlands	10	407,421	410,330	99	0	1,981	2,015	98
Norway	11	148,486	190,034	78	1	4,598	4,656	99
Poland	5	50,712	51,020	99	0	2,886	3,188	91
Portugal	5	53,888	57,470	94	0	–	–	–
Slovenia	1	10,720	10,788	99	0	–	–	–
Spain	24	321,651	326,001	99	2	19,510	20,041	97
Sweden	25	136,626	161,454	85	5	32,263	32,748	99
UK	40	526,869	555,662	95	5	33,009	33,708	98

Source: UK Department for Transport (DfT) (2005) (Annex D).
[a]In addition to ports that already meet the freight traffic threshold.

EU ties and commitments to accelerate national reforms. For them, over-coming deficiencies via industrial restructuring stands as more important than the implications of the specifics of the EU policy (Pallis & Vaggelas, 2005).

2.3. Port Level (Temporary?) Distortions

The specifics of the proposed directive bring forth practical implementation problems at port level. The negative effects are associated with bureaucracy, transformation of commercial into administrative relations and hampering of ongoing port level adjustments. Authorisations of a mandatory (rather than freely negotiated) nature might create bureaucracy. The lowering of maximum durations for authorisations[1] questions the potential for a normal return on investments in port facilities and equipment. Given that author-isations are not renewable, this might result in a counterproductive legis-lative bidding procedure.

This policy initiative might also undermine (or destabilise) the strategies of several port authorities that have concentrated on long-term cooperation with terminal operators. These relations might deteriorate if existing service providers see their investments cut short and face new tender procedures. The new regime might give rise to complaints about unfair competition: newcomers might take advantage of market opportunities developed by existing providers and win tenders without compensating the original en-trepreneur for the market risk and client establishment costs he had to bear. Moreover, the need for faster return on investments will strengthen the market power of large companies that offer services in EU ports, challeng-ing further the contestability of an already concentrated European cargo-handling market (six companies[2] control nearly 70 percent of the market). The new regime also risks higher prices and cargo concentration on fewer ports due to lack of stimulation from new entrants.

For those ports that will have to reintroduce intra-port competition, the situation can have negative effects on innovation and modernisation, un-dermining the potential of achieving the targeted efficiency. Still, due to the fact that the current performance in these European ports differs substan-tially (Table 21.2), their expectations regarding the directive, or the impli-cations of any other (de)regulatory policy reform, in improving efficiency vary considerably. As lower productivity levels are observed in the Med-iterranean region, where intra-port competition is limited, a certain degree of regulatory diversity should be allowed.

Table 21.2. Comparative Terminal Productivity.

Ports	Quay Length (m)	Terminal Area (ha)	1999 Throughput ('000 TEU)	TEU per Meter of Quay	TEU per Hectare
Felixstowe (UK)	2,523	137	2,697	1,069	19,684
Rotterdam (NL)	6,800	362	6,343	933	17,522
Thamesport (UK)	650	27	492	757	18,226
Bremerhaven (D)	3,000	167	2,181	727	13,060
Southampton (UK)	1,357	62	921	679	14,859
Hamburg (D)	5,700	265	3,738	656	14,107
Antwerp (B)	7,918	366	3,614	456	9,875
Zeebrugge (NL)	2,350	146	850	362	5,823
Le Havre (FR)	5,241	190	1,378	263	7,255

Source: UK Department for Transport (DfT) (2005) (Annex, D).

With respect to self-handling, this is more an issue that creates short- and medium-term social unrest. According to port authorities and operators, this issue should be settled locally. Shipowners favour the provision services on the ship by the crew, aiming to minimise operational costs and improve efficiency. Shippers advocate their right not to be obliged to contract with any third party that has the appropriate personnel and physical means, but enjoys a dominant position. Port workers express forceful objections, feeling that this rule will result in the employment of less qualified, cheaper labour, and lead to deteriorating safety standards. They continue to fight "a war on Europe's waterfront" (Turnbull, 2006). With the Commission believing that self-handling will enhance the presence of shipowners' capital in ports, but the European Parliament re-endorsing the view that this provision should be excluded from the directive, the prospects for establishing this right are still questionable.

3. REDEFINING INTER-PORT COMPETITION

3.1. Financing and Charging Port Operations

The port services directive is also associated with the creation of a level playing field for EU ports. Trends in port organisation and financing, particularly the increasing participation of the private sector, motivated EU institutions to consider ports as commercial operators that have to compete on a level playing field. Therefore, an EU framework regarding financial flows between public authorities, ports authorities, operators, users as well as the charging methods for port services lies at the centre of the EU interest for more than the past decade.[3]

Once more, the multiplicity of institutional structures and the difficulty of identifying the extent to which existing practices affect port competition present two major obstacles. To date, most member states intervene to ensure/facilitate port restructuring via financial assistance that assumes several forms, such as offset of operating losses, loans on privileged terms (a common practice in public or municipal ports), profits not subject to tax, leasing of land under privileged conditions, and free provision of administrative and technical services. Ports where investments and managerial decisions, including charging, are to various extents dependent on or influenced by public bodies, handle approximately 90 percent of the EU maritime trade (European Sea Ports Organization (ESPO), 2005).

The EU is trying to determine what, if any, form of state aid to the port sector is in Community interest. The transparency of public financial flows is a main restrictive factor in relation to the formulation, or even the rejection, of a specific policy. Ports are currently monitored by different accounting systems, and their financial accounts cannot provide aggregate information on the flow of public funds.

In order to implement its approach, the Commission initiated successive inventories. The first concluded in 2001 (CEU, 2001b), and the second will conclude in late 2005, despite limitations linked to limited information and confidentiality problems. According to the first inventory, public financing exhibits significant variation between EU port regions. In some cases it focuses on infrastructure modernisation, elsewhere on new infrastructure construction. The highest rate of public financing is observed in the North Sea, where almost 50 percent of the EU traffic is handled. The Mediterranean is experiencing high growth rates of public investment in port infrastructure. In the Baltic region there is a boom of financial support for more commercially oriented investments. The share of total public spending in Atlantic EU ports is the lowest in the EU, with a clear orientation towards support for investments in superstructure and typical start-up investments in services provision.

These differences render the design of such an EU policy impractical. Despite this setback, in the context of the latest port services directive draft, the Commission has undertaken the task of drawing up common guidelines for funding ports and indicating the cases in which funding of ports is compatible with the principles governing the internal European market.

Using the analogy of airport guidelines, any state aid is likely to be declared as incompatible with the European market and a distortion to competition. However, exemptions may be granted provided the relevant authority (i.e. European Commission) has been notified, and certain criteria have been met (e.g. financing infrastructure to be used by all rather than just some users).

The EU aims to introduce in parallel a common user-pay charging system, but this is still under development. The Commission is considering a charging system based on marginal costs, which would grant the full recovery of (direct and indirect) costs by placing on port users the full social marginal cost of infrastructure including operating, maintenance and external costs (CEU, 1997a, 1997b). Both theoretical (Bergantino, 2002) and practical problems (Meersman, Van de Voorde, & Vanelslander, 2003) lead to the argument that there is no single solution to the problem of port pricing (Pettersen-Strandenes & Marlow, 2000). Besides, although port tariffs in Europe are still based on a mix of pricing strategies, it is notable that most port authorities expect that the potential adoption of full recovery pricing would have a limited impact on pricing levels, and only for some cargoes (i.e. containerised and Ro-Ro; Haralambides, Verbeke, Musso, & Benacchio, 2001). At the moment, the EU policy is quite far from the objective of developing specific port pricing reforms in the EU, although the progress in the case of road and rail transport might eventually result in some proposals for discussion.

3.2. Ports and the Grand Transport Design

The goal of EU transport policies is higher efficiency in the transport system through interconnection of different modes. It also intends to eliminate barriers to uninterrupted network access via the interoperability of existing links through technical harmonisation. This design insists on the expansion of those modes that have the lowest energy demands and the least negative effects on the environment (i.e. shortsea shipping). By mobilising both direct and indirect (via public–private partnerships) public funding of some, rather than all EU ports, this project also has implications for port governance.

The initial plans for a unified trans-European transport network excluded ports. The potential plan of designated ports of EU interest was seen as a discriminatory practice against those ports that would not be included (Pallis, 1997). However, the need to integrate transport modes created de facto conditions for reversing this decision. For example, integrating a trans-European network of railways without including in the plan a list of ports was impossible (CEU, 1997b). Even though it affects the competitive conditions and the practices of stakeholders, this mobilisation is considered a way to overcome externalities and enhance, rather than cause distortions of, competition between transport modes.

Since 2001, international ports with annual traffic of less than 1.5 million tonnes or 200,000 passengers, and community ports with annual traffic of less than 500,000 tonnes or 100,000–199,000 passengers that have established intermodal links with the TEN-T might be financed as part of TEN-T plans. The same is true in the case of ports located on islands or in remote inland areas and which are necessary for the provision of steady connections with specific areas. Access to financial resources is gained due to the development of EU actions to support the shortsea shipping industry and intermodality (i.e. creation of Motorways of the Sea). The condition for funding support the economic viability of the specific project, assessed on the basis of financial and social cost/benefit analyses.

The effect is not limited to the generation of financial benefits for those that undertake specific projects. To put the concept into practice, ports have to join forces and discuss, together with administrators, investors, port users and supply chain parties, European projects that will help a modal shift and cohesion. This policy promotes cooperation strategies and networking between the various stakeholders.[4] However, there is the potential danger of EU ports being in a situation where national or supranational policy-makers will decide which is a strategic port that needs to be supported.

3.3. Ports, Safety, Security and Environmental Protection

The wide range of initiatives attempting to regulate a range of safety, security and environmental protection issues (Psaraftis, 2005) represents another array of proposals that affects the governance of European ports. Irrespective of whether they target port operations per se, or ships through inspections and detention, these measures emphasise the port-state (rather than the flag-state) role of member states and utilise port operations in order to implement a regulatory framework governing maritime traffic-related safety, security and environmental protection.

The EU aims at a uniform implementation of the existing international rules (i.e. Port State Control). Still, in some cases, e.g. security, the EU has adopted stricter applications than those detailed internationally. The transposition of the International Ship and Port Facility Security Code into EU law is more stringent in terms of its mandatory requirements than the original Code.[5] This measure has been accompanied by a proposal for a directive on port security to govern the ship/shore interface, while an additional proposal to regulate intermodal security is eminent. The trend of involving ports for the resolution of safety/security problems between the regulatory authorities and other maritime industries (shipping lines or shippers) implies an increase in the port industry's responsibility as well (i.e. the discussions of an EU rule establishing liability imposed on port authorities along with other stakeholders) in the case of ship-source pollution.

Although the effectiveness of maritime transport might be enhanced, all these changes entail considerable compliance costs and complicate port operations. Security costs have led to average increases of 5 percent for port tariffs (Containerisation International, 2003). For ports, the immediate challenge is to finance the compliance costs, and to incorporate them in pricing and marketing strategies while maintaining market share and achieving reasonable profit margins. Then, competition effects arise from the distribution of the compliance costs. When an EU rule imposes technical demands for port reception facilities for ship-generated waste and cargo residues, but does not specify cost-recovery systems, the impacts on port costs are visible: in the absence of common pricing rules there is the potential of undermining the ability of some ports to compete. The long-term challenge involves adjusting relations with suppliers and customers so as to ensure agile and competitive supply chains capable of overcoming risk and vulnerability threats while still delivering value to ultimate customers and users (Bichou, 2004).

4. CONCLUSIONS

During the last decade, a series of EU policy proposals have aimed to assist the adjustment of European ports to contemporary challenges. However, the produced policy output remains comparatively insignificant. A major reason is that any EU rules need to satisfy a sector of remarkably heterogeneous structures and policy traditions. The discussions for an EU regime governing competition *between* and *within* EU ports have hinted at contradictory potential impacts. Hence, a controversial decision-making process has yet to reach conclusions.

If adopted, EU measures on financing, charging and access to port services market will in the long-term favour a stable pan-European operating environment with well-defined and transparent rules. In the short and medium terms, however, the whole process produces a substantial dynamism and potential port-level distortions. This is due to the different operating and management practices of the ports located in various EU port regions. Complexities increase as the details of this still inconclusive policy-making process might raise negative implications. The latter might be strong enough to destabilise the investment climate and damage existing stakeholder relations. Dissimilar developments in national policy reforms generate variable implementation costs, thus further perplexing the operating environment. If not well considered, these two factors may turn the existing EU policy-making difficulties into long-term obstacles.

At the moment there is some contradictory evidence of a centralised administrative process (i.e. financing, authorisation for port services provision). However, the proposal for a port services directive explicitly advances a regime wherein decision-making and customisation are advanced in highly decentralised conditions.

Overall, the potential implications of the EU initiatives are two-fold. On the one hand, there are efforts to promote port efficiency, via the opening of market access to port services and the definition of a level playing field between ports, either via the transparency of financial accounts or the reconsideration of charging practices. Taking into account the services included in the port services directive, one might conclude that the latter measure puts emphasis on reforming solely the ship-port interface, rather than the entire logistics chain, within an efficiency improvement concept.

On the other hand, there is a clear focus on the effective operation of the port sector. The EU initiatives attempt to redirect port strategies by fostering specialisation and minimising any monopolistic practices in services provision. Policy-makers do not hesitate to intervene in port financing in order to promote the interconnection and interoperability of transport systems in Europe, in essence considering the competition implications of this decision to be of secondary importance. Furthermore, they endorse policies aimed at effective maritime transport services, in terms of safety, security and environmental protection, undermining to a certain extent the impact of these decisions on port costs and consequently port efficiency.

The spirit of the EPP is in line with the overall European integration policy and the liberal orientation of EU sectoral policies. Despite the variation observed in EU ports the analysed policies will progress, one way or another. In any case, port authorities will need to redirect their activities towards a new role, distancing themselves from the traditional role of subsidised port service providers that frequently demand excessive rent for the provided services. To fit in the new environment, port authorities have to redefine their relations with port users and service providers, occasionally rebuilding these relations from scratch (i.e. due to obligatory restarting or introduction of authorisations). Foremost, they will need to act as smart managers of complex port clusters that are integrated in logistics and multimodal transport chains. They will have to solve problems requiring collective action and advance effective networking between numerous actors involved in the port arena.

NOTES

1. The proposal suggests maximum durations of 8 years when there are no investments, 12 years in cases of significant investments in movable assets, and 30 years in cases of significant investments in immovable assets. In many member states, durations of 50, 75 or 99 years are common.

2. Hutchison Port Holding, Port of Singapore Authority Corp., APM Terminals, P&O Ports, Eurogate, and Hamburger Hafen und Logistik AG. (Drewry Shipping Consultants, 2003).

3. Meanwhile, the European Court of Justice has been examining issues of competition and finance during the allocation of concessions on a case-by-case basis since 1991 (European Court of Justice, 1991).

4. An example is the development of the Eurocoast Network Association (ENA) with partners in five member states.

5. *Regulation (EC) No. 725/2004* on enhancing ship and port security.

REFERENCES

Bergantino, A. S. (2002). The European Commission approach to port policy: Some open issues. *International Journal of Transport Economics, 31*(3), 337–379.

Bichou, K. (2004). The ISPS Code and the cost of port compliance: An initial logistics and supply chain framework for port security assessment and management. *Maritime Economics and Logistics, 6*(4), 322–348.

Bichou, K., & Gray, R. (2004). A logistics and supply chain management approach to port performance measurement. *Maritime Policy and Management, 31*(1), 47–67.

CEU. (1991). *New challenges for maritime industries.* Com (91)335, final. Brussels: European Commission.

CEU. (1995). *The development of short sea shipping in Europe: Prospects and challenges.* Com (95)317, final. Brussels: European Commission.

CEU. (1997a). *Green paper on seaports and maritime infrastructure.* Com (97)678, final. Brussels: European Commission.

CEU. (1997b). *Trans-European rail freight freeways.* Com (97)242, final. Brussels: European Commission.

CEU. (2001a). *Reinforcing quality services in seaports: A key for European transport.* Com (2001)35, final. Brussels: European Commission.

CEU. (2001b). *Report on public financing and charging practices in the community sea port sector.* Sec (2001)234, final. Brussels: European Commission.

CEU. (2004). *Proposal for a directive of the European Council and the European Parliament on market access to port services.* Com (2004)654, final. Brussels: European Commission.

Chlomoudis, C. I., Karalis, V. A., & Pallis, A. A. (2003). Port reorganisation and the worlds of production theory. *European Journal of Transport and Infrastructure Research, 3*(1), 77–94.

Chlomoudis, C. I., & Pallis, A. A. (2002). *European port policy: Towards a long-term strategy.* Cheltenham, UK: Edward Elgar.

Chlomoudis, C. I., & Pallis, A. A. (2004). Port governance and the smart port authority: Key issues for the reinforcement of quality services in European ports. *Proceedings of the 10th world conference on transport research,* Istanbul, June.

Containerisation International. (2003). *Cost of port security begins to show,* December, p. 29.

de Langen, P. W. (2002). Clustering and performance: The case of maritime clustering in the Netherlands. *Maritime Policy and Management, 29*(3), 209–221.

de Langen, P. W., & Pallis, A. A. (2006). The effects of intra-port competition. *International Journal of Transport Economics, XXXIII*(1), 69–85.

Drewry Shipping Consultants. (2003). *Annual review of global container terminal operators.* London: Drewry.

European Court of Justice. (1991). *Merci convenzionali Porto di Genova v. Siderurgica Gabrielli,* ECR I –5589.

European Sea Ports Organization (ESPO). (2005). *European ports – Factual report.* Brussels: ESPO.

ESPO, & FEPORT. (2002). *Joint statement on the European Commission's amended proposal for a directive on market access to port services,* 15 March, Brussels.

Farrell, S. (2001). If it ain't bust, don't fix it: The proposed EU directive on market access to port services. *Maritime Policy and Management, 28*(3), 307–313.

Fleming, D. K., & Baird, A. J. (1999). Some reflections on port competition in the United States and Europe. *Maritime Policy and Management, 26*(4), 383–394.

Goss, R. O. (1999). On the distribution of economic rent in Seaports. *International Journal of Maritime Economics, 1*(1), 1–9.

Haralambides, H., Verbeke, A., Musso, E., & Benacchio, M. (2001). Port financing and pricing in the European Union: Theory, politics and reality. *International Journal of Maritime Economics, 3*(3), 368–386.

Hooydonk, E. van. (2005). The European port services directive: The good or the last try? *Journal of International Maritime Law, 11*(3), 188–220.

McConville, J. (Ed.). (2002). Editorial: EU port policy. *Maritime Policy and Management, 29*(1), 1–2.

Meersman, H., Van de Voorde, E., & Vanelslander, T. (2003). Pricing considerations on economic principles and marginal costs. *European Journal of Transport Infrastructure Research, 3*(4), 371–386.

Notteboom, T., & Winkelmans, W. (2004). *Overall market dynamics and their influence on the port sector.* Factual report – Work package 1. Brussels: ESPO.

Orru, E. (2005). Current organization of port services in Europe. *International Workshop on the 2nd EU port package: The good or the last try?* University of Antwerp, 21 May.

Paixao, A. C., & Marlow, P. B. (2003). Fourth generation ports – a question of agility? *International Journal of Physical Distribution and Logistics Management, 33*(4), 355–376.

Pallis, A. A. (1997). Towards a common ports policy? EU Proposals and the industry's perceptions. *Maritime Policy and Management, 24*(4), 365–380.

Pallis, A. A., & Vaggelas, G. K. (2005). Port competitiveness and the EU port services directive: The case of Greek ports. *Maritime Economics and Logistics, 7*(2), 116–140.

Pettersen-Strandenes, S., & Marlow, P. B. (2000). Port pricing and competitiveness in short sea shipping. *International Journal of Transport Economics, 27*(3), 315–334.

Psaraftis, H. N. (2005). EU Ports policy: Where do we go from here? *Maritime Economics and Logistics, 7*(1), 73–82.

Robinson, R. (2003). Ports as elements in value-driven chain systems: The new paradigm. *Maritime Policy and Management, 29*(3), 241–255.

Slack, B., & Fremont, A. (2005). Transformation of port terminal operations: From the local to the global. *Transport Reviews, 25*(1), 117–130.

Turnbull, P. (2006). The war on Europe's waterfront – Repertories of power in the port transport industry. *British Journal of Industrial Relations, 44*(2), 325–326.

UK Department for Transport (DfT). (2005). *Initial regulatory impact assessment: Market access to port services directive.* London: DfT.

PART IV:
PORT PERFORMANCE

CHAPTER 22

PORT PERFORMANCE: AN ECONOMICS PERSPECTIVE

Wayne K. Talley

ABSTRACT

This chapter presents methodologies for evaluating the economic perform-ance of a port. This performance may be evaluated from the standpoint of technical efficiency, cost efficiency and effectiveness by comparing the port's actual throughput with its economic technically efficient, cost effi-cient and effectiveness optimum throughput, respectively. The port's eco-nomic performance may also be evaluated by comparing the actual values of its performance indicators to their standards (that satisfy an economic objective of the port). If the actual values approach (depart from) the standards over time, the port's performance with respect to its economic objective has improved (deteriorated) over time.

1. INTRODUCTION

The crucial question that arises in evaluating a port's performance is how to measure performance. Should a port's performance be evaluated relative to its performance over time (a single-port approach) or relative to the per-formance of other ports (a multi-port approach)? Should the performance be evaluated from an engineering or an economics perspective?

Devolution, Port Governance and Port Performance
Research in Transportation Economics, Volume 17, 499–516
Copyright © 2007 by Elsevier Ltd.
All rights of reproduction in any form reserved
ISSN: 0739-8859/doi:10.1016/S0739-8859(06)17022-5

Ports have traditionally evaluated their performance by comparing their actual and optimum throughputs (measured in tonnage or number of containers handled). If a port's actual throughput approaches (departs from) its optimum throughput over time, the conclusion is that its performance has improved (deteriorated) over time. Engineering optimum throughputs have typically been used in such evaluations, defined as the maximum throughput that a port can physically handle under certain conditions.[1]

In an environment in which ports have natural hinterlands and are not in competition with one another, an engineering performance evaluation methodology of comparing actual and engineering optimum throughputs may be appropriate. In an environment in which ports are in competition with one another (where shippers and carriers are part of the port-selection process), a port should not only be concerned with whether it can physically handle cargo, but also whether it can compete for cargo. In a competitive environment, port time-related costs in addition to port charges incurred by shippers and carriers are important determinants in port selection. Since port cargo remains in the shipper's inventory (assuming the shipper retains ownership), the shipper incurs time-related inventory (or logistics) costs in port; water and inland carriers also incur port time-related costs, e.g. depreciation and insurance costs on their ships and vehicles while in port. A port can reduce these time-related costs by reducing the time that the cargo of shippers and the ships and vehicles of carriers are in port, i.e. by improving the quality of its service.

If a port's performance in a competitive environment is evaluated by comparing actual and optimum throughputs, an economic (rather than the engineering) optimum throughput should be utilized (Talley, 1988a). A port's economic optimum throughput is that throughput for which the port achieves an economic objective, e.g. maximizing port profits.

An alternative methodology to that of comparing actual and optimum throughputs for evaluating the performance of a port is one that makes use of port performance indicators. From an economics perspective, port performance indicators are choice variables (i.e. variables whose values are under the control of port management) for optimizing the port's economic objective. If the economic objective is to maximize profits, port management would select values for the port indicators that would result in maximum profits for the port. These values of the performance indicators have been referred to in the literature as performance indicator standards (or benchmarks). If the actual values of the port's performance indicators approach (depart from) their respective standards over time, the port's performance – with respect to its economic objective – has improved (deteriorated) over time.

Note that the port performance evaluation methodologies, comparing actual and optimum throughputs and comparing actual values and standards of performance indicators, are consistent with one another. For example, if the port's economic objective function (e.g. to maximize profits) is known, the port's economic optimum throughput can be determined by substituting the standards for the performance indicators and solving.

The determination of a port's economic optimum throughput and its performance indicator standards (as defined above) requires knowing or having to estimate the port's economic objective function(s). For example, to obtain a port's profit function, the port's throughput demand function must be known or estimated. However, economic objective functions (i.e. their functional form and parameter values) are not likely to be known with certainty. Further, available data may not be sufficient to obtain reliable estimates of these functions, nor to obtain reliable estimates of the port's economic optimum throughput and performance indicator standards.

The following section presents a basic economic model of a port. The model provides the theoretical structure for understanding the approaches that have been used in evaluating the performance of ports. In Section 3 the single-port approach for port evaluation is discussed. This approach may involve evaluating a port's throughput as well as the values of its performance indicators over time. In Section 4 the multi-port performance evaluation approach and the frontier statistical models that have been used in multi-port technical performance evaluations are discussed, followed by a summary of the discussions in Section 5.

2. AN ECONOMIC MODEL OF A PORT

A port's economic production function represents the relationship between the port's maximum throughput and given levels of its productive resources, i.e.

$$\text{Maximum Port Throughput} = f(\text{Port Productive Resources}) \qquad (1)$$

where throughput may be the number of containers (measured in 20-foot equivalent units or TEUs) or tons of cargo handled and port productive resources include labor, immobile capital (e.g. berths and buildings), mobile capital (e.g. cranes and vehicles), fuel and ways (e.g. port roadways and railways). If the port achieves the maximum throughput for given levels of its resources, then it is technically efficient; otherwise it is technically inefficient.

A port's economic cost function represents the relationship between the port's minimum costs to be incurred in handling a given level of throughput, i.e.

$$\text{Minimum Port Costs} = g(\text{Port Throughput}) \qquad (2)$$

where the costs are those incurred in the use of the port's resources, e.g. wages paid to labor and vehicle fuel expenses. If the port provides throughput at a minimum cost (given the unit costs or resource prices to be paid), then it is cost efficient; otherwise it is cost inefficient.

In order for a port to be cost efficient, it must be technically efficient, i.e. the latter is a necessary condition for the former (Nicholson, 1992, Chapter 12). If a port is technically inefficient, it can handle more throughput with the same resources by becoming technically efficient. Further, given the same resources and thus the same resource costs, the average cost per unit of throughput will decline with the port becoming technically efficient. Alternatively, if the port is technically inefficient, it must follow that it is also cost inefficient.

A port, especially in a competitive environment, is concerned not only with whether it is efficient (technically and cost), but also with whether it is effective in providing throughput. Economic operating objectives of a port may be classified as either efficiency or effectiveness objectives. For example, port efficiency operating objectives include the technical efficiency objective of maximizing throughput in the employment of a given level of resources (exhibited by the port's economic production function) and the cost efficiency objective of minimizing cost in the provision of a given level of throughput (exhibited by the port's economic cost function). Effectiveness is concerned with how well the port provides throughput service to its users – shippers and carriers (ocean and inland). From the perspective of the port, this may be measured by its adherence to its effectiveness operating objective, e.g. maximizing profits. That is, the port's throughput service is at the level at which profits cannot be further increased.

Port effectiveness operating objectives will differ between privately-owned and government-owned ports. If the port is privately owned, its effectiveness economic operating objective might be to maximize profits or to maximize throughput subject to a minimum profit constraint. If the port is owned by government, its effectiveness economic operating objective might be to maximize throughput subject to a zero operating deficit (where port revenue equals cost) or subject to a maximum operating deficit (where port revenue is less than cost) that is to be subsidized by government.

In order for a port to be effective, it must be efficient – i.e. it must be cost efficient, which in turn requires that it must be technically efficient

(Nicholson, 1992, Chapter 13). For example, if a port has the effectiveness operating objective of maximizing profits and is cost inefficient, it can obtain greater profits for the same level of throughput service by lowering costs in becoming cost efficient. However, note that a port can be cost efficient without being effective.

A critical component of a port's effectiveness operating objective is the demand for its throughput services. A port's throughput demand function represents the relationship between the demand for the port's throughput services by its users and the generalized port price (per unit of throughput) incurred by these users, i.e.[2]

$$\text{Port Throughput} = h(\text{Generalized Port Price}) \qquad (3)$$

where

$$\begin{aligned}
\text{Generalized Port Price} = \ &\text{Port Price Charged} \\
&+ \text{Ocean Carrier Port Time Price} \\
&+ \text{Inland Carrier Port Time Price} \\
&+ \text{Shipper Port Time Price} \qquad (4)
\end{aligned}$$

The Port Price Charged per unit of throughput represents prices charged by the port for various port services, e.g. wharfage, berthing and cargo handling charges; the Ocean Carrier Port Time Price per unit of throughput represents the time-related costs incurred by ocean carriers while their ships are in port, e.g. ship depreciation, fuel and labor costs; the Inland Carrier Port Time Price per unit of throughput represents the time-related costs incurred by inland (rail and truck) carriers while their vehicles are in port, e.g. vehicle depreciation, fuel and labor costs; and the Shipper Port Time Price per unit of throughput represents the time-related costs incurred by shippers while their shipments are in port, e.g. inventory costs such as insurance, obsolescence and depreciation costs.

If a port seeks to maximize profits, its profit (or effectiveness operating objective) function may be written as

$$\text{Profit} = \text{Port Price Charged} * \text{Port Throughput - Minimum Costs} \qquad (5)$$

Substituting the port's throughput demand function (3) and economic cost function (2) into profit function (5) and rewriting, it follows that:

$$\begin{aligned}
\text{Profit} = \ &\text{Port Price Charged} * h(\text{Generalized Port Price}) \\
&- g(\text{Port Throughput}) \qquad (6)
\end{aligned}$$

Finally, substituting the economic production function (1) into profit function (6) and rewriting, it follows that:

$$\text{Profit} = \text{Port Price Charged} * h(\text{Generalized Port Price})$$
$$- g[f(\text{Port Resources})] \qquad (7)$$

The resources in profit function (7) in turn may be expressed as functions of the port's operating options and the amounts of given types of cargo (provided by carriers and shippers) to be handled by the port. A port's operating options are the means by which it can vary the quality of its throughput service. If the functions relate the minimum amount of a given resource employed by the port to its levels of operating options and amounts of given types of cargo to be handled, such functions in the literature have been referred to as resource functions (see Talley, 1988b):

$$\text{Minimum Port Resources} = j(\text{Port Operating Options};$$
$$\text{Amounts of Given Types of Cargo}$$
$$\text{Provided by Carriers and Shippers}) \qquad (8)$$

Substituting the resource function (8) into profit function (7) and rewriting, it follows that:

$$\text{Profit} = \text{Port Price Charged} * h(\text{Generalized Port Price})$$
$$- g\{f[j(\text{Port Operating Options};$$
$$\text{Amounts of Given Types of Cargo}$$
$$\text{Provided by Carriers and Shippers})]\} \qquad (9)$$

A port can differentiate the quality of its service with respect to such operating options as: (a) ship loading and unloading service rates – ship loading and unloading times incurred per port call, (b) ship berthing and unberthing service rates – ship berthing and unberthing times incurred per port call, (c) inland-carrier vehicle loading and unloading service rates – vehicle loading and unloading times per port call, (d) inland-carrier vehicle entrance and departure service rates – vehicle entrance and departure times per port call. Entrance (departure) time for an inland-carrier vehicle is the queuing time incurred to be cleared for entrance into (departure from) the port once arriving at the port's entrance (departure) gate.

What are the means by which port management can optimize its effectiveness-operating objective? That is to say, what are the choice variables to be utilized by port management in the optimization? For a variable to qualify as a choice variable, its value must be under the control of port management. Suppose the port's effectiveness operating objective is to maximize profits, where profits are expressed as in profit function (9). In this function, Port Price Charged is a choice variable, unless constrained by port competition. The other choice variables are the port's operating options. Changes in the values of operating options not only affect the level of resources used by the port, and thus port costs, but also the times incurred in port by ocean carriers' ships, inland carriers' vehicles and shippers' cargo. These times in turn affect the port time costs incurred by these port users – consequently, affecting the port's profits.

3. A SINGLE-PORT PERFORMANCE EVALUATION APPROACH

The single-port approach to port performance evaluation evaluates a port's performance from the perspective of its performance over time. This may be done by comparing the port's actual throughput to its optimum throughput or comparing the actual values of its performance indicators to the standards of these indicators over time.

3.1. Throughput Performance Evaluation

If a port's actual throughput departs from (approaches) its optimum throughput over time, one would conclude that its performance has deteriorated (improved) over time. While a port's optimum throughput may be a technically (or production) efficient, cost efficient or effectiveness optimum throughput, the technically efficient optimum throughput has typically been utilized in port throughput performance evaluations. A port's optimum throughput may also be an engineering or an economic optimum throughput.

A port's engineering production optimum throughput is the port's maximum throughput that physically can be handled by the port under certain conditions. This throughput has also been referred to as the port's capacity.[3] The port's engineering production optimum throughput may be measured theoretically or empirically.

The theoretical engineering production capacity (or optimum throughput) of a port has been classified as: (a) design capacity, (b) preferred capacity and (c) practical capacity (Chadwin, Pope, & Talley, 1990). A port's design capacity is its maximum utilization rate. For example, the design capacity of the storage area of a container port is the maximum number of containers that can physically be stored in the area. A port's preferred capacity is the utilization rate beyond which certain utilization characteristics or requirements cannot be obtained, e.g. the utilization rate beyond which port congestion occurs. Port congestion at the gate of a container port occurs when the waiting times for trucks to enter the gate increase beyond normal waiting times due to the increase in the number of trucks seeking entrance. A port's practical capacity is its maximum utilization rate under normal or realistic conditions. For example, the practical capacity for a container port's shipside crane is the maximum number of containers that the crane is expected to load and unload from a ship per hour under normal working conditions.

The empirical engineering production optimum throughput is the estimated maximum throughput for the port, usually based upon the actual throughput productions of several similar ports. One exception is found in a port handbook by Hockney and Whiteneck (1986). A modular method for estimating the capability of a given port is presented, where capability is defined as the maximum annual throughput (in tons of cargo) that a port can handle under normal working conditions. To determine the capability estimate for a given port, the handbook first estimates the maximum annual throughput for the various components of the port: ship-to-apron transfer capability, apron-to-storage transfer capability, yard storage capability, storage-to-inland transport transfer capability and inland transport unit processing capability. The port's capability estimate is that estimate with the lowest throughput value among the five estimates. The lowest throughput estimate is the constraining capability of the port (or choke point) and thus is selected as the maximum annual throughput that the port can handle under normal working conditions.

A port's economic optimum throughput is that throughput that satisfies an economic objective or objectives of the port. It may be either an economic: (a) technically efficient optimum throughput (based upon the port's economic production function), (b) cost efficient optimum throughput (based upon the port's economic cost function) or (c) effectiveness optimum throughput (based upon a port's effectiveness operating objective such as maximizing profits). The economic performance of a port may be evaluated from the standpoint of technical efficiency, cost efficiency and effectiveness by comparing its actual throughput with its economic technically efficient optimum throughput, cost efficient optimum throughput and effectiveness optimum throughput, respectively.[4]

A port's economic technically efficient optimum throughput is generally more difficult to determine than its engineering production optimum throughput, especially if the latter is the theoretical engineering production optimum throughput. In general, port economic optimum throughputs are based upon estimated economic objective functions, since the specific forms of these functions are generally unknown. Estimated port production and cost functions are found in studies by Kim and Sachish (1986) and De Neufville and Tsunokawa (1981).[5] To determine a port's economic effectiveness optimum throughput, not only must a port's economic cost function be modeled and estimated, but also its demand and revenue functions. Although port pricing has been investigated, little attention has been given to estimating port demand and revenue functions.

3.2. Indicator Performance Evaluation

3.2.1. Indicator Selection

Two contrasting methodologies have appeared in the transportation literature for selecting performance indicators for transportation firms – the operating objective specification methodology and the criteria specification methodology (Talley, 1986). The operating objective specification methodology requires the specification of an operating objective for the purpose of then selecting performance indicators. The criteria specification methodology specifies the criteria that selected performance indicators must satisfy.

With respect to an economics operating objective, a port's performance indicators are those variables whose values are under the control of port management (i.e. choice variables) for optimizing the operating objective. The values of these variables that optimize the economic objective are the indicators' standards (or benchmarks). If the actual values of the indicators approach (depart from) their perspective standards over time, the port's performance with respect to the given economic objective has improved (deteriorated) over time.

Criteria of the criteria specification methodology that may be used by a port for selecting performance indicators include: (a) conciseness, (b) consistency with objectives, (c) data availability, (d) data collection time and cost, (e) measurability, (f) minimization of uncontrollable factors, and (g) robustness (Talley, 1994). The conciseness criterion requires that the redundancy and overlap among selected indicators be limited. The consistency-objective criterion requires that the indicators be consistent with the port's operating objectives, i.e. they affect these objectives. In addition to the availability of data, the time and cost to be incurred in the collection of the indicator data should be considered in the selection of port performance indicators. The measurability criterion requires that the selected indicators be measurable, i.e. having a continuous as opposed to a discrete unit of measurement. The criterion, minimizing uncontrollable factors, requires that the values of the port's selected indicators be under the control of port management. The robustness criterion requires that the selected indicators allow for the port to be evaluated under various scenarios.

Port performance indicators selected from an economic operating objective specification perspective (i.e. choice variables for optimizing the given economic objective) in general satisfy the selection criteria of the criteria specification methodology. The conciseness criterion is specifically addressed, since choice variables in optimization theory are independent variables, i.e. not a function of one another. Thus, by definition, no redundancy and overlap among selected indicators exist. The selected indicators are consistent, since they are derived from a port's economic objective. Since the performance indicators are variables whose values are under the control of management, the criterion – minimization of uncontrollable factors – will be satisfied. Also, the criteria of data availability, measurability and robustness will likely be satisfied. Although the economic operating objective specification methodology does not specifically address the time and cost commitment to collecting indicator data, it does so indirectly by generally selecting a smaller number of indicators than would be selected by the criteria specification methodology.

In Section 2, it was noted that the port's choice variables with respect to its effectiveness economic operating objective, maximizing profits, are its prices charged to users for various port services as well as port operating options – choice variables by which the port can vary the quality of its services. These port choice variables may also be labeled as the port's effectiveness performance indicators with respect to maximizing profits.

In a model of a mixed-cargo port (handling bulk and container cargo) that seeks to maximize annual throughput subject to a profit constraint, Talley (1996) found the following choice variables or effectiveness port performance indicators with respect to this effectiveness operating objective:

1. Annual average port charge per throughput ton (for a given type of cargo).
2. Annual average ship loading service rate (for a given type of cargo), i.e. tons of cargo loaded on ships per hour of loading time.
3. Annual average ship unloading service rate (for a given type of cargo), i.e. tons of cargo unloaded from ships per hour of unloading time.
4. Annual average loading service rate (for a given type of cargo) for port vehicles of inland carriers, i.e. tons of cargo loaded on port vehicles per hour of loading time.

5. Annual average unloading service rate (for a given type of cargo) for port vehicles of inland carriers, i.e. tons of cargo loaded from vehicles per hour of unloading time.
6. Annual average daily percent of time that the port's channel adheres to authorized depth and width dimensions (a port channel accessibility indicator).
7. Annual average daily percent of time that the port's berth adheres to authorized depth and width dimensions (port berth accessibility indicator).
8. Annual average daily percent of time that the port's channel is open to navigation (port channel reliability indicator).
9. Annual average daily percent of time that the port's berth is open to the berthing of ships (port berth reliability indicator).
10. Annual average daily percent of time that the port's entrance gate is open to inland-carrier vehicles (entrance gate reliability indicator).
11. Annual average daily percent of time that the port's departure gate is open to inland-carrier vehicles (departure gate reliability indicator).
12. Annual expected probability of damage to ships while in port.
13. Annual expected probability of loss of ship property while in port.
14. Annual expected probability of damage to inland-carrier vehicles while in port.
15. Annual expected probability of loss to the property of inland-carrier vehicles while in port.
16. Annual expected probability of damage to shippers' cargo while in port.
17. Annual expected probability of the loss of shippers' cargo while in port.

In Australia, port performance indicators have been used to evaluate the performance of the country's ports for waterfront reform – both single-port and multi-port evaluations – comparing indicator values for a given port over time and across ports for a given time period, respectively. The selected indicators measure the change in the utilization and productivity of port resources; thus, they can be argued to be productivity (or technical) efficiency indicators. Their selection was based upon the criteria specification methodology, using selection criteria of stevedores, shipping lines and port authorities (Talley, 1994).

The selected stevedoring performance indicators measure the productivity and utilization of a port's equipment and labor resources across stevedoring operations. Port indicators from an equipment perspective include: (a) cargo handling rate (the rate at which ships are loaded and discharged), (b) number of ships and amount of cargo handled (an indicator of workload),

(c) containers handled per crane and (d) cargo handled per man-shift (total cargo handled divided by the number of man-shifts). The indicators from a labor perspective include: (a) number of employees, (b) average age of the labor force, (c) average hours worked per week by employees and (d) labor idle time (the percentage of time employees are available for work but are not required to work).

The selected shipping-line port performance indicators reflect ship delays: (a) average delay to ships waiting for berths and (b) average delay to ships while alongside berths. The port-authority indicators reflect port utilization and throughput: (a) cargo tonnage handled, (b) truck queuing times at port gates and (c) facility utilization (as a percentage of total available time).

Note that although the intent of the Australian criteria specification methodology was to select only port production efficiency indicators, several of the selected indicators are similar to those derived by Talley (1996) with respect to the port's effectiveness operating objective of maximizing throughput subject to a profit constraint: (a) cargo handling rate, (b) average delay to ships waiting berths and (c) truck queuing times at port gates. Hence, these indicators may be both efficiency and effectiveness indicators.

In 1988, the US Army Corps of Engineers introduced performance indicators for evaluating its operation and maintenance of the national navigational waterway system, i.e. in provision of navigational aids and new and maintenance dredging of channels. The indicator-selection process was imperfect in retrospect. Six years later the US Army Corps of Engineers' National Program Proponents Workshop - Navigation (1994) was held. Among the performance indicators selected at the workshop, four relate to harbor waterways: (a) percentage of annual days wherein designed vessels may operate within the harbor, unrestricted by wave, current or shoaled conditions (unrestricted accessibility indicator), (b) percentage of annual days wherein designed vessels may operate within the harbor, restricted by tides, wave actions and shoaled conditions (restricted accessibility indicator), (c) percentage of annual days where harbor operations cease due to scheduled navigational maintenance (scheduled-maintenance accessibility indicator) and (d) percentage of annual days where harbor operations cease due to unscheduled navigational maintenance (unscheduled-maintenance accessibility indicator). These indicators were selected based upon the criteria specification methodology. Specifically, the Army Corps of Engineers specified that the selected performance indicators be able to evaluate whether it is providing a given level of navigational service for the national waterway system at the least cost.

3.2.2. Evaluation in Practice

If the specific form of a port's economic objective function is not known (or a reliable estimate is not available), the port's performance over time with respect to the economic objective can still be evaluated by means of performance indicators: it can be evaluated by just knowing the actual values of its performance indicators. Specifically, if the direction of movement in these values over time moves the port nearer to (away from) achieving its economic objective, the conclusion is that the port's performance has improved (deteriorated) over time. For one indicator, a rising trend over time in its actual values might move the port nearer to achieving its economic objective; for another indicator, it might be a declining trend in its actual values.

An advantage to a port in having individual performance indicators to evaluate its performance over time is that the performance of its various services and service areas (e.g. the dock, entrance and departure gates, and the port channel) can be evaluated, thereby allowing for the detection of port activity centers where performance is improving or declining. However, a disadvantage is the problem of how to evaluate the net impact of changes in these indicators on the port's overall performance, given that the changes in some indicators may improve performance and changes in other indicators may negatively affect performance. What is needed is an overall (or single) port performance indicator that captures the net impact of the changes in the individual performance indicators on the port's performance.

In a study by Talley (1996), where the port's economic objective is to maximize annual throughput subject to a profit constraint, this overall performance indicator is the port's annual throughput per profit dollar (given its profit constraint). If this overall indicator is rising (declining) over time, it follows that the port's performance has been improving (declining) over time; furthermore, the net impact of the changes over time in the individual performance indicators on port performance has been positive (negative).

4. MULTI-PORT PERFORMANCE EVALUATION APPROACH

Although it is tempting to compare the performance of one port to that of another, such comparisons may be misleading. Ports operate in different economic, social and fiscal environments. For example, even if ports have the same economic objective of maximizing annual throughput subject to a profit constraint, the profit constraint is likely to differ among ports. Also, one port may have a negative profit (or deficit) constraint that is to be subsidized, while another port may have a positive or break-even profit constraint. Ports may also have different economic objectives (see Suykens, 1986). Thus, in a multi-port performance evaluation approach, where the performance of one port is compared to that of another, similar ports should be used in the comparisons. A principal component analysis for identifying similar ports in a group of ports is found in a study by Tongzon (1995).

In the literature, multi-port performance evaluations of the technical efficiency of ports generally rely upon frontier statistical models. These models utilize the throughputs (or outputs) and resources (or inputs) of a group of ports to investigate whether the ports are technically efficient – i.e. whether their throughputs are the maximum throughputs for given levels of resources (or on their production frontiers) – or technically inefficient, where their throughputs are less than their maximum throughputs for given levels of resources and therefore lie below their production frontiers.

Frontier statistical models used in multi-port technical performance evaluations generally utilize data envelopment analysis (DEA) techniques – non-parametric mathematical programming techniques for deriving the specification of the frontier model. DEA techniques derive relative efficiency ratings for the ports that are used in the analysis. Thus, the development of standards against which efficiency can be measured is not required, although such standards can be incorporated into the DEA analysis. DEA techniques make no assumptions about the stochastic properties of the data. When such assumptions are made, the frontier statistical model is referred to as a stochastic frontier model. For an in-depth discussion of these models, see Cullinane (2002).

In the Tongzon (2001) study, DEA is used to investigate the relative technical efficiency of 16 international (including four Australian) container ports for the year 1996. Initially, the investigation considered two-port output and six input variables. The output variables are the total number of containers in TEUs loaded and unloaded (cargo throughput) and the number of containers moved per working hour (ship working rate). The input variables include: (a) number of cranes, (b) number of container berths, (c) number of tugs, (d) delay time (the difference between total berth time plus time waiting to berth and the time between the start and finish in working a ship), (e) terminal area and (f) the number of port authority employees. Two versions of the DEA model were used in the investigation: the CCR version that assumes constant returns to scale in production and the Additive version that allows for variable returns to scale. See Cullinane (2002) for a discussion of these two DEA versions.

Given the small sample size in the Tongzon (2001) study, only one output – cargo throughput – was used in the final analysis. More ports were found to be technically inefficient based upon the CCR version than for the Additive version. This is not surprising, since the CCR version has the restrictive assumption of constant returns to scale. For both DEA versions, the ports of Melbourne, Rotterdam, Yokohama and Osaka were identified as technically inefficient and the ports of Hong Kong, Singapore, Hamburg, Keelung, Zeebrugge and Tanjung Priok were identified as technically efficient. Since a number of the ports within each group are quite different with respect to size and function (e.g. hub versus a non-hub container port), the results suggest that the technical efficiency of ports does not depend only upon port size or function. For example, in the technically inefficient group, Rotterdam is large relative to the port of Melbourne and is a hub container port as opposed to the ports of Melbourne, Yokohama and Osaka.

Multi-port technical-efficiency performance evaluation studies that utilize stochastic frontier models include studies by Notteboom, Coeck, and van den Broeck (2000), Coto-Millan, Banos-Pino, and Rodriguez-Alvarez (2000) and Cullinane, Song, and Gray (2002).

5. SUMMARY

Should a port's performance be evaluated from an engineering or an economics perspective? Should it be evaluated relative to its performance over time (a single-port approach) or relative to the performance of other ports (a multi-port approach)? Ports have traditionally been evaluated by the engineering single-port approach of comparing their actual and engineering optimum throughputs, i.e. the maximum throughputs or cargo tonnage that ports can physically handle under certain conditions. If a port's actual throughput approaches (departs from) its optimum throughput over time, the conclusion is that its performance has improved (deteriorated) over time.

A port's economic optimum throughput is that throughput that satisfies an economic objective of the port. It may be an economic: (a) technically efficient optimum throughput (based upon the port's economic production function, representing the relationship between a port's maximum throughput and given levels of its productive resources), (b) cost efficient optimum throughput (based upon the port's economic cost function, representing the relationship between a port's minimum costs to be incurred in handling a given level of throughput) or (c) effectiveness optimum throughput (based upon a port's effectiveness operating objective such as maximizing profit).

The economic performance of a port may be evaluated from the standpoint of technical efficiency, cost efficiency and effectiveness by comparing its actual throughput with its economic technically efficient optimum throughput, cost efficient optimum throughput and effectiveness optimum throughput, respectively.

In addition to comparing a port's actual throughput to its optimum throughput, the single-port approach for evaluating a port's performance may also involve comparing the actual values of a port's performance indicators (i.e. variables whose values are under the control of port management) to their standards. The latter are values of the performance indicators that satisfy an economic objective of the port. Thus, the standards may be technically efficient standards, cost efficient standards or effectiveness standards. If the actual values of the port's performance indicators approach (depart from) their respective standards over time, the port's performance – with respect to its economic objective – has improved (deteriorated) over time. If performance indicator standards are unknown, a port's performance can be evaluated just by knowing the actual values of its performance indicators. Specifically, if the direction of movement in these values over time moves the port nearer to (away from) achieving its economic objective, the conclusion is that the port's performance has improved (deteriorated) over time.

Methodologies for selecting performance indicators include the operating objective specification methodology and the criteria specification methodology. The operating objective specification methodology requires the specification of an operating objective for the purpose of then selecting performance indicators. The criteria specification methodology specifies the criteria that selected performance indicators must satisfy.

In the literature, multi-port performance evaluations of the technical efficiency of ports generally rely upon frontier statistical models that utilize DEA techniques – non-parametric mathematical programming techniques for deriving the specification of the production frontier model. DEA techniques derive relative efficiency ratings for the ports that are used in the analysis. These ports should be similar; otherwise, the efficiency ratings may be misleading.

NOTES

1. A modular method for estimating the engineering optimum throughput of a port is found in Hockney and Whiteneck's (1986) port handbook. The module (or component) of the port with the least capability is the constraining capability component (or "choke point") of the terminal, which thus serves as the estimate of the

maximum annual throughput (or capability) that the port can handle under normal working conditions.

2. The port throughput demand function is a function of the Generalized Port Price as opposed to just the port price (or prices charged by the port to its users). The Generalized Port Price includes the latter as well as the port time prices incurred by port users. The time prices are obviously other determinants of port throughput demand, but are also functions of the capital depreciation, fuel and labor costs of carriers and the logistics costs of shippers incurred in port. Thus, the latter are indirectly considered determinants of port throughput demand in port. Thus, the latter are indirectly considered determinants of port throughput demand.

3. For a discussion of capacity with respect to a port's infrastructure, see Jansson and Shneerson (1982). For a general discussion of economic capacity, see Wilson (1980).

4. For a discussion of effectiveness and efficiency in transit performance, see Talley and Anderson (1981).

5. A discussion of port cost functions is found in studies by Jansson and Shneerson (1982), Schonfeld and Frank (1984) and De Weille and Ray (1974).

REFERENCES

Chadwin, M. L., Pope, J. A., & Talley, W. K. (1990). *Ocean container transportation: An operational perspective*. New York: Taylor & Francis.

Coto-Millan, P., Banos-Pino, J., & Rodriguez-Alvarez, A. (2000). Economic efficiency in Spanish ports: Some empirical evidence. *Maritime Policy and Management, 27*, 169–174.

Cullinane, K. P. B. (2002). The productivity and efficiency of ports and terminals: Methods and applications. In: C. Grammenos (Ed.), *The handbook of maritime economics and business* (pp. 426–442). London: Lloyds of London Press.

Cullinane, K. P. B., Song, D.-W., & Gray, R. (2002). A stochastic frontier model of the efficiency of major container terminals in Asia: Assessing the influence of administrative and ownership structures. *Transportation Research A, 36*, 743–762.

De Neufville, R., & Tsunokawa, K. (1981). Productivity and returns to scale of container ports. *Maritime Policy and Management, 8*, 121–129.

De Weille, J., & Ray, A. (1974). The optimum port capacity. *Journal of Transport Economics and Policy, 8*, 244–259.

Hockney, L. A., & Whiteneck, L. L. (1986). *Port handbook for estimating marine terminal cargo handling capability*. Washington, DC: Maritime Administration, US Department of Transportation.

Jansson, J. O., & Shneerson, D. (1982). *Port economics*. Cambridge, MA: MIT Press.

Kim, M., & Sachish, A. (1986). The structure of production, technical change and productivity in a port. *Journal of Industrial Economics, 35*, 209–223.

Nicholson, W. (1992). *Microeconomic theory: Basic principles and extensions*. Philadelphia: Dryden Press.

Notteboom, T., Coeck, C., & van den Broeck, J. (2000). Measuring and explaining the relative efficiency of container terminals by means of Bayesian stochastic frontier models. *International Journal of Maritime Economics, 2*, 83–106.

Schonfeld, P., & Frank, S. (1984). Optimizing the utilization of a containership berth. *Transportation Research Record*, *984*, 50–63.

Suykens, F. (1986). Ports should be efficient (even when this means that some of them are subsidized). *Maritime Policy and Management*, *13*, 105–126.

Talley, W. K. (1986). A comparison of two methodologies for selecting transit performance indicators. *Transportation*, *13*, 201–210.

Talley, W. K. (1988a). Optimum throughput and performance evaluation of marine terminals. *Maritime Policy and Management*, *15*, 327–331. Reprinted in Brooks, M., Button, K., & Nijkamp, P. (Eds.). (2002). *Maritime transport* (pp. 511–515). Cheltenham, UK: Edward Elgar Publishing.

Talley, W. K. (1988b). *Transport carrier costing*. New York: Gordon and Breach Science Publishers.

Talley, W. K. (1994). Performance indicators and port performance evaluation. *Logistics and Transportation Review*, *30*, 339–352.

Talley, W. K. (1996). *Performance evaluation of mixed-cargo ports*. Alexandria, VA: Water Resources Support Center, Army Corps of Engineers, Unpublished report.

Talley, W. K., & Anderson, P. (1981). Effectiveness and efficiency in transit performance: A theoretical perspective. *Transportation Research A*, *15*, 431–436.

Tongzon, J. (1995). Systematizing international benchmarking for ports. *Maritime Policy and Management*, *22*, 171–177.

Tongzon, J. (2001). Efficiency measurement of selected Australian and other international ports using data envelopment analysis. *Transportation Research A*, *35*, 107–122.

US Army Corps of Engineers. (1994). *The national program proponents workshop – Navigation*. Institute for Water Resources, US Army Corps of Engineers, Fort Belvoir, VA. Unpublished Report.

Wilson, G. W. (1980). *Economic analysis of intercity freight transportation*. Bloomington: Indiana University Press.

CHAPTER 23

DATA ENVELOPMENT ANALYSIS (DEA) AND IMPROVING CONTAINER PORT EFFICIENCY

Kevin Cullinane and Teng-Fei Wang

ABSTRACT

This chapter analyses the relevance of Data Envelopment Analysis (DEA) to the estimation of productive efficiency in the container port industry. Following an exposition of the DEA methodology, the many previous applications of the technique to the port industry are reviewed and assessed. The DEA technique is illustrated through a detailed example application using sample data relating to the world's leading container ports. The different DEA models give significantly different absolute results when based on cross-sectional data. However, efficiency rankings are rather similar. The efficiency estimated by alternative approaches, therefore, exhibits the same pattern of efficiency distribution, albeit with significantly different means. An analysis of panel data reveals that container port efficiency fluctuates over time, suggesting that the results obtained from an analysis of cross-sectional data may be misleading. Overall, the results reveal that substantial waste exists in container port production. It is also found that the sample ports exhibit a mix of decreasing, increasing and constant returns to scale. The chapter concludes that the optimum efficiency levels indicated by DEA results might not be achievable in reality, because each individual port has its own specific and

Devolution, Port Governance and Port Performance
Research in Transportation Economics, Volume 17, 517–566
Copyright © 2007 by Elsevier Ltd.
ISSN: 0739-8859/doi:10.1016/S0739-8859(06)17023-7

unique context. Consequently, more singular aspects of individual ports should be investigated to determine the reasons that explain estimated efficiency levels.

1. INTRODUCTION

In an era characterised by the globalisation of production and consumption patterns, recent trends in international trade have led to the increasing importance of container transportation. This is largely because of the numerous technical and economic advantages it possesses over traditional methods of transportation. Standing at the interface of sea and inland transportation, container ports play a pivotal role in the container transportation process.

One distinctive feature of the contemporary container port industry is that competition between container ports is more intensive than has previously been the case. Port markets used to be perceived as monopolistic due to the exclusive and immovable geographical location of the port and the unavoidable concentration of port traffic. However, the rapid development of international container and intermodal transportation has drastically changed the market structure from one of monopoly to one where fierce competition is rife in many parts of the world. Many container ports no longer enjoy the freedom yielded by a monopoly over the handling of cargoes from their hinterland. Instead, they have to compete for cargo with their neighbouring ports (Cullinane, Wang, & Cullinane, 2004).

It is the intense competition which characterises the container port industry that has stimulated an overt interest in the efficiency with which it utilises its resources. Indeed, the analysis of the performance of individual container ports or terminals is of great importance for the health of the industry and the survival and competitiveness of its players. Not only can such an analysis provide a powerful management tool for port operators, but it also constitutes an important input for informing regional and national port planning and operations. Alluding to the significant level of competition in the industry provides one justification for undertaking an analysis of relative efficiency within it. In so doing, however, it is not necessarily the case that any direct causal relationship is proposed to exist between changing levels of efficiency over time and the degree of competition in the container port industry. Although this is certainly a hypothesis worthy of testing, particularly because of the potential impact of market

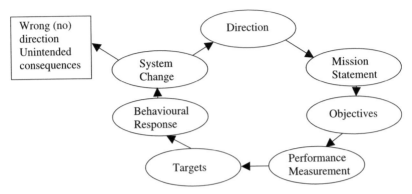

Fig. 23.1.   Performance Measures and Organisational Development. *Source:* Dyson (2000, p. 5).

contestability (Baumol, Panzer, & Willig, 1982), it is one of such complexity and import that it merits a separate study in its own right.

Performance measurement plays an important role in the development of a company (or any other form of organisation). Dyson (2000) claimed that performance measurement plays an essential role in evaluating production because it can define not only the current state of the system but also its future. Performance measurement helps move the system in the desired direction through the effect exerted within the system by the behavioural responses towards these performance measures. As highlighted in Fig. 23.1, however, mis-specified performance measures will cause unintended consequences, with the system moving in the wrong direction.

Performance measurement in the port industry has evolved in tandem with the evolution of the industry itself. The most straightforward and still widely used approach in this respect is to use multiple indicators of partial productivity measures (Ashar, 1997; Cullinane, 2002). This is because ports are essentially providers of service activities, in particular for vessels, cargo and inland transport. As such, it is possible that a port may provide an acceptable level of service to vessel operators on the one hand and unsatisfactory service to cargo or inland transport operators on the other. Therefore, port performance cannot normally be assessed on the basis of a single value or measure.

The main weakness of multiple indicators, partial productivity measures for example, lies in the difficulty of examining whether port performance has improved or deteriorated when changes in some indicators improve performance and changes in others affect it negatively. To overcome this, Talley (1994) attempted to build a single performance indicator – the shadow price of variable port throughput per profit dollar – to evaluate the performance of a port. Several methods have been suggested in a similar vein, such as the estimation of a port cost function (De Neufville & Tsunokawa, 1981), the estimation of the total factor productivity of a port (Kim & Sachish, 1986) and the establishment of a port performance and efficiency model using multiple regression analysis (Tongzon, 1995).

This chapter aims to analyse the applicability of Data Envelopment Analysis (DEA) to the port industry and to identify the economic implications that can be derived from its application. To this end, the fundamentals of the DEA technique are presented in Section 2 and previous applications of the technique to the port industry are reviewed in Section 3. A methodology for arriving at a definition of the input and output variables needed to drive the technique is provided in Section 4, together with a discussion of the most important theoretical and practical considerations to take into account in so doing. An illustrative example is provided to show how the DEA technique might be applied in practice and Section 5 details the sample data and the process by which they were collected. The example applies DEA to estimate the efficiency of the world's leading container ports. It utilises both cross-sectional and panel data, as well as a variety of different DEA approaches. The results of this analysis are presented and discussed in Section 6, with particular emphasis placed on the economic, policy and managerial implications of the outcomes of the DEA analysis. Finally, conclusions are drawn in Section 7.

Two points should be particularly emphasised at this stage. First, this chapter focuses more on the specifics of the DEA approach as applied to the container port industry, rather than on the more general topic of container port production itself. Readers interested in the latter are referred to Wang, Cullinane, and Song (2005) where container port production is given a more comprehensive treatment. Second, this chapter deals with the application of DEA exclusively to the container port industry. Given the different characteristics of production at ports that specialise in the handling of different cargo categories (e.g. general cargo and bulk), it is inadvisable to generalise aspects contained in this chapter directly to all types of ports. A thorough investigation of the characteristics of production at these different types of port and an assessment of the appropriateness of the approach are necessary before DEA can be logically applied.

2. DATA ENVELOPMENT ANALYSIS

2.1. Conceptual Exposition

Efficiency is a fundamental concept in the field of economics and has been variously defined in different textbooks. The concept is basically concerned with the economic use of resources (inputs) for production. Given the limited and finite nature of resources available to every productive process, the importance of studying efficiency is self-evident. In the very first sentence of his influential seminal paper, Leibenstein (1966, p. 392) goes so far as to state explicitly: 'At the core of economics is the concept of efficiency'.

DEA concerns itself with assessing the efficiency of an individual firm. This firm is the fundamental unit of analysis that, following aggregation, makes up the sample for analysis and is typically defined as the *Unit of Assessment* (Thanassoulis, 2001) or the *Decision Making Unit* (*DMU*) (Charnes, Cooper, & Rhodes, 1978). In either case, the terminology refers to the organisational entity responsible for controlling the process of production and for making decisions at various levels that may influence the productive process and, *ergo*, the level of efficiency associated with it. These include daily operational, short-term tactical and long-term strategic decisions. DEA can be employed to measure the holistic efficiency of a firm by comparing it with other homogeneous units that transform the same group of measurable positive inputs into the same types of measurable positive outputs. In fact, this homogeneity of both the inputs and outputs constitutes a fundamental underlying assumption upon which the veracity of DEA efficiency measures is based. In the absence of such an assumption, the relevance of measuring efficiency across any set of DMUs could undoubtedly be called into question.

The basic principle of utilising DEA to measure the efficiency level of firms within a given sample can be explained through the use of the example data presented in Table 23.1 and the conceptual illustration provided in

Table 23.1. Single Input and Single Output.

Terminal	T1	T2	T3	T4	T5	T6	T7	T8
Stevedores	10	20	30	40	50	50	60	80
Throughput	10	40	30	60	80	40	60	100
Productivity (throughput/stevedores)	1	2	1	1.5	1.6	0.8	1	1.25
Efficiency (%)	50	100	50	75	80	40	50	62.5

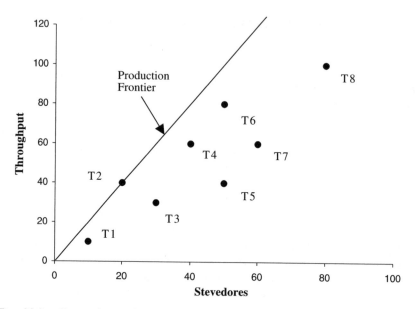

Fig. 23.2. In relation to assessing the validity of the homogeneity assumption, it is critically important that the quality of inputs utilised by the various ports across the sample is similar.

Together, Table 23.1 and Fig. 23.2 present a simplified vision of the production process of eight container terminals, with 'Stevedores' being the only input considered and 'throughput' (i.e. containers per unit time) the only output. The productivity of any entity is simply the absolute measure of outputs/inputs and, therefore, in Table 23.1 the productivity of each terminal is represented by the calculated ratio of 'throughput/stevedore'. It should be pointed out that this measure is equivalent to the slope of the line connecting each point to the origin in Fig. 23.2 and corresponds to the number of containers moved per stevedore per unit time. It is clear from Fig. 23.2 that T2 has the highest absolute level of productivity compared with the other points and is, therefore, the most efficient unit in the sample. As such, the line from the origin through T2 is deemed to define the production frontier, at least insofar as current relative information on a limited sample allows. All the other points are inefficient compared with T2 and are termed to be physically 'enveloped' by the production frontier. In the context of DEA, the relative efficiencies of all units of analysis other than T2 (as shown in the bottom line of Table 23.1) are measured by comparing their productivity with that of T2. The term 'Data Envelopment Analysis' stems directly from this graphical representation of a frontier 'enveloping' datapoints and of datapoints being 'enveloped' by a frontier.

The way in which efficiency is calculated in Table 23.1 and Fig. 23.2 is based on an assumption that production exhibits constant returns-to-scale. In other words, there are no (dis)economies of scale as the level of productive output changes. The DEA model corresponding to this assumption is termed the DEA-CCR model (named in recognition of the seminal contribution of Charnes, Cooper & Rhodes, 1978). Apart from the DEA-CCR model, the DEA-BCC model (similarly named in recognition of Banker, Charnes, & Cooper, 1984) as well as the Additive model, are the two other DEA models that are widely studied and applied. The main difference between these two models and the DEA-CCR model is that the former allows for perhaps a more realistic assumption of variable returns-to-scale, in contrast to the constant returns-to-scale assumed in the DEA-CCR model. Accordingly, the production frontier representations associated with these models are different. When either the DEA-BCC model or the Additive model are utilised, Fig. 23.3 shows the piecewise linear production frontier that would be estimated for the same data sample given in Table 23.1. The contrast with the production frontier in Fig. 23.2, where the DEA-CCR model is utilised, can be clearly seen. In Fig. 23.3, the datapoints defined by T1, T2, T6 and T8 are all located on the estimated production frontier and, therefore, are defined as efficient. In essence, therefore, given the validity of the variable returns-to-scale assumption, none of these sample datapoints can dominate each other. The other observations that are 'enveloped' by these *efficient* units are deemed to be *inefficient*.

The DEA-BCC model and the Additive model are identical in terms of the production frontiers that are estimated by applying each of the respective techniques. The main difference between them lies with the derivation of the projection path from each of the datapoints that represent the inefficient firms onto the production frontier. This is important since it is the proportionate distance that results from the projection path that impacts directly upon the efficiency estimate derived for a specific inefficient datapoint. For instance, the inefficient firm located at T3 can be projected to either T3I or T3O under the DEA-BCC model in terms of either the input or output orientation. On the other hand, it is projected to T2 in the Additive model. Ultimately, it is merely this different method of projection that yields the different estimates of the relative efficiencies for the inefficient firms.

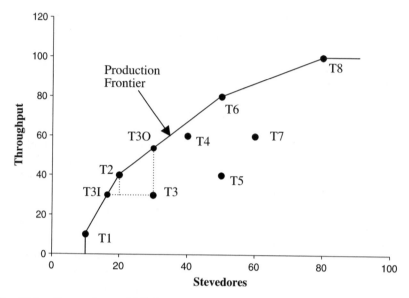

Fig. 23.3. Comparison of Efficiencies of Container Terminals (BCC and Additive Models). *Source:* Drawn by the Authors.

The basic information derived from the above three DEA models, i.e. the DEA-CCR model, the DEA-BCC model and the Additive model, is whether or not a firm can improve its performance relative to the set of firms to which it is being compared. A different set of firms is likely to provide different efficiency results because of the possible movement of the production frontier.

The lack of allowance for statistical noise is widely regarded as the most serious limitation of DEA (Ray, 2002), because this puts a great deal of pressure on users of this technique to collect data on all relevant variables and to measure them accurately. In consequence, a great deal of research effort has been exerted on the sensitivity of efficiency estimates to measurement error and to changes in the sample being analysed, particularly with respect to sample size and to changes in the number of input or output variables. Interested readers are referred to Charnes and Cooper (1968), Charnes, Cooper, Lewin, Morey, and Rousseau (1985), Charnes and Neralic (1992), Charnes, Haag, Jaska, and Semple (1992), Charnes, Rousseau, and Semple (1996), Charnes, Clark, Cooper, and Golany (1985), Thrall (1989), Wilson (1995), Banker, Chang, and Cooper (1996) and Neralic (1997). Work is also underway for facilitating formal significance testing and other tests of inference in order to address another major limitation of DEA. Koop, Osiewalski, and Steel (1999) are particularly critical of the technique in relation to how difficult it is to construct confidence intervals around the efficiency estimates that it yields. In this respect, Simar and Wilson (1998, 2000) and Simar and Zelenyuk (2003) have introduced bootstrapping techniques into the DEA framework to overcome this and other associated shortcomings.

2.2. Model Specification

DEA models can be distinguished according to whether they are input- or output-oriented. The former allows the study of how to minimise inputs while producing a given level of output, while the latter is concerned with how to maximise outputs while using no greater quantity of any of the individual inputs within a given set of inputs. Mathematically, inputs are denoted as $x_k = (x_{1k}, x_{2k}, \ldots, x_{Mk}) \in R_+^M$ and, following their combination within a production process, they produce outputs $y_k = (y_{1k}, y_{2k}, \ldots, y_{Nk}) \in R_+^N$. The row vectors x_k and y_k form the kth rows of the data matrices X and Y respectively. Let $\lambda = (\lambda_1, \lambda_2, \ldots, \lambda_K) \in R_+^K$ be a non-negative vector, which forms the linear combinations of the K firms. Finally, let $e = (1, 1, \ldots, 1)$ be a suitably dimensioned vector of unity values.

An output-oriented efficiency measurement problem can be written as a series of K linear programming envelopment problems, with the constraints differentiating between the DEA-CCR and DEA-BCC models, as shown in Eqs. (1)–(5). Under this specification, U is a scalar quantity such that $1 \leq U \leq \infty$ and $U-1$ yields the proportional increase in outputs that could be achieved by the ith DMU, with input quantities held constant.

$$\max_{U, \lambda} \quad U \tag{1}$$

Subject to

$$U y_{k'} - Y' \lambda \leq 0 \tag{2}$$

$$X' \lambda - x_{k'} \leq 0 \tag{3}$$

$$\lambda \geq 0 \ \text{(DEA-CCR)} \tag{4}$$

$$e \lambda' = 1 \ \text{(DEA-BCC)} \tag{5}$$

The combination of Eqs. (1)–(4) and (1)–(5) form the DEA-CCR and DEA-BCC models, respectively. The equivalent input-oriented models can be specified in a similar fashion. Interested readers may refer to Seiford and Thrall (1990), Ali and Seiford (1993) and Cooper, Seiford, and Tone (2000) for a more in-depth discussion on the logic, formulation and interpretation of the above models.

The output-oriented measure of the technical efficiency of the kth DMU, denoted by TE_k, can be computed by applying the following equation:

$$TE_k = 1/U_k \qquad (6)$$

The technical efficiency estimates that are derived from both DEA-CCR and DEA-BCC models are frequently utilised together to obtain a measure of scale efficiency for each of the K DMUs $(1...k)$. As shown by Cooper et al. (2000), this is achieved by applying the following equation:

$$k = U_{\text{CCR}\_k}/U_{\text{BCC}\_k} \qquad (7)$$

where SE_k denotes the scale efficiency of the kth DMU, while U_{CCR_k} and U_{BCC_k} are the estimated technical efficiency measures for DMU k that are derived from applying the DEA-CCR and DEA-BCC models, respectively, as specified in Eqs. (1)–(5). Under this metric, if $SE_k = 1$ then this indicates scale efficiency, while if $SE_k < 1$ then scale inefficiency is indicated.

Scale inefficiency occurs in situations where either increasing or decreasing returns to scale prevail. Which of the two circumstances actually exists can be determined by inspecting the sum of each of the weights, $e\lambda'$, which is derived from applying the specification for a DEA-CCR model. If this sum is equal to one, constant returns to scale prevail. Increasing returns to scale exist when the sum is less than unity and decreasing returns to scale prevail when the sum is found to be greater than one.

In considering the pivotal influence of time, let t denote the point in time when a particular observation is made and T denote the total number of time periods observed. Then, the input and output variables of firm k can be rewritten as $(x_{kt}) = (x_{1kt}, x_{2kt} \; ... \; x_{Mkt}) \in R_+^M$ and $(y_{kt}) = (y_{1kt}, y_{2kt} \; ... \; y_{Nkt}) \in R_+^N$, respectively.

In analysing cross-sectional data using DEA models, the focus of interest lies solely with the comparison of one firm with all other firms in the feasible dataset. Distinct from this approach, evaluating the efficiency of an individual firm on the basis of a set of panel data involves the selection of only alternative subsets, termed *reference observations subsets* (Tulkens & van den Eeckaut, 1995), rather than the full data set. Tulkens and van den Eeckaut suggest that each observation in a set of panel data can be characterised in efficiency terms vis-à-vis three different kinds of frontiers, specified as (i) to (iii) below. Alternatively, an approach originally developed by Charnes et al. (1985) and described as (iv) below can also be applied:

(i) *Contemporaneous*: involving the construction of a *reference observations subset* at each point in time, with all the observations made only at that point in time. Each of the different *reference observations subsets* for each point in time can therefore be denoted as

$$\{(x_{kt}, y_{kt}) | k = 1, 2, \ldots, K\} \quad \text{for } t = 1, 2, \ldots, T$$

Over the whole period of analysis, a sequence of T *reference observation subsets* are constructed, such that there exists one that pertains to each time period t.

(ii) *Intertemporal*: involving the construction of a single production set from the observations made over the whole period of analysis. In this case, the *reference observations subset* is denoted simply as

$$\{(x_{kt}, y_{kt}) | k = 1, 2, \ldots, K; \quad t = 1, 2, \ldots, T\}$$

(iii) *Sequential*: involving the construction of a *reference observations subset* at each point in time t but utilising all the observations made from some point in time $h = 1$ up until $h = t$. The *reference observations subsets* at each time $t = 1, 2, \ldots, T$ can therefore be denoted as

$$\{(x_{kt}, y_{kt}) | k = 1, 2, \ldots, K; \quad h = 1, 2, \ldots, t\}$$

Obviously, this method does suffer from the disadvantage that, as t moves towards T, it does lead to a certain imbalance in the number of observations over which a given estimate of average efficiency is calculated. Because of this problem, this approach is rarely applied in practice and, indeed, is omitted from any further consideration herein.

(iv) *Window analysis*: a time-dependent version of DEA. The basic idea is to regard each firm as if it were a different firm in each of the reporting dates. Then each firm is not necessarily compared with the whole dataset, but rather only with alternative subsets of panel data. If w is the window width which describes the time duration for the *reference observations subsets*, then a single window *reference observations subset* can be expressed as

$$\{(x_{kt}, y_{kt}) | k = 1, 2, \ldots, K; \quad h = t, t+1, \ldots, t+w; \quad t \leq T - w\}$$

Successive windows, defined for $t = 1, 2, \ldots, T-w$, yield a sequence of *reference observations subsets*.

Window analysis is based on the assumption that what was feasible in the past remains feasible forever. In addition, the treatment of time in window analysis is more in the nature of an averaging over the periods of time covered by the window. So that the efficiency estimates derived from *window analysis* reflect solely the difference between the actual level of production of a firm and the best level of contemporaneous production, the window width should ideally be defined to correspond to the standard cycle time between technological innovations in the industry being analysed. Even if such a thing should exist, the technological innovation cycle time within the port industry is difficult to observe in practice and, therefore, it is difficult to find more than an ad hoc justification for the size of the window, as well as for the fact that part of the past is ignored (Tulkens & van den Eeckaut, 1995).

The difference between *contemporaneous*, *intertemporal* and the *window* approaches (where the window width has been arbitrarily defined as 3, $w = 3$) is depicted conceptually in Fig. 23.4. In this illustrative example, an input-oriented model with two inputs that are used to produce the same amount of output(s) is assumed. Fig. 23.4 portrays the production of four firms A, B, C and D ($k = 1, 2, 3, 4$) that are shown over four time periods ($t = 1, 2, 3, 4$). Logically, this constitutes 16 observations, although not all are mapped in Fig. 23.4 because it is assumed that the individual observations that are not shown do not influence the production or cost frontier.

In Fig. 23.4 then, $D3$ refers to the production of firm D at time $t = 3$. *FI*, *FW* and *FC* in Fig. 23.4 denote each of the frontiers associated with the application of an *intertemporal* approach, *window* analysis and *contemporaneous* analysis. $A1$, $B2$ and $D4$ are the most efficient assuming a frontier derived from the application of *intertemporal* analysis and, thus, together with $(x_{111}, +\infty)$ and $(+\infty, x_{244})$ define the cost frontier for all 16 observations that make up the set of panel data. However, when applying *window* analysis, $D4$ is excluded and replaced by $D3$ in the first window $\{(x_{kh}, y_{kh}) | k = 1, 2, \ldots, 4; \quad h = 1, 2, 3\}$, such that the points $A1$, $B2$, $D3$, $(x_{111}, +\infty)$ and $(+\infty, x_{243})$ define the cost frontier for the first window.

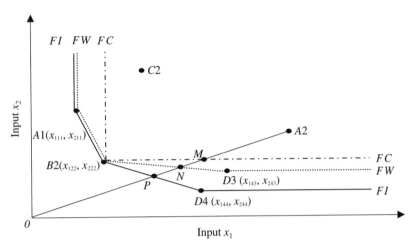

Fig. 23.4. Conceptual Illustration of Contemporaneous, Intertemporal and Window Analyses. *Source:* Drawn by the Authors.

Finally, when applying *contemporaneous* analysis when $t = 2$, the points $B2$, $(x_{122}, +\infty)$ and $(+\infty, x_{222})$ define the efficient frontier for the observations $A2$ and $C2$ which occur within the same time period. Thus, when applying *contemporaneous* analysis to this hypothetical data, the only observation at time $t = 2$ which is operating efficiently is $B2$. The production of DMUs A and C in time $t = 2$ is deemed inefficient in that observed values of their level of production in this time period do not lie on the efficient frontier. Focusing exclusively on the efficiency of $A2$ as estimated by the application of *contemporaneous* (U_c), *window* (U_w) and *intertemporal* (U_i) analyses, the resulting values are given by the ratios $0M/0A2$, $0N/0A2$ and $0P/0A2$, respectively. The efficiency of all other observations in this extremely simplified, but still complicated, example can be derived analogously.

Either by calculating the values of these ratios or by merely seeing the location of the observation $A2$ in relation to each of the efficient frontiers ascribed as *FI*, *FW* and *FC*, then it is patently obvious that $U_i \leq U_w \leq U_c$. In essence, *contemporaneous* and *intertemporal* analyses are two extreme situations of *window* analysis where, respectively, the window widths are specifically allocated the boundary values of $w = 1$ and $w = T$. This implies that the efficiency of every observation tends to decline as window width (w) increases. This can be rather simply explained by the fact that a firm in a small sample has fewer counterparts to be compared against and, therefore, the probability of being dominated by other observations is smaller. An alternative perspective is that there is a higher probability of it being classified as more efficient.

Observing the impact of changes in window width on efficiency estimates is of great significance. It can ultimately be explained by the relationship between growth in total factor productivity and its two components: technical progress and changes in technical efficiency. Technical progress relates to the shifting of the production frontier over time (i.e. as brought about by technological developments), while technical efficiency indicates the ability of firms to follow best-practice techniques so that they can continue to operate on the frontier at any point in time. In other words, to make best use of the technology that is available at any point in time. Alternatively, assessing the level of dynamic change and trends in technical efficiency involves the analysis of whether the gap between best-practice techniques and the methods of production actually employed has been diminishing or widening over time. It is relatively safe to assume that the level of cutting-edge technology that is available remains constant during the time period defined by a given window width w, at least and especially when w is reasonably small (e.g. $w = 1$). In such a case, estimates of technical efficiency that are produced by employing this set of DEA techniques that utilise panel data are actually measuring whether a firm is following best practice at any particular point in time. On the other hand, when window width is large, estimates of technical efficiency are measuring, both simultaneously and in combination, whether a firm is keeping abreast of both the latest technology and industry best practice in the production process. This differentiation has crucial policy implications in that it helps to identify the source of any technical inefficiency that is found. For instance, it would be a waste of precious and finite resources if a firm is simply unable to make efficient use of its existing technology and facilities, but instead attributes its inefficiency to technological shortcomings and erroneously embarks on a program of technological upgrading that unnecessarily absorbs valuable investment funds that have a high opportunity cost.

3. APPLICATIONS OF DEA TO THE PORTS SECTOR[1]

In recent years, DEA has been increasingly used to analyse port production. Compared with traditional approaches, DEA has the advantage that it can cater for multiple inputs to and outputs from the production process. This accords with the characteristics of port production, so that there exists, therefore, the capability of providing an overall summary evaluation of port performance.

- Roll and Hayuth (1993) probably represents the first work to advocate the application of the DEA technique to the ports context. However, it remains a purely theoretical exposition, rather than a genuine application.
- Martinez-Budria, Diaz-Armas, Navarro-Ibanez, and Ravelo-Mesa (1999) classified 26 ports into three groups, namely high-, medium- and low-complexity ports. After examining the efficiency of these ports using DEA-BCC models, the authors conclude that the ports of high complexity are associated with high efficiency, compared with the medium and low efficiency found in other groups of ports.
- Tongzon (2001) uses both DEA-CCR and DEA-Additive models to analyse the efficiency of four Australian and 12 other international container ports for 1996. The results suggest that Melbourne, Rotterdam, Yokohama and Osaka are the most inefficient ports in the sample, with enormous scope for improvement found to exist with respect to container berths, terminal area and labour inputs. Clearly, plagued by a lack of data availability and the small sample size (only 16 observations), more efficient ports than inefficient ports are naturally identified. Realising this serious drawback, the author concludes that further work should be done in collecting more observations to enlarge the sample analysed.
- Valentine and Gray (2001) apply the DEA-CCR model to 31 container ports out of the world's top 100 container ports for 1998. The objective of the authors' work is to compare port efficiency with a particular type of ownership and organisational structure to determine if there exists any relationship between them. The authors conclude that cluster analysis is a viable tool for identifying organisational structures and that the ports sector exhibits three structural forms that seem to have a relationship to estimated levels of efficiency.
- For the period 1990–1999, Itoh (2002) conducted a DEA window analysis using panel data relating to the eight international container ports in Japan. Tokyo was found to be consistently efficient in terms of its infrastructure and labour productivity over the whole period, while Nagoya performed well during the early part of the period covered by the analysis. At the other extreme, efficiency scores for Yokohama, Kobe and Osaka were found to be low throughout the duration of the period under study.
- Barros (2003a) applies DEA to the Portuguese port industry in 1999 and 2000. The motivation for the analysis is to determine what relationship exists between the governance structure that has been established for the

Portuguese port sector, the incentive regulation promulgated under this structure and the ultimate impact on port efficiency. The author concludes that extant incentive regulation has been successful in promoting enhanced efficiency in the sector, but that this could be improved upon by the implementation of recommendations aimed at redefining the role of Portugal's Maritime Port Agency, the regulatory body responsible for port matters.

- This time using data for 1990 and 2000, Barros (2003b) again applies DEA to the Portuguese port industry to derive estimates of efficiency that can then be utilised to determine the source of any inefficiency that may be identified. One of the results of the analysis is that while Portuguese ports have attained high levels of technical efficiency over the period covered by the analysis, the sector has generally not kept pace with technological change. The author concludes that the financial aids to investment that form part of the EU's Single Market Program have stimulated greater efficiency in the port sector, particularly as the result of the greater competition that is faced; a feature that is particularly relevant for Portuguese ports located near the border with Spain. Through the application of Tobit regression analysis, it is also found that container ports are more efficient than their multi-cargo counterparts (suggesting that there are diseconomies of scope in cargo handling), that efficiency is positively related to market share and, finally, that greater public sector involvement is negatively related to efficiency.

- In yet another extension of this work, Barros and Athanassiou (2004) apply DEA to the estimation of the relative efficiency of a sample of Portuguese and Greek seaports. The broad purpose of this exercise was to facilitate benchmarking so that areas for improvement to management practices and strategies could be identified and, within the context of European ports policy, improvements implemented within the seaport sectors of these two countries. The authors conclude that there are economic benefits from the implementation of this form of benchmarking and go on to evaluate their extent.

- Bonilla, Casasus, Medal, and Sala (2004) apply a version of DEA that includes a statistical tolerance for inaccuracies in input and output data to the investigation of commodity traffic efficiency within the Spanish port system. Their sample comprises 23 ports and annual data are collected for 1995–1998 inclusive. The analysis is unusual in that the sample ports handle a range of cargoes – solid bulk, liquid bulk and general break-bulk – rather than being restricted to a single form of cargo (most usually containers). Given a calculated high level of correlation between prospective input variables, a single input encapsulating infrastructure endowment is incorporated into the analysis. The most and least efficient Spanish ports are identified and, using an 'incidence analysis', the authors conclude by identifying which ports are most sensitive to variations in traffic volumes among the different types of cargo handled.

- Given the characteristics of the container port industry and the random effects associated with a single measured value of production for each port or terminal in a sample and the level of measured efficiency associated with it, Cullinane, Song, Ji, and Wang (2004) recognised that the analysis of cross-sectional data will inevitably provide inferior estimates of efficiency than those based on panel data. In seeking to allow for this potential, they applied alternative DEA approaches based on cross-sectional and panel data analysis, respectively. The authors conclude that by so doing, the development of the efficiency of each container port or terminal in a sample can be tracked over time and that this provides interesting and potentially useful insights for both policy formulations and management.

- Recognising the limitations in assessing the efficiency of ports solely on the basis of capital and labour inputs, Park and De (2004) develop what they refer to as a 'Four-Stage DEA Method'. This involves the disaggregation of the overall efficiency model into its constituent components, so that better insight can be gained into the real sources of efficiency. The model comprises individual DEA components that determine the respective efficiency related to productivity, profitability, marketability and overall. In applying their method to a sample of Korean ports, the authors conclude that improving the marketability of Korean seaports should be the utmost priority of port authorities.

- Turner, Windle, and Dresner (2004) applied DEA to the determination of changes in infrastructure productivity in North American ports over the period 1984–1997. They then went on to use the productivity estimates as the dependent variable within a Tobit regression model, which sought to determine the causal factors affecting the scores they derived. Perhaps most significantly, the authors conclude that there are significant economies of scale present within the North American sector, both at port and at terminal level – a finding that concurs with the outcomes of most research investigating economies of scale in the port sector. They also find that access to the rail network is a pivotal determinant of container port infrastructure productivity in North America, but that there is no evidence to suggest that specific investment in on-dock rail facilities is a productive use of the land-take involved.

- Cullinane, Ji, and Wang (2005) empirically examine the relationship between privatisation and relative efficiency within the container port industry. The sampling frame comprises the world's leading container ports ranked in the top 30 in 2001, together with five other container ports from the Chinese mainland. DEA is applied in a variety of panel data configurations to eight years of annual data from 1992 to 1999, yielding a total of 240 observations. The analysis concludes that there is no evidence to support the hypothesis that greater private sector involvement in the container port sector irrevocably leads to improved efficiency.
- Using cross-sectional data for 2002, Cullinane and Wang (2006) apply DEA to the derivation of estimates of relative efficiency for a sample comprising 69 of Europe's container terminals with annual throughput of over 10,000 twenty foot equivalent units (TEUs). The sample was distributed across 24 European countries. The main finding is that significant inefficiency pervades the European container handling industry, with the average efficiency of container terminals under study amounting to 0.48 (assuming constant returns to scale) and 0.42 (assuming variable returns to scale). Most of the container terminals under study exhibit increasing returns to scale, with large container terminals more likely to be associated with higher efficiency scores. A further conclusion was that there was significant variation in the average efficiency of container terminals located in different regions, with those in the British Isles found to be the most efficient and Scandinavia and Eastern Europe the least efficient.
- Cullinane, Wang, Song, and Ji (2005) apply both DEA and Stochastic Frontier Analysis (SFA) to the same set of container port data for the world's largest container ports and compare the results obtained. A high degree of correlation is found between the efficiency estimates derived from all the models applied, suggesting that results are relatively robust to the DEA models applied or the distributional assumptions under SFA. High levels of technical efficiency are associated with scale, greater private-sector participation and with transhipment, as opposed to gateway, ports. In analysing the implications of the results for management and policy-makers, this paper concludes that a number of shortcomings of applying a cross-sectional approach to an industry characterised by significant, lumpy and risky investments are identified and the potential benefits of a dynamic analysis, based on panel data, are enumerated.
- Similarly, using the same dataset for the world's most important container ports and terminals, Cullinane, Song, and Wang (2006) evaluate and compare relative efficiency estimates derived from the two alternative techniques of DEA and the Free Disposal Hull model. The results confirm

expectations that the available mathematical programming methodologies lead to different results and the conclusion is drawn that the appropriate definition of input and output variables is a crucial element in meaningful applications of such techniques.

4. THE DEFINITION OF INPUT AND OUTPUT VARIABLES

4.1. Approach

The precise definition of both input and output variables is crucial to the successful and meaningful application of DEA. This is because however elaborate the models employed may be, the erroneous or ill-defined specification of variables for collection, collation and subsequent analysis will inevitably lead to the derivation of results that are easily misinterpreted and possibly misleading. This will obviously negate the veracity of any conclusions drawn from them. In this respect, Norman and Stoker (1991) have provided some useful guidance on variable definition and justification when applying DEA. They suggest several steps for the appropriate implementation of efficiency measurement, as summarised in Fig. 23.5.

The most pertinent of the steps involved can be encapsulated as follows:

- To identify and define the role and objectives of the firms in the sample population. This involves the identification of where authority lies (and its limits), where responsibility lies (and its limits) and what resources (labour, capital, land, knowledge, etc.) are at the disposal of the firms. The role must be determined at the level of the whole organisation or service and the range of products/services it seeks to provide. Some of the most important questions in this respect include, why was this firm/organisation/unit set up, what does it do and whom does it serve? The definition of the role of the firm or organisation leads in turn to the identification of a range of objectives that it seeks to achieve.
- The second step involves conducting a pilot exercise and is mainly concerned with the number of firms to be measured. This issue is important because a small number of firms is more likely to generate a high proportion of efficient units, thus depriving the analysis of any opportunity for discriminating between the efficiency of units or to generate the rich variety of subtle information that is potentially obtainable. Norman and Stoker (1991) suggest that the minimum number of firms that should be

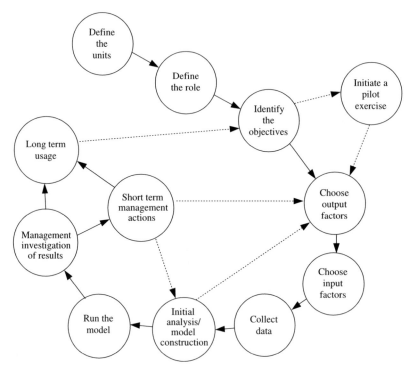

Fig. 23.5. A Performance Measuring System. *Source:* Norman and Stoker (1991, p. 168).

considered is 20 or, alternatively, that a general guideline for the minimum number of units that make up the sample for evaluation is at least twice the sum of the inputs and outputs. Cooper et al. (2000) adopt a more prudent approach in recommending a minimum sample size given by

$$N \geq \max \{m \times s, \ 3(m + s)\}$$

where N is the minimum sample size of DMUs, m the number of inputs and s the number of outputs. Since DEA is not founded on the principles and logic of statistics, there exists no possibility of drawing statistically valid inferences from the results of the analysis. However, in seeking to obtain as accurate a picture as possible of the distribution of relative efficiency across firms within the industry under study, efforts should be made to obtain data for, and subsequently analyse, a sample that is sufficiently large to be representative of the wider population of firms (DMUs) with which the analysis is concerned. Since the measures of technical efficiency that are derived from applying DEA are determined *relatively*, the use of as large a sample size as possible allows some loose, non-statistical, yet meaningful, generalisation of the results to the wider population and also enhances the accuracy of the estimates derived for individual DMUs.

- The third step is to choose output variables that reflect the provision of support to the achievement of the objectives of DMUs in the sample. These will be measurable quantities that refer to specific aspects of achievement. Norman and Stoker (1991) point out that the golden rule in selecting output variables is to be as inclusive and comprehensive as possible in covering the whole gamut of work (production) that the firms undertake. In certain circumstances, however, it should be recognised that this might be fairly difficult to achieve, not only in terms of the identification of all forms of output, but also in terms of collecting sufficiently accurate data that are appropriate to each of the identified forms. Norman and Stoker go on to recommend that the customers of the firms in the sample should first be identified and, in so doing, prime consideration should be given to two questions: (a) who buys or benefits from the goods or services produced by the firms under consideration, and (b) what metric can be applied to yield some quantity of the goods or services that are produced for consumption? Deducing the precise definitions of the output variables to be considered and for which data have to be collected can be based on the answers to these two questions.

It is important here to emphasise that the relationship between a firm's objective(s) and the choice of input and output variables is also discussed by Lovell (1993), who suggests that only *useful* input(s) and output(s) should be considered. If this were not the case, and all possible inputs and outputs were to be considered, then all firms would achieve the same unitary efficiency score since the model would then be overspecified. This is analogous to the overspecification which might occur in regression analysis whereby a very high coefficient of correlation that tends to unity (i.e. $r^2 = 1$) is easily achievable when a large number of independent (input) variables are regressed against a single dependent (output) variable, especially when the sample size is comparatively small. This analogy illustrates quite clearly that this problem of overspecification in the mathematical programming approach of DEA (i.e. an approach with non-statistical underpinnings) parallels a situation where *multi-collinearity* is present in regression analysis. From what has been said earlier, it is clear that this concept of usefulness is closely related to the objective(s) of the

DMUs in the population under study. Where these might differ within the population and, therefore, within the sample under study, it is potentially feasible that one product might be regarded as invaluable for one DMU and completely useless for another. In relation to this particular problem, Stigler (1976) observes that the measured inefficiency that results from the application of this sort of method may be a reflection of a failure to incorporate any one, or combination, of the right variable(s), the right constraints and the appropriate specification of the economic objective(s) of the firm. An empirical proof of this potential is provided by Kopp, Smith, and Vaughan (1982), who demonstrate the adverse consequences for efficiency measurements of failing to account for residual discharge constraints in steel production.

- The fourth step is to choose input variables. Norman and Stoker (1991) recommend that the normal practice in this respect should be to start with a list of factors that is impractically long, then review the list and eliminate obvious duplications or irrelevancies on an iterative basis.
- The fifth step in data collection is closely related to steps 3 and 4. In many cases, data are not actually available for all input and output variables for which it is desired to collect data. If this happens, Norman and Stoker (1991) recommend three alternative approaches to coping with this problem: (a) abridge the list of variables; (b) initiate data gathering and data mining exercises or (c) a mixture of (a) and (b).

4.2. Implementation

In line with the guidance given by Norman and Stoker (1991), a thorough discussion of variable definition for applications to container ports is provided in Wang (2004), but can be summarised as follows. The input and output variables should reflect actual objectives and the process of container port production as accurately as possible. As far as the former is concerned, the observed performance of a port might be closely related to its objective(s). For instance, a port is more likely to utilise state-of-the-art, expensive equipment to improve its productivity if its objective is to maximise cargo throughput. On the other hand, a port may be more willing to use cheaper equipment if its objective is simply to maximise profits.

The objectives of a port are, in fact, a crucial consideration in defining the variables for efficiency measurement. For instance, if the objective of a port is to maximise its profits, then employment or any information on labour should be counted as an input variable. However, if the objective of a port is to increase employment, then information on labour should be accounted for as an output variable.

In order to demonstrate how DEA may be applied in practice, this chapter will go on to apply the technique to a real set of data. To this end, the main objective of a port is assumed to be the minimisation of the use of input(s) and the maximisation of output(s). Because of the difficulty of obtaining confidential data that, in any case, are often only available on an inconsistent basis across different corporate entities, nations, etc., the respective prices of inputs or outputs are not taken into account in the illustrated example contained herein. In consequence, this assumed objective may not be entirely consistent with that of profit maximisation. The argument can be made, however, that the capital intensity of a container port, handling a largely homogeneous throughput with only small variation in the price charged per container, means that its profits do mainly stem from handling sufficiently large volumes to cover the significant fixed cost of investment.

The example application of DEA to container ports presented in this chapter is, therefore, exclusively concerned with the assessment of the *technical efficiency* of the DMUs that make up the sample. Because economic variables are not incorporated in the analysis and since the focus lies with assessing the extent to which physical resources and facilities (the inputs) are optimally utilised, it is quite feasible that any container port in the sample may be assessed as being technically inefficient while simultaneously achieving allocative efficiency. A practical situation where this might be the case is when, in a labour-abundant country, a port utilises a large volume of labour simply because it is less costly than capital. Of course, improving technical efficiency will improve overall economic efficiency.

This assumed objective of simultaneously seeking to minimise inputs while maximising outputs is justified not only by its analytical tractability but also by, inter alia, the facts that:

(i)   Contemporary container ports rely heavily on sophisticated equipment and information technology rather than being labour-intensive. In undertaking the illustrative empirical analysis contained herein to determine how well the assumed objective has been attained, the level of utilisation of state-of-the-art assets can be determined and, therefore, the overall quality of the management inferred. This has obvious implications for, and in all probability a high correlation with, the achievement of more orthodox corporate objectives such as profit maximisation.

(ii) Often because of investment in inland transport infrastructure and the development of enhanced logistics capabilities, the hinterlands of container ports have expanded quite rapidly in recent years and increasingly overlap (Cullinane & Khanna, 2000). This has led to more fervent competition between container ports. In consequence, achieving this assumed objective is likely to be more urgent than any other. Container ports compete on both their direct and indirect costs. In the case of the former, these are passed onto customers wherever possible. The latter are related to productivity levels and impact port customers through the value of customer time (Goss & Mann, 1974) and the generalised costs they face (Ortúzar & Willumsen, 2001). Given a standard unit cost, price competitiveness is undermined by the failure to minimise the use of inputs. Similarly, a failure to maximise outputs for a given input level will, irrespective of prices charged, undermine a port's ability to achieve maximum productivity through scale economies and through the inability to accumulate reserves for further investment;

(iii) This assumed objective also conforms to the findings of most research in the field (e.g., Edmond & Maggs, 1976; Gilman, 1983; Jansson & Shneerson, 1987).

Container throughput is the most important and widely accepted indicator of port or terminal output. Almost all previous studies treat it as an output variable because it closely relates to the need for cargo-related facilities and services and is the primary basis on which container ports are compared, especially in assessing their relative size, investment magnitude or activity levels. Another consideration is that container throughput is the most appropriate and analytically tractable indicator of the effectiveness of the production of a port.

On the other hand, container port production depends crucially on the efficient use of labour, land and equipment. Therefore, the total quay length, the terminal area, the number of quayside cranes, the number of yard gantry cranes and the number of straddle carriers constitute highly suitable elements to be incorporated into the models as input variables. In the light of the unavailability or unreliability of direct data, information on labour inputs for the illustrative example developed herein is derived from a predetermined and highly correlated relationship to terminal facilities. For future applications of DEA to container ports, it is very important to note that the predetermined relationship employed herein is not universally applicable to all types of ports possessing different production characteristics. Because of the different equipment and labour configurations that will be employed, it is also dangerous to apply this relationship to container ports operating at different scales of production.

5. THE DATA[2]

5.1. Specification of Samples

For the purpose of illustrating how the DEA technique may be applied to a sample of container ports, sets of both cross-sectional and panel data are analysed to allow the estimation of container port or terminal efficiency under a number of differing assumptions and model specifications.

Sample 1. This sample underpins the cross-sectional data analysis and is based on the world's leading container ports ranked in the top 30 in 2001. Of these 30, the Port of Tanjung Pelepas (PTP) in Malaysia and San Juan are excluded – the former because it did not officially open until 2000, and the latter because the required data are simply not available. Of the remaining 28 ports under consideration, some have multiple terminals for which the required data are available at the disaggregate level. Thus, the sample for analysis comprises a total of 57 observations that relate either to individual container ports themselves or to a number of individual container terminals within some of the container ports that are present in the sampling frame. Important descriptive statistics relating to sample 1 are summarised in Table 23.2.

Sample 2. This sample serves as the basis for the panel data analysis and also comprises the world's leading container ports ranked in the top 30 in 2001. Although ranked in the top 30, five container ports (Shenzhen, Gioia Tauro, PTP, Algeciras and San Juan) are finally excluded from the sample because either they have a shorter history than the study period or lack reliable data for analysis. Eight years of annual data from 1992 to 1999 are collected for each port. Thus, the sample for analysis comprises a total of 200 observations. The summary descriptive statistics for this sample are reported in Table 23.3.

5.2. Data Collection

The required secondary data are mainly taken from various issues of the *Containerisation International Yearbook* and *Lloyd's Ports of the World.* As

Table 23.2. Summary Statistics for Sample 1.

	Throughput (TEU[a])	Quay Length (m)	Terminal Area (ha)	Quayside Gantry (Number)	Yard Gantry (Number)	Straddle Carrier (Number)
Maximum	15,944,793	15,718	1,000	99	337	171
Minimum	204,496	305	6	3	0	0
Mean	2,293,516	3,208	139	22	32	27
Standard Deviation	2,625,305	3,125	171	20	55	40

[a]TEU is the abbreviation for "twenty foot equivalent unit", referring to the most common standard size for a container of 20 ft in length.

Table 23.3. Summary Statistics for Sample 2.

	Throughput (TEU[a])	Quay Length (m)	Terminal Area (ha)	Quayside Gantry (Number)	Yard Gantry (Number)	Straddle Carrier (Number)
Maximum	15,944,793	16,917	1,000	111	337	171
Minimum	33,705	142	4	1	0	0
Mean	2,455,824	4,609	194	29	39	31
Standard Deviation	2,459,148	4,022	206	25	61	40

[a]TEU is the abbreviation for "twenty foot equivalent unit", referring to the most common standard size for a container of 20 ft in length.

the publishers of these sources contact the ports under study every year, and the data are compiled based on their surveys, the data analysed are regarded as the most reliable and comprehensive available.

Based on the argument that container terminals are more suitable for one-to-one comparison than whole container ports (Wang, Song, & Cullinane, 2002), the question may be raised as to whether it is more appropriate to undertake an analysis of efficiency at the level of individual container terminals, rather than at the level of the port. While theoretically this may appear to be a better approach to the analysis, data sources often report the required data, especially container throughput, at the aggregate level of the whole port, rather than on the basis of the individual terminals that may make up each of the ports within the sample. In such cases, it proved necessary to define the input(s) and output(s) of a port as the aggregation of the input(s) and output(s) of individual terminals within the port.

6. EMPIRICAL RESULTS AND ANALYSIS

6.1. Cross-Sectional Data Analysis

The software DEA-Solver-PRO 3.0 (Cooper et al., 2000) was employed to solve the DEA models. Without precise information on the returns to scale of the port production function, two types of DEA models, namely the CCR and BCC models, are applied to analyse the efficiency of container terminals.

In line with Eqs. (1)–(7) and the relationship (discussed in Section 3) between scale inefficiency and the sum of weights, $e\lambda'$, under the specification of the CCR model, Table 23.4 reports the efficiency estimates for both the DEA-CCR and DEA-BCC models, an estimate of the scale efficiency and, based on this, the returns to scale classification of each container terminal (port).

It is clear from Table 23.4 that, as one would expect, the DEA-CCR model yields lower average efficiency estimates than the DEA-BCC model, with respective average values of 0.58 and 0.74, where an index value of 1.00 equates to perfect (or maximum) efficiency. Respectively, 9 and 22 out of the 57 terminals (ports) included in the analysis are identified as efficient when the DEA-CCR and the DEA-BCC models are applied. This result is not surprising since a DEA model with an assumption of constant returns to scale provides information on pure technical and scale efficiency taken together, while a DEA model with the assumption of variable returns to scale identifies technical efficiency alone. A Spearman's rank order correlation coefficient between the efficiency rankings derived from DEA-CCR and DEA-BCC analyses is 0.80. The positive and high Spearman's rank order correlation coefficient indicates that the rank of each terminal (or port) yielded by the two approaches is similar. This leads to the conclusion that the efficiency estimates yielded by the two approaches follow the same pattern across the sample of terminals (ports).

Table 23.4 also reports the scale properties of port production yielded by DEA. Of the 57 terminals (ports), 13 exhibit constant returns to scale, 10 exhibit increasing returns to scale, and 34 exhibit decreasing returns to scale. Among those terminals (ports) found to be scale-inefficient, all except for one of the *large* ports (classified as having annual container throughput of less than 1 million TEU) show decreasing returns to scale. On the other hand, scale-inefficient terminals (ports) having annual container throughput of less than 1 million TEU exhibit a mix of both increasing and decreasing returns to scale, but with almost four times as many of the former as the latter.

Table 23.4. Terminal Efficiency under the DEA-CCR and DEA-BCC Models.[a,b].

Port	Port/Terminal	Throughput	DEA-CCR	DEA-BCC	Scale Efficiency	Returns to Scale
Hong Kong	Hong Kong	10,121,971	0.6922	1.0000	0.6922	Decreasing
	HIT	5,236,594	0.8267	1.0000	0.8267	Decreasing
	MTL	2,594,000	0.4942	0.6940	0.7121	Decreasing
	Terminal 3	1,071,376	1.0000	1.0000	1.0000	Constant
	Cosco-HIT	1,220,001	0.6200	0.7510	0.8256	Decreasing
Singapore	Singapore	15,944,793	0.8999	1.0000	0.8999	Decreasing
Busan	Busan	4,074,087	0.2861	0.5435	0.5264	Decreasing
	Jasungdae	1,046,973	0.2505	0.3515	0.7127	Decreasing
	Shinsundae	1,196,207	0.3045	0.4536	0.6713	Decreasing
	Uam	368,765	0.3111	0.3185	0.9768	Decreasing
	Gamman_G	263,784	0.2686	0.6416	0.4186	Increasing
	Gamman_Hanjin	411,130	0.4187	1.0000	0.4187	Increasing
	Gamman_Hyundai	439,646	0.4477	1.0000	0.4477	Increasing
	Gamman_K	347,582	0.3339	1.0000	0.3339	Increasing
Taiwan	Kaohsiung	6,985,361	0.9959	1.0000	0.9959	Decreasing
Shanghai	Shanghai	4,210,000	0.7541	1.0000	0.7541	Decreasing
Rotterdam	Rotterdam	6,343,242	0.4601	0.8045	0.5719	Decreasing
	Home	959,450	0.7394	0.8318	0.8889	Increasing
Los Angeles	Los Angeles	3,828,852	1.0000	1.0000	1.0000	Constant
Shenzhen	Shenzhen	2,512,390	0.4513	0.6386	0.7067	Decreasing
	Yantian	1,588,099	0.6469	0.7578	0.8537	Decreasing
	Shekoui	574,100	0.4239	0.4437	0.9554	Decreasing
	Chiwan	350,191	0.2987	0.3067	0.9739	Decreasing
Hamburg	Hamburg	3,738,307	0.4774	0.7781	0.6135	Decreasing
	Burchardkai	1,880,000	0.7146	0.7561	0.9451	Decreasing
	Eurokai	1,085,906	0.6662	0.6688	0.9961	Decreasing
	TCT Tollerort	342,737	0.4789	0.7114	0.6732	Increasing
	Unikai	209,866	0.1907	0.1959	0.9735	Constant
Long Beach	Long Beach	4,408,480	1.0000	1.0000	1.0000	Constant

Antwerp	Antwerp	3,614,246	0.3310	1.0000	0.3310	Decreasing
	Europe Terminal	1,107,029	0.6424	0.6642	0.9672	Increasing
	Seaport	501,045	0.1545	0.1782	0.8670	Decreasing
	Noord Natie	448,500	0.5481	0.8411	0.6516	Increasing
	Noordzee	435,000	1.0000	1.0000	1.0000	Constant
Port Klang	Port Klang	2,550,419	0.2867	0.4596	0.6238	Decreasing
	Klang Container	938,934	1.0000	1.0000	1.0000	Constant
	Klang Port	810,439	0.2097	0.3148	0.6661	Decreasing
Dubai	Dubai	2,844,634	0.4447	0.5526	0.8047	Decreasing
New York/New Jersey	New York/New Jersey	2,863,342	0.6841	1.0000	0.6841	Decreasing
Bremen/Bremerhaven	Bremen/Bremerhaven	2,180,955	0.6094	0.9679	0.6296	Constant
Felixstowe	Felixstowe	2,696,659	0.3776	0.6056	0.6235	Decreasing
Manila	Manila	2,147,422	0.3748	0.5099	0.7350	Decreasing
	South Harbour	537,998	0.4027	0.4149	0.9706	Constant
	Manila International	865,816	0.2536	0.3306	0.7671	Decreasing
Tokyo	Tokyo	2,695,589	0.5119	0.6533	0.7836	Decreasing
Qingdao	Qingdao	1,540,000	0.4999	0.6303	0.7931	Decreasing
Gioia Tauro	Gioia Tauro	2,253,401	1.0000	1.0000	1.0000	Constant
Yokohama	Yokohama	2,172,919	0.3541	0.3796	0.9328	Decreasing
Laem Chabang	Laem Chabang	1,828,460	1.0000	1.0000	1.0000	Constant
Tanjunk Priok	Tanjunk Priok	2,273,303	0.5493	0.7771	0.7069	Decreasing
Algeciras	Algeciras	1,828,460	0.9022	1.0000	0.9022	Decreasing
Kobe	Kobe	2,176,004	0.2749	0.4694	0.5856	Decreasing
Nagoya	Nagoya	1,534,355	0.6037	0.6809	0.8866	Constant
	Kinjo Pier	204,496	0.4332	1.0000	0.4332	Increasing
	NCB	639,660	0.9267	1.0000	0.9267	Increasing
Keelung	Keelung	1,954,573	1.0000	1.0000	1.0000	Constant
Colombo	Colombo	1,732,855	1.0000	1.0000	1.0000	Constant
Average		2,293,516	0.5759	0.7382	0.7831	

[a] Alternative efficiency results and returns to scale are calculated by applying Eqs. (1)–(7). The estimates presented here are based on output-orientated DEA-CCR and DEA-BCC models.

[b] 1 = 'efficient'.

Although a rather arbitrary dichotomous classification of the sample has been made between *large* and *small* ports on the basis of a cut-off through-put of 1 million TEU per annum, these results do suggest an association between large ports and decreasing returns to scale and between small ports and increasing returns to scale. On the other hand, the proportion of *large* ports (or terminals) that exhibit constant returns to scale is similar to the proportion for the sub-sample of *small* ports (or terminals).

These findings are probably explained by a combination of the indivisible and lumpy nature of port investment, the consequent commercial risks involved and the level of competition in the market. The sample of *large* ports will probably have evolved as the result of successfully pursuing strategies aimed at attaining container hub status. This would inevitably mean that these ports have, over the years, invested heavily in expensive and ever more advanced equipment in order to attract new container shipping services to the port and enhance the technical efficiency of their operations.

Having achieved a certain level of operational scale, large ports are eventually faced with potential limits to their further growth. The level of investment required to further enhance the design capacity of what is already a large container port becomes very significant and may ultimately deter the port from taking the decision to expand. There may even be physical constraints such as the unavailability of land to facilitate any further expansion. At the very least, the decision to opt for further investment in throughput capacity is deferred until such point that all potential sources of improved technical efficiency have been utilised.

This has meant that many large ports typically operate at, or quite often even beyond, the level of throughput capacity for which existing facilities have been designed. With increasing concentration in the container shipping industry and the formation of strategic alliances between them, container shipping companies can exert considerable market power over the prices that ports or terminals are able to charge, especially where competition within the locale is intense and/or where a particular shipping company, or alliance of companies, is dominant in a port or terminal. Hence, competition between ports is rife. In consequence, there is very little opportunity for ports to avoid following this pattern if they wish to retain their competitiveness and maintain their hub port status.

At the other end of the scale, ports with lower throughput levels are also likely to have the objective of attaining or maintaining hub status. As implied above, this requires a certain minimum scale of operation, however, whereby network connectivity between mainline and feeder services can be facilitated (Yap, Lam, & Notteboom, 2006; Hoffmann, 2005). In consequence, *small* ports too are motivated to increase the scale of their operations. Since a larger scale of operation invariably means greater network connectivity, this is particularly the case given the level of competition in the market.

At the same time, *small* ports need not necessarily face any greater difficulty than *large* ports in gaining access to the requisite capital resources to make major investments in infrastructure. This is especially so given that the sums involved at this lower level of scale are also likely to be smaller. Hence, the risks associated with such investments are concomitantly less, even though they bring about significant proportionate growth in design capacity. Obviously, as is also the case with *large* ports, access to capital is even less of a problem where national or regional economic or social considerations are at stake (such as, for example, where local employment is an issue). Ports with lower levels of throughput are less likely to be faced with physical constraints on their expansion, especially since the *small* port sample is almost certain to contain new ports or terminals that are in the early stages of their evolution. On inspection of the samples, many relatively newly developed terminals have indeed been classified as *small* in throughput terms. The corollary of all this is that the throughput level of a *small* port is more likely to fall below design capacity than is the case for their *large* port counterparts and, as such, the potential exists for benefiting from increasing returns to scale.

6.2. Panel Data Analysis

As with the analysis of cross-sectional data using DEA approaches, in the absence of categorical empirical a priori evidence that the production function of container ports exhibits either constant or variable returns to scale, the DEA-CCR and DEA-BCC models were chosen from among several DEA models to analyse port production. Several alternative versions of DEA panel data analyses were implemented as part of this process. These included models that are integral to the *contemporaneous, intertemporal* and *window* approaches to the estimation of efficiency using panel data. While it is relatively straightforward to calculate efficiency estimates using *contemporaneous* and *intertemporal* analyses, caution should be exercised in defining the window width for conducting a *window* analysis. As mentioned earlier, it is difficult to find a justification for the choice of window size. In common with many previous studies of this kind, however, the length of the window used herein has been defined as three time periods. This is also consistent with the original work of Charnes et al. (1985). Six separate windows are represented as separate rows in Appendices A and B. The average of the 18 DEA efficiency scores and their associated standard deviations are presented in the columns denoted 'mean' and 'SD' The efficiency estimates are reported in Appendices A–F.

The approaches used in formulating Appendices A–F lend themselves to a study of 'trends' of efficiency over time. This is achievable through the adoption of a 'row view'. For instance, a cursory glance at Appendix A may prompt the inference that the efficiency of a container terminal differs significantly over time. Taking Hong Kong as an example, its efficiency varies from 0.74 in 1992 to 1 in 1999. Distinct from the other two approaches, *window* analysis also lends itself to the examination of the 'stability' of efficiency within windows by the adoption of a 'column view'. For instance, by adopting this perspective, it is possible to observe that the efficiency of a firm within the different windows can also vary substantially. The observation of 'trend' and 'stability' in *window* analysis reflects simultaneously both the absolute performance of a port over time and the relative performance of that port in comparison to the others in the sample. It is important to recognise, however, that since DEA measures *relative* efficiency, an upward (downward) trend in average scores does not necessarily imply any general technical advance (reduction) but rather a convergence (divergence) of performance within the industry.

Fig. 23.6 depicts the development of the year-by-year average efficiency of all of the container ports in the sample using *contemporaneous, intertemporal* and *window* analysis, assuming in each case both the CCR and BCC model forms. It is clear from Fig. 23.6 that the general trend of average efficiency for the results from applying *intertemporal* analysis during the study period is upward, compared with the downward and almost flat trends (with some fluctuations) observed in the average efficiency estimates derived from applying *window* and *contemporaneous* analyses, respectively. The former can be explained by the fact that long-term technological advancement and managerial development provide an important impetus for improving productivity and efficiency. Within a shorter time period, defined by a window width of a mere three years in this research, the different observations on the sample DMUs (it is important to recognise that the same port observed at different time periods is treated as being different DMUs) are more likely to use the same or similar technology and management. In such a case, estimated efficiency results are not greatly influenced by the technology and management utilised. Fig. 23.6 also shows that an advance in technology does not necessarily imply an overall improvement in efficiency; a feature

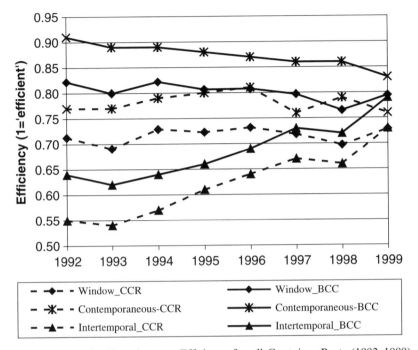

Fig. 23.6. Year-by-Year Average Efficiency for all Container Ports (1992–1999).

that can be shown by the decline in the average efficiency estimates from 1997 to 1998 when applying *intertemporal* analysis.

Fig. 23.6 also shows that the average efficiency reduces in turn for each of the *contemporaneous*, *window* and *intertemporal* analyses. This is not surprising; it is explained by the conceptual illustration shown in Fig. 23.4. In this illustrative example, each port is compared with 24 other counterparts in the same set for a *contemporaneous* analysis, with 74 counterparts for *window* analysis and 199 counterparts for *intertemporal* analysis. Clearly, the number of counterparts against which each port is compared will be negatively correlated to the level of average estimated efficiency.

7. CONCLUSIONS

As one of the most important concepts in the context of economics, efficiency is concerned with how to use limited resources more economically for any sort of production. As a benchmarking approach to study efficiency, DEA enables a port to evaluate its performance vis-à-vis its peers. In so doing, the possible waste of resources and the industry best practice can be identified.

This chapter has investigated the fundamentals of DEA and demonstrated how DEA can be applied to measure the efficiency of container ports. The most frequently used DEA models, including DEA-CCR and DEA-BCC models that respectively correspond to the assumptions of constant returns to scale and variable returns to scale of port production, are applied to analyse both cross-sectional and panel data related to port production.

Empirical results reveal that substantial waste exists in the production process of the container ports in the sample. For instance, the average efficiency of container ports using the DEA-CCR model and cross-sectional data amounts to 0.58. This indicates that, on average, the ports under study can dramatically increase the level of their outputs by 1.7 times as much as their current level while using the same inputs.

Empirical results also reveal that the terminals (ports) in the sample were found to exhibit a mix of decreasing, increasing and constant returns to scale at current levels of output. Such information is particularly useful for port managers or policy makers to decide on the scale of production.

The analysis of the efficiency yielded by two DEA models (CCR and BCC) using cross-sectional data confirms that the different mathematical programming methodologies tend to give significantly different results. On the other hand, Spearman's rank order correlation coefficient revealed that although the efficiencies estimated by the alternative approaches differ significantly from each other, the ranking of the sample ports using alternative approaches is rather similar. This leads to the conclusion that the efficiency estimated by alternative approaches exhibits the same pattern of efficiency distribution, albeit with significantly different means.

The static analysis of efficiency using cross-sectional data may be rather misleading. This is largely as a result of the lumpy investments that a port might make to achieve beneficial results in the future or, simply just as a result of random effects. In this sense, it is important to review the various methodological approaches to DEA using panel data. The alternative approaches comprising *contemporaneous*, *intertemporal* and *window* analyses are applied to panel data pertaining to the container port industry. The empirical results confirm that the efficiency of different container ports can fluctuate over time to different extents, sometimes even drastically. This consequently confirms the necessity of using alternative DEA panel data approaches.

In final conclusion, it is important to note that to estimate the efficiency of a container port is the beginning and not the end of any analysis. It is undoubtedly the case that each individual container port has its own specific and unique context within which it operates and which will contribute to its level of efficiency. Put differently, although DEA results provide important information on the port industry, they should be carefully interpreted as the ideal efficiency indicated by DEA results might not be achievable in reality for the ports or terminals under study. This implies that such an analysis needs to be supplemented by an investigation of other more singular aspects of individual ports on a case-by-case basis. It will then be useful to explore the more subtle reasons behind the degree to which each individual container port is (in)efficient.

NOTES

1. This section is based on Cullinane (2002), but has been supplemented by a review of more recent applications.
2. This illustrative example of a DEA application that is presented in the next few sections has been derived from Cullinane et al. (2005, 2006) for the analysis of the cross-sectional data and from Cullinane, Song et al. (2004) for the panel data analysis.

REFERENCES

Ali, A. I., & Seiford, L. M. (1993). The mathematical programming approach to efficiency analysis. In: H. Fried, C. A. K. Lovell & S. Schmidt (Eds), *The measurement of productive efficiency: Techniques and applications*. Oxford: Oxford University Press.

Ashar, A. (1997). Counting the moves. *Port Development International* (November), 25–29.

Banker, R. D., Chang, H., & Cooper, W. W. (1996). Simulation studies of efficiency, returns to scale and misspecification with nonlinear functions in DEA. *Annals of Operations Research, 66*, 233–253.

Banker, R. D., Charnes, A., & Cooper, W. W. (1984). Some models for estimating technical and scale inefficiencies in data envelopment analysis. *Management Science, 30*, 1078–1092.

Barros, C. P. (2003a). Incentive regulation and efficiency of Portuguese port authorities. *Maritime Economics and Logistics, 5*(1), 55–69.

Barros, C. P. (2003b). The measurement of efficiency of Portuguese seaport authorities with DEA. *International Journal of Transport Economics, XXX*(3), 335–354.

Barros, C. P., & Athanassiou, M. (2004). Efficiency in European seaports with DEA: Evidence from Greece and Portugal. *Maritime Economics and Logistics, 6*(2), 122–140.

Baumol, W. J., Panzer, J. C., & Willig, R. D. (1982). *Contestable markets and the theory of industry structure*. San Diego: Harcourt Brace Jovanovich.

Bonilla, M., Casasus, T., Medal, A., & Sala, R. (2004). An efficiency analysis of the Spanish port system. *International Journal of Transport Economics, XXXI*(3), 379–400.

Charnes, A., Clark, C. T., Cooper, W. W., & Golany, B. (1985). A developmental study of data envelopment analysis in measuring the efficiency of maintenance units in the U.S. Air Forces. *Annals of Operation Research, 2*, 95–112.

Charnes, A., & Cooper, W. W. (1968). Structural sensitivity analysis in linear programming and an exact product form left inverse. *Naval Research Logistics Quarterly, 15*, 517–522.

Charnes, A., Cooper, W. W., Lewin, A. Y., Morey, R. C., & Rousseau, J. J. (1985). Sensitivity and stability analysis in DEA. *Annals of Operations Research, 2*, 139–156.

Charnes, A., Cooper, W. W., & Rhodes, E. (1978). Measuring the efficiency of decision making units. *European Journal of Operational Research, 2*, 429–444.

Charnes, A., Haag, S., Jaska, P., & Semple, J. H. (1992). Sensitivity of efficiency calculations in the additive model of data envelopment analysis. *International Journal of System Sciences, 23*, 789–798.

Charnes, A., & Neralic, L. (1992). Sensitivity analysis of the proportionate change of inputs (or outputs) in data envelopment analysis. *Glasnik Matematicki, 27*, 393–405.

Charnes, A., Rousseau, J. J., & Semple, J. H. (1996). Sensitivity and stability of efficiency classifications in DEA. *Journal of Productivity Analysis, 7*, 5–18.

Cooper, W. W., Seiford, L. M., & Tone, K. (2000). *Data envelopment analysis: A comprehensive text with models, applications, references and DEA-solver software.* Boston: Kluwer Academic Publishers.

Cullinane, K. P. B. (2002). The Productivity and efficiency of ports and terminals: Methods and applications. In: C. T. Grammenos (Ed.), *Handbook of maritime economics and business* (pp. 803–831). London: Informa Professional.

Cullinane, K. P. B., Ji, P., & Wang, T.-F. (2005). The relationship between privatization and DEA estimates of efficiency in the container port industry. *Journal of Economics and Business, 57*(5), 433–462.

Cullinane, K. P. B., & Khanna, M. (2000). Economies of scale in large containerships: Optimal size and geographical implications. *Journal of Transport Geography, 8*, 181–195.

Cullinane, K. P. B., Song, D.-W., Ji, P., & Wang, T.-F. (2004). An application of DEA windows analysis to container port production efficiency. *Review of Network Economics, 3*(2), 186–208.

Cullinane, K. P. B., Song, D.-W., & Wang, T.-F. (2006). The application of mathematical programming approaches to estimating container port production. *Journal of Productivity Analysis, 24*(1), 73–92.

Cullinane, K. P. B., & Wang, T.-F. (2006). The efficiency of European container ports: A cross-sectional data envelopment analysis. *International Journal of Logistics: Research and Applications, 9*(1), 19–31.

Cullinane, K. P. B., Wang, T.-F., & Cullinane, S. L. (2004). Container terminal development in mainland China and its impact on the competitiveness of the port of Hong Kong. *Transport Reviews, 24*(1), 33–56.

Cullinane, K. P. B., Wang, T.-F., Song, D.-W., & Ji, P. (2005). A comparative analysis of DEA and SFA approaches to estimating the technical efficiency of container ports. *Transportation Research A: Policy and Practice, 40*(4), 354–374.

De Neufville, R., & Tsunokawa, K. (1981). Productivity and returns to scale of container ports. *Maritime Policy and Management, 8*(2), 121–129.

Dyson, R. G. (2000). Performance measurement and data envelopment analysis – Ranking are rank!. *OR Insight, 13*(4), 3–8.

Edmond, E. D., & Maggs, R. P. (1976). Containership turnaround times at UK Ports. *Maritime Policy & Management, 4*(1), 3–19.

Gilman, S. (1983). *The competitive dynamics of liner shipping.* Aldershot, Hants: Gower.

Goss, R. O., & Mann, M. C. (1974). *Cost of ships' time.* London: Government Economic Service Occasional Papers.

Hoffmann, J. (2005). Liner shipping connectivity. *UNCTAD Transport Newsletter*, First Quarter, *27*, 4–12.

Itoh, H. (2002). Efficiency changes at major container ports in Japan: A window application of data envelopment analysis. *Review of Urban and Regional Development Studies, 14*(2), 133–152.

Jansson, J. O., & Shneerson, D. (1987). *Liner shipping economics.* London: Chapman and Hall.

Kim, M., & Sachish, A. (1986). The structure of production, technical change and productivity in a port. *Journal of Industrial Economics, 35*(2), 209–223.

Koop, G., Osiewalski, J., & Steel, M. F. J. (1999). The components of output growth: A stochastic frontier analysis. *Oxford Bulletin of Economics and Statistics, 61*, 455–487.

Kopp, R. J., Smith, V. K., & Vaughan, W. J. (1982). Stochastic cost frontiers and perceived technical inefficiency. In: V. K. Smith (Ed.), *Advances in applied microeconomics, 2.* Greenwich, CT.: JAI Press.

Leibenstein, H. (1966). Allocative efficiency vs. "X-efficiency". *American Economic Review, 56*, 392–415.

Lovell, C. (1993). Production frontiers and productive efficiency. In: H. Fried, C. A. K. Lovell & S. Schmidt (Eds), *The measurement of productive efficiency: Techniques and applications* (pp. 3–67). Oxford: Oxford University Press.

Martinez-Budria, E., Diaz-Armas, R., Navarro-Ibanez, M., & Ravelo-Mesa, T. (1999). A study of the efficiency of Spanish port authorities using data envelopment analysis. *International Journal of Transport Economics, XXVI*(2), 237–253.

Neralic, L. (1997). Sensitivity in data envelopment analysis for arbitrary perturbations of data. *Glasnik Matematicki, 32*, 315–335.

Norman, M., & Stoker, B. (1991). *Data envelopment analysis, the assessment of performance.* Chichester: Wiley.

Ortúzar, J. de D., & Willumsen, L. G. (2001). *Modelling transport* (3rd ed.). Chichester: Wiley.

Park, R.-K., & De, P. (2004). An alternative approach to efficiency measurement of seaports. *Maritime Economics and Logistics, 6*, 53–69.

Ray, S. C. (2002). William W. Cooper: A legend in his own times. *Journal of Productivity Analysis, 17*, 7–12.

Roll, Y., & Hayuth, Y. (1993). Port performance comparison applying data envelopment analysis (DEA). *Maritime Policy and Management, 20*(2), 153–161.

Seiford, L. M., & Thrall, R. (1990). Recent development in DEA: The mathematical programming approach to frontier analysis. *Journal of Econometrics, 46*(1/2), 7–38.

Simar, L., & Wilson, P. (1998). Sensitivity of efficiency scores: How to bootstrap in nonparametric frontier models. *Management Science, 44*, 49–61.

Simar, L., & Wilson, P. (2000). A general methodology for bootstrapping in nonparametric frontier models. *Journal of Applied Statistics, 27*, 779–802.

Simar, L., & Zelenyuk, V. (2003). *Statistical inference for aggregates of Farrell-type efficiencies.* Discussion Paper 0324, Institut de Statistique, Université Catholique de Louvain, Belgium.

Stigler, G. J. (1976). The existence of X-efficiency. *American Economic Review, 66*(1), 213–216.

Talley, W. K. (1994). Performance indicators and port performance evaluation. *Logistics and Transportation Review, 30*(4), 339–352.

Thanassoulis, E. (2001). *Introduction to theory and application of data envelopment analysis.* Norwell, MA: Kluwer Academic.

Thrall, R. M. (1989). Classification of transitions under expansion of inputs and outputs. *Managerial and Decision Economics, 10*, 159–162.

Tongzon, J. (1995). Systematising international benchmarking for ports. *Maritime Policy and Management, 22*(2), 171–177.

Tongzon, J. (2001). Efficiency measurement of selected Australian and other international ports using data envelopment analysis. *Transportation Research Part A: Policy and Practice, 35*(2), 113–128.

Tulkens, H., & van den Eeckaut, P. (1995). Nonparametric efficiency, progress and regress measures for panel-data: Methodological aspects. *European Journal of Operational Research, 80*(3), 474–499.

Turner, H., Windle, R., & Dresner, M. (2004). North American containerport productivity: 1984–1997. *Transportation Research E, 40*, 339–356.

Valentine, V. F., & Gray, R. (2001). The measurement of port efficiency using data envelopment analysis. *Proceedings of the 9th world conference on transport research*, 22–27 July, Seoul.

Wang, T.-F. (2004). *Analysis of the container port industry using efficiency measurement: A comparison of China with its international counterparts.* Ph.D. thesis, Hong Kong Polytechnic University.

Wang, T.-F., Cullinane, K. P. B., & Song, D.-W. (2005). *Container port production and economic efficiency.* Basingstoke: Palgrave-Macmillan.

Wang, T.-F., Song, D. -W., & Cullinane, K. P. B. (2002). The applicability of data envelopment analysis to efficiency measurement of container ports. *Proceedings of the International Association of Maritime Economists Conference*, Panama, 13–15 November.

Wilson, P. W. (1995). Detecting influential observations in data envelopment analysis. *Journal of Productivity Analysis, 6*, 27–46.

Yap, W. Y., Lam, J. S. L., & Notteboom, T. (2006). Developments in container port competition in East Asia. *Transport Reviews, 26*(2), 167–188.

APPENDIX A

DEA-CCR Window Analysis for Container Port Efficiency[a].

Port	Efficiency Scores								Summary Measures	
	1992	1993	1994	1995	1996	1997	1998	1999	Mean	SD
Hong Kong	0.72	0.72	0.77						0.89	0.09
		0.72	0.77	1.00						
			0.94	0.97	1.00					
				0.89	0.94	1.00				
					0.92	1.00	1.00			
						0.87	0.89	0.93		
Singapore	0.95	1.00	0.69						0.89	0.10
		1.00	0.68	0.71						
			0.80	0.83	1.00					
				0.76	0.92	1.00				
					0.86	0.93	1.00			
						0.89	0.95	1.00		
Busan	0.79	0.92	1.00						0.79	0.13
		0.86	0.99	0.91						
			0.89	0.88	1.00					
				0.71	0.79	0.75				
					0.79	0.74	0.52			
						0.65	0.49	0.48		
Kaohsiung	1.00	0.94	1.00						0.95	0.04
		0.89	0.94	1.00						
			0.94	1.00	0.97					
				0.97	0.93	0.97				
					0.90	0.91	1.00			
						0.82	0.90	1.00		
Shanghai	0.75	0.94	1.00						0.83	0.11
		0.87	0.96	0.94						
			0.87	0.78	0.81					
				0.66	0.71	0.87				
					0.69	0.82	1.00			
						0.60	0.73	1.00		
Rotterdam	0.39	0.39	0.42						0.51	0.09
		0.37	0.40	0.48						
			0.40	0.48	0.65					
				0.46	0.64	0.57				
					0.61	0.55	0.64			
						0.51	0.59	0.56		
Los Angeles	0.98	0.96	1.00						0.98	0.02
		0.95	0.99	1.00						
			0.99	1.00	1.00					

APPENDIX A (*Continued*)

DEA-CCR Window Analysis for Container Port Efficiency[a].

Port	Efficiency Scores								Summary Measures	
	1992	1993	1994	1995	1996	1997	1998	1999	Mean	SD
				1.00	1.00	1.00				
					1.00	1.00	0.95			
						1.00	0.88	0.99		
Hamburg	0.47	0.49	0.55						0.54	0.04
		0.46	0.52	0.53						
			0.52	0.53	0.56					
				0.51	0.53	0.58				
					0.52	0.56	0.66			
						0.53	0.62	0.59		
Long Beach	0.61	0.56	0.68						0.68	0.11
		0.55	0.67	0.75						
			0.63	0.71	0.70					
				0.58	0.57	0.66				
					0.53	0.60	1.00			
						0.58	0.93	1.00		
Antwerp	0.34	0.35	0.41						0.33	0.04
		0.32	0.37	0.39						
			0.37	0.39	0.42					
				0.26	0.29	0.29				
					0.29	0.28	0.31			
						0.27	0.30	0.36		
Port Klang	1.00	0.64	0.79						0.56	0.19
		0.61	0.74	0.76						
			0.74	0.76	0.41					
				0.69	0.34	0.41				
					0.33	0.39	0.40			
						0.33	0.32	0.42		
Dubai	0.85	0.89	1.00						0.68	0.15
		0.83	0.93	0.50						
			0.93	0.54	0.67					
				0.49	0.58	0.59				
					0.56	0.58	0.62			
						0.54	0.59	0.57		
New_York/New Jersey	0.57	0.50	0.68						0.76	0.16
		0.46	0.63	1.00						
			0.63	1.00	0.98					
				0.90	0.83	0.98				
					0.79	0.93	0.60			
						0.90	0.60	0.76		

APPENDIX A (*Continued*)

DEA-CCR Window Analysis for Container Port Efficiency[a].

Port	Efficiency Scores								Summary Measures	
	1992	1993	1994	1995	1996	1997	1998	1999	Mean	SD
Bremen/Bremerhaven	0.82	0.85	0.99						0.80	0.06
		0.72	0.83	0.85						
			0.83	0.85	0.85					
				0.70	0.74	0.73				
					0.74	0.73	0.78			
						0.73	0.78	0.94		
Felixstowe	0.70	0.74	0.78						0.60	0.08
		0.64	0.67	0.72						
			0.58	0.59	0.64					
				0.50	0.54	0.58				
					0.52	0.56	0.63			
						0.45	0.52	0.52		
Manila	0.71	0.54	0.60						0.58	0.09
		0.54	0.58	0.79						
			0.50	0.69	0.81					
				0.55	0.64	0.51				
					0.64	0.51	0.42			
						0.51	0.42	0.47		
Tokyo	0.50	0.43	0.48						0.52	0.05
		0.40	0.45	0.53						
			0.48	0.58	0.62					
				0.54	0.56	0.54				
					0.55	0.52	0.57			
						0.45	0.49	0.65		
Qingdao	0.68	0.54	0.88						0.77	0.15
		0.51	0.82	1.00						
			0.53	0.74	1.00					
				0.58	0.79	1.00				
					0.79	1.00	0.78			
						1.00	0.55	0.70		
Yokohama	0.57	0.50	0.50						0.48	0.05
		0.47	0.47	0.59						
			0.47	0.59	0.50					
				0.53	0.46	0.45				
					0.45	0.43	0.46			
						0.39	0.42	0.36		
Laem Chabang	0.11	0.69	0.47						0.66	0.20
		0.55	0.47	0.71						
			0.36	0.55	1.00					

APPENDIX A (*Continued*)

DEA-CCR Window Analysis for Container Port Efficiency[a].

Port	Efficiency Scores								Summary Measures	
	1992	1993	1994	1995	1996	1997	1998	1999	Mean	SD
				0.44	0.85	1.00				
					0.62	0.64	1.00			
						0.55	0.85	1.00		
Tanjung Priok	0.97	0.80	1.00						0.91	0.08
		0.69	0.86	1.00						
			0.86	1.00	0.97					
				0.96	0.93	1.00				
					0.93	1.00	0.95			
						0.94	0.72	0.86		
Kobe	0.43	0.44	0.41						0.38	0.06
		0.41	0.38	0.21						
			0.41	0.22	0.40					
				0.21	0.36	0.48				
					0.36	0.47	0.47			
						0.36	0.37	0.38		
Nagoya	0.99	1.00	1.00						0.94	0.06
		0.95	0.94	1.00						
			0.94	1.00	0.99					
				1.00	0.99	0.96				
					1.00	0.96	0.73			
						0.97	0.74	0.83		
Keelung	1.00	0.96	1.00						0.96	0.04
		0.96	0.94	1.00						
			0.94	1.00	0.97					
				1.00	0.97	1.00				
					1.00	1.00	0.86			
						1.00	0.86	0.84		
Colombo	0.94	1.00	1.00						0.94	0.08
		1.00	1.00	0.86						
			1.00	0.77	1.00					
				0.64	0.82	1.00				
					0.82	1.00	1.00			
						1.00	1.00	0.99		

[a] 1 = 'efficient'.

APPENDIX B

DEA-BCC Window Analysis for Container Port Efficiency[a].

Port	Efficiency Scores								Summary Measures	
	1992	1993	1994	1995	1996	1997	1998	1999	Mean	SD
Hong Kong	0.72	0.72	0.77						0.91	0.12
		0.72	0.77	1.00						
			0.94	0.99	1.00					
				0.92	0.97	1.00				
					0.94	1.00	1.00			
						1.00	1.00	1.00		
Singapore	0.95	1.00	1.00						0.94	0.08
		1.00	0.96	1.00						
			0.80	0.83	1.00					
				0.76	0.92	1.00				
					0.86	0.93	1.00			
						0.89	0.95	1.00		
Busan	0.81	0.94	1.00						0.87	0.16
		0.89	1.00	0.97						
			0.89	0.93	1.00					
				0.93	1.00	0.93				
					1.00	0.91	0.61			
						0.71	0.53	0.54		
Kaohsiung	1.00	0.95	1.00						0.96	0.05
		0.89	0.94	1.00						
			0.94	1.00	0.97					
				1.00	0.97	1.00				
					0.95	0.91	1.00			
						0.82	0.90	1.00		
Shanghai	0.76	0.96	1.00						0.86	0.12
		0.87	0.96	0.95						
			0.87	0.85	0.92					
				0.76	0.80	0.93				
					0.72	0.82	1.00			
						0.60	0.73	1.00		
Rotterdam	0.82	0.82	0.88						0.85	0.06
		0.77	0.83	0.91						
			0.83	0.91	0.95					
				0.84	0.88	0.95				
					0.80	0.86	0.85			
						0.77	0.77	0.80		
Los Angeles	1.00	0.99	1.00						0.99	0.03
		0.97	0.99	1.00						
			0.99	1.00	1.00					

APPENDIX B (*Continued*)

DEA-BCC Window Analysis for Container Port Efficiency[a].

Port	Efficiency Scores								Summary Measures	
	1992	1993	1994	1995	1996	1997	1998	1999	Mean	SD
				1.00	1.00	1.00				
				1.00	1.00	0.95				
				1.00	0.89	1.00				
Hamburg	0.75	0.80	0.81						0.73	0.06
		0.75	0.76	0.81						
			0.66	0.71	0.75					
				0.64	0.68	0.74				
					0.64	0.70	0.81			
						0.66	0.77	0.73		
Long Beach	0.80	0.83	1.00						0.85	0.10
		0.65	0.80	0.89						
			0.80	0.88	0.89					
				0.84	0.84	0.98				
					0.72	0.79	1.00			
						0.73	0.93	1.00		
Antwerp	0.77	0.78	0.92						0.71	0.10
		0.73	0.86	0.90						
			0.61	0.65	0.72					
				0.56	0.62	0.66				
					0.61	0.64	0.71			
						0.62	0.69	0.76		
Port Klang	1.00	0.71	0.86						0.61	0.23
		0.75	0.91	0.89						
			0.79	0.78	0.41					
				0.70	0.38	0.42				
					0.36	0.40	0.40			
						0.34	0.34	0.46		
Dubai	0.89	0.89	1.00						0.72	0.17
		0.89	1.00	0.51						
			0.96	0.54	0.68					
				0.52	0.61	0.71				
					0.57	0.66	0.71			
						0.57	0.61	0.57		
New_York/New Jersey	0.91	0.85	0.95						0.92	0.09
		0.79	0.88	1.00						
			0.66	1.00	1.00					
				0.95	0.85	1.00				
					0.85	1.00	1.00			
						1.00	0.90	1.00		

APPENDIX B (*Continued*)

DEA-BCC Window Analysis for Container Port Efficiency[a].

Port	Efficiency Scores								Summary Measures	
	1992	1993	1994	1995	1996	1997	1998	1999	Mean	SD
Bremen/Bremerhaven	0.87	0.90	1.00						0.90	0.09
		0.89	0.98	1.00						
			0.97	0.99	1.00					
				0.77	0.78	0.86				
					0.78	0.86	0.91			
						0.78	0.83	1.00		
Felixstowe	0.89	0.94	1.00						0.71	0.13
		0.66	0.70	0.77						
			0.64	0.71	0.75					
				0.63	0.68	0.73				
					0.61	0.67	0.74			
						0.50	0.56	0.58		
Manila	0.82	0.57	0.62						0.67	0.13
		0.56	0.59	0.82						
			0.54	0.71	0.82					
				0.69	0.80	0.88				
					0.79	0.78	0.58			
						0.54	0.44	0.51		
Tokyo	0.50	0.44	0.49						0.53	0.06
		0.41	0.46	0.54						
			0.48	0.58	0.62					
				0.56	0.59	0.56				
					0.55	0.52	0.58			
						0.47	0.49	0.65		
Qingdao	1.00	0.61	1.00						0.83	0.17
		0.61	1.00	1.00						
			0.53	0.74	1.00					
				0.58	0.79	1.00				
					0.79	1.00	0.78			
						1.00	0.68	0.87		
Yokohama	0.59	0.53	0.55						0.50	0.07
		0.49	0.52	0.61						
			0.51	0.61	0.52					
				0.57	0.49	0.48				
					0.45	0.44	0.48			
						0.40	0.44	0.38		
Laem Chabang	0.15	1.00	0.49						0.74	0.28
		1.00	0.52	0.78						
			0.37	0.57	1.00					

APPENDIX B (*Continued*)

DEA-BCC Window Analysis for Container Port Efficiency[a].

Port	1992	1993	1994	1995	1996	1997	1998	1999	Mean	SD
				0.46	1.00	1.00				
					1.00	0.64	1.00			
						0.55	0.85	1.00		
Tanjung Priok	1.00	0.80	1.00						0.92	0.10
		0.69	0.86	1.00						
			0.86	1.00	0.97					
				0.96	0.93	1.00				
					0.93	1.00	0.96			
						1.00	0.73	0.88		
Kobe	0.52	0.51	0.49						0.43	0.11
		0.48	0.47	0.24						
			0.49	0.26	0.41					
				0.24	0.39	0.63				
					0.36	0.52	0.52			
						0.42	0.44	0.45		
Nagoya	1.00	1.00	1.00						0.96	0.09
		1.00	1.00	1.00						
			1.00	1.00	0.99					
				1.00	0.99	0.96				
					1.00	0.96	0.74			
						1.00	0.74	0.83		
Keelung	1.00	0.96	1.00						0.96	0.05
		1.00	0.94	1.00						
			0.94	1.00	0.97					
				1.00	0.97	1.00				
					1.00	1.00	0.86			
						1.00	0.86	0.84		
Colombo	1.00	1.00	1.00						0.99	0.05
		1.00	1.00	1.00						
			1.00	0.80	1.00					
				1.00	1.00	1.00				
					1.00	1.00	1.00			
						1.00	1.00	0.99		

[a]1 = 'efficient'.

APPENDIX C

DEA-CCR Contemporaneous Analysis of Container Port Efficiency[a].

	1992	1993	1994	1995	1996	1997	1998	1999	Average
Hong Kong	0.79	0.72	1.00	1.00	1.00	1.00	1.00	0.93	0.93
Singapore	1.00	1.00	1.00	0.96	1.00	1.00	1.00	1.00	1.00
Busan	0.91	1.00	1.00	1.00	1.00	0.75	0.64	0.49	0.85
Kaohsiung	1.00	1.00	1.00	1.00	1.00	0.99	1.00	1.00	1.00
Shanghai	0.88	1.00	1.00	0.94	0.86	0.87	1.00	1.00	0.94
Rotterdam	0.41	0.42	0.44	0.48	0.68	0.57	0.70	0.56	0.53
Los Angeles	1.00	1.00	1.00	1.00	1.00	1.00	0.96	1.00	1.00
Hamburg	0.48	0.53	0.56	0.53	0.57	0.58	0.72	0.64	0.58
Long Beach	0.68	0.58	0.69	0.88	0.71	0.66	1.00	1.00	0.77
Antwerp	0.36	0.39	0.41	0.39	0.43	0.29	0.36	0.43	0.38
Port Klang	1.00	0.70	0.80	0.76	0.41	0.42	0.40	0.43	0.61
Dubai	0.93	0.98	1.00	0.55	0.68	0.59	0.74	0.57	0.76
New York/New Jersey	0.60	0.55	0.68	1.00	1.00	0.98	0.72	0.92	0.81
Bremen/Bremerhaven	0.83	0.91	0.99	0.85	0.86	0.73	0.95	1.00	0.89
Felixstowe	0.96	0.91	0.78	0.73	0.65	0.58	0.71	0.53	0.73
Manila	0.75	0.56	0.65	0.89	0.81	0.51	0.58	0.50	0.66
Tokyo	0.58	0.48	0.53	0.58	0.63	0.55	0.59	0.67	0.58
Qingdao	0.79	0.63	0.88	1.00	1.00	1.00	0.82	0.70	0.85
Yokohama	0.62	0.56	0.51	0.59	0.52	0.45	0.48	0.37	0.51
Laem Chabang	0.15	0.86	0.48	0.85	1.00	1.00	1.00	1.00	0.79
Tanjung Priok	1.00	1.00	1.00	1.00	1.00	1.00	1.00	0.87	0.98
Kobe	0.48	0.49	0.45	0.23	0.41	0.48	0.49	0.38	0.43
Nagoya	1.00	1.00	1.00	1.00	1.00	0.97	0.89	1.00	0.98
Keelung	1.00	1.00	1.00	1.00	1.00	1.00	1.00	1.00	1.00
Colombo	1.00	1.00	1.00	0.91	1.00	1.00	1.00	1.00	0.99
Average	0.77	0.77	0.79	0.80	0.81	0.76	0.79	0.76	0.78
Number of efficient ports	8	9	11	9	12	8	9	10	

[a] 1 = 'efficient'.

APPENDIX D

DEA-BCC Contemporaneous Analysis of Container Port Efficiency[a].

	1992	1993	1994	1995	1996	1997	1998	1999	Average
Hong Kong	0.80	0.72	1.00	1.00	1.00	1.00	1.00	1.00	0.94
Singapore	1.00	1.00	1.00	1.00	1.00	1.00	1.00	1.00	1.00
Busan	0.95	1.00	1.00	1.00	1.00	1.00	0.65	0.54	0.89
Kaohsiung	1.00	1.00	1.00	1.00	1.00	1.00	1.00	1.00	1.00
Shanghai	0.88	1.00	1.00	0.95	0.95	0.93	1.00	1.00	0.96
Rotterdam	0.95	0.86	0.90	0.91	0.99	0.95	0.85	0.80	0.90
Los Angeles	1.00	1.00	1.00	1.00	1.00	1.00	1.00	1.00	1.00
Hamburg	0.77	0.87	0.81	0.81	0.77	0.74	0.94	0.79	0.81
Long Beach	0.87	0.96	1.00	0.89	0.95	1.00	1.00	1.00	0.96
Antwerp	0.80	0.86	0.92	0.90	0.73	0.66	0.91	0.88	0.83
Port Klang	1.00	0.78	0.94	0.90	0.42	0.42	0.42	0.46	0.67
Dubai	0.96	1.00	1.00	0.55	0.69	0.71	0.76	0.59	0.78
New York/New Jersey	0.95	0.94	0.95	1.00	1.00	1.00	1.00	1.00	0.98
Bremen/Bremerhaven	1.00	1.00	1.00	1.00	1.00	0.86	1.00	1.00	0.98
Felixstowe	1.00	1.00	1.00	0.78	0.75	0.74	0.74	0.64	0.83
Manila	0.89	0.62	0.65	0.90	0.82	0.90	0.59	0.51	0.74
Tokyo	0.58	0.49	0.53	0.58	0.64	0.56	0.60	0.68	0.58
Qingdao	1.00	1.00	1.00	1.00	1.00	1.00	1.00	1.00	1.00
Yokohama	0.66	0.59	0.55	0.61	0.54	0.48	0.49	0.38	0.54
Laem Chabang	1.00	1.00	0.56	1.00	1.00	1.00	1.00	1.00	0.95
Tanjung Priok	1.00	1.00	1.00	1.00	1.00	1.00	1.00	1.00	1.00
Kobe	0.60	0.53	0.54	0.27	0.43	0.63	0.53	0.50	0.50
Nagoya	1.00	1.00	1.00	1.00	1.00	1.00	1.00	1.00	1.00
Keelung	1.00	1.00	1.00	1.00	1.00	1.00	1.00	1.00	1.00
Colombo	1.00	1.00	1.00	1.00	1.00	1.00	1.00	1.00	1.00
Average	0.91	0.89	0.89	0.88	0.87	0.86	0.86	0.83	0.87
Number of efficient ports	12	14	15	13	13	13	14	14	

[a] 1 = 'efficient'.

APPENDIX E

DEA-CCR Intertemporal Analysis of Container Port Efficiency[a].

	1992	1993	1994	1995	1996	1997	1998	1999	Average
Hong Kong	0.61	0.62	0.66	0.76	0.79	0.87	0.89	0.93	0.77
Singapore	0.92	1.00	0.65	0.68	0.81	0.89	0.95	1.00	0.86
Busan	0.54	0.63	0.67	0.68	0.77	0.65	0.47	0.46	0.61
Kaohsiung	0.76	0.67	0.76	0.86	0.83	0.82	0.90	1.00	0.82
Shanghai	0.39	0.49	0.54	0.57	0.48	0.60	0.73	1.00	0.60
Rotterdam	0.34	0.33	0.35	0.42	0.56	0.51	0.59	0.56	0.46
Los Angeles	0.78	0.79	0.96	0.99	1.00	1.00	0.88	0.99	0.92
Hamburg	0.41	0.41	0.46	0.47	0.49	0.53	0.62	0.59	0.50
Long Beach	0.43	0.38	0.46	0.52	0.51	0.58	0.93	1.00	0.60
Antwerp	0.20	0.21	0.24	0.25	0.28	0.27	0.30	0.36	0.26
Port Klang	1.00	0.54	0.66	0.61	0.29	0.33	0.32	0.42	0.52
Dubai	0.69	0.72	0.81	0.40	0.49	0.54	0.59	0.57	0.60
New York/New Jersey	0.40	0.39	0.44	0.90	0.77	0.90	0.60	0.76	0.64
Bremen/Bremerhaven	0.60	0.62	0.69	0.70	0.74	0.73	0.77	0.93	0.72
Felixstowe	0.38	0.40	0.43	0.40	0.43	0.45	0.52	0.52	0.44
Manila	0.48	0.36	0.39	0.53	0.62	0.51	0.42	0.47	0.47
Tokyo	0.37	0.33	0.37	0.43	0.47	0.45	0.49	0.65	0.44
Qingdao	0.43	0.26	0.42	0.58	0.79	1.00	0.55	0.70	0.59
Yokohama	0.46	0.38	0.38	0.46	0.40	0.39	0.42	0.36	0.41
Laem Chabang	0.06	0.36	0.25	0.37	0.55	0.55	0.85	1.00	0.50
Tanjung Priok	0.58	0.61	0.77	0.90	0.87	0.94	0.72	0.86	0.78
Kobe	0.34	0.34	0.32	0.17	0.28	0.36	0.37	0.38	0.32
Nagoya	0.90	0.95	0.94	1.00	0.99	0.96	0.73	0.82	0.91
Keelung	0.99	0.95	0.94	1.00	0.97	1.00	0.86	0.84	0.94
Colombo	0.71	0.76	0.75	0.63	0.81	1.00	1.00	0.99	0.83
Average	0.55	0.54	0.57	0.61	0.64	0.67	0.66	0.73	0.62
Number of efficient ports	1	1	0	2	1	4	1	5	

[a] 1 = 'efficient'.

APPENDIX F

DEA-BCC Intertemporal Analysis of Container Port Efficiency[a].

	1992	1993	1994	1995	1996	1997	1998	1999	Average
Hong Kong	0.64	0.65	0.71	0.88	0.93	1.00	1.00	1.00	0.85
Singapore	0.92	1.00	0.65	0.68	0.81	0.89	0.95	1.00	0.86
Busan	0.55	0.65	0.69	0.74	0.80	0.71	0.50	0.54	0.65
Kaohsiung	0.76	0.67	0.76	0.87	0.83	0.82	0.90	1.00	0.83
Shanghai	0.49	0.62	0.70	0.58	0.49	0.60	0.73	1.00	0.65
Rotterdam	0.59	0.59	0.63	0.69	0.72	0.77	0.77	0.80	0.69
Los Angeles	0.80	0.80	0.97	1.00	1.00	1.00	0.89	1.00	0.93
Hamburg	0.49	0.52	0.53	0.57	0.60	0.66	0.77	0.73	0.61
Long Beach	0.44	0.45	0.55	0.62	0.64	0.73	0.93	1.00	0.67
Antwerp	0.43	0.44	0.51	0.54	0.60	0.62	0.69	0.76	0.58
Port Klang	1.00	0.59	0.72	0.62	0.30	0.33	0.33	0.45	0.54
Dubai	0.72	0.75	0.84	0.40	0.49	0.57	0.61	0.57	0.62
New York/New Jersey	0.53	0.50	0.58	0.95	0.85	1.00	0.90	1.00	0.79
Bremen/Bremerhaven	0.63	0.65	0.72	0.73	0.76	0.78	0.83	1.00	0.76
Felixstowe	0.39	0.41	0.44	0.43	0.46	0.50	0.56	0.58	0.47
Manila	0.48	0.36	0.41	0.53	0.63	0.54	0.44	0.51	0.49
Tokyo	0.37	0.33	0.37	0.44	0.48	0.46	0.49	0.65	0.45
Qingdao	1.00	0.26	0.42	0.58	0.79	1.00	0.66	0.84	0.69
Yokohama	0.46	0.38	0.39	0.47	0.40	0.40	0.44	0.38	0.42
Laem Chabang	0.15	1.00	0.25	0.38	0.57	0.55	0.85	1.00	0.59
Tanjung Priok	0.88	0.66	0.82	0.96	0.93	1.00	0.73	0.88	0.86
Kobe	0.37	0.37	0.37	0.20	0.31	0.42	0.44	0.45	0.37
Nagoya	0.93	0.98	0.96	1.00	0.99	0.96	0.74	0.83	0.92
Keelung	1.00	0.96	0.94	1.00	0.97	1.00	0.86	0.84	0.95
Colombo	1.00	1.00	1.00	0.65	0.82	1.00	1.00	0.99	0.93
Average	0.64	0.62	0.64	0.66	0.69	0.73	0.72	0.79	0.69
Number of efficient ports	4	3	1	3	1	7	2	9	

[a]1 = 'efficient'.

CHAPTER 24

REVIEW OF PORT PERFORMANCE APPROACHES AND A SUPPLY CHAIN FRAMEWORK TO PORT PERFORMANCE BENCHMARKING

Khalid Bichou

ABSTRACT

Most practical and theoretical approaches to port performance measurement benchmarking are reducible to three broad categories: performance metrics and index methods, economic impact studies and efficiency frontier approaches. However, despite the plethora of performance models and measurement systems, an integrative benchmarking approach is seldom adopted and performance measurements are often fragmented and biased towards sea access. This paper proposes an integrative framework to port performance by conceptualising ports from a logistics and supply chain management (SCM) approach. The model was tested and improved in a reported survey inquiry. Examples and hypothetical case studies are also presented for illustration.

1. INTRODUCTION

Over the last three decades or so, there has been a growing amount of both theoretical and practical work on port performance measurement and

Devolution, Port Governance and Port Performance
Research in Transportation Economics, Volume 17, 567–598
Copyright © 2007 by Elsevier Ltd.
All rights of reproduction in any form reserved
ISSN: 0739-8859/doi:10.1016/S0739-8859(06)17024-9

benchmarking. Many authors have studied individual performance metrics, performance measurement frameworks and the relationship between performance systems and the port environment. All too often, though, relevant work on mechanisms and techniques of port performance and efficiency has taken place at different disciplinary levels, yet with fragmented layers of operational, functional and spatial port systems. Fundamental differences between these conflicting approaches and their proposed methodologies meant that, despite the variety of tools and instruments available, no consensus on a single framework for port performance benchmarking has been established. Examples of such core differences include

- Fundamental differences on the definition and taxonomy of performance (efficiency, productivity, utilisation, effectiveness, etc.), their individual and combined applications in measurement and benchmarking contexts;
- Perceptual differences among multi-institutional port stakeholders (regulator, operator, user/customer, etc.) and the resulting impact on the objective, design and implementation of performance frameworks and analytical models;
- Boundary-spanning complexities of seaport operational (types of cargo handled, ships serviced, terminals managed, processes/systems operated, etc.) and spatial (cluster, port, terminal, quay system, yard system, etc.) dimensions bring confusion not only on what and how to measure, but also on what to benchmark against; and
- Dissimilarity, in both space and time, between world ports' operational structures, functional scopes, institutional models and strategic orientations.

This paper provides a critical review of existing frameworks and models of port performance benchmarking in the light of supply chain management (SCM) considerations. It identifies the main theoretical approaches to port performance and demonstrates that the current knowledge and techniques are incompatible with recent developments in the port industry, particularly with regard to global logistics integration, institutional restructuring and new supply chain typologies of modern ports and terminals. The paper develops a supply chain framework for port operations and management, proposes a corresponding model for port performance benchmarking and concludes with a series of lessons highlighting further research and development needs.

2. REVIEW OF THE LITERATURE ON PORT PERFORMANCE AND BENCHMARKING

Most practical and theoretical approaches to benchmarking performance in ports can be grouped into three broad categories: individual metrics and indices, economic impact studies and frontier approaches. Studies using simulation techniques for the purpose of operational optimisation do not fall under the subject of performance benchmarking, and are therefore not covered in this paper.

2.1. Performance Metrics and Index Methods

Like most other operating and management systems, performance measurement in seaports and terminals starts with individual metrics at each functional or operational level. A performance measure or metric is presented numerically to quantify one or many attributes of an object, product, process or any other relevant factor and must allow for the comparison and evaluation vis-à-vis goals, benchmarks and/or historical figures. A performance metric can fall within one or a combination of three main categories, namely input measures (e.g., time, cost and resource), output measures (e.g., production/throughput, profit) and composite measures (productivity, efficiency, profitability, utilisation, effectiveness, etc.). The latter are normally presented in the form of output/input ratios, with the objective of maximising the former and/or minimising the latter. Furthermore, each composite index may be broken down into two or more components depending on the approach, typology and dimensions of performance. For instance, in the engineering and manufacturing literature, efficiency encompasses at least two dimensions: cost efficiency (low production costs) and capital efficiency (low investments) (Wheelwright, 1978), while in the field of production economics, efficiency is usually decomposed into technical efficiency, allocative efficiency and distributional efficiency. Technical efficiency indicates the ability to produce the maximum level of output from a given set of inputs (output oriented) or to reduce the input to the minimum given the same output (input oriented). Allocative efficiency reflects the ability to optimally allocate inputs at a minimum cost of outputs, for a given set of input prices and technology. Unlike technical efficiency, where costs or profits are not considered, allocative efficiency studies the costs of production given that the information on prices and a behavioural assumption, such as cost minimisation or profit maximisation, is properly established. When combined, the two measures are referred to as total economic efficiency. Distributional efficiency, on the other hand, is related to consumer choice or welfare optima. Further discussions on performance concepts and their analytical applications in the context of port operations and management are provided in Section 3 of this paper.

In the seaport literature, the lack of uniformity on what constitutes a standard industry practice has shifted the focus of port performance measurement away from the effectiveness and utilisation dimensions to the efficiency dimension, although financial and utilisation metrics are widely used in the profession. An efficiency measure can be loosely defined as the ratio of actual (current) output quantity to the actual quantity of input. Depending on the range and nature of the inputs/outputs selected and the methodology used to calculate them, existing ratio measures for ports can be divided into three types of measurement: financial productivity measures, single and partial factor productivity indicators (SFP, PFP), and total factor (or multi-factor) productivity indices (TFP, MFP).

2.1.1. Financial Metrics and Financial Productivity Measures

Financial metrics use ratios applied in costing and management accounting systems in an approach similar to that of physical indicators, with the difference of using monetary values for input and output data. Financial performance measurement is close to the concept of profitability, usually defined as being the ratio between revenue and cost.

In the port industry, financial ratios are used widely with the most cited and comprehensive study being the annual survey of financial performance of US public ports undertaken by the US Maritime Administration (MARAD) (see, for instance, MARAD, 2003, financial report). Common measures for financial performance in the industry include return on investment, return on assets, capital structure and short-term liquidity.

Conventional financial ratios are inappropriate for performance measurement and benchmarking for a number of reasons. A significant issue is that financial performance can have little correlation with the efficient and effective use of resources. For instance, higher profitability may be driven by price inflation and other external conditions rather than by efficient productivity or utilisation. Kaplan (1984) argues that superior financial performance may be attributable to the use of 'novel financing and ownership arrangements' rather than to efficient operating and management systems. Examples include the arbitrary allocation of overhead costs on a volume basis (e.g., labour or machine hours) and the calculation of depreciation using traditional accounting systems whereby written-down facilities tend to be favoured over new ones. Holmberg (2000) indicates that the main bias of financial techniques is that they show the results of past actions and that they are designed to meet external evaluators' needs and expectations. Vitale and Mavrinac (1995) criticise financial ratios because they are incapable of assessing the contribution of intangible activities such as innovation and development programmes.

In ports, the focus of financial ratios on short-term profitability is inconsistent with the nature and objectives of long-term port investments. Dissimilarity between various costing and accounting systems is equally a major problem when one tries to compare ports from different countries. Even within a single country, port financing and institutional structures (private, landlord, tool, etc.) are hardly comparable. Many other aspects influence port financial performance including price and access regulation, market power, statutory freedom and access to private equity. For these reasons and others, physical productivity measures are regarded as more reliable performance indicators than financial measures.

2.1.2. Physical Productivity Measurements

A single productivity indicator or SFP is the ratio of a measure of a single output quantity to the quantity of a single factor input. The latter is typically based on an input resource (labour, land or capital), while the output quantity is usually based on the cost drivers of the activity or resource being measured. However, the data on cost drivers in ports are often unavailable and physical productivity measures are used instead. The concept behind a partial productivity indicator or PFP is similar to that of SFP, with the difference that the former seeks to compare a subset of outputs to a subset of inputs when multiple inputs and outputs are involved, such as in the case of seaports. The objective is to construct performance measures that compare one or more outputs to one or more inputs. Examples of PFP ratios in ports include crane throughput per machine hour, berth or quay throughput per square-metre capacity and worker or gang output per man-hour. SFP and PFP indicators seek to capture a change in productivity prompted respectively by a single factor or by a subset of factors. Thus while they are quite easy to calculate, their 'subjective' construction and narrow focus on a single/partial form of output constitute a major bias.

Much of the conventional port literature falls under the SFP/PFP category (see, for example, Bendall & Stent, 1987; De Monie, 1987; Frankel, 1991; Fourgeaud, 2000; Talley, 1988; UNCTAD, 1976), and so do many industry and professional publications (ports' statistics, trade journals, market reports, etc.). Nevertheless, many performance metrics used in the literature only provide 'snapshot' measurements, such as for a single port operation (loading, discharging, storage, distribution, etc.) and/or facility (crane, berth, warehouse, etc.). Annual container throughput in 20-foot equivalent units (TEUs) is a typical example of such measures, and is widely, and quite misleadingly, used to rank world container ports and terminals. A common mistake is to equate production with productivity or efficiency since the latter is relative rather than absolute concepts. Sometimes, composite metrics are used for specific performance and benchmarking purposes, for example, the number of containers per hour versus ship's size (Drewry Shipping Consultants, 1997) and the net crane rate by liner shipping trade (Australian Productivity Commission, 1998). Coordination with land modes of transport may also feature in a port productivity indicator, for instance, cargo dwell time or the time elapsed from when the cargo is unloaded from a ship until she leaves the port. The latter is usually used in conjunction with time-based utilisation metrics such as berth occupancy rate and average ship's service time. An utilisation ratio typically compares the input actually used against that of available resources, for example, working time versus service time. However, utilisation metrics share similar shortcomings with single productivity indices and may not be appropriate for port performance benchmarking. In a typically complex port operating and management system, SFP and PFP indicators may be simply described as incomplete measures of performance.

2.1.3. Multifactor or Total Factor Productivity Measurements

Multifactor productivity (MFP) and total factor productivity (TFP) indicators incorporate multiple inputs and outputs through the use of aggregated index methods or estimated indices from specified cost or production functions. The TFP concept seeks to provide an aggregate indication of productivity using a measure of total input/output quantity, but can be decomposed by introducing statistical effects of model decomposition. The method behind TFP is to synthesise a productivity index by assigning weights that reflect the relative importance of its cost and production components. Primarily, a TFP index can be constructed direct from data without the need for statistical estimation of a production or cost function, but this requires information on output and input data categories such as price, cost and revenue shares. When these data are not available, it is possible to estimate the weights from econometric cost or production functions.

Few efficiency studies have estimated or used a TFP index in the context of port performance. Early attempts were undertaken by Kim and Sachish (1986) and Bendall and Stent (1987). In Kim and Sachish, the composed TFP index consisted of labour and capital expenditure as input and throughput in metric tonnes as output. Later, Talley (1994) suggests a TFP index using a shadow price variable, while Sachish (1996) refers to a weighting mechanism of partial productivity measures. Lawrance and Richards (2004) developed a decomposition method for the Törnqvist index to investigate the distribution of benefits from productivity improvements in an Australian container terminal.

The main advantage of MFP/TFP indices is that they reflect the joint impacts of the changes in combined inputs on total output. Such a feature is not accounted for when using single or partial productivity indicators. However, TFP results depend largely on the technique used and the definition of weights, and thus different TFP indices may yield different efficiency results.

2.2. Economic Impact Studies

Port impact studies have emerged as an area of applied research that can bridge trade with wider regional economic impacts, but have been criticised because of the selection of limited industrial categories and the approach undertaken to compare 'ports as regions' rather than 'ports as firms' (Bichou & Gray, 2005). The literature on port impact studies depicts two separate lines of research: port economic impacts and port trade efficiency.

(A) The first category may be considered as a branch economic geography, extended to the field of urban planning and environmental economics due to the increasing importance of the port–city interface. Port impacts on the economy are measured to assess the economic and social impacts (direct, indirect and induced) of ports on their respective hinterlands or forelands. In this approach, ports are seen as economic catalysts for the regions they serve, where the aggregation of services and activities generates benefits and socio-economic wealth. Here, the performance of a port is depicted in terms of its ability to generate maximum or optimal output and economic wealth. Relevant conceptual work in the field can be found in the AIPV (International Association of Ports and Cities, 2005; www.aipv.orgwww.aipv.org) references as well in related academic literature (e.g., De Langen, 2002; Rodrigue, Slack, & Comtois, 1997).

On the other hand, much of the applied research on the subject is based on input–output models. An input–output model is a set of linear equations in which the outputs of various branches in the economy are calculated based on an empirical estimation of inter-sector transactions. The US MARAD's Port Economic Impact Kit (PortKit) is probably the most comprehensive and regularly updated input–output port model. Since its first publication in the mid-1970s, it has become the standard model for assessing economic impacts of US ports. The latest PortKit version was released in December 2000 in the form of PC-based software comprising a 30-sector table. Hamilton, Ramsussen, and Zeng (2000) developed similar software versions for US inland ports. Outside the US, input–output port models have been used to assess the impacts of existing port facilities (Moloney & Sjostrom, 2000) or to justify future port investments (e.g., Le Havre Port, 2000).

An alternative method to assess port economic impacts is through the use of computable equilibrium models. In a unique application in ports, Dio, Tiwari, and Itoh (2001) use a computable general equilibrium (CGE) model to analyse the impacts of port efficiency improvements on the Japanese economy. The objective is to analyse the relationship between an assigned or given size 'shock' to productivity growth on the GDP of a region, country or group of countries. CGE models have gained more popularity in the last decade or so, with applications across different sectors including for quantifying the benefits of improved port efficiency on trade facilitation (APEC, 1999). A good reference to CGE models and their applications in trade reform policies is provided by Devarajan and Rodrik (1991).

Gravity models analysing the relationship between geographical distance and trade flows have also been used to investigate the impacts of selected trade facilitation indicators including port efficiency (Wilson, Mann, & Otsuki, 2003). CGE and gravity models are both separate branches of theoretical econometrics and deserve a thorough analysis beyond the scope of this paper. What one needs to note, though, is that CGE data and model equations are calibrated to national accounts and input–output table details. However, since each port-country depicts a separate economic structure and inter-sectoral configuration, both CGE and input–output models would prove inadequate for international benchmarking of ports. Other limitations of the general equilibrium theory include its reliance on many constraints, some of which are inconsistent with the structure of the port industry, for instance, the assumptions of perfect competition, constant returns to scale and that both labour and capital move freely between sectors.

(B) The second category of economic impact analysis assesses port efficiency in relation to transport and logistics costs. This part of the economic impact literature is rapidly establishing itself as a 'separate' branch due mainly to the recent emphasis on the role of ports in trade facilitation.

Research on trade facilitation is, however, still at its infancy as the definition of the concept and the approach to it have not stabilised yet. Relevant port literature in the field includes the works of Sanchez et al. (2003) and De and Ghosh (2003), both employing principal component analysis (PCA). Previous work on the relationship between port efficiency and waterborne transport costs used proxies such as GDP per capita (Fink, Mattoo, & Neagu, 2000), perception surveys (Hoffmann, 2001) and infrastructure indicators (Micco & Perez, 2001). Beyond the limitations of applying PCA and regression techniques to multi-input/multi-output port production systems, the existing literature on the relationship between port efficiency and trade facilitation depicts inherent discrepancies at both conceptual and analytical levels, including for using throughput, rather than traffic, indicators to analyse trade-related issues!

2.3. Frontier Approaches

The frontier concept denotes the lower or upper limit to a boundary-efficiency range. Unlike the typical statistical central tendency approach where performance is evaluated relative to an average firm or unit, the frontier approach measures the efficiency in relation to the calculation or estimation of a frontier. Under this approach, a firm is defined as efficient when it operates on the frontier and inefficient when it operates away from it (below it for a production frontier and above it for a cost frontier). The frontier can be either absolute or relative (i.e., best practice) depending on the method of parameter construction, i.e., parametric versus non-parametric methods.

2.3.1. Parametric Approaches
Parametric, or econometric, methods require a functional form whereby a set of input and output observations can be statistically estimated. Parametric models refer to the calculation or estimation of a cost or a production frontier function from the relevant input–output data, meaning that parameter values can be statistically inferred from data observations. The parametric representation can, however, be either deterministic or stochastic, depending on whether or not certain assumptions are made regarding the data used. A detailed review of the applications of cost and production functions in ports is provided by Tovar, Jara-Diaz, and Trujillo (2003). Note that recent applications have estimated multi-output cost functions for ports (see Gonzales & Trujillo, 2005; Jara Díaz, Martinez-Budría, Cortes, & Basso, 2002), which seems to overcome the problem of using a single measure of product

or output technology. However, the main argument against the use of parametric models stems from the deterministic requirement of a functional specification, which does not allow for relative comparisons with the best multi-factor practice. Furthermore, parametrical approaches may not be suitable for international port benchmarking. As pointed out by Braeutigam, Daughety, and Turnquist (1984) and Kim and Sachish (1986), the structure of port production may limit the econometric estimation of a cost or production function to the level of a single port or terminal.

2.3.2. Non-Parametric Approaches
Unlike econometric models, non-parametric approaches do not require a pre-defined functional formulation but use linear programming techniques to determine rather than estimate the efficiency frontier. Much of the recent research in the field involves the use of data envelopment analysis (DEA). The methodology works by solving a series of linear programming problems and selecting the optimal solution that maximises the efficiency ratio of weighted output to weighted input for each DMU (decision-making unit). The rationale behind DEA is that, in seeking to solve the issue of DMUs assigning different value weights to their respective inputs/outputs, each DMU is allowed to set a combination of weights that shows it in the most favourable position vis-à-vis other DMUs. DEA applications in ports are quite recent with the first attempt being attributed to Roll and Hayuth (1993). A detailed review of the use of DEA techniques in ports is provided by Estache, González, and Trujillo (2002), although since then many studies have been published on the subject. The literature in the field may be divided in terms of four broad criteria. First, between DEA–CCR applications (Valentine & Gray, 2001) and DEA–BCC applications (Martínez-Budría, Díaz-Armas, Navarro-Ibañez, & Ravelo-Mesa, 1999), although some studies use both models (Tongzon, 2001). Second, between input-oriented models (Barros, 2003) and output-oriented models (Wang & Cullinane, 2005). Third, between applications looking at aggregate port operations (Barros & Athanassiou, 2004) and those focusing on a single port function (Cullinane, Song, & Wang, 2004). Last, but not least, between studies relying on DEA results solely and those complementing DEA with a second-stage analysis (Bonilla, Medal, Casaus, & Salas, 2002; Turner, Windle, & Dresner, 2004). An additional categorisation may be made between studies using cross-sectional data versus those using panel data, although there is a general tendency to use the former.

The DEA approach to efficiency analysis has many advantages over parametric approaches. The methodology accommodates multiple inputs and outputs, and provides information about the sources of their relative (factor specific) efficiency. Under DEA, there is no necessity to pre-define relative weight relationships, which should free the analysis from subjective weighting and randomness. Similarly, DEA neither imposes a specification of a cost/production function nor requires an assumption about the technology. Moreover, firms or DMUs are benchmarked against an actual 'best' firm rather than against a statistical measure, an exogenous or average standard. All such features and others make DEA particularly attractive for port-related efficiency studies, which explain the increasing number of academic publications on the subject.

On the other hand, one could argue that the same features that make DEA a powerful tool also create major limitations. For instance, DEA does not allow for stochastic factors and measurement errors and there is no information on statistical significance or confidence intervals. Although a second-stage regression analysis is sometimes used to solve this, regression assumes data interdependency and requires the imposition of a functional form, which deprives DEA of its major advantage. Another major drawback of DEA stems from the sensitivity of efficiency scores to the choice and weights of input–output variables. This is of major concern because a DMU can appear efficient simply because of its patterns of inputs and outputs. In the port literature most DEA applications assume constant efficiency over time, which overlooks the incremental nature of port investment, and hence favours ports that are not investing in new facilities or equipment at the time of the investigation. Moreover, input (output) saving (increase) potentials identified under DEA are not always achievable in port operational settings, particularly if this involves small amounts of an indivisible input or output unit.

Other comments on the port frontier literature (both parametric and non-parametric) are highlighted in the following:

(a) In none of the studies mentioned above were the determinants (inputs/outputs) of port performance formally linked to or justified by a correspondingly valid empirical analysis. Variables were selected either subjectively or at best from previous literature, much of which was, in fact, based on subjective and arbitrary appraisal.

(b) Many port researchers have each applied parametric and non-parametric methods on different occasions, which implies that no consensus has been reached on a single and consistent approach that best analyses port performance.

(c) The above partly explains why some of the findings of the port frontier literature provide inconsistent results, for instance, when analysing the relationships between size and efficiency (Martínez-Budría et al., 1999 versus Coto-Millan, Banos-Pino, & Rodriguez-Alvarez, 2000), ownership structure and efficiency (Notteboom, Coeck, & Van-Den Broeck, 2000 versus Cullinane, Song, & Gray, 2002) and locational/logistical status and efficiency (Liu, 1995 versus Tongzon, 2001).

(d) Finally, most frontier applications in ports focus solely on sea access, which in fact is a common feature of much of the literature on port performance. The emphasis on quayside operations overlooks other processes of the port operating system and ignores the interests of other members of the port's supply chain network.

The failure to link quayside operations with landside systems also under-lines a major gap in the port literature. The structure of much of the port production and operating systems is marked by the existence of many critical processes or bottlenecks whereby the performance and capacity of one stage (or sub-system) is a binding constraint for that of the next stage, and sub-sequently for the total performance (e.g., output) of the whole port system, extended to that of the overall port supply-chain network. An illustration of bottleneck problems in a container port operating and management system is depicted in Fig. 24.1.

The process in Fig. 24.1 implies a dual relationship between (a) dispro-portionate performance and capacity levels at the internal logistics level

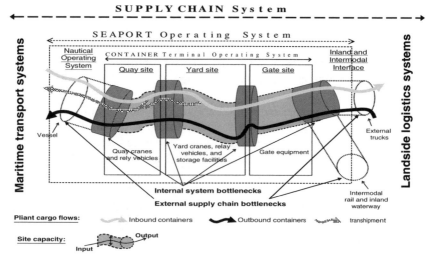

Fig. 24.1.   Illustration of Bottleneck Problems in a Container Port Operating System.

(e.g., a sub-system working fully, while concurrent ones remain under-utilised) and (b) demand and supply variability scenarios at the external supply chain level (e.g., uncertainty of vessel schedules, changes in trade patterns, landside logistics disruptions). In the context of this paper, the latter aspect is of particular importance as it shifts the focus of port performance from the traditionally fragmented internal efficiency to the contemporary integrated supply-chain efficiency. Following a brief section on the criteria of performance assessment and evaluation, the remainder of this paper seeks to outline and illustrate the basis and benefits of the logistics and supply chain approach to port performance benchmarking.

3. CRITERIA OF ASSESSMENT AND EVALUATION

Issues in developing a valid and integrative performance framework involve major gaps and discrepancies at both analytical and methodological levels.

3.1. Performance Taxonomy and Measurement Gaps

Performance measures are primarily designed to capture the performance of an activity or a transformational process. Performance is a broad concept that covers almost any objective of operational, management and competitive excellence of a firm and its activities. A sample of the taxonomy of performance measurement dimensions is depicted below. The list in Fig. 24.2 is for illustration only, and is neither authoritative nor exhaustive.

The main problem with performance measurement systems is that while they depict various performance dimensions, their definitions and specific

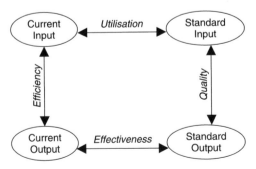

Fig. 24.2. Basic Matrix of Performance-Ratio Dimensions.

applications are not always consistent between researchers or fields. For instance, the term 'productivity' may be interpreted differently depending on the approach used. Ghobadian and Husband (1990) suggest that there are at least three broad categorisations of productivity: the economic concept (efficiency of resource allocation), the technological concept (relationship between ratios of output to the inputs used in its production) and the engineering concept (relationship between the actual and the potential output of a process). Edwards (1986) points out that even within the narrow context of management accounting, productivity is a misused concept and can mean anything from cost reduction and efficient utilisation to work measurement and programme evaluation.

Performance concepts are a source of debate as to which dimension or combination of dimensions most accurately reflects a firm's performance. In the port literature, the underlying relationship between variations in the indicators and performance has been difficult to establish. Typically, performance measurement systems in ports are split between measuring either internal efficiency or external effectiveness, but are hardly used to capture both dimensions. A single focus on either efficiency or effectiveness does not seem to be the only way to increase performance, as there are many examples of ports around the world that operate effectively but are still inefficient, and vice versa. To close these measurement gaps, a multi-dimensional performance system that encompasses both operational tasks and strategic activities is required.

3.2. Comprehension, Consistency and Usefulness

A comprehensive performance measurement system involves (a) capturing all appropriate activities in the process and (b) incorporating the interests of all relevant members and stakeholders. In the context of port performance, this is very difficult to achieve due to the complex interactions of port missions, institutions and functions, which raises the question of whose perspective or standpoint (regulator, operator, customer, etc.) one has to consider when undertaking performance measurement and benchmarking. Conventional frameworks tend to favour the regulator's (e.g., port authority) interest, but even within this context, port authorities may have different, sometimes conflicting, missions and objectives. A further complexity arises when an outside institution performs a port function, for instance, when a shipping line or its subsidiary acts as a port operator. In such cases, a port's performance is often equated to ship's efficiency at berth or in port (i.e., minimising service time or time in port), hence blurring the boundaries between the objectives of the shipping line as a customer and those of the port as a service provider. On the other hand, a port where job generation and environmental sustainability are the primary missions may find its performance manifesto and objectives being fundamentally different from those of a dedicated port or terminal. In this respect, it is worth noting that many performance studies tend to overlook this dimension when undertaking benchmarking exercises of ports with different missions or functions. A thorough discussion of port attributes and the relationship with operational and management decisions is provided by Bichou and Gray (2005). The argument they advocate is particularly relevant to port performance in that it acknowledges the role of a port's major attribute, or decisive factor, in shaping the approach to performance benchmarking (see Table 24.1). In the next sections, I outline the benefits of supply chain thinking in overcoming common obstacles of role, functional and operational inadequacies within and among ports so as to ensure a comprehensive performance measurement system.

A second criterion is the consistency of the performance system. Consistency implies coherence with other approaches and performance systems. For instance, the results of frontier benchmarking exercises should be broadly consistent with financial analysis and investor perceptions. Central to consistency is the requirement for compatibility and that the performance system must be designed in line with a firm's objectives and future orientations. This is particularly relevant in the context of port performance benchmarking as recent structural changes of the industry suggest a shift of emphasis from measuring the internal efficiency of local stevedores to analysing the supply chain efficiency of global operators.

Finally, a performance system should be designed to guide and influence the decision-making process. To achieve this, performance models need be useful and avoid over-complexity, otherwise they may end up being ignored or discarded. The criterion of usefulness must not conflict with the validity and robustness of the benchmarking process, but rather ensures that the selected techniques can be implemented in practice while the results can still demonstrate reasonable stability over time and space.

3.3. Benchmarking Issues

Performance benchmarking is the merging of two methodologies: performance measurement, i.e., selecting measurement metrics and designing measurement systems, and benchmarking against a comparable group. In the port

Table 24.1. The Role of Decisive Factors in Determining Port Performance Approaches and Frameworks of Analysis.

Examples of Approaches to Ports in the Literature		Decisive Factors				Corresponding Performance Approaches
		Missions	Assets/facilities	Functions	Institutions	
Macro-economic approaches	Economic catalyst	Major				Economic impact analysis
	Job generator	Major				
	Trade facilitator	Major				
Institutional/ organisational models	Private/public	Minor			Major	Index methods/ frontier approaches
	Landlord/tool/service		Major		Minor	Economic impact analysis
Geographic and spatial approaches	Port-city	Major				
	Waterfront estate	Minor	Major			
	Sea/shore interface	Minor		Major		
	Logistics centre	Minor		Major		
	Clusters				Major	
	Trade and distribution centres	Major		Minor		
	Free zones and trading hubs	Minor		Major		
Hybrid approaches	UNCTAD generations (1st–4th)	Major		Major	Minor	Index methods
	World Bank 'Port Authority' model	Major			Major	
Alternative new approaches	Combinative strategies (cargo–sea–land, supply–demand led)	Major		Major		Index methods/ frontier approaches
	Logistics/production systems (Tele-port, trade port, etc.)	Major		Major		
	Business units (production, marketing, pricing, etc.)	Minor		Major		

Source: Extracted and adapted from Bichou and Gray (2005, p. 84).

literature, most available approaches seem to undertake either exercise, but not both. A major weakness in this respect is that port performance focuses largely on competitive benchmarking (usually against a neighbouring port but rarely against other direct product competitors such as regional distribution centres (RDCs) and to a lesser extent on internal benchmarking (e.g., between two warehouses or container terminals within the same port). The other two missing levels of benchmarking are (a) process benchmarking used to compare operations and business processes (e.g., process re-engineering) with the best practice in the industry such as for terminal design and layout, and (b) generic benchmarking, which seeks best practices irrespective of the industry (for instance, benchmarking port operations against similar processes such as those found in manufacturing and distribution environments).

3.4. Multi-Dimensionality

Although extensive literature has addressed theories and practices in port performance measurement, little has emerged on linking and integrating operations, design and strategy within the multi-institutional and cross-functional port context. In particular, the association of ports with performance benchmarking is likely to imply some methodological difficulty, with the major obstacles being identified as:

- *Multi-firm dimensions.* Identifying and accessing the wide range of members working in and across port supply chains (shippers, ocean carriers, port operators, logistics providers, freight forwarders, public authorities, etc.).
- *Cross-functional dimensions.* Recognising and minimising differences of operational/strategic viewpoints in a traditional port setting often typified by institutional fragmentation and conflict over channel control and management.
- *Inter-disciplinary dimensions.* Understanding the interdisciplinary scopes of port practice and research, the first extending across manufacturing, trade and service industries, while the second intersects wide subjects ranging, inter alia, from engineering and operations research to marketing and quality management.

A normative framework is therefore required to integrate the various internal functions and processes and link them to those of the external supply chain members. This paper proposes that this can be done by adopting a logistics and supply chain orientation to ports. The essence of logistics and SCM is an

Table 24.2.   Purpose and Relationships between Performance
Evaluation Criteria and Key Concepts of Logistics and SCM.

Criterion	Key Logistics and SCM Concepts
Multi-dimensional	Cross-functionality and multi-firm perspective
Comprehensive	Systems approach, integrative framework, trade-off analysis
Consistent	Customer focus, future-oriented
Robust	Multi-disciplinary perspective, process approach, value-chain orientation
Useful	Practical orientation, action-oriented

integrative approach to the interaction of different functions and processes within a firm extended to a network of organisations for the purpose of cost reduction and customer satisfaction (Stank, Keller, & Daugherty, 2001). The key concepts behind this definition are time compression, cost reduction, process integration, planning and control, collaborative arrangements and trust relationships, customer focus, value creation and value chain analysis and a system's perspective. Table 24.2 shows how logistics and SCM concepts can satisfy the criteria of performance evaluation and assessment.

4. CONCEPTUAL MODEL AND FRAMEWORK OF ANALYSIS

4.1. Channel Orientation to Port Operations and Management

Increasing recognition of seaports as logistics centres requires them to be conceptualised from a logistics and SCM perspective. Traditional port management is often typified by institutional fragmentation and conflict with other members of the logistics channel, whereas the SCM philosophy advocates process integration and partnership. Nevertheless, although there is a widespread recognition of the potential of ports as logistics centres, little has emerged on supply chain mapping, design and orientation.

This paper distinguishes between logistics, trade and supply channels. The logistics channel consists primarily of specialists (e.g., shipping lines, freight forwarders) that facilitate the efficient progress of cargo through, for example, warehousing and transportation. Both the trade channel and supply channel are associated with ownership of goods moving through the system, with the difference that the trade channel is normally perceived to be at the level of the sector or industry (e.g., the oil trade) and the supply channel at the level of the firm.

At the operational level, this approach is particularly useful for designing, managing and integrating the various port functions and activities for the purpose of overall flow and process optimisation. The conceptualisation in terms of channel management translates various movements and operations into flows and processes related to a network of activities and institumshy;tions brought together for the purpose of the chain. A channel is a pathway tracing the movement of a cargo-shipment across a typology of multi-institutional and cross-functional cluster alignments, while flows are the derived transactions (business interactions) between various functional-institutions within each channel. Thus, the nature and number of channels and flows will depend on each port or terminal, and may change over time and space.

At the strategic level, the approach allows ports to formulate long-term strategies (e.g., traffic forecasting, strategic planning) in relation to channel typologies of external members. Traditionally, the emphasis has been on the nature, origins and destinations of freight movements that disintegrate port operations and management from supply chain structures. We believe that this dimension only justifies part of the evolution of port operating and management systems, and that the full explanation lies in understanding the logistics and SCM configurations of a port's external members. For instance, bulk maritime transport has traditionally been analysed by trade (crude oil, iron ore, etc.), and this dimension still explains much of distribution patterns of bulk commodities, including the location, management and operations of bulk ports and terminals. Nevertheless, the trade-channel perspective does not always explain port operational flows and processes involving other types of commodities. For example, freight distribution patterns of manufactured cars seem to follow a supply-chain orientation, whereby manufacturing firms' (i.e., car makers), organisational structures (e.g., location of branch offices), industrial capabilities (e.g., ownership or operation of car-carriers) and relationships with suppliers, retailers and service providers are the most decisive factors behind port choice and terminal operational features. In a similar vein, many container maritime flows, including transshipment and management of empty containers (reverse logistics) are also explained by the strategic orientations of ocean carriers rather than only by the structure of trade patterns between containerised markets. In this case, the emphasis on the logistics channel suggests that the selection of a transport network in container shipping, including port choice, is undertaken at the level of the logistics provider (e.g., the shipping line) rather than at the level of the firm (shipper, manufacturer) or the industry (container trade).

Ports have an important role to play in the integration of all three types of channels. Furthermore, they offer a unique location in which members of different channels (trade, logistics and supply channels) can meet and interact:

- From a logistics channel standpoint, the port is a very important node since it serves as an intermodal/multimodal transport intersection and operates as a logistics centre for the flows of goods (cargo) and people (passengers).
- From a trade channel perspective, the port is a key location whereby channel control and ownership can be identified and/or traded.
- From a supply channel approach, the port not only links outside flows and processes but also creates patterns and processes of its own. Robinson (2002) argues that ports should be seen as key elements in value-driven chain systems. He claims that ports contribute to supply chains through the creation of competitive advantage and value-added delivery. At the supply chain level, ports are one of the very few networking sites that can bring together various members in the supply channel.

Fig. 24.3 depicts the interactions between channels and flows in a typical port supply chain network. Channel and flow configurations are portrayed in terms of linear path combinations for the purpose of simplification, although a better illustration would be in terms of web-type network re-lationships. Note the fundamental distinction between institutions and functions in international shipping and logistics. Often, a single institution undertakes various functions and thus can be part of more than one chan-nel, e.g., a manufacturer or shipper acting as an ocean carrier (e.g., indus-trial shipping, bareboat chartering), or a carrier (e.g., shipping line) acting as a port operator or as a logistics provider. The interaction between functions and institutions is often overlooked in international shipping and logistics, which may constitute a fundamental bias when designing supply chains or locating decisions spots.

The conceptualisation of ports from a logistics and supply chain approach has proven to be constructive on more than one level, including recog-nising and integrating the multi-institutional and cross-functional dimen-sions of ports. This is particularly the case for measuring port performance. Common obstacles of role, functional and operational inadequacies are also to be overcome as the segmentation of the port business in terms of the trade, logistics and supply channels typology will greatly help in identifying which institution is performing what function to achieve which ultimate mission. Similarly, performance measurement and management will be analysed, val-ued and assessed in terms of a port's contribution to the overall combined

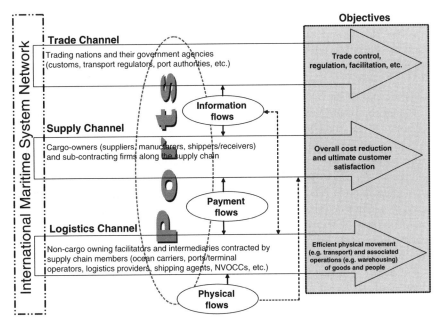

Fig. 24.3. Channel Typologies and Components of the Port Network System.

channel added-value, and thus port competition will shift from the institutional, functional and/or spatial levels to the channel management level.

The logistics approach regularly adopts a cost trade-off analysis between functions, processes and even supply chains (Rushton, Oxley, & Croucher, 2000). This could be beneficial to port efficiency by directing port operations towards relevant value-added logistics activities, while overcoming bottleneck obstacles through process integration. Logistics performance is based on the notion that sub-optimisation at one point is permitted as long as it contributes to overall performance. In a supply chain context, this is expanded across functional and organisational interfaces thanks to lean and agile practices (Christopher & Towell, 2000; Paixao & Marlow, 2003).

4.2. An Illustration of Internal Logistics Measurements and Costing in Ports

The logistics and supply chain literature offers an abundance of performance measures and measurement frameworks ranging from bottom-up approaches such as costing techniques, e.g., activity-based costing (ABC), total cost analysis (TCA) and direct product profitability (DPP) to top-down models such as the balanced scorecard (BSC) and the supply chain operations reference (SCOR) model (see Kaplan & Norton, 1992; Supply Chain Council Website, 2002). A detailed review of the literature on port performance, logistics and SCM measurements is provided by Bichou and Gray (2004).

Under the traditional cost accounting system, every department or activity in ports is regarded as a cost centre (e.g., maintenance, warehousing, administration), which makes it difficult to identify the different costs that result from servicing customers with a particular product mix. Logistics costing techniques that have gained recognition from both logistics professionals and academics include ABC and TCA. The former proposes an evaluation of the costs of a firm's activities based on the actual resources and time consumed to perform them (Liberatore & Miller, 1998), whereas the latter proposes a trade-off analysis among different internal functions to minimize the total cost while at the same time maintaining customer satisfaction (Bowersox & Closs, 1996). The principles behind ABC and TCA are also valid in a supply chain context. Other relevant methods of logistics and supply chain costing include DPP, mission costing and customer profitability analysis.

This section demonstrates why and how the logistics costing approach can be relevant to port performance, and uses ABC as an example. ABC does not allocate direct or indirect costs based on volume alone, but determines which activities are responsible for these costs, and associates them with their respective portion of overhead costs. ABC works on a two-stage procedure for assigning cost-to-cost object. The first stage focuses on determining the costs of activities within the system, while the second stage allocates activity costs to port cost centres consuming the work performed. Fig. 24.4 illustrates a basic application of ABC in a typical port setting.

Consider, for instance, the costing of a public port warehouse working under the traditional approach of allocating costs based on total output shipped to each customer. Port warehouse's customers (shipping lines, land transport carriers, freight forwarders, etc.) and products (container boxes, break bulk consignments, etc.) do not consume warehouse resources (labour, machine time, fuel and electricity, etc.) proportionally to their value or weight volume. An ABC view of costs at this facility would not only allow for a better allocation of costs, but also provide a clearer identification of performance deficiencies and ways to improve them.

Performance measures appear as a logical consequence of an ABC system. Activity description includes financial information (cost, profit, etc.) and non-financial information (time, quality, etc.), whereby the port management can

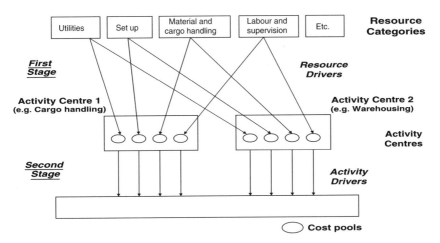

Fig. 24.4. A Simplified Example of a Two-Stage ABC Cost Allocation in a Port Setting.

develop a set of performance indicators based on ABC results. The technique also supports continuous improvement by identifying where incremental improvements at the activity level can improve the overall port performance. Finally, ABC can be extended across the port external system so as to identify opportunities for eliminating redundant (waste) activities within the supply chain, identifying port members with excessive resource consumption, and providing attractive alternatives to existing channel structures.

4.3. Towards an Integrated Framework to Port Performance Benchmarking

In the previous sections, the limits of conventional models in both their approach and methodology for benchmarking port performance were stressed. Many frameworks in the field are oriented towards fragmented aspects of port operations and management (e.g., efficiency versus effectiveness, seaside versus landside, operator versus regulator), which may reduce their universality in the context of performance benchmarking of world ports and terminals. Similarly, many approaches display inherent limitations for conducting benchmarking exercises, particularly in complex and multi-faceted settings such as ports and terminals.

Despite these difficulties, significant academic work on the subject has been undertaken in the last two decades or so, with many noteworthy attempts to adapt and improve models applied elsewhere. In a similar effort, this chapter attempts to analyse the subject of port performance from the perspective of logistics and SCM. It argues that a supply chain framework can prove particularly useful in overcoming complexities and obstacles described earlier in the chapter, with a view to developing an integrative approach to performance benchmarking of ports.

The next section shows how an integrated logistics and SCM framework to port performance benchmarking can be designed and successfully implemented. The model presented is extracted and adapted from a previous inquiry on the same subject. Although the overall research was much wider, it is partly reported here, which restricts the discussion to the conceptual and analytical elements. The methodology and full results of the inquiry can be found in Bichou and Gray (2004).

The approach and methods reported both in this chapter and in the framework described below are neither inclusive nor exhaustive of all concepts and techniques developed in the field of SCM. Many SCM topics that are also relevant to port operations and management are not discussed here, including the aspects of network topologies, queuing and scheduling models, demand forecasting and amplification (e.g., the bullwhip effect), inventory control models, collaborative and contractual relationships, vulnerability and risk pooling, and system's agility and lean responsiveness. The same applies to performance models for supply chain optimisation and benchmarking; in particular, those applying quantitative techniques such as linear and dynamic programming, enterprise modelling and simulation.

4.3.1. Initial Investigation and Model Development

The methodology used was to present port managers and a panel of experts and academics with an interim framework of port performance for examination and assessment by them leading to an improved model. Participants in the study were clustered into three focus groups or panels.

1. *Ports Panel (1)*. 45 port managers drawn from a sample of 60 ports worldwide to represent different regions and continents.
2. *International Institutions Panel (2)*. 14 employees of international institutions (World Bank, UNCTAD, IMO, etc.) drawn from a sample of 17 originating from 11 countries.
3. *Academics and Consultants Panel (3)*. 14 academics and other port experts drawn from a sample of 17 academics, 3 consultancy firms and 3 independent (freelance) consultants originating from 11 countries.

The initial model was the result of diagnostic work undertaken through an online questionnaire designed exclusively for port managers (Panel 1) in order to investigate both their perception of logistics concepts and the methods used by them to measure port performance and efficiency. In some cases, surveys were conducted through face-to-face interviews or administered over the telephone. Tables 24.3 and 24.4 report on the performance techniques used by the port participants, while Fig. 24.5 informs about their perception and ranking of supply chain members.

Table 24.3. Respondents' Cumulative Ranking of Performance Indicators by Frequency of Use.

	Internal Performance		External Comparison		Cumulative Responses	
	Positive ranking	% of total	Positive ranking	% of total	Compiled	% of grand total
Financial indicators	44	38	35	25	79	31
Throughput indicators	37	32	43	31	80	31
Productivity indicators	24	21	35	25	59	23
Economic-impact indicators	7	6	25	18	32	13
Others	3	3	1	1	4	2
Total	115	100	139	100	254	100

Table 24.4. Satisfaction with Current Performance Indicators versus Frequency of Use of Logistics Techniques.

Satisfaction with the Indicators Currently Used			Scale	Use of Logistics Techniques		
	Replies	%			Replies	%
Very satisfied	5	11	5	Always	2	4
Satisfied	3	7	4	Often	7	16
Neither one nor the other	13	29	3	Sometimes	11	25
Dissatisfied	16	35	2	Very few times	15	33
Very dissatisfied	8	18	1	Never	10	22
Total	45	100		Total	45	100

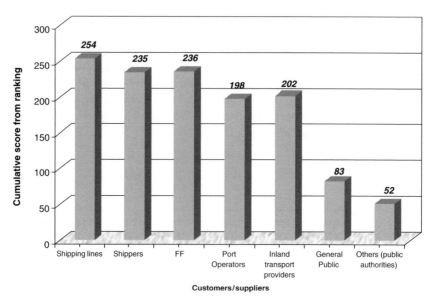

Fig. 24.5. A Rating Scale Analysis of Participants' Main Customers and Suppliers (Supply Chain Members). *Note:* FF, Freight Forwarders.

4.3.2. Testing, Results and Analysis

Fig. 24.6 depicts a framework for integrative port performance and monitoring based on logistics and SCM concepts. The model links different components for the port performance system throughout a chain of influence and relationships. Note the links between logistics and operations on the one hand and SCM and strategy on the other. In both the upper and lower parts of the model, performance measurement starts with activity and process mapping based on a proper channel typology as illustrated in Fig. 24.4 and in the example in Fig. 24.5. It then traces the contribution of the derived internal and external performances to form an aggregate port performance index. The model was sent to and discussed with participants from the three panels to assess its validity and feasibility. An explanatory note of the techniques and concepts proposed by the model was enclosed in the initial model sent to port participants.

Responses from the port's panel participants varied in many aspects, although most port respondents considered the model valid as a 'first initiative' that looks at port efficiency from the perspective of logistics and SCM. Almost all port participants have measured efficiency in a way similar to that of performance 1 in the model. However, about half mentioned the

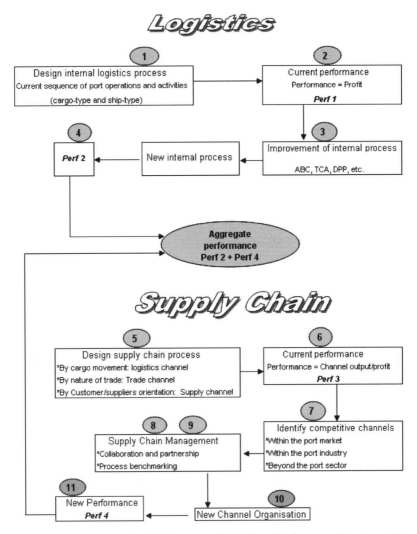

Fig. 24.6. An Integrated SCM Framework for Port Performance Benchmarking.

problems of accountability and process continuity in performance monitoring. An important issue is that many port participants indicate that they
do not know where their logistics process (or activity sequencing) starts and
particularly where it ends, simply because many activities are at the interface
between the port area and the outside world (e.g., inland logistics centres).

At the supply chain level, they also admit facing great difficulty in understanding and designing their respective channel typologies. Some ports attribute these limitations to the lack of reliable information, while others relate this to the complexity of channels and the confusion that surrounds their categorisation. However, nearly all ports appreciated the concepts of activity sequencing, channel integration and collaborative arrangements, with as many as 90 per cent of participants (40 ports) intending to use logistics costing techniques to improve performance and visibility.

Responses from other panels were even more supportive to the application of a logistics and supply chain approach to port performance. They particularly valued the association of internal process mapping with external channel design. The remainder of the two panels' comments and feedbacks focused on the detailed aspects of the framework rather than its methodological or conceptual orientation. In particular, there has been a widespread call for the quantification of the model.

5. CONCLUSION

The aim of this paper was to demonstrate that conceptualising the port system from the perspective of logistics and SCM can be relevant to port performance and benchmarking, where conventional approaches often provide a fragmented framework of analysis.

After reviewing the applications and limitations of the various approaches to port efficiency, a supply chain framework was proposed through linking internal processes with external channel orientations, analysing and integrating different performance dimensions. A subsequent interim model was designed and tested by adopting a structured approach involving a wide range of participants and interest groups, including ports. The results show a common interest in logistics and SCM across the various panels, although port respondents have shown a lack of familiarity and understanding of the techniques and concepts involved.

Finally, the framework and methods given in this chapter are primarily illustrative and not intended to be conclusive. Both this chapter and the reported inquiry stand as a first and modest initiative and more research is needed particularly with regard to the quantification of the model and the concepts associated with it. Further empirical and analytical investigations are also required so as to allow a more comprehensive analysis and generalise the scope and scale of the study to a large sample of world ports and terminals. The results of the present study could serve as a basis to point the way forward and highlight the potential for integrative and collaborative logistics and SCM, with a view to designing and managing valid frameworks for port performance measurement and benchmarking.

REFERENCES

Asia-Pacific Economic Cooperation (APEC). (1999). *Assessing APEC trade liberalisation and facilitation: 1999 update*. APEC Economic Committee Publication #99-EC-01.1. Online, available at www.apec.org/content/apec/publications/all_publications/economic_committee.html

Association Internationale des Ports et Villes (AIPV/IAVP). (2005). Online, available at www.aivp.org/infos.html

Australian Productivity Commission. (1998). *International benchmarking of the Australian waterfront*. Research Report. Canberra: Australian Government Productivity Commission.

Barros, C. P. (2003). Incentive regulation and efficiency of Portuguese port authorities. *Maritime Economics and Logistics, 5*, 55–69.

Barros, C. P., & Athanassiou, M. (2004). Efficiency in European seaports with DEA: Evidence from Greece and Portugal. *Maritime Economics and Logistics, 6*, 122–140.

Bendall, H., & Stent, A. (1987). On measuring cargo handling productivity. *Maritime Policy and Management, 14*, 337–343.

Bichou, K., & Gray, R. (2004). A logistics and supply chain management approach to port performance measurement. *Maritime Policy and Management, 31*(1), 47–67.

Bichou, K., & Gray, R. (2005). A critical review of conventional terminology for classifying seaports. *Transportation Research A, 39*, 75–92.

Bonilla, M., Medal, A., Casaus, T., & Salas, R. (2002). The traffic in Spanish ports: An efficiency analysis. *International Journal of Transport Economics, 19*(2), 237–253.

Bowersox, D. J., & Closs, D. J. (1996). *Logistical management: The integrated supply chain process*. New York: McGraw-Hill.

Braeutigam, R., Daughety, A., & Turnquist, M. (1984). A firm specific analysis of economies of density in the US railroad industry. *Journal of Industrial Economics, 33*, 3–20.

Christopher, M., & Towell, D. R. (2000). Supply chain migration from lean and functional to agile and customised. *Supply Chain Management, 5*(4), 206–213.

Coto-Millan, P., Banos-Pino, J., & Rodriguez-Alvarez, A. (2000). Economic efficiency in Spanish ports: Some empirical evidence. *Maritime Policy and Management, 27*(2), 169–174.

Cullinane, K. P. B., Song, D. W., & Gray, R. (2002). A stochastic frontier model of the efficiency of major container terminals in Asia: Assessing the influence of administrative and ownership structures. *Transportation Research Part A, 36*, 743–762.

Cullinane, K. P. B., Song, D. W., & Wang, T. (2004). An application of DEA windows analysis to container port production efficiency. *Review of Network Economics, 3*(2), 186–208.

De, P., & Ghosh, B. (2003). Causality between performance and traffic: An investigation with Indian ports. *Maritime Policy and Management, 30*(1), 5–27.

De Langen, P. W. (2002). Clustering and performance: The case of maritime clustering in the Netherlands. *Maritime Policy and Management, 29*(3), 209–221.

De Monie, G. (1987). Measuring and evaluating port performance and productivity. In: *UNCTAD Monographs on Port Management, No. 6, International Association of Ports and Harbours* (pp. 2–11). Geneva: UNCTAD.

Devarajan, S., & Rodrik, D. (1991). Pro-competitive effects of trade reform: Results from CGE model of Cameroon. *European Economic Review, 35*, 1157–1184.

Dio, M., Tiwari, P., & Itoh, H. (2001). A computable general equilibrium analysis of efficiency improvements at Japanese ports. *Review of Urban & Regional Development Studies, 13*(3), 187–206.

Drewry Shipping Consultants. (1997). *World container terminals 1997*. London: Drewry Market Report.

Edwards, J. B. (1986). *The use of performance measures*. Montvale, NJ: National Association of Accounts.

Estache, A., González, M., & Trujillo, L. (2002). Efficiency gains from port reform and the potential for yardstick competition: Lessons from México. *World Development, 30*(4), 545–560.

Fink, C., Mattoo, A., & Neagu, I. C. (2000). *Trade in international maritime services: How much does policy matter?* Washington, DC: The World Bank.

Fourgeaud, F. (2000). *Measuring port performance*. Washington, DC: The World Bank.

Frankel, E. G. (1991). Port performance and productivity measurement. *Ports and Harbours, 36*(8), 11–13.

Ghobadian, A., & Husband, T. (1990). Measuring total productivity using production functions. *International Journal of Production Research, 28*(8), 1435–1436.

Gonzales, M. M., & Trujillo, L. (2005). *Reforms and infrastructure efficiency in Spain's container ports*. World Bank Research Policy Paper No. 351. Washington, DC: The World Bank.

Hamilton, G. L., Ramsussen, D., & Zeng, X. (2000). *Rural inland waterways economic impact kit: User guide*. Institute for Economic Advancement, University of Arkansas at Little Rock. Online, available at www.ntl.bts.gov/data/user-guide1.pdf

Hoffmann, J. (2001). Latin American ports: Results and determinants of private sector participation. *International Journal of Maritime Economics, 3*, 221–230.

Holmberg, S. (2000). A systems perspective on supply chain measurements. *International Journal of Physical Distribution and Logistics Management, 30*(10), 47–68.

Jara Díaz, S., Martinez-Budría, E., Cortes, C., & Basso, L. (2002). A multi-output cost function for the services of Spanish ports' infrastructure. *Transportation, 29*(4), 419–437.

Kaplan, R. S. (1984). Yesterday's accounting undermines production. *Harvard Business Review, 62*(4), 95–101.

Kaplan, R. S., & Norton, D. P. (1992). The balanced scorecard: Measures that drive performance. *Harvard Business Review, 70*(1), 71–79.

Kim, M., & Sachish, A. (1986). The structure of production, technical change and productivity in a port. *Journal of Industrial Economics, 35*(2), 209–223.

Lawrance, D., & Richards, A. (2004). Distributing the gains from waterfront productivity improvements. *Economic Record, 80*, 43–52.

Le Havre Port. (2000). *Impact socio-economique du port 2000*. Online, available at www.havre-port.net/pahweb.html

Liberatore, M. J., & Miller, T. (1998). A framework for integrating activity-based costing and the balanced scorecard into the logistics strategy development and monitoring process. *Journal of Business Logistics, 19*(2), 131–154.

Liu, Z. (1995). The comparative performance of public and private enterprises: The case of British ports. *Journal of Transport Economics and Policy, 29*(3), 263–274.

MARAD. (2003). *Public port finance survey for the financial year 2001.* Washington, DC: US Maritime Administration.

Martínez-Budría, E., Díaz-Armas, R., Navarro-Ibañez, M., & Ravelo-Mesa, T. (1999). A study of the efficiency of Spanish port authorities using data envelopment analysis. *International Journal of Transport Economics, 26*(2), 237–253.

Micco, A., & Perez, N. (2001). *Maritime transport costs and ports efficiency.* Santiago: Inter-American Development Bank.

Moloney, R., & Sjostrom, W. (2000). *The economic value of the Port of Cork to Ireland in 1999: An input–output study.* Report to the Irish government. Cork: National University of Ireland.

Notteboom, T., Coeck, C., & Van-Den Broeck, J. (2000). Measuring and explaining the relative efficiency of container terminals by means of Bayesian stochastic frontier models. *International Journal of Maritime Economics, 2*(2), 83–106.

Paixao, A. C., & Marlow, P. B. (2003). Fourth generation ports – a question of agility. *International Journal of Physical Distribution and Logistics Management, 33*(4), 355–376.

Robinson, R. (2002). Ports as elements in value-driven chain systems: The new paradigm. *Maritime Policy and Management, 29*(3), 241–255.

Rodrigue, J. P., Slack, B., & Comtois, C. (1997). Transportation and spatial cycles: Evidence from maritime systems. *Journal of Transport Geography, 5*(2), 87–98.

Roll, Y., & Hayuth, Y. (1993). Port performance comparison applying data envelopment analysis (DEA). *Maritime Policy and Management, 20*(2), 153–161.

Rushton, A., Oxley, J., & Croucher, P. (2000). *The handbook of logistics and distribution management.* London: Institute of Logistics and Transport.

Sachish, A. (1996). Productivity functions as a managerial tool in Israeli ports. *Maritime Policy and Management, 23*(4), 341–369.

Sanchez, R. D., Hoffmann, J., Micco, A., Pizzolitto, G. V., Sgut, M., & Wilsmeier, G. (2003). Port efficiency and international trade: Port efficiency as a determinant of maritime transport costs. *Maritime Economics and Logistics, 5,* 199–218.

Stank, T. P., Keller, S. B., & Daugherty, P. J. (2001). Supply chain collaboration and logistics service performance. *Journal of Business Logistics, 22*(1), 29–47.

Supply Chain Council Website. (2002). Online, available at www.supply-chain.org/

Talley, W. K. (1988). Optimum throughput and performance evaluation of marine terminals. *Maritime Policy and Management, 15*(4), 327–331.

Talley, W. K. (1994). Performance indicators and port performance evaluation. *Logistics and Transportation Review, 30*(4), 339–352.

Tongzon, J. L. (2001). Efficiency measurement of selected Australian and other international ports using data envelopment analysis. *Transportation Research A, 35*(2), 113–128.

Tovar, B., Jara-Diaz, S., & Trujillo, L. (2003). *Production and cost functions and their applications to the port sector: A literature survey.* World Bank Policy Research Working Paper 3123, pp. 1–31. Washington, DC: The World Bank.

Turner, H., Windle, R., & Dresner, M. (2004). North American container-port productivity: 1984–1997. *Transport Research E, 40,* 339–356.

United Nations Conference on Trade and Development (UNCTAD). (1976). *Port performance indicators.* Geneva: UNCTAD.

Valentine, V. F., & Gray, R. (2001). The measurement of port efficiency using data envelopment analysis. *Proceedings of the 9th world conference on transport research,* Seoul.

Vitale, M. R., & Mavrinac, S. C. (1995). How effective is your performance measurement system? *Management Accounting, 77*(2), 43–47.

Wang, T. F., & Cullinane, K. B. P. (2005). Measuring the economic efficiency of Europe's container terminals. In: *Proceedings of the international association of maritime economists annual conference*, Cyprus, 30 June–2 July.

Wheelwright, S. C. (1978). Reflecting corporate strategy in manufacturing decisions. *Business Horizons, 21*(1), 57–66.

Wilson, J., Mann, C., & Otsuki, T. (2003). *Trade facilitation and economic development: Measuring the impact*. Washington, DC: The World Bank Research Group.

CHAPTER 25

ISSUES IN MEASURING PORT DEVOLUTION PROGRAM PERFORMANCE: A MANAGERIAL PERSPECTIVE

Mary R. Brooks

ABSTRACT

One of the premises of devolution is that productivity gains will result from better management of the resources of the entity devolved. The success of a devolution program, however, is difficult to measure and may be a product of many variables. A review of the literature on port devolution indicates mixed results. The literature on port performance measurement provides a myriad of efficiency measures with little emphasis on whether these ports are effective or meet customer or stakeholder needs. This paper examines the literature on performance measurement, both at the firm level (public and private) and at the program level, and explores what constructs may be suitable to measure the performance of devolution programs for ports from a strategic management perspective.

1. INTRODUCTION

One of the premises of devolution is that productivity gains will result from better management of the resources of the entity devolved. However, it is

Devolution, Port Governance and Port Performance
Research in Transportation Economics, Volume 17, 599–629
Copyright © 2007 by Elsevier Ltd.
ISSN: 0739-8859/doi:10.1016/S0739-8859(06)17025-0

difficult to measure the "success" of devolution programs, because success
will be dependent on the objectives of devolution and most governments do
not attempt to quantify what they mean by "better allocation of resources" or
"better management."[1] Success may be a product of many factors, only some
of which may result from altering the governance system. A survey of the
literature on port reform outcomes results in more questions than answers.
There is certainly no shortage of research on measuring efficiency in ports;
neither is there a lack of data on the available capacity of ports. A report
detailing available capacity may tell the reader of the opportunity available,
but growth in available capacity may or may not be the planned product of a
government devolution program. A report indicating the efficient use of that
capacity may lead to the assumption that the port is effective in the conduct of
its business, but is year-over-year improvement in the efficiency of a nation's
ports the only means by which a government can determine that its port
reform program is a success? This paper explores the topic of performance
measurement at both the firm and the program level, and draw conclusions
about how ports may measure success and how governments may choose to
define program success, beyond the measurement of efficiency.

2. THE LITERATURE ON PORT DEVOLUTION OUTCOMES

Devolution programs are founded on the belief that the private market is the
most efficient means of allocating resources. However, a number of port
studies suggest that the desired "efficient markets" ideal does not materialize.
Anecdotes of greed among privatized boards with large salaries paid to staff
and perquisites of cars and padded expense accounts abound (e.g. Thomas,
1994; Baird, 1995). Other studies note significant improvements in port services post-privatization (e.g. Shashikumar, 1998; Serebrisky & Trujillo, 2005)
or mixed outcomes (Estache, Gonzáles, & Trujillo, 2002; Pestana Barros,
2003). This conflict raises the question, why does the unhappiness reported on
post-privatization performance occur if the outcome is greater efficiency?

It is quite clear that some of the unhappiness stems from the failure of
privatized entities to pass on the benefits achieved, in the form of reduced
charges or improved services, to customers of the entity. Saundry and Turnbull
(1997), for example, noted the large profits made by former managers when a
number of UK trust ports were privatized. They attributed improvements in
economic performance primarily to the abolition of the National Dock Labour
Scheme. Privatization, they argued, did not transform the financial and

economic performance of UK trust ports sufficiently to justify the private gains of port management shareholders, and represented a "huge public loss." However, it is not clear if the outcome would have been better if the UK government had provided greater regulatory oversight post-privatization.

Similarly, studies of Australian port reform, while finding improved productivity at the stevedore level (Bureau of Transport and Communication Economics, 1995; Gentle, 1996), did not wholeheartedly endorse the devolution program as successful. Dick and Robinson (1992) concluded that the reform program failed to alter the fundamental structure of port authorities, and that the culture within port authorities discouraged initiative and innovation.

The assessment of privatization program outcomes in Latin America provides some insight into public dissatisfaction with devolution. The Inter-American Development Bank Research Department (IADB, 2002a) examined four indicators of privatization success (profitability, operational efficiency, output and employment) in seven countries (Argentina, Bolivia, Brazil, Chile, Colombia, Mexico and Peru). In all countries, profitability, operational efficiency and output were improved. Success of the privatization initiatives was reflected (IADB, 2002b) in an increase in foreign investment that, in many countries, had previously been deterred by restrictions on the inflow of foreign capital. However, in four countries – Argentina, Colombia, Mexico and Peru – short-run employment suffered. Most interesting, however, was the public opinion survey finding that, in the eyes of the general public, privatization had not been of benefit. The survey showed that "from 2000 to 2001 the percentage of Latin Americans who said privatization did not benefit their countries rose from 57% to 64%" (IADB, 2002b, p. 2). The IADB speculated that this gap between perception and reality was tied to the short-run loss of employment that can accompany privatizations, in spite of evidence that long-run employment may rise as new jobs are created in some sectors. So how should the performance of devolution programs be measured? A starting point would be to understand how firm-level "performance" is measured first.

3. MEASURING PERFORMANCE: A FIRM-LEVEL PERSPECTIVE

Benchmarking, or measuring a *company*'s performance against others of its type or with similar processes, is a key strategic activity. By reference to the successful practices and outcomes of other companies, the organization is able to compare its own performance against others, make internal adjustments, monitor the outcome and explore further changes in the way it conducts its business. Benchmarking is not a new management tool. It has been associated with quality initiatives (e.g. the Deming Award, the Malcolm Baldridge National Quality Award) and certification processes (e.g. ISO, 9000) for more than two decades.

While benchmarking has its critics who focus on hidden costs (Elnathan, Lin, & Young, 1996) or associated strategic risks (Cox & Thompson, 1998), the practice of benchmarking by the international airport community appears to be further advanced than that of the port community.[2] Research of the type reported by Francis, Humphreys, and Fry (2002a, b), Gillen (2001) or the Airports Council International in its reports to member airports is not present in the port literature. The port literature has focused more narrowly on identifying criteria and measuring efficiency (several chapters in this book have reported on that approach). The broader subject of benchmarking as done by airports will be discussed later.

One schema well known by practising managers for benchmarking performance is that proposed by Kaplan and Norton (1996) in their *Balanced Scorecard* approach. It is an integrated approach, taking four areas and assessing the performance of each in a cohesive manner. These areas – financial, customer, internal business processes, learning and growth – which implement strategy and vision in a top-down manner, translate the philosophy of the organization into a set of objectives, measures, targets and initiatives. In this approach, managers evaluate existing company performance and set objectives for future improvement and growth, and then measure progress against the base. The company is then set on a cycle that clarifies its strategy, links it to the objectives and measures, supports the plan with targets and strategic initiatives, and then enhances performance through feedback and organizational learning. Key to the implementation of performance measures in the supply chain, of which ports are critical players, are the principles of (a) tying performance measures to strategy, (b) focusing on the critical few performance measures (not trying to measure too many), and (c) taking a process view of performance measurement (Keebler, Mandrodt, Durtsche, & Ledyard, 1999).

Gillen, Henriksson, and Morrison (2001, p. 63), in their study of performance measurement in airports, provided a rationale for why performance measurement in airports is important. These can be restated to suggest reasons for ports to do the same. First, carriers may view ports either as a cost center or as a strategic business partner and want value for money; they assess cost efficiency or service delivery against that offered by competing ports. (This applies to arm's length carrier-port relationships and not those between related companies.) Second, ports are increasingly required to go to external capital markets and they require some means of judging the port's future performance, likely credit-worthiness and the risk of the investment. Container terminals, according to Drewry (1998, p. 6), find it extremely difficult to access private capital because of the high returns required for a private operator to finance the risk of the market; in Drewry's assessment, for example, a greenfield site to handle the transhipment market is unbankable without subsidy.

The area of container-*terminal* efficiency in particular is well studied (e.g. Cullinane, 2002). This has been a long-standing area of research since well before Talley's definitive work on container terminals (Talley, 1988, 1994). Some authors (e.g. Turner, 2000; Chen, 1998) have examined productivity or terminal efficiency quite specifically, with leasing or investment in mind. Others are interested in identifying the most efficient ports or in benchmarking their productivity (e.g. Tongzon, 1995). The two preceding chapters have detailed much of this existing terminal efficiency-focused research so it will not be repeated here, as the purpose of this chapter is to examine performance at the port authority and government program levels.

At the port level, it is important to compare or benchmark with like, and to find appropriate ports to benchmark against. For example, the Drewry studies (Drewry, 1998, 2002) examined container port performance in an effort to provide guidance to terminal operators; they found it important to segment ports into similar groups. For Drewry (1998) size was a clear, distinguishing feature and comparisons were made within two groupings: medium-sized terminals (handling about 210,000 TEU per year) and large terminals (handling about 600,000 TEU per year). In its latest report, Drewry (2002), the benchmark categories were adjusted to those with less than one million and those with more than one million TEU per year; the study also provided further disaggregation of standards by geographical region, arguing that Asian operations are sufficiently different from those in North America and Europe that geographically distinct operating standards should exist. The critical point is that, while benchmarks exist for container terminal operations in ports, albeit for purchase, benchmarks for other types of port activities do not seem to be readily available.

This raises the question, What should be measured? In examining performance measures for airports, Gillen (2001) notes that indicators chosen must not only be useful for inter-facility comparisons but for intra-facility comparisons. However, Gillen (2001), Gillen et al. (2001) and Francis et al. (2002a) all stop short of comparisons beyond system (financial and non-financial) and operational (or functional) internal firm factors. According to

Best (2005, pp. 38–40), performance should also be measured against external factors. While Gillen (2001) notes the importance of a customer orientation, he fails to propose an approach other than one of operational measurement. This chapter proposes a broader perspective, one that explores both internal measures (system financial and non-financial as well as operational) and external customer perspectives in order to ensure that performance can be measured from both efficiency and effectiveness perspectives.

This broad perspective ties in neatly with Talley's (1996) argument that performance must be put into the context of the particular port's objectives. He suggested that port objectives may be to maximize port profit or market share (in the case of a private port), or maximize throughput subject to a zero-profit constraint (in the case of a non-profit public port), or to maximize throughput subject to a maximum allowable operating deficit (in the case of a subsidized port). He also noted that mixed-cargo ports are more difficult to evaluate and proposed that the optimization model be limited to factors under a port manager's control (e.g. prices per ton of cargo, throughput per profit dollar before depreciation adjusted for inflation, ship and vehicle loading and unloading rates by cargo type, channel accessibility and reliability, berth accessibility and reliability, entry gate accessibility and reliability, probabilities of ship, vehicle or cargo damage and loss or theft of each).

Therefore, it is more than just the measurement of efficiency that needs to be contemplated by port managers. From this perspective, the studies of Roll and Hayuth (1993), Hop, Lea, and Lindjord (1996), Tongzon and Ganesalingam (1994), Friedrichsen (1999) and Marlow and Paixão (2003) are particularly relevant. Hop et al. (1996) attempted to broaden the port performance debate beyond production efficiency to measure port market effectiveness. They noted that some of the measures of port performance are really customer-oriented measures, in particular the following four:

• price (loading/unloading costs, terminal charges, commodity taxes, port expenses, crane rental);
• time (ship turnaround time, cargo dwell time is replaced by departure cut-off times and hours to availability after arrival, as some cargo interests will want in-transit storage to be available);
• availability (destinations served and departure frequency; number of arrivals by vessel type and market origin);[3] and
• reliability.[4]

Roll and Hayuth (1993), in their examination of port efficiency using data envelopment analysis were also concerned about effectiveness, including a measure of users' satisfaction (on a scale of 1–10 from a user survey). Friedrichsen (1999), in his study of eight Danish ports, noted that the best practices of efficient ports include external surveys of customers' perceptions of the port.

Tongzon and Ganesalingam (1994) divided efficiency measures into two broad categories: operational efficiency and customer-oriented measures. The former includes commonly used measures such as containers per net crane hour or TEUs per crane or TEUs per berth meter. The latter includes reliability and ship's waiting time. However, the problem is that not all the customer-oriented metrics are either easily or consistently measured, making comparison difficult. For example, the "minimization of delay" is an objective, not a metric. It can be measured at both the port and at the terminal level, although the former may require the cooperation of the latter depending on the processes for vessel management within port boundaries. To measure delay, a port could decide to use the metric "vessel hours waiting (at anchor) after scheduled arrival," for example, and the decrease in delay from one reporting period to the next would be seen as performance improvement in a service-oriented port or terminal. The port could also compare delay metrics between terminals as part of its marketing informatics activities.

Finally, Marlow and Paixão (2003) pointed to the importance of measuring port effectiveness in the context of their work on lean ports and the need for leanness and agility in a highly competitive environment. Their focus on effectiveness is supportive of the concept of measuring more than just financial and operating performance and of tying performance outcomes to customer requirements.

Deiss (1999) noted that key performance factors to be measured for railways are of five types: (1) financial performance; (2) efficiency or productivity; (3) asset utilization; (4) reliability (operator view) and service (customer view); and (5) safety. For the reliability and service constructs (effectiveness factors), Deiss relies on choice criteria, such as timeliness, delay and incidents, but does not develop constructs to measure these. Problematically, however, the customer view fails to examine satisfaction, a critical factor in the measurement of effectiveness. The measurement of service quality and satisfaction has been deeply examined in the last three decades of marketing research and constructs for its measurement are well developed and so will not be discussed further here. The key point is that they should not be overlooked; the tie between effectiveness and profitability is clear and strong (Narver & Slater, 1990; Heskett, Jones, Loveman, Sasser, & Schlesinger, 1994).

The differing types of performance measurement at the firm level have been illustrated in Fig. 25.1. While sample indicators are given in the table below the figure, the specific indicators are decided by directors as those most appropriate for the objectives (and strategy) of the individual firm. These indicators would then be used internally to monitor performance improvement objectives and for inter-port comparison in future. Of these, it would be instructive if an agreed subset of indicators could be identified for third-party evaluation.

Therefore, reviewing the literature at the firm level and the classification system offered in Fig. 25.1, it appears that the measurement of performance in ports has focused on internal system performance, primarily financial,

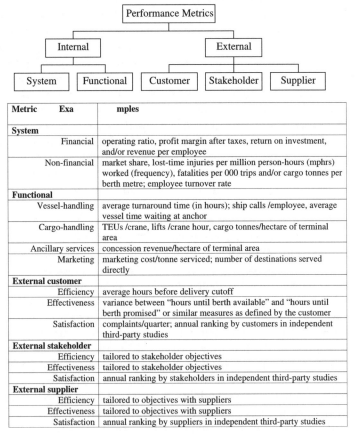

Metric	Exa	mples
System		
	Financial	operating ratio, profit margin after taxes, return on investment, and/or revenue per employee
	Non-financial	market share, lost-time injuries per million person-hours (mphrs) worked (frequency), fatalities per 000 trips and/or cargo tonnes per berth metre; employee turnover rate
Functional		
	Vessel-handling	average turnaround time (in hours); ship calls /employee, average vessel time waiting at anchor
	Cargo-handling	TEUs /crane, lifts /crane hour, cargo tonnes/hectare of terminal area
	Ancillary services	concession revenue/hectare of terminal area
	Marketing	marketing cost/tonne serviced; number of destinations served directly
External customer		
	Efficiency	average hours before delivery cutoff
	Effectiveness	variance between "hours until berth available" and "hours until berth promised" or similar measures as defined by the customer
	Satisfaction	complaints/quarter; annual ranking by customers in independent third-party studies
External stakeholder		
	Efficiency	tailored to stakeholder objectives
	Effectiveness	tailored to stakeholder objectives
	Satisfaction	annual ranking by stakeholders in independent third-party studies
External supplier		
	Efficiency	tailored to objectives with suppliers
	Effectiveness	tailored to objectives with suppliers
	Satisfaction	annual ranking by suppliers in independent third-party studies

Fig. 25.1.   Performance Metrics at the Firm Level.

and on the provision of vessel-handling and cargo-handling capacity as a proxy for the assessment of internal functional activities. The link between the provision of capacity and the efficient and effective use of that capacity is not clearly made. As for the external measures, individual ports may collect external measures (and not report them) but, other than the Drewry reports on container terminals (Drewry, 1998, 2002), there appears to be no systematic, independent third-party assessment of port performance. This is in contrast to the existence of systematic third-party assessment of airport performance established by the airport industry.

According to the marketing literature, the important issues include not only how efficiently you use what you have, but also how effective you are in its use. From this perspective, there are two motivations for improving port performance – attracting customers and retaining customers. There is a substantial literature on port selection, most of it of poor quality and not worth referencing here.[5] The exception is Lirn, Thanopoulou, Beynon, and Beresford (2004). They have developed and validated a transhipment port selection model for the container trades; its relevance here is that it clearly demonstrates that perceived port performance (quality perceptions of four key port dimensions – physical and technical infrastructure, location, management and administration, and carrier's terminal costs) is the final step in the transhipment port selection process. In this model, the majority of port performance indicators found in the literature are classified as management and administration factors. How a potential customer views the port's capacity to deliver an expected level of service (across a broad range of performance variables) is a determinant in attracting customers.

The second critical area from a strategic management perspective is retaining the customers you have already acquired. This is because, according to the service-profit chain (Heskett et al., 1994), effective efficiency[6] results in a satisfied, loyal customer, and given the costs of acquiring new customers, loyal customers are more profitable even though the net present value of a customer and the retention rate vary by industry (Reichheld & Teal, 1996). Using cargo-handling operations as an example, if, in its efforts to improve both efficiency (by increasing lifts per crane hour) and effectiveness (by securing lower terminal handling charges), a terminal increases the number of customers served (so as to improve berth utilization and spread berth costs over more customers), the result may be an average increase in delay times, and a reduction in customer satisfaction. In other words, the terminal may have improved its efficiency at the expense of the port authority's objective to improve customer satisfaction; terminals and port authorities do not always have the same port performance objectives.

While, loyalty is not solely dependent on satisfaction, customer dissatisfaction is a major cause of customer defection (Buttle & Burton, 2002). If
delay is increased but the service failure is adequately addressed by a recovery
plan, defection may not happen. Therefore, efficiency and effectiveness (including satisfaction) must be measured in order to ensure that the port's
strategy results in sustainable high performance at the firm level. Otherwise,
short-term pro-efficiency initiatives may prove to be sub-optimal in the long-
term when their impacts on effectiveness, satisfaction and, ultimately, on
revenue are realized. Likewise, a focus on effectiveness may also have unintended consequences for efficiency. The key issue then is the strategy chosen
by the port, and whether it is primarily focused on efficiency or effectiveness.

Once the strategic approach has been decided and the general goals for
performance determined, the port specifically defines what external performance metrics it will measure and how it will measure them. The measurement of customer satisfaction in service companies is well established in
the marketing literature and need not be repeated here. However, there is
very little written in the port literature other than what has been noted
above on the external measurement side. Thinking more broadly of the
transport industry, Brooks (2000), in her assessment of carrier performance,
noted that external metrics measurement is undertaken differently depending on the type of company; manufacturers measure carrier performance
differently than do carriers themselves and than do third-party logistics
suppliers. Common methods of measuring performance included third-
party audits, process reviews, customer satisfaction surveys and measuring
customer complaint levels. How the port industry currently benchmarks
performance will be the focus of a later section.

4. PERFORMANCE MEASUREMENT AT AIRPORTS: ARE THERE LESSONS FOR PORTS?

The academic community has long had an interest in airport performance;
not unlike the academic community in ports, the scholarly focus has been on
airport productivity. The Air Transport Research Society took the issue
under its wing and produced a multi-country, multi-airport study of total
factor productivity at airports (Oum, Yu, & Fu, 2003, reported on the 2002
study of 50 airports). The next year, Oum and Yu (2004), noting that
ancillary services vary considerably at airports, extracted the ancillary or
non-aeronautical service outputs so as to place the study of airports on a
more appropriate base for comparison using variable-factor productivity

rather than total-factor productivity. This 2003 study included 76 airports in Asia, North America and Europe, and, based on 2000 and 2001 data, examined the problems of performance measurement across a wide range of variables, ignoring the problems of accurately accounting for capital inputs through the use of physical inputs as proxies. Neither study attempted to correlate operating efficiency with profitability or financial structure, but the 2002 study concluded that an airport's ownership structure does not appear to have any effect on its total factor productivity performance.[7]

Most important, both studies incorporated the IATA Global Airport Monitor satisfaction rating so that variables under the control of management through resource allocation (to meet desired service levels) were represented in the analysis. Two findings are of interest. (1) Overall passenger satisfaction does not have a statistically significant effect on variable-factor productivity, implying that raising customer satisfaction may not cost more but can often be achieved with existing resources. From a market share (output) perspective, this finding is important as firms may influence their industry environment and their relative performance without significant resource investment. (2) The study found that capacity-constrained airports are likely to have higher variable-factor productivity because they impose delays and other negative externalities on customers. Therefore, while they may be more efficient from an operating perspective, they may not exhibit effective efficiency.

From a strategic management perspective, such studies are only half the feedback necessary for continuous improvement. In the airport industry, more than a decade ago, the Airports Council International teamed up with the International Air Transport Association to provide third-party independent customer surveys (the Global Airport Monitor) to identify airport performance in terms of how well they serve their customers (passengers but not freight). Building on the success of the Global Airport Monitor, the Airports Council International and the International Air Transport Association launched a joint program in December 2003, AETRA (ACI, 2005). The program is a customer satisfaction benchmarking initiative with 66 airport members in 2004 and an additional 11 as of 30 June 2005. The instrument measures 23 elements of a passenger's journey through the airport, including items such as ease of wayfinding, availability and cleanliness of washrooms, walking distance, availability of baggage carts and so on, as well as asking customers to indicate their overall level of satisfaction with the facility on a Likert 5-point scale (the measure used as an input in Oum & Yu, 2004). Airports participate in three size categories and six regional categories, and both domestic and international passengers are approached in the departure lounge. The results are consolidated and provided to the participating airports quarterly so they can see their relative performance against all the others. No such independent assessment process has been identified for ports.

Best of all, the AETRA program sets the scene for an individual airport to undertake its own performance gap analysis (Fig. 25.2) after Hooley, Saunders, and Piercy (1998). Through its own surveys of customer importance of the various elements of the service package, the airport is able to identify where best to allocate its resources in terms of improvement in both efficiency and effectiveness. It is the combination of importance-performance assessment and comparative benchmarking that enables the planning of both strategic initiatives and continuous improvement in operating activities.

Finally, there is also the question of what airports do in terms of other methods of measuring performance at the firm level. Francis et al. (2002b) completed a very detailed examination of performance measurement at airports (see Table 25.1). Using this as a base, the port performance pesearch network[8] asked 42 ports in nine countries (Argentina, Australia, Belgium, Canada, Chile, Italy, Turkey, United Kingdom and United States) what measures they used and these are reported in Table 25.1.

From Table 25.1 it seems clear that ports do not exhibit the same pattern of performance-monitoring activity that airports do. The benchmarking and

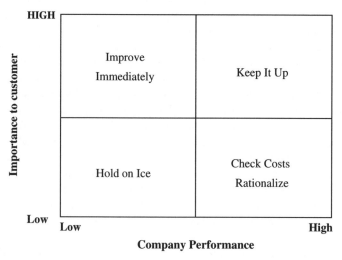

Fig. 25.2. Performance Gap Analysis.

Table 25.1. Performance Management Techniques Used.

Performance Management Technique	Percent Use by Airports (1)	Percent Use by Ports (2)
Activity based costing	36	24
Balanced scorecard	25	17
Best practice benchmarking	46	24
Business process re-engineering	23	10
Environmental management systems (e.g. ISO 14000)	27	44
Quality management systems (e.g. ISO 9000; BS5750 or similar)	23	49
Total quality management (TQM)	41	12

Note:
(1) As identified by Francis et al. (2002b, Table 4, p. 103).
(2) From Questionnaire A of the Port Performance Research Network.

total quality management techniques that are dominant in airports are not as prominent in ports; on the other hand, ports are more focused on ISO-type programs for evaluation. It is not that performance measurement is not done, but that it is approached in a different way. That does not mean that ports cannot learn from airport experience, it is just that the lessons will not always be easily transferable.

5. PERFORMANCE MEASUREMENT: THE GOVERNMENT PERSPECTIVE

From the well-developed area of firm performance assessment, we now move to the area of program performance assessment. Governments are interested in program performance in general as part of the process of new public management, a theme of which is greater efficiency. Performance assessment is often now entrenched in legislation, such as the *Government Performance and Results Act of 1993* in the US and the *Citizen's Charter* in the UK. Governments are interested in port performance because well-functioning infrastructure bodes well for national competitiveness, and the country that trades efficiently grows its economy. Today, the port is seen as one link in a global supply chain and so efficiency concerns are broader (Bichou & Gray, 2004). This book is focused on ports and so the comments presented here focus only on the efficiency and effectiveness of ports on the whole.

In addition to their work on airport performance, Gillen et al. (2001, p. 63) provided reasons why performance measurement in airports (read ports) would be important to governments. Governments, as they implement policies, need to see whether the intent of the policy has been realized and what changes might be needed; this is as true of port reform as it is of airport reform. Measures of fiscal effort and fiscal capacity provide a means of assessing whether port management is being efficient and rigorous in raising revenue, while measures of traffic volume for the given capacity provide an indication of whether government financial support is being well spent.

A focus on program performance goes beyond financial support for infrastructure and the wise use of fiscal capital. The following arguments are often heard:

- Successful program delivery raises national competitiveness internationally.
- Successful program delivery is not only a function of efficiency but also a function of effectiveness and taxpayer satisfaction (although performance in terms of efficiency and financial results is what is usually measured).
- Performance outcomes become inputs to the next performance period in terms of expectations.

Benchmarking is also an important means by which *government* can assess program success. In the case of measuring program performance, the government compares what is happening system-wide with the performance outcomes of individual entities. As one example, the *Canadian Transportation Act Review Panel* sought to examine the outcome of its port reform program by contracting a study that measured the performance of Canadian ports (Boucher, 2001). As another Deiss (1999) reported on the competitiveness indicators, governments usually use to "benchmark performance," citing as illustrations the International Institute for Management Development *World Competitiveness Yearbook* (IIMD, annual) and the United Nations Development Program (UNDP) Human Development Index among many.

However, the approach taken by Gillen et al. (2001) in measuring airport performance for airport program evaluation focused primarily on 25 system measures of financial and non-financial airport performance (Table 25.2), recognizing that one area not adequately evaluated is efficiency. The author would respectfully argue that not only is efficiency missing, but so are external measures of effectiveness and, as indicated by the IADB (2002b), so is satisfaction. Like firm-level performance evaluation, effectiveness and satisfaction play a role beyond what can be gained from solely a focus on efficiency.

Table 25.2. Range of Performance Indicators for Airports.

Performance Indicator	Measured As
Total revenue per Air Transport Movement (ATM)	(Aeronautical revenue + commercial revenue + other revenue)/ATM
Total revenue per passenger	(Aeronautical revenue + commercial revenue + other revenue)/passengers
Total revenue per employees	(Aeronautical revenue + commercial revenue + other revenue)/employees
Aeronautical revenue per ATM	Aeronautical revenue/ATM
Aeronautical revenue as a % of total cost	[Aeronautical revenue/(operating costs + personnel costs + depreciation)] × 100
Aeronautical revenue as a % of total revenue	[Aeronautical revenue/(aeronautical revenue + commercial revenue + other revenue)] × 100
Aeronautical revenue per passenger	Aeronautical revenue/passengers
Commercial revenue per passenger	Commercial revenue/passengers
Commercial revenue as a % of total revenue	Commercial revenue/(aeronautical revenue + commercial revenue + other revenue)
Total costs per ATM	(Operating costs + personnel costs + depreciation)/ATM
Total cost per passenger	(Operating costs + personnel costs + depreciation)/passengers
Operating costs per passenger	Operating costs/passengers
Staff costs as a % of operating and staff costs	Personnel costs/(operating costs + personnel costs)
Staff costs per passenger	Personnel costs/passengers
Staff costs as a % of turnover	(Personnel costs/turnover) × 100
Operating profit	Revenue – expenses
Return on capital employed	Operating profit/total assets
Operating profit per passenger	Operating profit/passengers
Passengers per employee	Passengers/employees
Passengers per ATM	Passengers/ATM
Total assets per passenger	Total assets/passengers
Assets per employee	Assets/employees
Capital expenditure per passenger	Actual expenditure on infrastructure or equipment/passengers
Capital expenditure as a percentage of turnover	(Actual expenditure on infrastructure or equipment/turnover) × 100
Net cash generation per passenger	Operating profit + non-cash depreciation – capital expenditure

Source: Gillen et al. (2001, p. 66, Table 4–1).

Saundry and Turnbull (1997) measured UK port performance using capital expenditure as a percentage of gross revenue, finding that there was no evidence that private ports had a superior investment record. Furthermore, they examined growth in annual-operating profit to determine if profitability improved after privatization, and did not find supporting evidence here either. What they did find was that employment fell significantly, attributable to the abolition of the National Dock Labour Scheme, and privatization was not accompanied by the creation of new jobs. Saundry and Turnbull also expected that privatization would yield more aggressive management, and that this would be reflected in improvements in individual port market shares; however, there was little difference in performance between those ports privatized and those not. While this study provided a major contribution to the assessment of strategic outcomes at the firm level from program changes, it did not fully measure program success.

Turner, Windle, and Dresner (2004) used data envelopment analysis (DEA) to examine productivity in container ports in the US and Canada and draw conclusions for policy makers from a structure and conduct perspective about whether existing investments in infrastructure were being used productively. While they incorporated variables such as dedicated quay length as a percentage of total quay length as a measure of port conduct, for example, they concluded that size matters (inferring that governments should primarily invest in the largest), but they did not provide guidance to either governments or firms about what should be measured for other relevant policy decisions.

Noting the global drive to privatization, Kent and Hochstein (1998) highlighted reform/privatization initiatives in three Latin American countries where dominant ports faced limited competition and handled comparatively low volume. They questioned whether the absence of competition would limit the benefits of port reform, concluding that three-tiered volume thresholds might improve competition by setting the levels at thresholds that would encourage private sector participation in intra-terminal, inter-terminal and inter-port activities. They did not, however, tie performance measures to program success but did use them to set thresholds for government action.

In the 1990s, the Australian government stood alone, according to Lawrence, Houghton, and George (1995), in developing a benchmarking program to evaluate both the efficiency and the effectiveness of its transport sector, among other infrastructure-based service industries. The Bureau of Industry Economics (1993), in addition to examining the charges the customer faced, framed port performance in terms of the desired customer service characteristics of timeliness (measured as cargo dwell time, ship turnaround time and truck and rail turnaround times) and reliability, as measured by delays and vessel days lost. The government used these to monitor the impact of labor reform on the waterfront. The Bureau of Transportation and Communication Economics (1995) analyzed the change in port productivity factors and examined the extent to which these were passed onto port users (ship operators and shippers), deeming the program to be successful as the users acquired the financial benefits anticipated from the program of reform. The government imposed a temporary levy on ship operators to pay for the costs of laying off stevedores, and reduced the revenue the terminals would receive from handling charges. The offset was not complete, as the terminals used some of the difference to modernize. Shippers gained benefits in the form of reductions in freight rates and decreased transit times (as a result of improved vessel turnaround time) with consequent reductions in inventory carrying costs and truck demurrage charges (Gentle, 1996).

Likewise, Boucher (2001) evaluated, for the Canadian government, the performance of key Canadian ports. His primary focus was the relative performance of each, based on borrowing power calculated as the discounted value of future revenues. Comparing financial indicators (net income, operating expenses) at the firm level, he drew conclusions about port reform in Canada. However, his assessment, found in Table 25.3, was limited to examining financial variables and was remiss in not looking more broadly at system and network service factors or external effectiveness in meeting market needs. It would seem logical that devolution program performance would be best measured over time, assessing performance before the reform is implemented (establishing a baseline) and comparing results after an acceptable period has passed, as change is a long process to implement.

Government port reform has been fraught with adverse stakeholder reaction. Ircha (1997) conducted a set of surveys of government officials, port managers and stakeholders to draw conclusions about their perception of the environment and drivers of change; this is as close as it appears any researchers have been in focusing on the perceptions of stakeholders in assessing reform program variables. Morris (2000) provided a review of the situation with Australia's port reform throughout the 1990s, through the period of Waterfront Reform (as already noted above) to the 1998 strike and lockout, one of the worst in Australia's history. He underscored the notion that governments will not always fully satisfy stakeholders in any effort at reform. He noted that reform was aimed at both improving productivity and profitability, with national competitiveness as a program output.

Table 25.3. Financial Profile of Seven Canada Port Authorities and the Entire National Port System, 1999, in Millions of Dollars.

Item (1)	Vancouver	Montreal	Halifax	Quebec City	Saint John	Fraser River	Toronto	Total for 7	Total for Network
OR	76.8	58.9	15.8	11.7	10.4	15.2	12.2	201	240.0
OE	54.8	57.7	12.8	11.9	9.5	14.3	17.4	161	212.1
OI	22.1	1.2	3.0	(0.2)	0.9	0.9	(5.2)	22.7	27.9
NI	17.9	4.5	3.4	2.6	1.6	0.8	(3.4)	36	36.0
NFI	399.0	160.7	82.6	44.2	57.7	104.8	43.2	807	1,105.4
NAA	5.4	9.4	7.6	9.4	1.8	15.2	2.3	51.1	62.9
TA	453.5	273.1	88.7	74.9	72.4	130.1	63.0	1,155.7	1,470.1
Equity (%)	381.0	250.9	76.2	24.6	68.8	87.5	55.4	944.4	1,220.9

Note: Abbreviations stand for operating revenue (OR), operating expenditures (OE), operating income (OI), net income (NI), net fixed assets (NFI), net acquisition of assets (NAA), total assets (TA) and the equity (Equity).

Source: Database from Economic Analysis, Transport Canada as cited by Boucher (2001, p. 92, Table 5).

Therefore, stakeholder satisfaction is an important metric in evaluating program performance. As already noted, if taxpayers believe that privatization has resulted in the squandering of public funds or the loss of employment opportunities for those in the industry, they may judge the outcome negatively when, on efficiency and effectiveness grounds, it has succeeded. The result may be a negative political backlash. (Being able to document reform impacts has the positive side benefit of helping government confront stakeholder dissatisfaction through information provision.) If governments choose to measure output only by surveying the post-event status of firms involved, or by conducting studies that measure efficiency but not effectiveness or satisfaction, the result is only a partial picture, and therefore one that could be misleading.

As noted previously by Deiss (1999), governments tend to measure their own success by their ranking on one of the many competitiveness indices. However, in the marine mode, the *World Competitiveness Yearbook 2001* (IIMD, annual) is unable to measure maritime transport performance as other modes are measured, and must resort to an opinion measure by survey as to whether the system is adequate to meet needs (Table 25.4). This is a very serious failing, as these are indicators of capacity not measures of the performance of that capacity, and the opinion survey of business leaders is a second best approach to actual measures of efficiency. The gap between what should be measured in the marine mode and what is measured is wider than for other transport modes. This confirms the author's contention that a more holistic view is required. Taking the firm-performance concept, and adapting it, the resulting scheme is presented in Fig. 25.3.

In Fig. 25.3, the internal metrics that government could decide to collect from all ports in its system for port monitoring purposes look very similar to those that would be (or are currently) collected by ports in Fig. 25.1. The only difference is that governments do not need to monitor the marketing aspects of ports, as these metrics are really only of use to the individual port and have no relevance to the effectiveness of a national system. External metrics focus on national stakeholders and in particular the suppliers and users of the facilities. This means that governments can choose to put in place a national system of data collection, one that is more than just the internal-system metrics that most use now, or support a multilateral approach, which could be a third-party industry-led one like the AETRA for airports (led by the Airports Council International) or one led by a multilateral agency like the International Maritime Organisation (IMO). As illustrated in Fig. 25.3, this could be extended by government to a program of international comparison across more than just the port system to one

Table 25.4.   Measuring Infrastructure Provision.

Performance Measure	Description
Output Data	
Roads	Density of the network in km per square km
Railroads	Density of the network in km per square km
Air transportation	Number of passengers carried by main companies
Opinion Data (Input)	
Water transportation	Meets business requirements
Values of the society	Support competitiveness

Note: These variables include three output measures for road, rail and air modes, and two input opinion data elements that reflect the missing marine mode and the values that might support infrastructure investment.
Source: IIMD (2001).

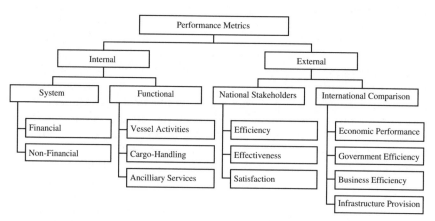

Fig. 25.3.   Measuring Performance of a Port Devolution Program.

that looks at the whole of the transport sector and its function within, for example, the IIMD variables of economic performance, government efficiency, business efficiency and infrastructure provision (IIMD, annual). This fourth branch of metrics is most useful once the government has adopted a performance management ethic throughout all its programs, not just those involving the port sector. It is the author's contention that metrics across competing entities like ports are best implemented by independent third-party audits of the industry-led variety.

6. PERFORMANCE METRICS AND PRACTICES IN USE

During 2004 and 2005, the port performance research network also asked 42 ports in 10 countries about the performance measures they currently collect and whether those data are collected for internal management purposes or more widely distributed to customers, government or stakeholders. Each port was given a list of performance metrics and asked about its performance metrics practices. Table 25.5 presents what was found.

Given a selection of 10 (internal system) financial measures, six vessel operations measures, 13 container operations measures, and a mix of nine other internal and external metrics, 38 ports chose to answer this part of the survey and share their usage patterns for these 38 metrics. Four ports did not collect any metrics, and the remaining ports collected between three and 30 metrics each, with a mean of 14. While three metrics are probably too few to gain sound insight into the performance of a port (particularly one with multiple types of markets served), 30 is certainly too many. The selection of metrics, tied to the port's strategy and objectives, is a key component of process improvement and the number of metrics must be manageable.

More than half of the ports report collecting financial system measures and they generally group them into those publicly reported, such as ancillary revenue as a percentage of gross revenue, debt to equity ratio and growth in profit before taxes, and those in a second group that are used for internal purposes only and seldom reported, such as average days accounts receivable and return on capital employed. These five are the most frequently collected, but the one most used internally is terminal charges as a percent of gross revenue.[9] It is not the financial system measures that are the focus of this chapter as these are the most likely metrics to be already in use. This leads to the question: what non-financial metrics do ports collect and how are they used?

Positive, published vessel operations metrics indicate to customers and stakeholders that port management and operations are effective. In addition, vessel operations metrics can be used by the port authority and/or terminal operators for marketing purposes, so it is not surprising that there is a pattern of reporting these to customers and to government (although not to the general public). Metrics like equipment downtime and revenue per tonne handled are less likely to be reported to customers than used internally. Still, whatever their use, the activity of collecting and using metrics internally is a good first step. Of the metrics proposed, average vessel calls

Table 25.5. Measures Collected and Their Use by Ports (1).

Performance Measure	Currently Collect	Report Externally to (2)			Internal Use Only
		Customers	Public	Government	
Financial Measures n = 30					
Ancillary revenue as % of gross revenue	22	6	11	12	10
Average days accounts receivable	19	1	2	3	13
Capital expenditure as % of gross revenue	13	3	6	4	6
Debt: equity ratio	23	5	7	8	6
Growth in profit (before taxes)	25	4	9	11	9
Interest coverage ratio	21	2	4	9	10
Port-related profit as % of port-related revenue	19	5	4	7	9
Return on capital employed	21	2	5	6	13
Terminal charges as a % of gross revenue	19	3	4	4	14
Yield % on shares, if publicly traded	10	2	1	1	7
Vessel Operations n = 34					
Average turnaround time/per vessel (in hours)	24	10	1	4	10
Average vessel calls per week	29	8	14	7	9
Average vessel waiting time at anchor	22	10	1	4	10
Hours of equipment downtime per month	14	3	1	2	10
Length of quay in metres (as a capacity measure)	22	6	6	9	13
Revenue per tonne handled	19	4	4	7	9
Container Operations n = 30					
20' TEU as a % of total TEU for year	17	2	5	2	7
Average revenue per TEU	9	2	1	2	7
Average vessel turnaround time/per 100 lifts (in hours)	4	2	1	2	1
Average yard dwell time in hours (3)	10	4	1	1	3
Container port throughput (TEU/metre of quay/year)	18	3	3	1	11
Departure cut-off time (hours)	3	0	0	0	3

Growth in TEU throughput	19	9	7	2	3
Import containers as a percent of total containers	17	3	8	2	7
Lifts per crane hour	12	2	3	1	7
Percent of containers grounded (ship to rail ops only)	5	3	3	2	1
Reliability (qualitative factor)	1	1	1	1	0
Transhipment (as % of total throughput)	9	3	6	0	0
Yard hectares to quay metres (3)	5	0	1	0	2
Other Measures n = 34					
Customer complaints per month	15	0	0	3	12
Destinations served this year	21	6	13	5	5
Employee turnover rate	14	1	2	3	9
Employment (full-time equivalents) per tonne handled	7	2	3	4	2
Employment (full-time equivalents) per TEU handled	5	1	2	2	3
Invoice accuracy percent	7	1	1	1	4
Number of customers served	18	6	10	5	5
Overall customer satisfaction	15	6	2	3	3
Stakeholder satisfaction	7	1	0	1	6

Note:

(1) Only 38 of the 42 ports chose to answer the question on the measures collected and who they were reported to, and of these, 34 collect performance metrics, hence *n* = 34. Of the 35 ports reporting container operations, 31 chose to answer the question on measures collected and 30 collect performance metrics, hence *n* = 30 for these measures.

(2) The respondent could choose as many of these as applicable.

(3) While ports indicated they collected this data, they did not indicate its use in every case.

per week (a measure more of connectivity than of efficiency) was the most commonly collected and frequently reported to the public.

Of the responding ports, 35 in the group had container operations and 30 reported collecting metrics, so there are slightly fewer of these to examine. Here the focus on reliability in the literature as a "must measure" option is not borne out in practice (it was only collected by one port). Growth in TEU throughput is widely reported (not surprising given the focus of the media and governments on this) and most frequently collected but seldom used internally. The measure of port productivity most used internally is TEU per metre of quay per year. This is a factor that one would expect to see in a third-party audit of ports. In fact, given the limited external reporting in this group of measures, it is not clear how ports are able to benchmark their activities without a third-party audit process.

Ports were given a list of other measures that were a mix of internal and external measures not in the financial system and functional performance areas. It is a positive sign that a substantial number of ports (although not as many as there should be) track external factors like customer complaints and overall customer satisfaction as well as non-financial system measures like employee turnover or safety (incidence and frequency of accidents). Destinations served was the metric most frequently collected; this was expected, as it is good politically for a port to be perceived as hub or gateway and "connected" to the world.

Finally, those responding to the survey were asked the five most important measures. While many chose the ones in Table 25.5, and generally reinforced the financial and productivity measures most commonly cited, there were some obvious ones not listed that were held to be of substantial importance in some ports – those relating to worker safety and port security. There were also others of interest that had not been contemplated in the testing of the survey: passengers per cruise ship, number of new customers, and dollar value of grants, to name a few.

Not all ports are in a position to collect operating metrics as leased terminals may choose not to share their commercial data. The key to acquiring data in these circumstances is tied to one of two scenarios: (1) the port–terminal relationship is not adversarial and data sharing can be negotiated for mutual benefit; or (2) government may impose data reporting requirements tied to devolution program monitoring initiatives. Measures collected for internal use by terminals may be used productively by port authorities if shared confidentially in a trust-based, cooperative relationship.

This leads to the question: which of these measures could or should be used by ports to assess their performance and which could be adopted by government in its port devolution program assessment? They should not be completely mutually exclusive. These questions can and should be asked as a next step in using metrics to improve port performance and to evaluate devolution program outcomes. Hauser and Katz (1998) provide an excellent resource in answering these questions. As they note, choosing the appropriate metrics is critical. Not all measures are created equal; some are more important than others and dictated by either firm strategy (in the case of a port's continuous improvement) or government goals (in the case of a government's review of regulatory reform).

7. CONCLUSIONS

In their assessment of ports in supply chain performance, Bichou and Gray (2004, p. 47) noted that port "organizational dissimilarity constitutes a serious limitation to enquiry, not only concerning what to measure but how to measure." This chapter has presented a framework for the measurement of port efficiency and effectiveness at the firm level to enable ports to, individually, benchmark their performance against others. For inter-port comparisons, it is, as Drewry (2002) has found, important to measure against those of similar size. It is also important to benchmark against those with similar goals and objectives; to use a port focused on growth in local employment as a benchmark also for one seeking technical operating efficiency is not appropriate.

Moreover, it has been noted that there is significant overlap in measures that governments may use to measure port devolution program performance, particularly in the area of internal metrics. Only if the measures are tied to the government's objectives in devolving the program can the government be assured that efforts in improvement will be targeted in the appropriate way and a better outcome result. The linkage between objectives (strategy as input) and performance (output) must be made.

This chapter has also provided some food for thought on the potential to learn from the airport industry. Like ports, airports provide passenger and cargo handling services to both large global transport companies and small local customers, and have both core and ancillary services that may be provided through a wide variety of governance models (public, private and mixed). In particular, airport performance measurement has moved into the domain of third-party customer satisfaction measurement to the benefit of those airports that seek to follow a customer-driven service strategy. (Airports following an operational efficiency strategy can also use the third-party performance measurement program for benchmarking, but the inclusion of satisfaction measurement broadens the usefulness of the program to a wider range of airports.) Ports have not yet reached this stage; Drewry's studies of container ports is a first step but much less inclusive. This chapter has noted that if a port authority association, such as the American Association of Port Authorities, were to be interested, it is possible to develop a common instrument, not unlike the Air Transport Research Society initiative in the airport industry, to provide a third-party independent assessment of performance that can be used to support port management initiatives and strategic development.

Finally, this chapter has presented the possibility that there is much in common between that measured by ports for their own continuous improvement activities and that needed by government to monitor the progress of reform (devolution). The measures may differ, but the process and approach are similar. The key to all of these is benchmarking similar types of organizations.

Tovar, Jara-Díaz, and Trujillo (2003, p. 26), in their study of cost and production functions in ports, conducted a literature review and concluded that

> applied research still has a long way to go in the sector. … If governments are serious about their commitment to improve the competitiveness of their countries, they will have to ensure that port costs and hence rents are minimized and this can only be done if costs are measured and assessed properly. One of the regulator's tasks should be to help obtain relevant information from adequate sources for the performance of these types of studies. One way to do that is that the obligation to submit such information could be placed with regulated firms.

It can be argued that this is just not possible at the country level. The example of airports provided earlier makes the case for a larger, unbiased view. By developing an industry-led third-party initiative, ports could collect information for government to use in program evaluation and each port would have the added benefit of being able to use the metrics of its relative position for its own operating processes and strategy evaluation. Then ports would be encouraged not only to improve competitiveness to meet their own objectives, however they may be defined, but also to improve national competitiveness in line with national objectives. The intersection of the two would bring any conflicts between firm and program objectives to negotiated settlement, with both sides having the information they need to make rational decisions rather than instinctive decisions. The outcome is that you can manage that which you can measure for the desired balance of effectiveness and efficiency.

NOTES

1. Furthermore, Newcomer (2001) notes that politics plays a role here. "The circuitous route through which evaluation professionals may contribute to improving our collective understanding of the operations and results of public programs is fairly treacherous because of the political origins of demands for performance data" (Newcomer, 2001, np).

2. Almost one-half (46%) of airports studied by Francis et al. (2002b) used some form of benchmarking technique.

3. UNCTAD (2004) has established a better term for this – connectivity.

4. There is considerable disagreement in the performance measurement circles about whether "reliability" can be measured, and whether it is a measure of efficiency or effectiveness. From an internal efficiency perspective, standard deviation around arrival and departure times, loading and unloading rates are proxies for reliability; from an external effectiveness perspective, it is a subjective measure of whether customer expectations of service have been met. For some customers, the missing of a delivery window by more than four hours is indicative of an unreliable supplier, while for another a day late is still reliable. Hop et al. (1996) were in the second group, and considered reliability a qualitative measure of effectiveness.

5. One category of port selection literature is usually founded on a survey of exporters, and most exporters are port-blind (Brooks, 1983), resulting in the choice model being conceptually flawed. The second category of port selection literature sees port choice as a route optimization problem for a carrier, but route characteristics are only some of the inputs to the carrier's port choice decision. Shipping lines in the container trades choose the ports that will provide them with the perceived best service package to their target customers. The only cases where cargo interests play a dominant port selection role are the tanker, dry bulk and neo-bulk trades.

6. Effective efficiency is being both highly efficient and effective, resulting in satisfied customers at low costs (American Marketing Association, 1995). Baltazar and Brooks in Chapter 17 note that the strategic management literature suggests high performance in one (effectiveness or efficiency) and satisfactory performance in the other as the goal to be met. As the American Marketing Association no longer lists the concept of effective efficiency in its online dictionary, it is likely that the need for a primary focus has been realized and adopted in practice. From this point onwards, the concept of effective efficiency used in this paper means high performance in one and above average performance in the other. As Katz and Kahn (1978) suggest, the best performing organizations tend to be concerned with both.

7. As the majority of airports are publicly owned, this finding may not be applicable to ports where the devolution of terminal facilities within one port is common. In other words, the management and ownership of "ports" presents a very different pattern than exists with "airports."

8. Members of the network who assisted with data collection were Khalid Bichou (Imperial College London), Mary R. Brooks (Dalhousie University), Dionisia Cazzaniga Francesetti (Universita Di Pisa), A. Güldem Cerit (Dokuz Eylül University), Peter W. de Langen (Erasmus University Rotterdam), Sophia Everett (Melbourne University), James A. Fawcett (University of Southern California), Ricardo Sánchez

(Austral University/ECLAC), Marisa Valleri (University of Bari), and Thierry Vanelslander (Universiteit Antwerpen).

9. This is hardly surprising as any landlord operation, be it a port, an airport, a shopping centre or a residential housing complex, needs to have a good knowledge of its rental or lease revenue in its total revenue picture if revenue management practices are to be optimized.

REFERENCES

Airports Council International. (2005). www.aci-na.org, accessed August 22.

American Marketing Association. (1995). *Marketing encyclopedia*. Chicago: American Marketing Association.

Baird, A. J. (1995). Privatisation of trust ports in the United Kingdom: Review and analysis of the first sales. *Transport Policy, 2*(2), 135–143.

Best, R. J. (2005). *Market-based management: Strategies for growing customer value and profitability* (4th ed.). Upper Saddle River, NJ: Prentice-Hall.

Bichou, K., & Gray, R. (2004). A logistics and supply chain management approach to port performance measurement. *Maritime Policy and Management, 31*(1), 47–67.

Boucher, M. (2001). Newly commercialized transport infrastructure providers: Analysis of principles of governance, accountability and performance. www.tc.gc.ca/ctar/

Brooks, M. R. (1983). *Determinants of shipper's choice of container carrier: A study of eastern Canadian exporters*. Ph.D. thesis, University of Wales.

Brooks, M. R. (2000). Performance evaluation of carriers by North American companies. *Transport Reviews, 20*(2), 205–218.

Bureau of Industry Economics. (1993). *International performance indictors – Waterfront*. Research report 47. Canberra, Australia.

Bureau of Transport and Communication Economics. (1995). *Review of the waterfront industry reform program (Report 91)*. Canberra: Australian Government Publishing Service.

Buttle, F., & Burton, J. (2002). Does service failure influence customer loyalty? *Journal of Consumer Behaviour, 1*(3), 217–227.

Chen, T. (1998). Land utilization in the container terminal: A global perspective. *Maritime Policy and Management, 25*(4), 289–304.

Cox, A., & Thompson, I. (1998). On the appropriateness of benchmarking. *Journal of General Management, 23*, 1–20.

Cullinane, K. P. B. (2002). The productivity and efficiency of ports and terminals: Methods and applications. In: C. T. Grammenos (Ed.), *Handbook of maritime economics and business* (pp. 803–831). London: Informa Publishing.

Deiss, R. (1999). Benchmarking European transport. In: *Transport benchmarking: Methodologies, applications and data*. European conference of ministers of transport: proceedings of the Paris conference, November, pp. 35–82.

Dick, H. & Robinson, R. (1992). Waterfront reform: The next phase. Presented to the national agriculture and resources outlook conference, Australia.

Drewry Shipping Consultants Ltd. (1998). *World container terminals: Global growth and private profit*. London: Drewry.

Drewry Shipping Consultants Ltd. (2002). *Global container terminals: Profit, performance and prospects*. London: Drewry.

Elnathan, D., Lin, T. W., & Young, S. M. (1996). Benchmarking and management accounting. *Journal for Management Accounting Research, 8*, 37–54.

Estache, A., Gonzáles, M., & Trujillo, L. (2002). Efficiency gains from port reform and the potential for yardstick competition: Lessons from Mexico. *World Development, 30*(4), 545–560.

Francis, G., Humphreys, I., & Fry, J. (2002a). The benchmarking of airport performance. *Journal of Air Transport Management, 8*(4), 239–247.

Francis, G., Humphreys, I., & Fry, J. (2002b). International survey of performance measurement in airports. *Transportation Research Record, 1788*, 101–108.

Friedrichsen, C. (1999). Benchmarking of ports: Possibilities for increased efficiency of ports. In: *Transport Benchmarking: Methodologies, Applications and Data. European Conference of ministers of transport: Proceedings of the Paris conference*, November, pp. 159–168.

Gentle, N. (1996). The distribution of the benefits of waterfront reform. *Maritime Policy and Management, 23*(3), 301–319.

Gillen, D. (2001). Benchmarking and performance measurement: The role in quality management. In: A. M. Brewer, K. J. Button & D. A. Hensher (Eds), *Handbook of logistics and supply chain management* (pp. 325–338). Oxford: Pergamon Press.

Gillen, D., Henriksson, L., & Morrison, W. (2001). *Airport financing, costing, pricing and performance: Final report to the Canadian Transportation Act Review Committee.* www.tc.gc.ca/ctar/

Hauser, J., & Katz, G. (1998). Metrics: You are what you measure. *European Management Journal, 16*(5), 517–528.

Heskett, J. L., Jones, T. O., Loveman, G. W., Sasser, W. E., Jr., & Schlesinger, L. A. (1994). Putting the service-profit chain to work. *Harvard Business Review, 72*(2), 164–174.

Hooley, G., Saunders, G., & Piercy, N. (1998). *Marketing strategy & competitive positioning* (2nd ed.). London: Prentice Hall.

Hop, Ø., Lea, R., & Lindjord, J.-E. (1996). Measuring port performance and user costs of small ports. Presented to international association of maritime economists annual conference, Vancouver.

IIMD (annual). *World competitiveness yearbook.* Lausanne: International Institute for Management Development. www.imd.ch

Inter-American Development Bank (IADB). (2002a). The privatization paradox. *Latin American Economic Policies, 18*, 1–3. www.iadb.org/res

Inter-American Development Bank (IADB). (2002b). The privatization boom in Latin America. *Latin American Economic Policies, 18*(2). www.iadb.org/res

Ircha, M. C. (1997). Reforming Canadian ports. *Maritime Policy and Management, 24*(2), 123–144.

Kaplan, R. S., & Norton, D. P. (1996). *The balanced scorecard: Translating strategy into action.* Boston: Harvard University Press.

Katz, D., & Kahn, R. L. (1978). *The social psychology of organizations.* New York: Wiley.

Keebler, J. S., Mandrodt, K. B., Durtsche, D. A., & Ledyard, D. M. (1999). *Keeping score: Measuring the business value of logistics in the supply chain.* Oak Brook, IL: Council of Logistics Management.

Kent, P. E., & Hochstein, A. (1998). Port reform and privatisation in conditions of limited competition: The experience in Colombia, Costa Rica and Nicaragua. *Maritime Policy and Management, 25*(4), 313–333.

Lawrence, D., Houghton, J., & George, A. (1995). International comparisons of Australia's infrastructure performance. *Journal of Productivity Analysis, 8*(4), 361–378.

Lirn, T. C., Thanopoulou, H. A., Beynon, M. J., & Beresford, A. K. C. (2004). An application of AHP on transhipment port selection: A global perspective. *Maritime Economics and Logistics, 6*(1), 70–91.

Marlow, P., & Paixão Casaca, A. C. (2003). Measuring lean ports' performance. *International Journal of Transport Management, 1*(4), 189–202.

Morris, R. (2000). A watershed on the Australian waterfront? The 1998 stevedoring dispute. *Maritime Policy and Management, 27*(2), 107–120.

Narver, J., & Slater, S. (1990). The effect of market orientation on business profitability. *Journal of Marketing, 54*, 20–35.

Newcomer, K. E. (2001). Tracking and probing program performance: Fruitful path or blind alley for evaluation professionals? *American Journal of Evaluation, 22*(3), 337–341.

Oum, T. H., & Yu, C. (2004). Measuring airports' operating efficiency: A summary of the 2003 ATRS global airport benchmarking report. *Transportation Research Part E: Logistics and Transportation Review, 40*(6), 515–532.

Oum, T. H., Yu, C., & Fu, X. (2003). A comparative analysis of productivity performance of the world's major airports: Summary report of the ATRS global airport benchmarking report – 2002. *Journal of Air Transport Management, 9*(5), 285–297.

Pestana Barros, C. (2003). Incentive regulation and efficiency of Portuguese port authorities. *Maritime Economics and Logistics, 5*(1), 55–69.

Reichheld, F. F., & Teal, T. (1996). *The loyalty effect.* Boston: Harvard Business School Press.

Roll, Y., & Hayuth, Y. (1993). Port performance comparison applying data envelopment analysis (DEA). *Maritime Policy and Management, 20*(2), 153–162.

Saundry, R., & Turnbull, P. (1997). Private profit, public loss: The financial and economic performance of U.K. ports. *Maritime Policy and Management, 24*(4), 319–334.

Serebrisky, T., & Trujillo, L. (2005). An assessment of port reform in Argentina: Outcomes and challenges ahead. *Maritime Policy and Management, 32*(3), 191–207.

Shashikumar, N. (1998). The Indian port privatization model: A critique. *Transportation Journal, 37*(3), 35–48.

Talley, W. K. (1988). Optimum throughput and evaluation of marine terminals. *Maritime Policy and Management, 15*, 327–331.

Talley, W. K. (1994). Performance indicators and port performance evaluation. *Logistics and Transportation Review, 30*, 339–352.

Talley, W. K. (1996). Performance evaluation of mixed cargo ports. Presented to international association of maritime economists annual conference, Vancouver.

Thomas, B. J. (1994). The privatization of United Kingdom seaports. *Maritime Policy and Management, 21*(2), 135–148.

Tongzon, J. (1995). Systematizing international benchmarking for ports. *Maritime Policy and Management, 22*(2), 171–177.

Tongzon, J., & Ganesalingam, S. (1994). Evaluation of ASEAN port performance and efficiency. *Asian Economic Journal, 8*(3), 317–330.

Tovar, B., Jara-Díaz, S., & Trujillo, L. (2003, August). *Production and cost functions and their application to the port sector: A literature survey.* Policy Research Working Paper 3123. World Bank Institute, Washington, DC.

Turner, H. S. (2000). Evaluating seaport policy alternatives: A simulation study of terminal leasing policy and system performance. *Maritime Policy and Management, 27*(3), 283–301.

Turner, H. S., Windle, R., & Dresner, M. (2004). North American containerport productivity: 1984–1997. *Transportation Research Part E: Logistics and Transportation Review, 40*(4), 339–356.

UNCTAD. (2004). *Efficient transport and trade facilitation to improve participation by developing countries in international trade (TD/B/COM.3/67), 7 December.* Geneva: UNCTAD.

CHAPTER 26

CONCLUSIONS AND RESEARCH AGENDA

Mary R. Brooks and Kevin Cullinane

ABSTRACT

This chapter reviews the literature on devolution outcomes from a broader spectrum of industries. Having derived reasonably conclusive evidence in Chapter 18 as to the inadequacy of existing models of port governance, a more appropriately complex model has been proposed. This model exhibits greater detail in the definition of groups of port activities for which governance decisions must be taken and a much greater refinement of the range of governance typologies available to decision-makers. Drawing conclusions from the content of previous chapters, an agenda for future research into devolution, governance and port performance is set out. The key objective lies with investigating, and possibly identifying, the nature of the relationship between this newly expanded range of port governance configurations and port performance. Finally, this chapter proposes a conceptual framework for analyzing the firm–governance relationship and how this affects the performance of both ports and devolution programs.

1. INTRODUCTION

The preceding chapters have provided a very full discussion of the complexity of port policies, governance models and resulting outcomes from

Devolution, Port Governance and Port Performance
Research in Transportation Economics, Volume 17, 631–660
Copyright © 2007 by Elsevier Ltd.
ISSN: 0739-8859/doi:10.1016/S0739-8859(06)17026-2

more than 10 different countries or geographic regions. Chapters 3–16 discussed these qualitatively while Chapters 17, 18 and 25 used quantitative data to examine them in more detail. While there have been attempts to simplify the issues in these chapters, the overwhelming conclusion that should be reached is a simple one: while governments may have had the best of intentions in establishing a more commercialized footing for port operations, the outcomes have not always delivered the full benefits sought. Some experiments have worked better than others. This chapter intends to examine the experiences reported more generally, draw conclusions about the models imposed, and about the resulting performance outcomes. Its purpose is to inform and to present a working framework for a future research agenda.

It begins with a recap of prior research on devolution and outcomes, both generally and specifically related to ports, and the debate about appropriate governance models for the desired performance outcomes. It then takes the opportunity to elaborate on the Baltazar and Brooks' Matching Framework presented in Chapter 17, and question what is still missing from our knowledge of port devolution and its link to performance. The authors propose the use of a third-party port performance instrument for future research and guidance to ports and governments. Using the Matching Framework, the authors then examine what is missing from the performance–governance debate and map a future research agenda to assist government in determining port policies designed to achieve the government's strategies.

2. DEVOLUTION AND PERFORMANCE

There is no shortage of research on the benefits of privatization and economists to date have disagreed on the appropriate approach to privatizing state-owned enterprises (SOEs). Parker (2000), for example, laid out over several pages the benefits and detractions of privatization, particularly for SOEs. (In many countries, ports have been SOEs until recently.) Gillen and Cooper (1995) summarized the rationale for port privatization, pointing out that governments believe it encourages and improves efficiency, makes industry more responsive to the demands of customers, reduces public debt and forces management (and unions) to face the realities of the marketplace. Furthermore, they noted that Boardman and Vining (1989) had found that, in terms of profitability, public sector firms perform substantially worse than private sector firms. While this earlier research records the incentive for port privatization, success stories have been difficult to hear over the voices of the detractors of devolution programs. Likewise, the earlier chapters of this book highlight that while there have been benefits to government devolution programs, in the end the authors often add caveats to their evaluations. We can only conclude that none of the port reform programs examined can be called an unqualified success.

Much of the literature assumes that the primary governance issue is one of ownership, and that private sector ownership is the appropriate model to gain market competitiveness and that superior firm performance will result. Yet, it is the devolution program established by government that sets the model or models that will be imposed by government on port authorities. As seen in Chapters 3–16, seldom was outright privatization pursued. In the port industry, governments seemed unwilling to fully privatize the industry, preferring to retain some components and some control over what is deemed by many to be a strategic national asset.

Is ownership necessary for superior performance outcomes? According to Backx, Carney, and Gedajlovic (2002), in the airline industry the relationship between ownership structure and airline performance is complex and it is difficult to make international generalizations. They pointed to evidence that making such generalizations is fraught with "counterfactual" problems, such as whether there was a fundamental restructuring in the business prior to its privatization (Jenkinson, 1998), or whether government competition and regulatory policy directly impacted post-privatization financial performance (Vickers & Yarrow, 1989), or whether the privatizations were staged and therefore the mixed ownership could be transient. Backx et al. (2002) concluded that performance cannot be isolated from government policy issues.

Everett and Robinson (1998) agreed with this perspective; they noted that privatization is often the outcome of a focus on improving port efficiency. They concluded that, in the case of port reform in Australia, the government failed to remove non-core assets from ports or fund public service obligations outside of port budgets, or to control staffing in excess of that necessary for core port activities. In these cases, they argued, the ownership of the port is not relevant to the performance outcome. They also noted that governments are not just seeking port efficiency but also national competitiveness. In their subsequent work in Chapter 12, they assessed that fault for failure in Australian port reform lies at the state government level, as the nature and form of the legislation enacted to facilitate corporatization were neither adequate nor appropriate in all states. In particular, they noted that Tasmania's governance model worked better than those implemented elsewhere in Australia. We can draw from this and some of the other experiences, for example

Canada (Chapter 11), that the "devil is in the details" and the details of the governance model do matter. Furthermore, these chapters in Part 2 of the book indicated that the state of the industry prior to changes in the imposed governance regime also mattered, as did what the government intended to have as a result.

Devolution is often accompanied by deregulation, and deregulation of the transport sector has been an almost global phenomenon. Research in other industries supports the premise that deregulation has a negative impact on efficiency in the short term, as there are costs incurred in going from a regulated to a deregulated environment. This support comes from the US banking industry (Mukherjee, Ray, & Miller, 2001; Grabowski, Rangan, & Rezvanian, 1994), US electric power utilities (Delmas & Tokat, 2005) and natural gas distribution utilities (Hollas, MacLeod, & Stansell, 2002). Delmas and Tokat (2005), in particular, noted that policy-makers often underestimate the time required and costs likely to be incurred in the transition period, and that the period between fully regulated and deregulated states may be more than 10 years (citing Dyner & Larsen, 2001, p. 1153). Perhaps it is worth considering that, for many devolved ports, the difficulties they have in measuring their performance arise from their relatively new "devolved status" and the absence of a culture of performance measurement prior to transition. In addition, any attempt to derive objective evidence on the impact of devolution on the performance of ports is complicated by the usually simultaneous deregulation of the sector. The period of transition for the world's port industry is likely not yet over, as many have entered their new governance environment within the past decade. The timing for a research agenda focused on devolution, governance and performance could not be better.

As for government policy intentions, these were presented in Chapter 1 as three possible directions: adoption of a policy of local responsiveness (as opposed to one of executing national policy, thereby resulting in decentralization as a government's approach to governance); a desire to improve efficiency and responsiveness of public sector ports (thereby resulting in commercialization or corporatization being chosen as the approach to governance); or adoption of the policy that private sector-led efficiency and market competition are best (thereby resulting in privatization as the preferred governance model to be imposed, along with the simultaneous introduction of measures aimed at deregulating the market). In the case of the latter, suboptimal outcomes can be expected if privatization is not accompanied by deregulation. The governance models resulting from these three government policy objectives, in terms of ownership, control, responsibility and accountability, were outlined in Table 1.1.

The chapters subsequent, and in particular Chapter 18, confirmed that the governance models appropriate to a particular government's objectives were not often imposed as anticipated. The experiences recounted in Part 2 reflected widely diverse approaches to new public management implementation. It is only speculation as to the causes of such diversity in implementation; this may have occurred for one or more of several reasons: (1) the government did not have a ports policy and therefore no policy principles to follow (the US comes to mind); (2) the government saw more than one objective as desirable and so failed to have a clear vision of what it wanted to accomplish with devolution; (3) the design of the governance model to be imposed was compromised by the political process of developing the implementation of legislation and regulation; and (4) the execution of the model was flawed. To examine those cases where "failure" has occurred, it is really not important as to why the models (from Table 1.1) failed to deliver an effective port reform program (if indeed that is the conclusion drawn in a particular country), but it is important that governments now examine the performance of their port reform program and see this as a new opportunity for improvement. Fig. 26.1 recreates Fig. 1.1 and adds a feedback loop through evaluation to the contemplation of opportunities for additional port reform.

As an example of the feedback on government reform initiatives, Gillen and Morrison (2004, pp. 60–61) noted that a competitive environment is not

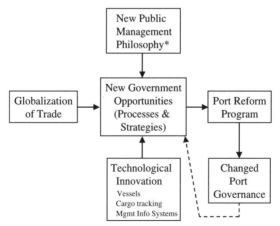

* Decouple management from ownership and/or regulation

Fig. 26.1.   Drivers of Port Reform.

the same as increased competition, and that, for the reform of Canada's airport policy, it was the former that was desirable. They concluded that the not-for-profit commercialized environment of Canada's airports meant that once airport authorities (e.g., managers) reached a certain threshold of revenue generation, they no longer had the incentive to cut costs or avoid gold-plated investments. Their evaluation of airport reform provided the feedback that there is currently an opportunity for the Canadian government to reform Canada's airport governance model in favor of one that would yield a more competitive environment, arguing that private shareholders would provide discipline against over-investment in airport facilities. There is widespread recognition in Canada that the airports policy originally sought investment in aging facilities and achieved that goal through a successful commercialization process; Gillen and Morrison's evaluation now suggests it is time to move further along the devolution continuum.

The key finding to draw from this section is that if devolution has not worked well, it is possible, through program review and evaluation, to determine how best to realign ports policy and the accompanying imposed governance model(s) with the desired objectives, to reconfigure the port reform program to achieve the desired performance. To do this, performance must be measured in the context of those objectives.

3. GOVERNANCE AND OUTCOMES

New public management is not just about imposing a governance model to meet devolution objectives; it is also about ensuring that the port performs optimally to achieve the program's objectives. Governance occurs in two contexts – at the firm level and at the government level. We begin with the former and end with the latter.

Yoshikawa and Phan (2003) looked at internal governance factors and their effect on firm performance. They concluded, based on empirical research of corporate governance reform in Japanese corporations, that governance rules such as the number of outside directors, the separation of board members and executive officers, and a reduction in the number of board directors (all factors common to Anglo-Saxon board reform proposals) were *not* related to firm financial performance. So while the devil may be in the details, it is not these particular details upon which our research agenda will need to focus, but rather those larger issues noted in Table 1.1.

In his discussion of corporate governance in multinationals (e.g., governance at the firm level), Luo (2005) noted that governance is the relationship between the corporation and its shareholders (or more broadly its stakeholders), and that it specifies the distribution of rights and responsibilities among the parties (board members, shareholders, executives and so on) and the rules and procedures for decision-making. Corporate governance provides the framework for setting objectives, the strategy for achieving the objectives and guidelines for monitoring performance. What makes Luo's perspective interesting is that (a) he assumed that the governance model provides guidelines for performance monitoring (which goes beyond that noted in other definitions of governance provided in Chapter 1) and (b) he extended the boundaries of decision-making as being throughout the various globally dispersed businesses that a corporate entity may have (concluding that governance also directs the distribution of rights, responsibilities and power internationally among these parties). Is this perspective applicable?

First, as we have seen in this book, not all ports engage in performance monitoring (perhaps because not all ports managers are directed to do so by those who hold the responsibility for risk and the meeting of statutory obligations). Second, as *ports* are seldom multinationals, there is no requirement for reconciliation across national boundaries of the rights, responsibilities and power of the parties affected by these decisions; the same will not be true at the *terminal* level. As noted by de Langen's examination of seaport clusters and their stakeholders (Chapter 20), many of the parties to a national port's governance system will have multinational concerns rather than local concerns. Therefore, we must not ignore the dilemma faced by modern *port authorities* that the delivery of local benefits desired of the authorities by government may not match the objectives of those who win terminal concession rights. This is particularly true in the container segment of the market, where *terminal operations* are consolidating in the hands of a small number of global companies, as noted by Notteboom (Chapter 2).

Moving to the government level, Starkie and Thompson (1985, p. 33) expressed considerable concern that the privatization of British airports would lead to cross-subsidization and premature development. They argued, in the case of airports, that each should be a separate financial subsidiary with ring-fencing of accounts. Ring-fencing occurs in other industries to prevent funds acquired for one purpose from being used for another. It is a governance mechanism imposed by government on devolved entities to ensure that the undesired activities do not take place. A variation on the ring-fencing that occurred in UK privatizations was proposed in the unsuccessful legislation for reforming the governance of Canada's airports: it was planned that they would be restricted from developing subsidiaries to perform activities not envisaged by the government's airport devolution program.

The previous chapters have noted that privatization is a popular topic when governments contemplate changing the governance system of ports through a program of port reform. The benefits of privatization are well documented in specific cases. For example, Laurin and Bozec (2001) documented the significant improvements in productivity (in terms of network density and tonne-kilometers per employee) when Canadian National (Canada's largest railway) was privatized; these productivity gains have continued and the company is now the most efficient Class 1 railroad in North America under a fully private, publicly traded corporate governance model. It should be noted that the Canadian government retains no shareholding and no role in operations.[1]

Both strategic management and agency theory suggest that different firm (port) performance will result because different owners pursue different goals and possess different incentives. Backx et al. (2002) noted the extensive research on incentives; briefly stated, private ownership alone does not necessarily provide strong incentives for performance, as strong owners may have motives other than value maximization (e.g., Demsetz & Lehn, 1985); widely dispersed owners produce free rider problems (e.g., Shleifer & Vishny, 1997) and, while the threat of takeover may provide discipline in developed countries (e.g., Hartley & Parker, 1991), in developing countries, where capital markets are weak, the threat of takeover as a source of discipline may be significantly less. Galal, Jones, Tandon, and Vogelsang (1994) examined privatizations and outcomes; they drew the policy conclusions that countries were better off when the government divested its public enterprises.[2] There were simply many more successes than failures. They concluded that all managers are fallible, and that both private and public sector managers can make bad decisions.

The above indicates to us that there is no clear research to tie appropriate governance models arising from devolution to performance, and that a research agenda to do this is needed. This does not mean that it has not been done in other industries. Perhaps most interesting of the studies in other industries is that by Delmas and Tokat (2005); they examined the choice of governance structure and tied it to performance in the gas distribution utility sector. They found that the greatest efficiencies came from structures that were either mostly vertically integrated (generating the power they sell) or mostly vertically disintegrated (buying the power they sell); both of these were found to be more efficient than firms with a hybrid structure (both generating and buying their power).

> Our findings show that both governance structures are efficient, albeit through different mechanisms. Transaction costs economics and the theories of flexible adaptation refer to different types of adaptation. The first is adaptation through hierarchy. That is to say, the firm "insulates" itself from market transactions and therefore uncertainty. The second is adaptation through market mechanisms where firms specialize in dealing with complex transactions and avoid the costs of organizational slack. Our findings are important because they suggest that both structures can be efficient in the same environment; they just represent different strategies. (Delmas & Tokat, 2005, p. 457)

In the port industry, Valentine and Gray (2001) pursued the topic of port efficiency and ownership structure, noting the divergent views held (in an excellent review of the literature on the topic). They produced one of the few studies in the port sector that attempted to examine the link between governance model and performance. Using cluster analysis, confirmed by discriminant analysis, they classified ports prior to assessing the three types of container ports (public, private and mixed)[3] for their performance outcomes in terms of efficiency. They concluded, based on a study of 31 ports, that there was no correlation between ownership structure and efficiency.[4]

Cullinane, Song, and Gray (2002) also examined the link between governance models (administrative and ownership structures) and the efficiency of major Asian container terminals. They found that the largest were the most productive, and that greater privatization and deregulation of the market were associated with better productive efficiency, but were not entirely satisfied given the anomalies found. Coto-Millán, Banos-Pino, and Rodriguez-Alvarez (2000), in their study of Spanish ports (using panel data from 1985–1989 and estimating a frontier cost function) concluded that smaller Spanish ports under central control are more economically efficient than those that had greater autonomy in management. It is clear to us that there is no consensus on the links between governance and performance. Performance in the port business, while there are economies of scale to be exploited, is not all about bigger and privately run being more efficient.

Heaver, Meersman, Moglia, and Van De Voorde (2000) noted that port efficiency is not only the product of the port authority's efforts, but also the efforts of other players such as stevedores, inland carriers, forwarders and agents. He concluded that "success" will be tied to the type of objectives the port authorities have. The objectives included in Heaver et al.'s study were cost minimization (type of port unspecified), maximization of cargo handling (public companies) and maximization of profits (in the case of private companies). This paves the way for Baltazar and Brooks' argument

(Chapter 17) that performance (success) is a product of the fit between the structure (the governance model established by a particular port to execute its strategy), the environment (including the governance model imposed by government on its relationship with the port) and the port's strategy (including objectives). We concur with Heaver's assertion that success (a positive performance outcome) will be defined differently for different port objectives.

Vickers and Yarrow (1989) are firm supporters of private ownership; they conclude that ownership of a firm will have significant impact on its performance given that ownership rights modify the structure of incentives available to decision-makers in the firm. According to the Matching Framework (Chapter 17), this does not mean that there is only "one best way" – privatization. What is necessary is internal consistency (called fit) between the environment, a government's port strategy and goals, and the structure and systems put in place at the time of devolution. According to configuration theory, a model that is optimal for the environment in which the port operates and the strategy port management seeks to implement (either mandated by government or adopted by managers) must fit with the appropriate governance structure to gain the desired optimal performance outcome. This leads to a two-phase approach: (1) identifying appropriate governance structures for the existing port strategy and environment and (2) determining what performance outcomes are linked to the combination. This then informs the future choice of governance structures (there may be more than one appropriate structure for different port types), matching expectations of outcome with the inputs (including imposed governance model).

As already noted, the work of Backx et al. (2002), while studying a quite different industry, suggests that measuring performance cannot be done in isolation from the variables of ownership and regulatory climate. If configuration theory as posited by Baltazar and Brooks in Chapter 17 is accepted as valid for ports, the way is paved for the argument that successful outcomes may differ not just by governance type, but also by the broader concept of configuration fit. The key point is that structure, if imposed by the regulator, can support or undermine performance if not matched to the Strategy and Environment of the port (according to Baltazar and Brooks' Matching Framework).

As noted by Porter (1995), the reduction of the role of the nation state has heightened the interest of local economic development interests in mobilizing indigenous assets to maximize local competitiveness. In assessing the port of Liverpool and the airport at Manchester as two case studies in local governance, Evans and Hutchins (2002) noted that the existing literature on governance is more focused on explaining its origins and characteristics than on measuring its impact or performance. They found, in these two case studies, that governance was of "middling rather than major importance to the competitiveness" of these entities (Evans & Hutchins, 2002, p. 437). They found that market factors and national government decisions had a greater role to play than local governance. As we have seen in this book, the impact of governance on performance for the most part has not been closely examined, and so its importance in port performance is not clear. As noted by Wolf (2004), it is so important to get the institutions right; that does not appear to have happened. Each governance model has its objectives and its built-in incentives; when governments impose internally inconsistent models on ports, we cannot expect that performance will be optimal, whatever that performance is intended to achieve.

4. KEY FINDINGS

Countries usually (but not always) develop an explicit policy regarding ports; this policy is the product of a political process and is implemented by bureaucrats as agents of the politicians. Are governments in this industry equivalent to owners, while port managers act as agents to deliver the benefits sought by government? In the case of a port policy focused on privatization and the objective of delivering private sector-led efficiency and market competition (far right column in Table 1.1), this is not true. Where government maintains its desire to have a modicum of control in the implementation of a port policy intended to deliver efficiency and local responsiveness, the port managers are treated as agents of the government and are expected to deliver results as if they were agents of owners. Lines of responsibility and accountability can become confused. When the outcome achieved is not the one expected by government, the temptation to interfere grows. When the local population as taxpayers and citizens do not like what they see, the pressure on politicians to interfere grows.

The issue of political interference is one raised in a few of the chapters and in prior research. It is a reported problem in Australia. "Ports and other government corporatized entities have good reason to be frustrated by the fact that ports remain subject to ongoing political and bureaucratic control" (Everett, 2003, p. 218). She concluded that corporatization does not sufficiently extract the ports from the interference of the Minister. Pallis (Chapter 7) reinforced this view with respect to the Greek ports; while they may have been privatized, the retention of ownership of the majority of shares by the national government explicitly indicated that the government sought to retain control of the assets, and therefore that political interference could be expected. Notteboom and Winkelmans (2001, p. 254) reported that political interference is widespread in Europe and may impinge on the flexibility ports need to respond to structural change in the global economy; they viewed politicization as a major drawback of government-funded infrastructure development.

It may be better to have commercialization (where it is clear that the government retains asset ownership and seeks better management of the assets via concession or management contracts) than to have corporatization with the government retaining a majority shareholding, as the government has only to exercise the property rights inherent in its majority shareholding to take back the control it should have relinquished. Whether this is the correct conclusion with respect to governance models remains to be validated in any future port performance research agenda.

Part 2 of this book also illustrated the wide variety of changes in government policy execution. These did encompass the full range. The UK (Baird & Valentine, Chapter 3) is the best example of a policy focused on delivering market competition and private sector-led efficiency, with the example of Hong Kong coming in close proximity for a country retaining control of the land and water assets but giving free rein on all other assets and their management. At the other end of the continuum, some ports in Turkey and Italy have retained a predominantly public approach. Particularly interesting among all of these country studies is the US. There has not been much change in US port policy since the formation of the country in the 1700s; it has always been a decentralized model with control at the state or local level (Fawcett, Chapter 10). In some US states, authorities have adopted a more corporatized approach, including setting up subsidiaries charged with the concessioning of many of the operating activities. The US approach is particularly diverse and not because of any change in policy. Throughout Part 2, the variety of examples of corporatization and commercialization illustrated the benefits and challenges of a mixed governance approach: neither strong government nor strong business, but a vision of something in between. The problem with something in between is getting the balance of incentives and risks right.

Part 3 introduced the Matching Framework (Baltazar & Brooks, Chapter 17) as one means of examining port performance and port competitiveness. The framework provides *managers* with a means to examine the components of a port's performance and determine if alignment of strategy and structure

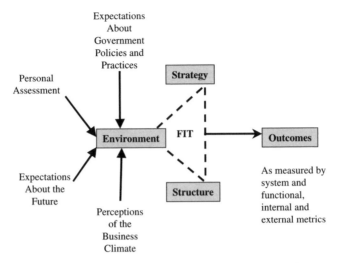

Fig. 26.2. Matching Framework (Firm Level) with Examples of Inputs. *Source:* Adapted from Brooks (2002).

matches with the environment, both task and operating, to achieve the desired performance outcome, be it efficiency- or effectiveness-focused. For managers, the benefit of applying the framework is its focus on aligning strategy and objectives with the achievement of desired performance outcomes, then measuring those outcomes and making adjustments to strategy, structure and the task environment, as illustrated in Fig. 26.2.

More than that, it is clear that government has a role in influencing the variables of the Matching Framework (the antecedents of port performance) in the way it establishes the business climate for port firms (e.g., an environment with clear codes of conduct for businesses), perhaps imposes a preferred governance model on port firms (structure may not be delineated by the firm but prescribed by legislation, regulation or letters patent, by-laws and so on), and even in some cases dictates the objectives that the port strategy must achieve. Therefore, from a regulator's viewpoint, the framework enables *government* to examine the alignment or fit between the various policy and regulatory elements that it imposes and their impacts on the resulting performance of the program (Fig. 26.3) and eventually on the port through a revised government impact on the port's task environment (Fig. 26.4, discussed later). A misalignment is likely to result in sub-optimal performance, be the outcome one focused on market competitiveness and return to the Treasury (or owners) or on local responsiveness and local

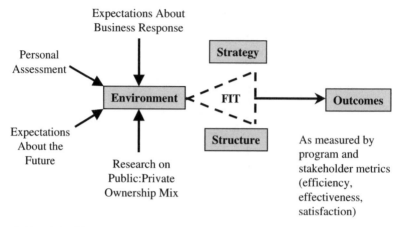

Fig. 26.3. Matching Framework at the Program Level (with Examples of Inputs).
Source: Adapted from Brooks (2002).

benefits. This process too requires government to decide what aspects of performance need to be measured to know if its objectives (e.g., local benefits, market competitiveness, etc.) are being met.

As suggested by Brooks in Chapter 25, the benchmarking of performance is an integral component of that application and that both internal and external measures are needed. The choice of measures must fit with the strategic objectives being sought; those that the firm chooses to measure will not necessarily be similar to those the government will seek to measure. The alignment is likely to be greater in public ports than in ports where government maintains a national strategic interest but has fully privatized the management and operations of the facilities. Brooks illustrated that while there were more commonly used measures of port performance, the choice of measures did not necessarily match the strategic intent of the port. While throughput was commonly measured, few ports reported collecting data on local benefits as part of the performance measurement process.

We would be remiss if we did not discuss the *limitations* of the research to date. Of greatest importance, the Port Performance Research Network only approached ports meeting a minimum throughput of two million tons with 40 percent international traffic. These two thresholds were arbitrary, again under the premise that you have to start somewhere. This gives rise to three issues. First, as Baltazar and Brooks note in Chapter 17, most ports of this size engage in multi-product strategies. Few large ports offer services of only one type. Governance issues and performance issues may be quite different

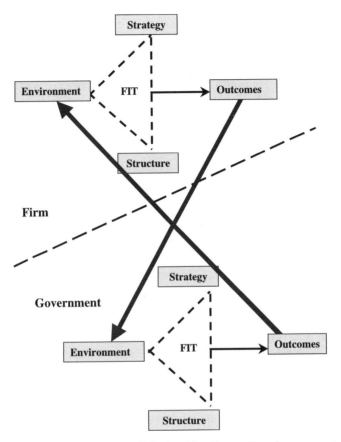

Fig. 26.4. The Firm–Government Relationship. *Source:* Based on conceptual work by Brooks (2002).

for single-product or single-user ports. Second, by only approaching the largest, those ports in remote areas or thin markets have been excluded. The performance issues and governance models will likely need to be different for these ports, and this research provides little contribution to the concerns of government or their managers. Third, tonnage is not a measure of value added by the port, so the relative importance of the port in the economy of the country or region might well be understated. Finally, the imbalance of responses across the countries has likely skewed the results. The support of UK, Italian, Turkish and US ports overshadows the contributions to our understanding made by those in other countries.

Based on the preceding chapters and this discussion, it is our assessment that there are two steps next in the process of examining and securing improved port performance. Both can be undertaken concurrently and adjusted dynamically. The first is the development of a research agenda for the validation of the relationships posited by the revised government–firm Matching Framework in Fig. 26.4. The second, at the firm and program levels, is the development of related performance-assessment instruments to evaluate outcomes in each of these areas. We address the second first.

5. WHAT NEXT? PERFORMANCE INSTRUMENTS

The development of instruments to measure port performance and agreed plans for their implementation does not need to wait until further research is complete. Instruments can be dynamically adjusted as more research becomes available. Two instruments are required. As illustrated by Brooks (Chapter 25), existing performance data collection by ports does not serve all purposes. What is reported to government is not entirely compatible with what the port needs for its own continuous improvement. The instrument developed to assess program performance needs to be related to the objectives of the government's program of port reform, while the instrument for use by ports internally needs to reflect its objectives, and the two sets of objectives may not be similar in scope or emphasis.

At the port or firm level, a third-party performance instrument not unlike the AETRA survey offered by the Airports Council International (discussed in Chapter 25) would be a good first step in the development of both internal and external measures of port performance. There is not a shortage of suitable agencies that could undertake this proposition on behalf of the port industry; coming quickly to mind are the International Association of Ports and Harbors, the American Association of Port Authorities and the European Sea Port Organization. The membership of the former two is broader than the last, but a collaborative approach would benefit all the members, increasing the likelihood of survival of an instrument in the longer term. Another independent, respected third party comes to mind: Lloyd's Register–Fairplay Ltd., an information and publication company. It assigns port numbers and tracks port compliance with the International Ship and Port Facility Code, is already active (recognized by the US authorities for this task and IMO-backed) and could extend its activities to include port performance benchmarking. Unlike the existing Drewry assessment of container ports, the subgroups, according to this research, should focus on reporting performance not only by size and type categories, but also by objectives so that those seeking local development are not compared with those having a profit motive. The features of the instrument must include customer evaluations to be truly as useful as the AETRA product is to the airport industry.

Any future research agenda will need to develop the actual constructs of performance assessment. As noted by Brooks in Chapter 25, many ports currently collect more performance measures than are manageable, while others do not collect any. Performance measures serve three constituencies – the port itself (measures used internally for performance improvement plus measures reported externally for marketing or public affairs purposes), government and the broader stakeholder community (measures required by law or requested by government), and the port's customers or potential customers (measures used by them for evaluating the services received or likely to be received). Those collected by the port for internal use only are not the focus of any future research agenda. What is of interest here is the externally reported measures, including those reported to government.

Brooks found that the measures reported to government often comprise only financial measures, with much less emphasis placed on the other categories of performance measurement. If a government is interested in local responsiveness, how should this be measured? Is this a measure suited for third-party collection or should the government determine its needs and legislate those measures that it desires to use to evaluate program success? The area of how to measure success in program performance in non-financial terms is ripe for further development.

6. WHAT'S NEXT? VALIDATING THE LINKAGES

The firm–government Framework shown in Fig. 26.4 combines both firm- and government-level use of the Matching Framework to illustrate that the two do not function in isolation. The dynamic nature of the relationship is illustrated by the large solid arrows crossing the boundary between government and firm, creating a Mobius strip of continual change. It is this complex set of linkages, with differing perspectives on port performance, that is ripe for further research. We need to examine both inputs and outputs to understand the Mobius strip of effective new public management. While Chapter 18 focused on developing proxies for the imposed governance structure, and validating the nature of strategy (as defined by the port's strategic intent), there remains a need to address the third input variable of government business climate and policies for corporate regulation and the output variables of port performance at both levels.

6.1. Key Input Variable: Strategy (Objectives)

Port objectives vary, and because port performance should be measured in the context of objectives, these need further discussion. Baltazar and Brooks in Chapter 17 proposed that the strategic intent of ports could be divided into two groups, those with economic objectives and those seeking non-economic ones, as a means to evaluate performance. In Chapter 18, Brooks and Cullinane suggested that there are three logical groups: (1) those that have solely non-economic objectives, including wider economic benefits, such as local economic development, cluster development (as per de Langen's Chapter 20) or not, and so on; (2) those that have strictly economic objectives, such as profit maximization and/or maximization of return on investment; and (3) those that have a mixture of both economic and non-economic objectives. While Brooks and Cullinane concluded in Chapter 18 that the way port authorities structure their cargo-handling activities and the strategic purposes they are trying to achieve are not related, that is, that ports do not appear to match methods of governing this activity with primary purpose, this conclusion needs to be validated with a much larger sample and a streamlined research instrument.

On the government side of the boundary in Fig. 26.4, the purpose of the government devolution program also needs to be confirmed. This assessment does not need to be sequenced but can be concurrent.

6.2. Key Input Variable: Structure (Imposed by Government, Adopted by Port)

The second set of variables that arise from the papers in this book concern the issue of governance type. Chapter 18 focused on trying to validate existing governance models, primarily in terms of centralization or decentralization of management and control for ports and in terms of shared or sole provision of mission-critical activities (internal port collaboration to achieve desired outcomes). While we initially identified five states of governance (A–E in Table 26.1), Chapter 18 found that it could neither validate existing oversimplified models nor find replacements that form a tractable basis for empirical analysis. The reality is a seemingly infinite hodgepodge of models

Table 26.1. Proxies for the Governance of Port Activities.

Port Activity	*Governance Proxy*
Services to vessels or terminals	Line handling for arriving/departing vessels
	Provision of fire protection services
	Terminal security
	Provision of pilotage services
Other activities	Port security
	Customs and immigration services
Policies, regulation and planning for the port	Setting and controlling port tariffs
	Determining applicable port safety and security policies
Type of Governance	*Description*
Public	Includes: (A) Central government owned with central government management and control and (B) Government owned but management and control are decentralized to a local government body
Mixed	Includes: (C) Government owned (federal, regional or municipal) but managed and controlled by a corporatized entity; and (D) Government owned but managed by a private sector entity via a concession or lease arrangement, or owned and managed via a public–private partnership agreement
Private	(E) Fully privately owned, managed and controlled

Source: As determined by Cullinane and Brooks in Chapter 18.

executed in almost unique ways. We felt compelled to reduce the five categories to three, even though we would lose the distinctions between centralized and decentralized public models and between commercialized versus corporatized mixed models.

Chapter 18 also focused on identifying the most representative individual port activity from each of the nine groupings of activities that reflected governance options, in order to act as a proxy for the group as a whole. These proxies (for the activities under these models) have streamlined the variables for future research into governance and performance, and are also summarized in Table 26.1.

6.3. Key Input Variable: Environment

The devolution–governance options defined in Table 1.1 (Chapter 1) confer varying degrees of flexibility on the devolved entity to respond to a changing industry and firm environment. For example, the entity need not own the land to have control of it for port purposes; such control may be granted by the terms and conditions of the governance system imposed by government. Therefore, the regulatory climate is, perhaps, more critical in conferring on a port the ability to respond strategically; its key role is defining the limits that are placed on that response. This would imply that an antecedent of port performance is the operating environment.

As a result, the next task is to define a proxy or proxies for the government "business climate." To develop appropriate indicators for this business climate variable does not require starting from scratch. As already suggested by Brooks (Chapter 25) and used by Sánchez and Wilmsmeier (Chapter 9), the *World Competitiveness Yearbook* (IMD, annual) supplies secondary attitudinal data collected from multinationals (business leaders) on an annual basis. This reflects, therefore, external perceptions of the *business controls* in place in a country and their impact on the ability of a business to develop efficient and effective operations within that country. Furthermore, IMD collects external perceptions of the country environment that result from *government policies and practices*, which also impact on the ability of business to develop efficient and effective operations within that country. Both types of factors could be used as input variables in the revised Matching Framework (Fig. 26.4) to determine which of these provides a proxy (or proxies) for the environment in which ports make strategy decisions and adapt to imposed governance models, in a manner similar to that used in Chapter 18 to generate proxies for governance approaches. A selected subset is proposed in Table 26.2 and could be used in future research focused on identifying which ones would best serve in assessing the environmental impacts on firm-level activities.

6.4. Key Output Variable: Performance (Efficiency or Effectiveness?)

In addition to the variables noted above, we see the operationalization of a future research agenda as having different sets of performance measures, reflecting different strategic purposes. Port performance should be measured in the context of the port authority's objectives, be they established to reflect a mandate from government or a requirement of owners or shareholders, or, in the case of commercialized or corporatized ports, a hybrid.

Finally, there is the choice of performance variables. Chapter 25 noted what gets measured and how it is used and distributed. This provides a starting point for the development of the port performance variables. From

Table 26.2. Environment Factors.

Perceptions of Business Controls
Legal framework (1 = detrimental to competitiveness)
Competition laws (10 = prevent unfair competition)
Price controls (10 = do not affect pricing)
Legal regulation of financial institutions/rights and responsibilities of shareholders (10 = well-
   defined)
Foreign and domestic companies (10 = treated equally)
Access to local capital markets (10 = not restricted for foreign companies)
Access to foreign capital markets (10 = not restricted for domestic companies)
Calculated (from above): ratio of access local: foreign capital markets

Perceptions of Government Policies and Practices
Tax evasion (common practice = 1)
Central bank policy (positive = 10, negative = 1)
Exchange rate policy (10 = supports competitiveness)
Consensus about policy direction inside the cabinet (10 = high)
Legislative activity of the parliament (10 = supports competitiveness)
Government economic policies (10 = adaptive)
Government decisions (10 = effectively implemented)
Political parties (10 = is well-adapted)
Transparency (10 = government communicates well)
Public service (10 = immune from political interference)
Bureaucracy (10 = does not hinder development)
Bribing and corruption (10 = do not exist in public sector)
Customs' bureaucracy (10 = efficient transit of goods)
Personal security & private property (10 = protected)
Protectionism (10 = open entry)

Note: All variables use a scale of 1–10.
Source: Selected variables from those included in IIMD (undated).

a firm perspective, while it was clear that many ports do not collect all the possible variables proposed, many collect more than was anticipated or possibly required for good management practices. At the other extreme, some conduct very little, if any, measurement of their own performance. The starting point for benchmarking is the externally reported measures, as these influence perceptions about the port's performance. It would be very useful to undertake further research on the measures desired by customers and expand this limited content further in anticipation of the creation of a third-party performance instrument (TPI).

Building on the research conducted by the Port Performance Research Network in Chapter 25, Table 26.3 is developed to illustrate how the measures might be used to develop performance instruments tied to objectives, be they firm or program ones. The table retains the measures in the order they

Table 26.3. Development of Performance Measures.

Measure	Firm Objective 1	Firm Objective 2[a]	Internal Management	Customer	External Public & Government	Government Objective 1	Government Objective 2[a]
Financial measures							
Ancillary revenue as % of gross revenue	Y	MIS	F		PAI	Y	Y
Average days accounts receivable	MIS	MIS	F				
Capital expenditure as % of gross revenue	Y		F		PAI	Y	Y
Debt: equity ratio	Y		F		PAI	Y	
Growth in profit (before taxes)	Y		F		PAI	Y	
Interest coverage ratio	MIS	MIS	F				
Port-related profit as % of port-related revenue	Y	Y	F		PAI	Y	Y
Return on capital employed	Y		F		PAI	Y	
Terminal charges as a % of gross revenue	Y		F				
Yield % on shares, if publicly traded	Y		F				
Vessel operations							
Average turnaround time per vessel (in h)	MIS	MIS	S	E, TPI			
Average vessel calls per week	Y	Y	S	E, TPI	PAI	Y	Y
Average vessel waiting time at anchor	MIS	MIS	S	E, TPI			
Hours of equipment downtime per month	MIS	MIS	F, S				
Length of quay in meters (as a capacity measure)	MIS	MIS	F, S		PAI	Y	Y
Revenue per tonne handled	Y		F	E	PAI	Y	Y
Container operations							
20′ TEU as a % of total TEU for year	MIS	MIS	F, S		PAI	Y	Y
Average revenue per TEU	Y	MIS	F				
Average vessel turnaround time per 100 lifts (in h)	MIS	MIS	S	E, TPI			
Average yard dwell time (in h)	MIS	MIS	F, S	E, TPI			
Container port throughput (TEU/meter of quay/year)	MIS	MIS	F, S	E, TPI	PAI	Y	Y
Departure cut-off time (h)	Y	Y	S				
Growth in TEU throughput	Y	Y	F, S	E	PAI	Y	Y
Import containers as a percent of total containers					PAI	Y	Y
Lifts per crane hour	MIS	MIS	F, S		PAI	Y	Y

Percent of containers grounded (ship to rail ops only)	Y	Y	S				
Reliability (qualitative factor)[a]	Y	Y	S	E, TPI		Y	Y
Transshipment (as % of total throughput)	MIS	MIS	F		PAI	Y	Y
Yard hectares to quay meters	Y	Y	F		PAI	Y	Y
Other measures							
Customer complaints per month	Y	Y	S				
Destinations served this year	Y	Y	S	E	PAI	Y	Y
Employee turnover rate	MIS	MIS	S				
Employment (full-time equivalents) per tonne handled	MIS	MIS	F, S		PAI	Y	Y
Employment (full-time equivalents) per TEU handled	MIS	MIS	F, S		PAI	Y	Y
Invoice accuracy percent	MIS	MIS	F, S				
Number of customers served			S		PAI		Y
Overall customer satisfaction		Y	S	E, TPI	PAI		Y
Stakeholder satisfaction	MIS	Y	S	TPI	PAI		Y

Notes: Objective 1 = This group has strictly economic objectives, such as profit maximization and/or maximization of return on investment. On the government program side, this would include those governments that hold port investments as part of a financial portfolio.

Objective 2 = This group is focused on wider benefits, such as local economic development, cluster development and so on.

Y = Yes, appropriate to the objective; may or may not be suitable for the imposed governance structure.

MIS = This information should be collected as part of an internal management information system and is an input to effective and efficient day-to-day operational decision-making.

F = Financial performance improvement.

S = Service performance improvement.

E = Customer evaluation of ports to be served.

TPI = Suitable for third-party performance instrument.

PAI = Program assessment information (e.g., input to program assessment).

[a]Further development needed.

appeared in Table 25.5; to assess their use, they could be reordered depending on how they might be used. For example, those external measures
in Table 26.3 that are marked as E (for use by customers to evaluate potential ports of call) could provide a starting point for the development of a
customer-oriented performance instrument, or those marked TPI could
provide a starting point for the development of that instrument.

On the internal side, and based on what ports already collect as found in
Chapter 25, Table 26.3 notes two types of measures: those that should be
collected by any management team as part of its internal performance assessment program (marked MIS for management information system) and
those that are appropriate to the objectives of the port (marked Y for yes, an
appropriate measure). The table also divides these measures into those focused on financial improvement (F) and those focused on service improvements (S). Some measures have financial impacts that will be directly linked
to service improvement and so are marked F, S. The ones focused on service
improvements may or may not be used by customers for evaluation. Again,
these could be validated as part of developing a TPI.

Measures used by ports to improve their financial performance tend to be
useful in measuring government program success, but only if improved financial performance is the intention of the government program.

Finally, it should be clear from Table 26.3 that more work is needed to
develop measures that will assess the performance of ports with broader local
benefits as their principal objective; the measures developed for the Port
Performance Research Network do not serve that strategic purpose well.

To conclude the firm perspective, each port will choose the measures by
which it wishes to assess its own performance; those of a service nature or
those used by customers to evaluate potential ports of call are suitable for
inclusion in a third-party port performance instrument. It is not necessary to
wait until the research is completed. New metrics in a third-party instrument
may be more readily agreed once there is some experience with existing ones.
While efficiency is reasonably well defined, effectiveness is less so. As effectiveness is tied to the perception of whether the port has delivered what its
customer was seeking, it is best to measure this by an external third-party
instrument (as noted previously).

From a government perspective, the measures should match the government's program objectives. If the government desires maximization of economic performance (profitability), it might choose to collect, or ask ports to
provide, the ones with Y in the column headed Government Program Objectives 1. If maximization of local benefits is desired, then those measures
marked Y from column 2 are appropriate. Problematically, as in the case of

ports seeking more local benefits, the list of measures to assess program outputs is not well developed and needs further research. Those marked PAI (for program assessment information) will provide useful feedback for a government attempting to determine program outcomes. It remains unclear, for example, what the best way would be to measure local responsiveness through other than a TPI.

7. CONCLUSIONS

In tandem with the wider contents of the book as a whole, this chapter has graphically illustrated that port devolution programs are prevalent throughout the world and are likely to continue to be so, not least because they tend to be phased in over a long period, with the process of transition from initial conditions to steady state potentially taking a number of years.

While governments have clearly had the best of intentions in (typically) seeking to place their nation's port operations on a more commercial footing, the outcomes of such policies and/or the way they have been implemented have proved inconsistent in delivering the full benefits that were sought. This chapter has shown that this conforms with research findings in other industries and can be at least partially explained by the complexity of the relationship between governance structure, performance outcomes and simultaneously implemented intervening phenomena, such as industrial restructuring (either driven from within the industry or externally by government), policies of market deregulation or the prevailing commercial environment, any or all of which may either reinforce or undermine the desired performance outcomes (government objectives) of devolution programs.

This chapter espouses the validity of the assertions made by Heaver et al. (2000) that success (which can clearly be considered a positive performance outcome) will be defined differently, depending upon the objectives of a port. Furthermore, each governance model has its own implicit objectives and incentives so that if governments impose internally inconsistent models on ports, performance simply cannot be optimal, irrespective of whatever performance outcomes it intended to achieve. Such a context provides clear support for the contention of Baltazar and Brooks (Chapter 17) that successful performance outcomes are a product not just of governance type, but also of the fit between the structure (the governance model established by a particular port to execute its strategy), the environment (including the governance model imposed by government on its relationship with the port) and the port's strategy (including objectives). The key point then becomes that structure can either support or undermine performance if not matched to the port's strategy and the environment in which the port operates. This is particularly germane if the government has a role in influencing the ante-cedents of port performance by conditioning the business climate in which ports operate (e.g., by fostering an environment with clear codes of conduct for businesses) or perhaps through the imposition of a preferred governance model on ports and even, in some cases, by dictating the objectives that port strategy must achieve.

In light of the relationships that have just been alluded to, it is critically important that governments constantly review and evaluate the perform-ance of any port reform program in terms of the objectives that the policy is seeking to achieve. This feedback process should be perceived as a contin-uing opportunity for tweaking, realigning or changing ports policy and the accompanying imposed governance model(s) to achieve the desired per-formance outcomes and associated policy objectives.

Chapter 18 revealed significant evidence of the oversimplified and inad-equate nature of existing port governance models. A new model for port governance was expounded that is characterized by both greater detail in the definition of the port activities for which governance decisions must be taken and a much more refined range of governance typologies that may be se-lected by decision-makers. This final chapter has elaborated what constitutes a complex future research agenda, with the key objective being the unrave-ling of the nature of the relationship between this newly expanded range of port governance configurations and the performance of ports. As suggested above, this must be achieved for each of a variety of contexts that impact not only port performance, but also the definition of performance per se.

With the achievement of this objective, however, the opportunity exists for the port governance model developed and presented in Chapter 18 to be transformed from a mere portrayal of available decision alternatives (in the sense of positive economics) to a much more normative tool that may be practically and usefully applied to the delineation of specific governance models (or sets of them) that are directly and explicitly linked to perform-ance outcomes that are flexibly defined.

Another of the conclusions reached in this chapter is that ports do not currently appear to match methods of governing cargo-handling activities with their primary strategic purpose or intent, through the adoption of an optimal governance–performance configuration. There remains, however, the need to validate this assertion using a much larger sample and a stream-lined research instrument, preferably one that does not rely on performance data collected by ports. For a variety of reasons addressed earlier in this chapter, mere internal information would simply not be fit for the purpose of assessing the performance of port reform programs as they relate to the objectives that have been established for them.

The final conclusion to be drawn from the contents of this chapter is that if the performance outcomes of port reform programs are to be improved through the linking of those outcomes to internal governance structures, the environment in which the port operates (including the external governance structure) and the objectives for the program within a holistic, more appropriately specified and (understandably) complex port governance model, then one of the best ways of achieving practical improvements in the sector will be through the use of a port performance instrument that has been objectively formulated by some credible third party. This can then form the basis for not only future research, but also as a benchmarking tool for both ports and governments. The authors avow that this instrument should incorporate both internal and external measures of port performance, including evaluations of both government and customer perceptions. The inclusion of the latter is not only conceptually logical, it also has the potential benefit of prompting buy-in on the part of the port sector in light of the marketing benefits that might accrue from access to this objective third-party information. It is important, however, that the instrument accounts for differences in size and type of port, as well as different objectives, so that those seeking local development are not compared with those having a profit motive. Only in this way can individual ports benchmark themselves against others on a like-by-like basis.

Through the development and widely adopted use of this instrument, the authors propose that government will be greatly helped in determining port policies and strategies that are designed to achieve their objectives for port devolution and reform. Given the oversimplified and, as shown in Chapter 18, fundamentally inadequate nature of existing port governance models, what has been proposed as a future research agenda in this chapter has the potential to lead to the development of models and tools that are applicable in practice and likely to make a significant contribution to improving the specification of objectives for the reform of port policies and of sensible performance outcomes that provide logical and realistic support to the achievement of these objectives. In addition, the elaborated future research agenda will prompt the development and international recognition of benchmarks that will increase the importance attached to the monitoring of internal and external performance measures, thereby providing a feedback loop that will influence the constant reformulation and realignment of specified objectives, governance structures and performance outcomes. With worldwide adoption and appropriate deployment, significant and ongoing improvements in the quality of international port services will result. Given the importance of the port sector to the smooth workings of international supply chains, the facilitation of trade and national and regional development, this is an outcome that can only serve to enhance global economic welfare in both developed and developing nations.

NOTES

1. It does impose statutory obligations – in terms of safety, security and competition, which are applicable to all transport entities, and fiduciary requirements – in terms of corporate practices imposed on all publicly traded companies.

2. Based on a study of 12 cases where the government privatized a public-owned enterprise, they found that world welfare change was positive in 11 of the 12. Workers as a class did not lose; in 10 cases workers were better off. Consumers were worse off in five cases as a result of the movement towards efficiency prices. The fiscal outcome was positive in nine of the 12. In particular, when poorly managed public firms were converted into well-managed private firms, productivity gains were evident.

3. Their classifications are not directly comparable to the ones used in this research as they follow Cass (1996), whose three-tier typology considers concessions, contracting out, public–private partnerships and corporatizations as private while we would argue that these fall into the mixed categories as specified in Table 1.1.

4. The authors concluded that sample size was small and so further confirmatory research was needed; we have reached the same conclusion based on 42 ports in this study using Port Performance Research Network data.

REFERENCES

Backx, M., Carney, M., & Gedajlovic, E. (2002). Public, private and mixed ownership and the performance of international airlines. *Journal of Air Transport Management*, *8*(4), 213–220.

Boardman, A. E., & Vining, A. R. (1989). Ownership and performance in competitive environments: A comparison of the performance of private, mixed and state owned enterprises. *Journal of Law and Economics*, *32*(1), 3–33.

Brooks, M. R. (2002). Measuring port performance. Presentation to the Workshop on Corporate Governance in the Port Sector, Organized by M. R. Brooks and R. Baltazar, Panama City, Panama, November 12.

Cass, S. (1996). *Port privatization*. London: Cargo Systems, IIR Publications Ltd.

Coto-Millán, P., Banos-Pino, J., & Rodriguez-Alvarez, A. (2000). Economic efficiency in Spanish ports: Some empirical evidence. *Maritime Policy and Management*, *27*(2), 169–174.

Cullinane, K. P. B., Song, D.-W., & Gray, R. (2002). A stochastic frontier model of the efficiency of major container terminals in Asia: Assessing the influence of administrative and ownership structures. *Transportation Research A*, *36*(8), 743–762.

Delmas, M., & Tokat, Y. (2005). Deregulation, governance structures, and efficiency: The U.S. electric utility sector. *Strategic Management Journal, 26*, 441–460.

Demsetz, H., & Lehn, K. (1985). The structure of corporate ownership: Causes and consequences. *Journal of Political Economy, 93*, 1155–1177.

Dyner, I., & Larsen, E. R. (2001). From planning to strategy in the electricity industry. *Energy Policy, 29*(13), 1145–1154.

Evans, S. R., & Hutchins, M. (2002). The development of strategic transport assets in Greater Manchester and Merseyside. *Regional Studies, 36*(4), 429–438.

Everett, S. (2003). Corporatization: A legislative framework for port inefficiencies. *Maritime Policy and Management, 30*(3), 211–219.

Everett, S., & Robinson, R. (1998). Port reform in Australia: Issues in the ownership debate. *Maritime Policy and Management, 25*(1), 41–62.

Galal, A., Jones, L., Tandon, P., & Vogelsang, I. (1994). Divestiture: Questions and answers. In: *Welfare consequences of selling public enterprises: An empirical analysis* (pp. 3–9). Oxford: Oxford University Press.

Gillen, D. W., & Cooper, D. (1995). Public versus private ownership and operation of airports and seaports in Canada. In: F. Palda (Ed.), *Essays in Canadian surface transport*. Vancouver: Fraser Institute (www.fraserinstitute.ca/publications).

Gillen, D. W., & Morrison, W. (2004). Airport pricing, financing and policy: Report to the National Transportation Act Review Committee. In: D. Forsyth, D. W. Gillen, A. Knorr, O. G. Mayer, H.-M. Niemeier & D. Strakie (Eds), *The economic regulation of airports* (pp. 45–62). Aldershot, UK: Ashgate.

Grabowski, R., Rangan, N., & Rezvanian, R. (1994). The effect of deregulation on the efficiency of U.S. banking firms. *Journal of Economics and Business, 46*(1), 39–54.

Hartley, K., & Parker, D. (1991). Privatization: A conceptual framework. In: A. F. Ott & K. Hartley (Eds), *Privatization and economic efficiency* (pp. 11–26). Aldershot, UK: Edward Elgar.

Heaver, T., Meersman, H., Moglia, F., & Van De Voorde, E. (2000). Do mergers and alliances influence European shipping and port competition? *Maritime Policy and Management, 27*(4), 363–373.

Hollas, D. R., MacLeod, K. R., & Stansell, S. R. (2002). A data envelopment analysis of gas utilities' efficiency. *Journal of Economics and Finance, 26*(2), 123–138.

Jenkinson, T. (1998). Corporate governance and privatization via initial public offering. In: *Corporate governance, state-owned enterprises and privatization* (pp. 87–118). Paris: Organisation for Economic Co-operation and Development.

Laurin, C., & Bozec, Y. (2001). Privatization and productivity improvement: The case of Canadian National. *Transportation Research E: Logistics and Transportation Review, 37*(5), 355–374.

Luo, Y. (2005). Corporate governance and accountability in multinational enterprises: Concepts and agenda. *Journal of International Management, 11*(1), 1–18.

Mukherjee, K., Ray, S. C., & Miller, S. M. (2001). Productivity growth in large U.S. commercial banks: The initial post-deregulation experience. *Journal of Banking and Finance, 25*(5), 913–939.

Notteboom, T. E., & Winkelmans, W. (2001). Reassessing public sector involvement in European seaports. *International Journal of Maritime Economics, 3*(2), 242–259.

Parker, D. (2000). Introduction. In: D. Parker (Ed.), *Privatization and corporate performance* (pp. xiii–xxviii). Cheltenham, UK: Edward Elgar.

Porter, M. (1995). Competitive advantage, agglomeration economies and regional policy. *International Regional Science Review, 19*, 85–90.

Shleifer, A., & Vishny, R. W. (1997). A survey of corporate governance. *Journal of Finance, 52*, 737–783.

Starkie, D. N. M., & Thompson, D. J. (1985). The airports policy white paper: Privatisation and regulation. *Fiscal Studies, 6*(4), 30–41.

Valentine, V. F., & Gray, R. (2001). The measurement of port efficiency using data envelopment analysis. *Ninth world conference on transport research, Seoul, Korea*, 22–27 July.

Vickers, J., & Yarrow, G. (1989). *Privatization: An economic analysis*. Cambridge, MA: MIT Press.

Wolf, H. (2004). Airport privatization and regulation: Getting the institutions right. In: D. Forsyth, D. W. Gillen, A. Knorr, O. G. Mayer, H.-M. Niemeier & D. Strakie (Eds), *The economic regulation of airports* (pp. 201–211). Aldershot, UK: Ashgate.

Yoshikawa, T., & Phan, P. H. (2003). The performance implications of ownership-driven governance reform. *European Management Journal, 21*(6), 698–706.

SUBJECT INDEX